D1209448

The Cambridge Encyclopedia of Stars

This unique encyclopedia provides a fascinating and fully comprehensive description of stars and their natures and is filled with beautiful color images. The book begins with ancient astronomy – with constellations and star names – and then proceeds to modern coordinate systems. Further chapters explain magnitudes, distances, motions, and the Galaxy at large. Double stars, clusters, and variables are introduced, and, once the different kinds of stars are in place, later chapters examine stellar evolution, beginning with the interstellar medium and star formation, proceeding to our Sun and its characteristics, and ending with the ageing processes of solar-type and high-mass stars. The book ends by showing how this information can be combined into a grand synthesis. Detailed cross-referencing enables the reader to explore topics in depth and makes this an invaluable work both for beginners and for those with a more advanced interest in stars and stellar evolution.

DR. JAMES B. KALER is Professor Emeritus of Astronomy at the University of Illinois at Urbana-Champaign. His research involved dying stars, specifically the graceful shells and rings of gas ejected in stellar death called "planetary nebulae." Long interested in science education and popularization, Dr. Kaler has written for a variety of magazines that include *l'Astronomia* (Italy), *Astronomy*, *Sky and Telescope*, *StarDate*, and *Scientific American*. He was a consultant for Time-Life Books on their "Voyage Through the Universe" series, has appeared frequently on regional Illinois television and radio, and lectures widely. He received the 1999 Armand Spitz Memorial Lectureship from the Great Lakes Planetarium Association for his work, has written several books including *Stars and their Spectra*, *Extreme Stars: At the Edge of Creation*, and *The Ever-Changing Sky: A Guide to the Celestial Sphere*, all published by Cambridge University Press. Dr. Kaler has an asteroid named after him in honor of his outreach work.

THE CAMBRIDGE
ENCYCLOPEDIA OF
STARS

James B. Kaler

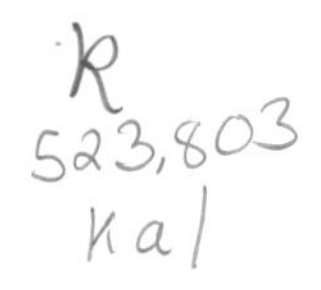

CAMBRIDGE UNIVERSITY PRESS
Cambridge, New York, Melbourne, Madrid, Cape Town, Singapore, São Paulo

Cambridge University Press
The Edinburgh Building, Cambridge CB2 2RU, UK

Published in the United States of America by Cambridge University Press, New York

www.cambridge.org
Information on this title: www.cambridge.org/9780521818032

First published 2006

Printed in the United Kingdom at the University Press, Cambridge

A catalogue record for this publication is available from the British Library

ISBN-13 978-0-521-81803-2 hardback
ISBN-10 0-521-81803-6 hardback

To my Family

Contents

Preface

In the summer of 1947, when I was eight years old, I wrote a five-page "book" called "Things That I Know About Stars." It calls out some of the constellations I had found. And it contains a few gems: "The Sun may expand to about 3 or 4 tims it's size"; "There are two kinds of stars (dwarfs and gaints)"; "A white dwarf may enter the solar system"; [The Sun] "is a fixed star traveling at about 12 miles per secend [sic]." I have no idea where these came from: eight-year-olds are not much into citations. And I clearly could have used an editor. But the statements are, even after nearly 60 years, all basically true.

I'm still writing the same book. It's just become a little bigger. I also found an editor. And it is not what I know, but what generations of astronomers have learned and allowed us all to know about the stellar science. Decades of research have not only supported the knowledge of the 1940s and earlier, but have added to it to an astonishing degree, to the point at which we have a good understanding of the beasts of the stellar zoo, and how they all link together. That is hardly to say that we have found all the answers. Indeed, for every discovery, there seems to unfold another mystery.

The Cambridge Encyclopedia of Stars is not laid out in traditional encyclopedic – alphabetic – format. Nor is it designed as a pedagogical textbook in which one chapter leads smoothly and logically into the next. It falls instead somewhere in between. Chapters concentrate on particular aspects of stellar astronomy – distances, double stars – and are meant to stand alone. At the same time, the reader is led continuously deeper and deeper into the subject. Discussion in one chapter assumes knowledge to be found in other chapters. Continuity is achieved by using both forward and backward referencing. For example, the chapter on distances discusses the Cepheid variable method, but does not treat either their creation or structure, which are reserved for later chapters and sections, and which are referenced with square brackets. The book is designed such that a reader can pursue it either from beginning to end or can dip into it at any point and read any particular chapter, in either case referring to the cross-referencing as needed.

The Cambridge Encyclopedia of Stars begins with two chapters on astronomical fundamentals that discuss constellations, star names, and coordinates. These are followed by another pair that examine stellar fundamentals: magnitudes and distances. An interlude places stars within the context of the Galaxy by looking at solar and stellar motions. The next pair attack the spectral sequence, the HR diagram, the various kinds of stars, and physical properties such as temperature, chemical composition, and the like. A trio then discusses binary systems, clusters, and variable stars. With these stellar characteristics in place, the book concludes with a set of four chapters on stellar structure and evolution that examine the interstellar medium and star formation, the Sun and the main sequence, the evolution of lower mass stars to white dwarfs, the evolution of higher masses to neutron stars and black holes.

Though the format is largely descriptive, equations are used where they enhance clear discussion. Calculus is not employed except by discussion of it in the section on stellar structure, in which the concepts are developed. Background and peripheral material is placed into boxed sidebars. The biggest problem, as always, was the selection of what to include within the limits of space allotted. I have tried to bring a balance to the book by maintaining what I consider to be consistent levels from one chapter to the next to give an overall uniform view of what is known about stars, and equally important, what is not known. I hope you enjoy the results.

Acknowledgements

As in any page of acknowledgements, especially for a book of this scope, there are many sets of people to thank. Foremost are the research astronomers who through their hard and dedicated work gleaned the knowledge that is presented here. The data in the many tables that appear in the text and appendices are a synthesis of a vast number of compilations and research papers by generations of scientists. Deep thanks and appreciation to them all.

Next are the individuals who made this labor of love possible. First to thank is Simon Mitton, who was responsible for getting the project started in the first place at the January 2001 meeting of the American Astronomical Society in San Diego. We were discussing the Cambridge encyclopedia projects, and I said "How about one on stars?". Four years later, the work was finally complete. To Simon, thanks. Following closely are my editor, Jacqueline Garget, who has patiently followed my progress through the forest of stellar astronomy, and has guided the book to fulfillment. Thanks also to editors Vince Higgs and Jo Bottrill, copy editor Peter Sinclair, and to designer Chris McLeod. I express great thanks as well for the fine new graphics rendered by Precision Graphics of Champaign, and for their diligence in preserving the old ones.

Third are the many professional and amateur astronomers who have allowed me the use of their wonderful imagery and who have helped make the book come alive. Following are the various observatories and organizations (and their personnel) who have been kindly cooperative and that include the American Association of Variable Star Observers (AAVSO), the Anglo-Australian Observatory, Big Bear Solar Observatory, the California Institute of Technology (Mt. Wilson and Palomar), the Chandra X-ray Observatory, the Institute for Solar Physics, Royal Swedish Academy of Sciences, Lick Observatory, Lowell Observatory, the National Optical Astronomy Observatories (NOAO), the Okayama Astrophysical Observatory, the Space Telescope Science Institute (Hubble Space Telescope, HST), the Subaru Telescope, and Yerkes Observatory. Thanks also to Sky Publishing Company and to the Special Collections and Rare Book Room of the University of Illinois. A very special thanks goes to Adam Black and to Addison-Wesley for allowing the use of numerous graphics from my textbook *Astronomy!*.

Last and hardly least are the many astronomers who taught me their craft, patiently answered my questions, and set me on corrected paths. Thanks to you all.

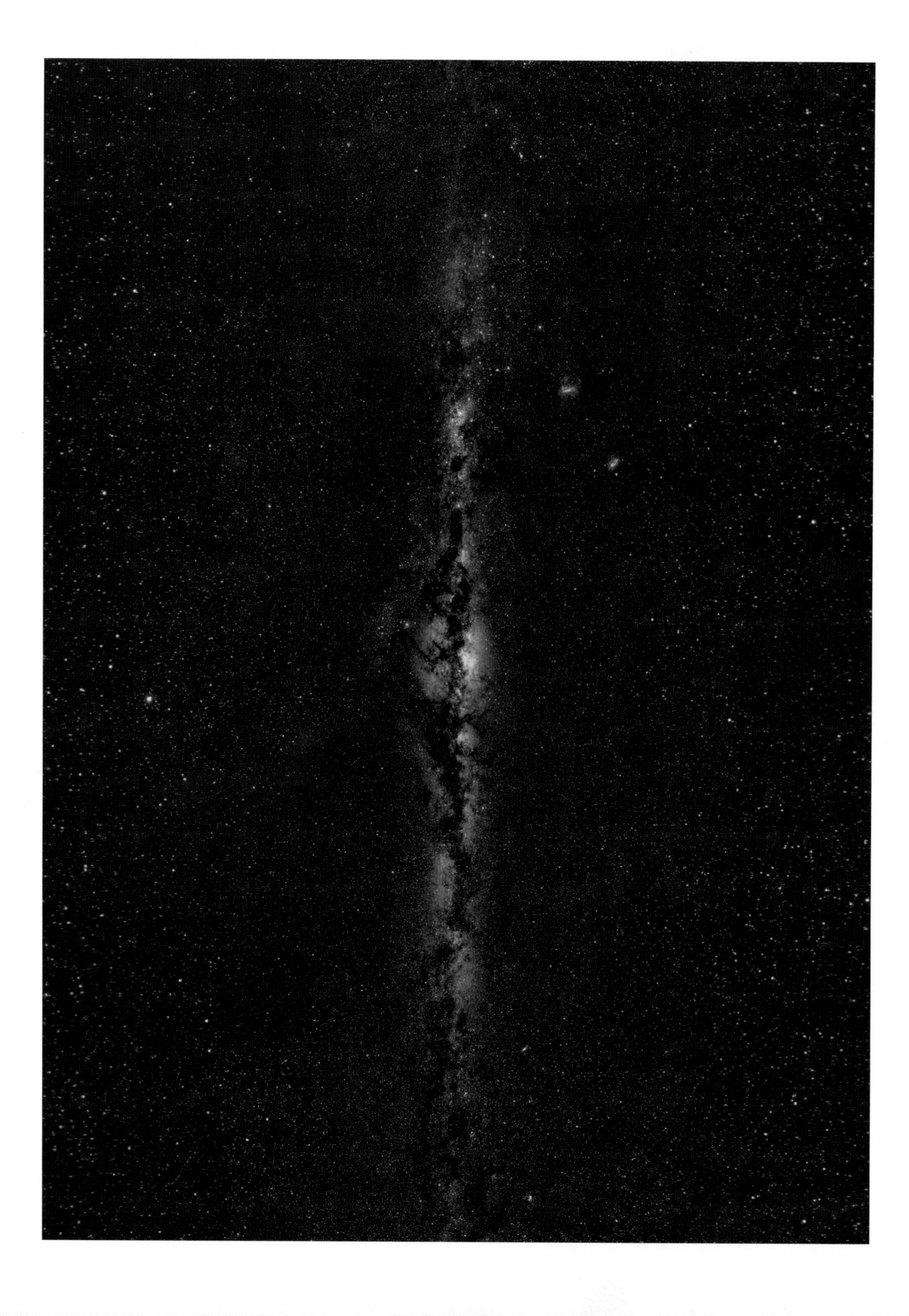

Chapter One

Stars and constellations

Fig. 1.1 **The Milky Way** A spectacular mosaic of the entire length of the Milky Way, focused on the center of the Galaxy in Sagittarius, reveals awesome complexity, all the constellations, millions of stars of all different kinds, and a great dark rift of interstellar dust in which stars are being born. In the middle is the Galactic bulge, while to the right are Orion and Canis Major, with Sirius, the brightest star of the sky. At center-left are Taurus and Auriga, where we find the Galaxy's anticenter [Appendix 5]. Ursa Major with its Big Dipper (or Plough) is above them. The two Magellanic Clouds, nearby irregular galaxies, are down from right center. See the frontispiece for an expanded view. (Courtesy of Axel Mellinger.)

Fig. 1.2 **The Pleiades** This most prominent of all open clusters has meant many things to many people, to some evoking sisters, to others brothers, a dipper, a basket, and much more. Note the dusty interstellar cloud lit by the hot blue stars. (Courtesy of Mark Killion.)

From the deep dark of the country, stars seem everywhere, bright ones set like colored jewels against a matrix of diamond dust. Learning them may seem a hopeless task. Closer examination, however, begins to show some order. The first sign of organization is their intense concentration toward the encircling band of the Milky Way, the visual manifestation of the disk of our Galaxy (Fig. 1.1). Superimposed upon this distribution are distinctive groupings (some physically real, others accidental) set into patterns that the human mind quickly recognizes and names, and in earlier times thought spiritually or otherwise significant. Thus were born the constellations. The moving Sun, Moon, and planets, embodying restless gods, resided within different starry "homes," and perhaps an understanding of where they seemed to be could tell something of human fate. If nothing else, the animals, heroes, and artifacts that in the ancient human mind populated the sky could be used to tell exciting stories as the stars wheeled overhead and the night played out its darkness.

In our own times, from the advent of our understanding of what the stars really are, the constellations have slowly lost their supernatural meanings (though not even now to those still enamored of astrology). Yet these figures retain their charm, their history, even considerable usefulness to modern astronomical nomenclature through their ability to parcel out the sky into manageable segments. The first step in our starry tale is therefore to explore what a newcomer to astronomy perhaps most wants to know: what does Orion look like, where can he be found, what is the name of that star over there, and why does it sound so strange to the ear, the subject inevitably leading back to physical reality and to the organization of the heavens.

All societies have to some degree explored the sky overhead. After all, it is half one's world. Though views overlap, every group has seen the sky through different eyes and has named different patterns of stars according to different rules. The ancient Greeks had 12 constellations in their zodiac, the Chinese, 30. The Greeks' Pleiades (Fig. 1.2) is the representation of the Seven Sisters, but to the Incas this lovely cluster was "cholca," a basket (which it indeed resembles). "Our" constellations, the ones that are used in "western" lore and are now adopted worldwide, are but one set; there are countless others. And they are neither ours nor western, as the oldest go back thousands of years to the lands of the middle east, with influences from the neighboring areas. Some figures actually look like something familiar. In most cases, however, the constellations' patterns were meant to represent, not portray, most looking little or nothing like the objects or people they were named after.

The constellations are commonly and neatly divided into "ancient" and "modern," as well as into informal "asterisms" and the mercifully "rejected." Such separation is rather artificial. One might rather think of constellation invention as a human desire that spans the ages and is chaotically continuous, each era providing its own richness to the mix. Still, the traditional groupings provide a good place to start.

1.1 Classical constellations

For a millennium, from Homer to Ptolemy, the astronomers and poets of the developing and mature Greek civilizations codified and expanded the myths of sky-figures that were handed down from forgotten times (Fig. 1.3). In his *Syntaxis*, the great work of mathematical astronomy known in Arabia as the *Almagest*, Ptolemy finally included 48 figures, the "ancient 48," which encompass the bright and obvious patterns (Table 1.1). The remainder of the starry heavens were referred to as "amorphous" (Greek, "amorphotoi") or "scattered," and presented a canvas left for others to paint on at later times. The old figures therefore have been termed the "ancient Greek constellations," though they long predate ancient Greece. The Romans, conquering Greece but also being conquered by its culture, gave to the figures the Latin names by which we know them today.

Fig. 1.3 **The northern ecliptic hemisphere** Constellation figures crowd the sky in an early eighteenth-century map. Centered on the north ecliptic pole, it features the ecliptic and the constellations of the Zodiac around the circular periphery. (Andraæ Cellarii, *Harmonia Cosmographia*, Amsterdam, edition 1708, courtesy of the Rare Book and Special Collections Library, University of Illinois.)

1.1.1 The Zodiac

Among the oldest and most revered figures were those that mark the apparent passages of the Sun, Moon, and planets, the street of the seven classical moving bodies of the sky, from which in part come our seven days of the week. The *Zodiac* (from Greek "zodia," for "animals") is filled with living fertility symbols that mark the time for planting, harvesting, and other human activities. In early classical times, the Sun passed through Aries – the Ram – on the first day of spring, through Virgo the Virgin and her golden sheaf of wheat as autumn approached, and into the "wet quarter" of Capricornus (the Water Goat), Aquarius (the Water Bearer), Pisces (the Fishes) as a prelude to spring. All these constellations are best seen six months opposite the time of solar passage.

From these dozen constellations are taken the occult astrological "signs" of the Zodiac. At the time of their invention, the signs (uniformly 30° wide in spite of the varying sizes of the astronomical constellations) overlapped the constellations and took their "magical" powers from them and their qualities. They no longer overlap because of precession [2.4], the 26 000-year wobble of the Earth's axis that carries the equinoxes and solstices [2.3.3], to which the signs are attached, westward through the constellations. Thus the sign of Libra sits on Virgo, the Virgin's sign on Leo, and so on. There are twelve because, to the nearest whole number, the Moon goes through its phases a dozen times per year. Therefore, every time the Moon passes a particular phase, the Sun is found in the next constellation (or, rather, sign) to the east. The constellation Ophiuchus (the Serpent Bearer) traditionally stands on, or crosses, the ecliptic, the Sun and its attendants actually passing through 13 constellations rather than 12. There is no "sign" for Ophiuchus, however, and he is a part of neither the astrological Zodiac nor of the classical astronomical Zodiac.

Today, the vernal equinox, the summer solstice, the autumnal equinox, and the winter solstice, are respectively held by Pisces, Gemini (technically, Taurus), Virgo, and Sagittarius. Two to three millennia ago, however, they were contained by Aries, Cancer, Capricornus, and Libra; hence the ram's horn ♈ as symbolic of the vernal equinox, the names of the northern and southern limits to the zenithal Sun (the Tropics of Cancer and Capricorn), and the non-animalistic invention of Libra, the Balance (which also marks the claws of Scorpius, making the two a double constellation). Quite by coincidence, the Milky Way cuts across the sky perpendicular to the ecliptic and almost along the solstitial colure (the great circle that runs through the celestial poles and the solstices), intersecting the Zodiac in Gemini and Sagittarius (Figs. 1.4 and 1.5).

As a set, the Zodiacal constellations are different from the other constellation groupings. Over the rest of the sky, the ancients generally singled out only the most visually prominent of patterns. Because of the great spiritual significance

Name	Meaning	Genitive	Abbr.	Loc.[a]	Luminary	Remarks
Andromeda	Chained Lady	Andromedae	And	EN–NP	Alpheratz = Alpha Mirach = Beta	Perseus myth
Antlia	Air Pump	Antliae	Ant	ES	Alpha	Modern
Apus	Bird of Paradise	Apodis	Aps	SP	Alpha	Modern
Aquarius	Water Bearer	Aquarii	Aqr	E–ES	Sadalsuud = Beta	Zodiac; Water Jar
Aquila	Eagle	Aquilae	Aql	E–EN	Altair = Alpha	Summer Triangle
Ara	Altar	Arae	Ara	SP	Beta	
Aries	Ram	Arietis	Ari	EN	Hamal = Alpha	Zodiac
Auriga	Charioteer	Aurigae	Aur	EN–NP	Capella = Alpha	
Boötes	Herdsman	Bootis	Boo	EN	Arcturus = Alpha	
Caelum	Engraving Tool	Caeli	Cae	ES	Alpha	Modern
Camelopardalis	Giraffe	Camelopardalis	Cam	NP	Beta	Modern
Cancer	Crab	Cancri	Cnc	EN	Al Tarf = Beta	Zodiac
Canes Venatici	Hunting Dogs	Canum Venaticorum	CVn	EN–NP	Cor Caroli = Alpha-2	Modern
Canis Major	Larger Dog	Canis Majoris	CMa	ES	Sirius = Alpha	Winter Triangle
Canis Minor	Smaller Dog	Canis Minoris	CMi	EN	Procyon = Alpha	Winter Triangle
Capricornus	Water Goat	Capricorni	Cap	ES	Deneb Algedi = Delta	Zodiac
Carina	Keel	Carinae	Car	SP	Canopus = Alpha	Argo
Cassiopeia	Queen	Cassiopeiae	Cas	NP	Shedar = Alpha	Perseus myth; Andromeda's mother
Centarus	Centaur	Centauri	Cen	ES–SP	Rigil Kentaurus = Alpha	Nearest star
Cepheus	King	Cephei	Cep	NP	Alderamin = Alpha	Perseus myth; Andromeda's father
Cetus	Whale, Sea Monster	Ceti	Cet	E	Deneb Kaitos = Beta	Perseus myth
Chamaeleon	Chameleon	Chamaeleontis	Cha	SP	Alpha	Modern
Circinus	Compasses	Circini	Cir	SP	Alpha	Modern

Table 1.1 **The constellations**

Name	Meaning	Genitive	Abbr.	Loc.[a]	Luminary	Remarks
Columba	Dove	Columbae	Col	ES	Phact = Alpha	Modern
Coma Berenices	Berenice's Hair	Comae Berenices	Com	EN	Beta	"Modern" but old; north Galactic pole
Corona Australis	Southern Crown	Coronae Australis	CrA	ES–SP	Alfecca Meridiana = Alpha, Beta equal	
Corona Borealis	Northern Crown	Coronae Borealis	CrB	EN	Alphecca = Alpha	
Corvus	Crow, Raven	Corvi	Crv	ES	Gienah = Gamma	
Crater	Cup	Crateris	Crt	ES	Delta	
Crux	Southern Cross	Crucis	Cru	SP	Acrux = Alpha	Modern
Cygnus	Swan	Cygni	Cyg	EN–NP	Deneb = Alpha	Northern Cross; Summer Triangle
Delphinus	Dolphin	Delphini	Del	EN	Rotanev = Beta	
Dorado	Swordfish	Doradus	Dor	SP	Alpha	Modern; south ecliptic pole
Draco	Dragon	Draconis	Dra	NP	Eltanin = Gamma	north ecliptic pole
Equuleus	Little Horse	Equulei	Eql	EN	Kitalpha = Alpha	
Eridanus	River	Eridani	Eri	ES–SP	Achernar = Alpha	
Fornax	Furnace	Fornacis	For	ES	Alpha	Modern
Gemini	Twins	Geminorum	Gem	EN	Pollux = Beta	Zodiac; Summer Solstice
Grus	Crane	Gruis	Gru	ES–SP	Al Nair = Alpha	Modern
Hercules	Hero, Hercules	Herculis	Her	EN	Kornephoros = Beta	"The Kneeler"
Horologium	Clock	Horologii	Hor	ES–SP	Alpha	Modern
Hydra	Water Serpent	Hydrae	Hya	E–ES	Alphard = Alpha	
Hydrus	Water Snake	Hydri	Hyi	SP	Beta	Modern
Indus	Indian	Indi	Ind	SP	The Persian = Alpha	Modern
Lacerta	Lizard	Lacertae	Lac	EN–NP	Alpha	Modern
Leo	Lion	Leonis	Leo	EN	Regulus = Alpha	Zodiac
Leo Minor	Smaller Lion	Leonis Minoris	LMi	EN	Praecipua = 46	Modern

Table 1.1 ***(cont.)***

Name	**Meaning**	**Genitive**	**Abbr.**	**Loc.**[a]	**Luminary**	**Remarks**
Lepus	Hare	Leporis	Lep	ES	Arneb = Alpha	
Libra	Scales	Librae	Lib	ES	Zubeneschamali = Beta	Zodiac
Lupus	Wolf	Lupi	Lup	ES–SP	Alpha	
Lynx	Lynx	Lyncis	Lyn	EN–NP	Alpha	Modern
Lyra	Lyre	Lyrae	Lyr	EN	Vega = Alpha	Summer Triangle
Mensa	Table	Mensae	Men	SP	Alpha	Modern
Microscopium	Microscope	Microscopii	Mic	ES	Gamma	Modern
Monoceros	Unicorn	Monocerotis	Mon	E	Beta	Modern
Musca	Fly	Muscae	Mus	SP	Alpha	Modern
Norma	Square	Normae	Nor	ES–SP	Gamma-2	Modern
Octans	Octant	Octantis	Oct	SP	Nu	Modern; south celestial pole
Ophiuchus	Serpent Bearer	Ophiuchi	Oph	E	Rasalhague = Alpha	With Serpens
Orion	Hunter, Orion	Orionis	Ori	E	Rigel = Beta	Winter Triangle
Pavo	Peacock	Pavonis	Pav	SP	Peacock = Alpha	Modern
Pegasus	Winged Horse	Pegasi	Peg	EN	Enif = Epsilon	Perseus myth; Great Square
Perseus	Hero, Perseus	Persei	Per	EN–NP	Mirfak = Alpha	
Phoenix	Phoenix	Phoenicis	Phe	ES–SP	Ankaa = Alpha	Modern
Pictor	Easel	Pictoris	Pic	NP	Alpha	Modern
Pisces	Fishes	Piscium	Psc	E-EN	Kullat Nunu = Eta	Zodiac; Vernal Equinox; Circlet
Piscis Austrinus	Southern Fish	Piscis Austrini	PsA	ES	Fomalhaut = Alpha	
Puppis	Stern	Puppis	Pup	ES–SP	Naos = Zeta	Argo
Pyxis	Compass	Pyxidis	Pyx	ES	Alpha	Modern
Reticulum	Net	Reticulii	Ret	SP	Alpha	Modern
Sagitta	Arrow	Sagittae	Sge	NE	Gamma	

Table 1.1 ***(cont.)***

Name	Meaning	Genitive	Abbr.	Loc.[a]	Luminary	Remarks
Sagittarius	Archer	Sagittarii	Sgr	ES	Kaus Australis = Epsilon	Zodiac; Winter Solstice; Little Milk Dipper; Teapot; Galactic center
Scorpius	Scorpion	Scorpii	Sco	ES	Antares = Alpha	Zodiac
Sculptor	Sculptor	Sculptoris	Scl	ES	Alpha	Modern; south Galactic pole
Scutum	Shield	Scuti	Sct	ES	Alpha	Modern
Serpens	Serpent	Serpentis	Ser	E	Unukalhai = Alpha	Two parts; with Ophiuchus
Sextans	Sextant	Sextantis	Sex	E	Alpha	Modern
Taurus	Bull	Tauri	Tau	EN	Alebaran = Alpha	Zodiac; Hyades; Pleiades Galactic anticenter
Telescopium	Telescope	Telescopii	Tel	ES–SP	Alpha	Modern
Triangulum	Triangle	Trianguli	Tri	EN	Beta	
Triangulum Australe	Southern Triangle	Trianguli Australis	TrA	SP	Alpha	Modern
Tucana	Toucan	Toucanae	Tuc	SP	Alpha	Modern
Ursa Major	Greater Bear	Ursae Majoris	UMa	NP	Alioth = Epsilon	Big Dipper/Plough; UMa cluster
Ursa Minor	Smaller Bear	Ursae Minoris	UMi	NP	Polaris = Alpha	Little Dipper; north celestial pole
Vela	Sails	Velorum	Vel	ES–SP	Regor = Gamma	Argo
Virgo	Maiden	Virginis	Vir	E	Spica = Alpha	Zodiac; Autumnal Equinox
Volans	Flying Fish	Volantis	Vol	SP	Gamma	Modern
Vulpecula	Fox	Vulpeculae	Vul	EN	Anser = Alpha	Modern

[a] Approximate location as follows:
E: Equatorial; lying on the celestial equator.
EN: Equatorial north; between the celestial equator and 45 degrees north declination.
NP: North polar; north of 45 degrees declination.
ES: Equatorial south; between the celestial equator and 45 degrees south declination.
SP: South polar; south of 45 degrees south declination.
Intermediate positions are indicated by combining location codes.

Fig. 1.4 **Gemini** The most northerly constellation of the Zodiac, lying in a faint part of the Milky Way, Gemini features two bright stars, Castor (the fainter) and Pollux. (Courtesy of Akira Fujii.)

of the Sun and planets, such was not true in the Zodiac. There had to be 12 constellations (or signs) whether the stars were bright or not. Thus we see quite a mixture, from brilliant Taurus [Fig. 4.7], Gemini, Scorpius [Fig. 9.9], and Leo (which really rather do look like what their names suggest) to dim figures like Cancer and Pisces that anywhere else in the heavens may well have been among the amorphotoi.

The Zodiac hosts a half-dozen prominent stars, the first magnitude [3.1] class K giants [6.2, 6.3] Aldebaran of Taurus and Pollux of Gemini (Fig. 1.4), the class B dwarfs Regulus of Leo and Spica of Virgo, the red supergiant [14.1] Antares of Scorpius, and second-magnitude sextuple class A star Castor (also of Gemini). The Zodiac's stars are also special because they are regularly occulted (or covered) by the Moon, which allows a measure of their angular diameters [7.2.2].

Several asterisms, or informal patterns, dot the Zodiac. Here we find the Pleiades and Hyades of Taurus, both of which are nearby open clusters [9.2], the latter making the Bull's thrusting head. Aldebaran, lying in the Hyades line of sight, burnishes the cluster, but is not a part of it. Within dim Cancer lies one of the few Christian allusions of the sky, the four-star "manger" that encloses the Praesepe (Beehive) open cluster, the two brightest stars "the asses." Leo is known for the "Sickle" that makes his head, Scorpius for its stellar "arteries" that flank Antares, Sagittarius for the southward-pouring "Little Milk Dipper" (which itself is part of the "Teapot," Fig. 1.5). Moving back northward, you can find Aquarius's prominent four-star "Water Jar" and Pisces' "Circlet."

1.1.2 The rest

Away from the Zodiac, the constellations take on more of a story-telling, rather than mystical, role. A great many fall into mythological, or at least logical, groupings. Perhaps the greatest of them, since it is seen from nearly all latitudes of the Earth (winter in the far northern hemisphere, summer in the southern), is centered on Orion [Figs. 3.2, 3.4], the Hunter, in mythology lover of Diana, goddess of the Moon (and in one story accidentally killed by her). Few figures are as arresting. Like Scorpius (which was Orion's killer in another myth), Orion is composed of related stars, the former from the Scorpius–Centaurus OB association complex [9.3], Orion from the Orion OB1 association (which averages about 450 parsecs [4.2.1] away). As such, the figures are filled with brilliant, massive blue-white O and B stars, many multiple

Fig. 1.5 **Sagittarius** In mythology, an archer and centaur, Sagittarius is better known for its delightful informal figures, the obvious "Little Milk Dipper," its handle stuck in the Milky Way, and the "Teapot," of which the Milk Dipper is the left-hand side. The Milky Way shines at right. The bright red glow at right center is the Lagoon Nebula. (J.B. Kaler.)

Fig. 1.6 **The Big Dipper or Plough** The best-known of all asterisms, the Big Dipper (the Plough in Britain), glides across the open slit of the University of Arizona's 2.3-meter telescope atop Kitt Peak. Mizar and Alcor, a naked-eye double star, are toward the top right. (J.B. Kaler.)

[8.1]. Here in Orion we find two first magnitude supergiants, blue class B Rigel and the red class M variable Betelgeuse, which closely match each other in visual brightness.

Accompanying the Hunter are two faithful companions and a rather sad bit of prey. To the southeast, following Orion across the sky, is majestic Canis Major, the Larger Dog, with the sky's stellar luminary, white class A Sirius, the "Dog Star" that reaches almost to minus second magnitude. Few stars have been so important, as the "heliacal rising" of Sirius (its first visibility in the glow of morning twilight) once announced the rising of the Nile. Below Sirius is a bright triangle topped at the northern end by Adhara (Epsilon Canis Majoris), which is either the faintest first magnitude star or the brightest of second magnitude (just topping Gemini's Castor), depending on which list you use. To the northeast lies little Canis Minor (the Smaller Dog). Though it consists of but two stars, the brighter, class F Procyon, provides first magnitude luster. Beneath the Hunter is Lepus the Hare, whose brightest star ranks only third magnitude. Obviously named by northerners, Procyon, Betelgeuse, and Sirius make the "Winter Triangle," which is seen in the southern hemispheres' summer heat. Other obvious asterisms include Orion's jeweled three-star "Belt," whose western star closely marks the celestial equator, and the dangling "Sword," which encloses the glowing Orion Nebula [11.1.3].

Three other classical groupings rival Orion and his family for fame, two in the northern celestial hemisphere, one in the southern. Among the most beloved of figures is Ursa Major, the Great Bear, outlined primarily by the striking seven-star asterism known as either the "Big Dipper" or "Plough" (Fig. 1.6). All but one of its stars second magnitude, the middle five are related, and are actually the pinnacle of a cluster some 25 parsecs [4.2.1] away. The second star in from the end of the Dipper/Plough's handle and its naked-eye companion, Mizar and Alcor, the Arabs' "horse and rider," vie for notoriety with the "Pointers." These rather well point toward Polaris, the second magnitude (and Cepheid variable [10.3]) North Star, which lies within a degree of the north celestial pole [2.2]. In parallel with Ursa Major, Polaris stands at the end of the handle of the Little Dipper, which is the major asterism of Ursa Minor, the Smaller Bear [Fig. 10.4]. Following along behind the Great Bear as he rounds the pole is the Herdsman, Boötes, which is anchored by the orange class K giant star Arcturus, the luminary (though not by much) of the northern hemisphere.

Seen in the spring and summer polar skies of the northern hemisphere, the Great Bear treads opposite the large Perseus group, which is led in story by Cassiopeia [Fig. 14.5], the Queen, her "W" figure making most of her uncomfortable-looking "Chair." To the west lies her husband, dim Cepheus the King, notable more for Delta Cephei, the prototype of the Cepheid variables, than for any great kingly deed. They are parents to Andromeda, made of two graceful streams of stars that emanate from Pegasus, the Flying Horse (both south of Cassiopeia), upon which Perseus rode after slaying Cetus the Sea Monster [Fig. 10.1] before it devoured the maiden (Fig. 1.7). Pegasus is renowned for its Great Square, Perseus for Algol (Beta Persei), the paradigm of the eclipsing variable stars [8.4], and Cetus for Mira (Omicron Ceti), which heads the list of the long period variables [10.5.1].

To the southern hemisphere belongs the great ship Argo, which sails on northern winter – southern summer – seas to the south and east of Canis Major, forever carrying Jason on his quest of the Golden Fleece. Almost unmanageably huge as a sector of the sky, Argo was separated in the nineteenth century into three portions that are now formal constellations in their

Fig. 1.7 **Pegasus, Andromeda, and Cassiopeia** Three of the constellations of the Perseus myth rise above the eastern horizon, the "Great Square" of Pegasus at right, Andromeda at lower center, the "W" of Cassiopeia at left. The Andromeda Galaxy, M31, is just below the center. (J.B. Kaler.)

own right: Puppis (the Stern), Carina (the Keel), and Vela (the Sails), giving us a modern count of ancient constellations of 50. Set into a stunning portion of the Milky Way, Carina contains the second brightest star, Canopus (Alpha Carinae), the extensive Carina Nebula, and one of the best of all supernova candidates [14.2], Eta Carinae.

Loosely associated with Argo are Centaurus the Centaur (Chiron, Jason's mentor, to the east of Argo), Hydra (the Water Serpent, and the longest constellation in the sky, seen to the north of Argo), and even mighty Hercules. Centaurus is at the heart of its own group, the Centaur seen strangling Lupus the Wolf upon Ara the Altar. This large constellation holds the closest and third brightest star (Alpha Centauri), as well as another of first magnitude (the B dwarf Hadar, Beta Centauri), and the grandest of all globular clusters [9.4], Omega Centauri. Hercules, which lies far to the north and contains the major northern globular cluster M13, is – in addition to belonging loosely to the Argo group – also a distinctly singular constellation, the descendent of the mysterious "Kneeler," the hero usually depicted upside down.

And again Ophiuchus rises both as himself and with further Argonian connection. Set north of Scorpius, the sprawling figure represents Asclepias, the physician aboard the Argo and the healer of the Trojan Wars, whose snake-wrapped body descended to us to become the physician's symbol. The Serpent is the only constellation entirely divided into two parts: Serpens Caput (the Serpent's Head, to the west of Ophiuchus) and Serpens Cauda (the Tail, to the east). It is nevertheless still treated as a single constellation, the two parts making individual asterisms.

Single constellations are scattered across the sky, some bearing loose relation to each other or to larger groups. Among them we watch three birds. Cygnus (the Swan) [Fig. 4.3] and Aquila (the Eagle) fly the northern Milky Way with the first magnitude A supergiant Deneb and the A dwarf Altair in tow, while Corvus the Crow boldly inspects the terrain to the west of Virgo. Tip Cygnus upside down and you find the informal Northern Cross. Each hemisphere also contains a crown – Corona Borealis (the Northern Crown, worn by Ariadne) and Corona Australis (the Southern Crown, perhaps belonging to Sagittarius). Between the two Bears winds long Draco the Dragon, and to the west of Orion runs even longer Eridanus, the River, which ends in the blue B dwarf Achernar. Within the set there is yet another wet-quarter fish (Piscis Austrinus, the Southern Fish, with the first magnitude A dwarf Fomalhaut), Orpheus's harp (Lyra, with the luminous A dwarf Vega), an arrow (Sagitta), a dolphin (Delphinus), a triangle (Triangulum, near Aries), a little horse (Equuleus, nicely near Pegasus), an amazingly dim cup (Crater, to the west of Corvus and on top of Hydra), and quite wonderful Auriga, the Charioteer, who rides to the north of Orion holding the double G giant Capella, the "she-goat," under his powerful arm. Arcturus, Vega, and Capella, of nearly equal brightness, somewhat tripartite the northern sky, while Vega, Deneb, and Altair make the northern "Summer Triangle," within which begins the Great Rift of the Milky Way, the two white celestial rivers flowing south to Scorpius and Sagittarius and beyond to the wonders of the deep southern sky.

1.2 Modern constellations

Fig. 1.8 **Centaurus and the Southern Cross** A modern invention in the middle of the bright Milky Way, with two stars of first magnitude; Crux (the Southern Cross) precedes Alpha and Beta Centauri (Alpha to the left) across the southern-hemisphere sky. A dark cloud of interstellar dust, the "Coalsack," lies to the left of the Cross. (Courtesy of Akira Fujii.)

Giving a whole new meaning to the concept of "modern," the modern constellations go back four centuries, and in one case much farther. After Ptolemy put the imprimatur on the old figures, locking them into place in the *Syntaxis*, constellation invention went into a long hiatus, their memory and Ptolemy's great work kept alive and intact by the great Arabian astronomers, whose status peaked in the tenth to twelfth centuries. Not until the end of the sixteenth century, when the excitement of scientific revolution that had begun with the Copernican realization of an Earth that orbited the Sun (instead of vice versa), did astronomers look skyward again to fill in the blanks (and at the same time to try to make themselves famous). Thus began a 200-year period of furious invention of obscure constellations made of mostly faint stars, plus those that surround the south celestial pole that were invisible from classical lands.

Though many astronomers – including Edmund Halley – tried to get into the act, most of the Greek amorphotoi were carved into constellations by Johann Hevel (1611–1687), better known as Hevelius, in his *Uranographia* of 1687. Here we find five animals and two artifacts that include well-known Canes Venatici (the Hunting Dogs, tucked into the curve of the Big Dipper's handle) and Scutum (the Shield, in the heart of the Milky Way, south of Aquila and known best for a bright patch of stars). The southern hemisphere was fertile territory for, among others, Nicolas de Lacaille, who some 50 years later introduced over a dozen then-modern artifacts that include the south celestial pole's Octans (the Octant), Antlia (the Air Pump), and Fornax (the Furnace). Others found the south polar region a fine place for new creatures, including Pavo (the Peacock), the eponymous Phoenix, Volans (the Flying Fish), and Musca Australis (the Southern Fly), which, because of the fortunate swatting of Musca Borealis (the Northern Fly), simply became known as "Musca."

Many of the southern moderns are far from obscure. They include one of the heavens' brightest and most famed constellations, Crux (the Southern Cross), which was stolen from the feet of Centaurus, the Centaur (Fig. 1.8). Another beauty, Coma Berenices (Berenice's Hair), spans the times. Due south of the Dipper's handle and Canes Venatici lies a lacy sprawl of fourth and fainter magnitude stars that belong to the Coma Berenices open star cluster, which lies 80 light years away (and is not to be confused with the vastly more distant Coma Berenices cluster of galaxies). The formal constellation actually extends beyond the cluster and includes stars that are not part of it. Though of ancient origin and alluded to by Eratosthenes in the third century BC, Coma Berenices finally achieved constellation stardom in the sixteenth century from none other than the best of the pre-telescopic observers, Tycho Brahe, thus connecting the ancient constellations with the moderns.

A great many modern constellations did not make the final cut for entry into the celestial pantheon. Edmund Halley invented Robur Carolinium ("Charles's Oak," honoring King Charles II), Maximillian Hell created Telescopium Herschelli (Herschel's Telescope), and Johannes Bode (he of Bode's law fame that gives the distances of planets from the Sun), that all-time champion of obsolete constellations, came up with – among many others – Officina Typographica (the "Printing Office") and Machina Electrica (the "Electric Machine"). Bode's stunning *Uranographia* of 1801, an atlas so good it could be used by backyard observers today, contains nearly 100 figures. Around 1930, the organization of professional research astronomers, the International Astronomical Union (IAU), put an end to the madness by selecting 38 modern figures, establishing the official number at 88, and by drawing rectangular boundaries around them all, thus relegating each star permanently to a proper place (see Table 1.1). The IAU also adopted simple three-letter abbreviations, Orion called Ori, Ursa Major UMa, and so on.

1.3 Common star names

Over the whole dark sky, with no moonlight or artificial lighting, nearly 10 000 stars are visible to the youthful unaided eye, stars of all different colors, sizes, masses, ages. To study the stars, to understand them and their life processes, we must first have some way of keeping them straight, of naming them. A remarkable variety of schemes is in place, some general, other specific to certain types. Most brighter stars can have dozens of names, which, given the number of stars, is enough to keep an accountant happy for ages.

1.3.1 Proper names

Start with proper names, names like those given to us: Mary, Bob, Fiona, Nigel. Stars carry proper names from a variety of languages, first (though not in number) Greek. "Sirius," for the Dog Star, means – rather obviously – "searing," or "scorching," certainly appropriate for the visually brightest star in the sky, one whose low altitude [2.3.1] above the horizon from northern lands allows the Earth's turbulent atmosphere to make it flash like the finest diamond. "Procyon," "before the Dog" in Greek ("pro-kion"), rises before Sirius (and like a heavenly butler announces one of the great stars of the sky), and "Canopus" – the second brightest star – refers to a man. Next the Romans speak with Latin, "Regulus" in Leo meaning "the Little King," "Polaris" the "pole star" (Table 1.2).

The great prize, however, goes to the Arabs of the middle ages. Both the Romans and those of the Arabic empire held Greek culture in high esteem; after all, the Romans adopted not only the Greek constellations, giving them Latin names, but some of Greek mysticism as well. The ancient Arabs had their own constellation lore, which can still be seen today in asterisms such as "the Virgins," the triangle of bright stars beneath Sirius. While some extolled Greek stellar virtues, others sacked the library at Alexandria, where Greek literature and knowledge were stored. With great fortune, however, much of the Greek writings – including those of Ptolemy – had been translated into Arabic.

Holding the "Greek" constellations in high regard, the Arabians gave names to many of the stars according to their places within them. Many Arabic star names are recognizable because they begin with the article "the," or "al." Thus "Algieba" means "the side", while "Algedi" refers to "the kid" for its place in Capricornus, the (Sea) Goat. Mixed into this system are the indigenous Arabic stars. The star at the end of Ursa Major's tail is "Alkaid," actually "Alkaid al Benatnasch," which, while obscure, means something like "the chief of the daughters of the bier"; in the Dipper (Plough), the Arabs saw the symbolism of a funeral.

The results of this astronomical exploration leaked back to the Europeans, who then translated much of it. Imagine the confusion, rather like Proust being translated by a first-year student of French. Mistranslations, corruptions, and contractions dominate star names, though not all: "deneb" simply means "tail," and was assigned chiefly to the star at the tail of Cygnus the Swan, our Deneb. Among the most famous misrepresentations is "Betelgeuse" (referring to the bright red supergiant in Orion), which is commonly given as the "armpit of the giant," but really refers to "yad al-jauza," the "hand of the central one," having nothing whatever to do with either a hunter or a giant ("al-jauza" seeming to refer to a female figure). And these examples are only a start.

Worse, Arabic names can refer to more than one star (or the wrong stars altogether). Thus "Deneb" also refers to stars in Leo ("Denebola," the tail of the Lion), Cetus ("Deneb Kaitos," the tail of the Whale), and "Deneb Algedi" (the tail of Capricornus). "Gienah" is the wing of both Cygnus the Swan and of Corvus the Crow! Worse yet, 2000 star names are mightily hard to recall. The authors of the Yale *Bright Star Catalogue*, who give a near-complete compilation of proper names, remarked about the list that "the very bulk of names should go far to discourage their use." Yet we love them still, and the names of the brightest stars – the 20 or so of first magnitude (plus a few others) – are used even in the professional literature, providing a connection and a compliment to the ancients.

1.3.2 Letters and numbers

A better system is clearly required. In the years preceding the sixteenth century, Tycho Brahe, a Danish nobleman, became deeply interested in astronomy, much of it fired by his observation of a brilliant "new star," then called a "nova" (now known to be a supernova) in the constellation Cassiopeia. The Danish King awarded Tycho an island off the mainland for an observatory. As there were yet no telescopes, the observatory consisted of angle-measuring devices, including a large quadrant, plus many assistants to help with the work. With his devices, Tycho constructed a catalogue of 777 stars (plus the observations of the planets that allowed Johannes Kepler to establish the laws of planetary motion, which in turn allowed Newton to formulate the law of gravity, which with Galileo's work on inertia led to the birth of physics . . .).

Name[1]	Derivation	Greek Letter Name	Meaning of Original Name[2]
Sirius	Greek	Alpha Canis Majoris	Scorching One
Canopus	Greek	Alpha Carinae	Proper name
Rigil Kentaurus	Greek	Alpha Centauri	Centaur's Foot
Arcturus	Greek	Alpha Boötis	Bear Watcher
Vega	Arabic	Alpha Lyrae	Swooping Eagle
Capella	Latin	Alpha Aurigae	She-Goat
Rigel	Arabic	Beta Orionis	The Foot of *al-jauza*[3]
Procyon	Greek	Alpha Canis Minoris	One Preceding the Dog
Achernar	Arabic	Alpha Eridani	River's End
Betelgeuse	Arabic	Alpha Orionis	Hand of *al-jauza*[3]
Hadar	Arabic	Beta Centauri	Untranslated
Altair	Arabic	Alpha Aquilae	Flying Eagle
Aldebaran	Arabic	Alpha Tauri	Follower (?)
Acrux	. . .		
Antares	Greek	Alpha Scorpii	Like Ares (Mars)
Spica	Latin	Alpha Virginis	Ear of Grain
Pollux	Latin[4]	Beta Geminorum	Proper name
Fomalhaut	Arabic	Alpha Piscis Austrini	Mouth of the Southern Fish
Deneb	Arabic	Alpha Cygni	Hen's Tail
Mimosa	. . .	Beta Crucis	Unknown significance
Regulus	Latin	Alpha Leonis	Little King
Adhara	Arabic	Eta Canis Majoris	Virgins
Castor	Greek	Alpha Geminorum	Proper name
Bellatrix	Latin	Gamma Orionis	Female Warrior
Alphard	Arabic	Alpha Hydrae	Solitary One
Polaris	Latin	Alpha Ursae Minoris	Of the Pole
Mizar	Arabic	Zeta Ursae Majoris	Groin
Saiph	Arabic	Kappa Orionis	Giant's Sword
Rasalhague	Arabic	Alpha Ophiuchi	Head of the Serpent Collector
Algol	Arabic	Beta Persei	Demon's Head
Denebola	Arabic	Beta Leonis	Lion's Tail
Zubeneschamali	Arabic	Beta Librae	Northern Claw (of the Scorpion)
Zubenelgenubi	Arabic	Alpha Librae	Southern Claw (of the Scorpion)
Vindemiatrix	Greek	Epsilon Virginis	Grape Gatherer
Alcyone	Greek	Eta Tauri	Proper name (one of the Pleiades)
Cor Caroli	Latin	Alpha Canum Venaticorum	Charles' (II) Heart
Porrima	Latin	Gamma Virginis	Proper name (Roman goddess)
Albireo	Latin	Beta Cygni	From "the bird"[5]
Rasalgethi	Arabic	Alpha Herculis	Kneeler's Head
Thuban	Arabic	Alpha Draconis	Serpent's Head
Alcor	Arabic	80 Ursae Majoris[6]	Black Horse
Merope	Greek	23 Tauri[6]	Proper Name (one of the Pleiades)
Pleione	Greek	28 Tauri[6]	Proper Name (one of the Pleiades)

[1] Names of first magnitude stars and others for which proper names are commonly used; presented in order of apparent brightness.
[2] Translations taken from *Short Guide to Modern Star Names and their Derivations*, P. Paul Kuntizsch and Tim Smart, Wiesbaden, Otto Harrassowitz, 1986.
[3] *al-jauza* in some way refers to a feminine "middle" figure.
[4] Originally derived from Greek.
[5] Named from a commentary on a mistranslation of a Greek word.
[6] Flamsteed number.

Table 1.2 **Commonly used proper names of bright stars**

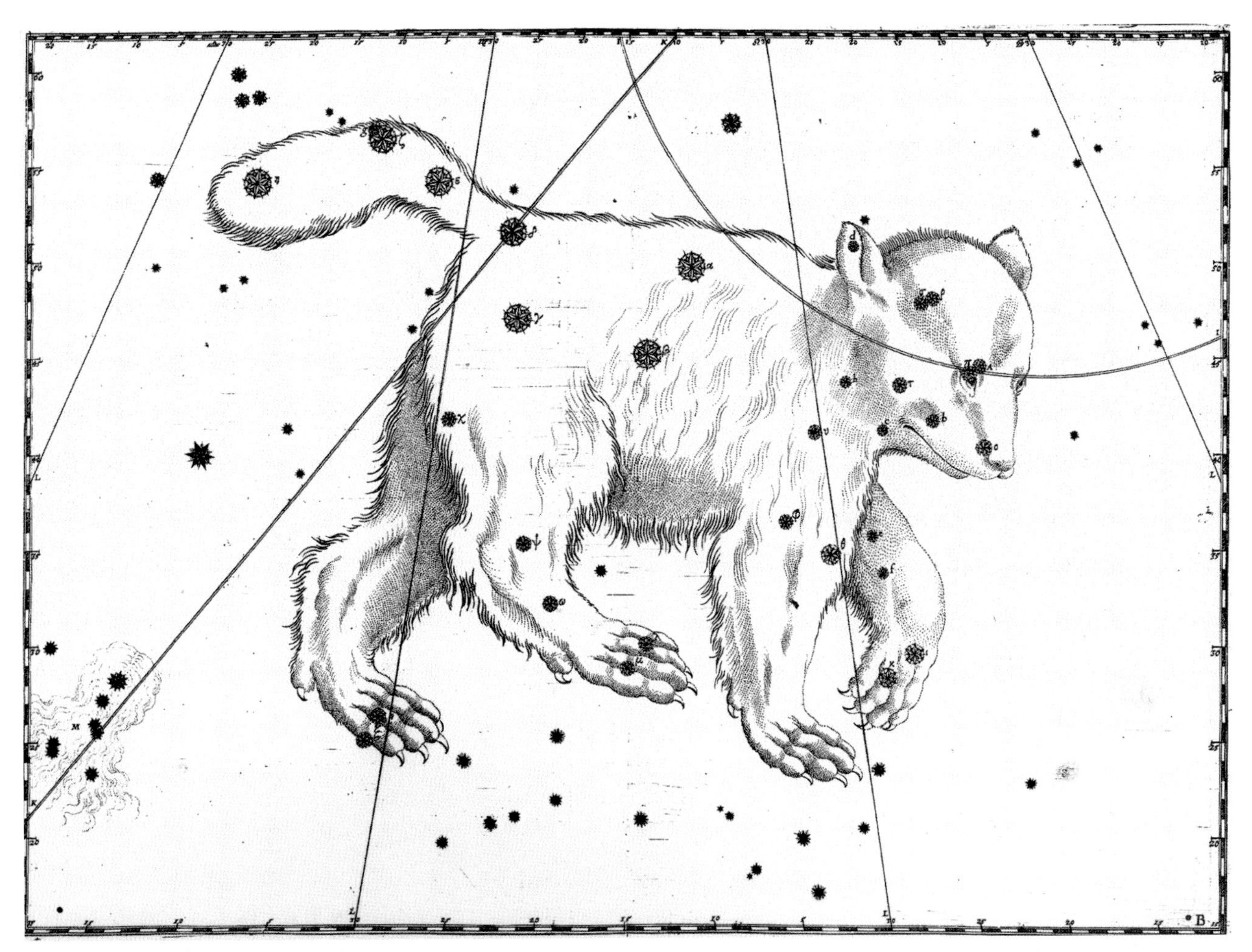

Fig. 1.9 **Bayer's Ursa Major** Ursa Major, the Great Bear, is seen walking beneath the north celestial pole. The Dipper/Plough stars are given Greek letters in sequential order, the rest more or less in order of brightness. Pairs of stars that mark three of the bear's feet were called "leaps of the gazelle" by the Arabs. Coma Berenices appears below and to the left. (Johannes Bayer, *Uranometria*, 1603, courtesy of the Rare Book and Special Collections Library, University of Illinois.)

Tycho's observations were the best by far. With them (and with other data of lesser quality), Johannes Bayer, working in today's Germany, constructed the foundation stone of the modern star atlas, his *Uranometria* ("measure of the heavens") of 1603. Each constellation recognized by Ptolemy has its own page (Fig. 1.9). Within each, he labelled the stars with lower-case Greek letters, "alpha" (α), "beta" (β), and so on. The first rule was to call out a constellation's stars according to visual brightness. Hence Sirius became the alpha star of Canis Major. But in English, the constellations are in Latin. Hence "Alpha of Canis Major" is implied by a Latin possessive case ending, and becomes "Alpha Canis Majoris." Literate astronomers therefore need learn not just 88 constellations and their abbreviations, but 88 genitives (possessives) as well. To shorten star names, astronomers commonly use the constellation abbreviation, which stands for both the nominative case (the actual constellation name) and the genitive, rendering Alpha Ursae Majoris (the Arabic "Dubhe") Alpha UMa or α UMa, Beta Canis Majoris (Mirzam) Beta CMa or β CMa, and so on.

Unfortunately, Bayer had other criteria than brightness in mind and did not bother to tell anyone what they were. The brightest star in a constellation, especially if of the first magnitude, is indeed commonly "Alpha" (though there are many exceptions). But number four is rarely "Delta." In some constellations, the Greek letters go from one end to the other, the seven bright stars of Ursa Major given Alpha through Eta from west to east. In other figures he clearly labelled within internal groups. Beta Canis Majoris, near Sirius at the constellations's northern end, is not even close to being the next-brightest star after Sirius, that title instead going to Adhara – Eta CMa – far down below. The Greek letter system is in constant use in the professional and amateur literature. Outside of the first magnitude few, Greek letters are in fact the default. So to the 88 constellation names, 88 abbreviations, and 88 possessives, add the 24 Greek letters.

Twenty-four is not large compared with the number of visible stars. The Greek letters can be enhanced by using superscripts for close groupings (for example, Pi-1 through Pi-6 Orionis, commonly written as π^1, π^2, etc.), usually numbered west to east, but that scheme has its limitations. To increase the number of star names, when Bayer was done

Fig. 1.10 **Orion and Taurus** Orion the Hunter and Taurus the Bull battle each other in a page from Flamsteed's atlas, which is marked with both equatorial and ecliptic coordinates. The Flamsteed numbers were not applied until much later. (John Flamsteed, *Atlas Coelestis*, edition of 1729, courtesy of the Rare Book and Special Collections Library, University of Illinois.)

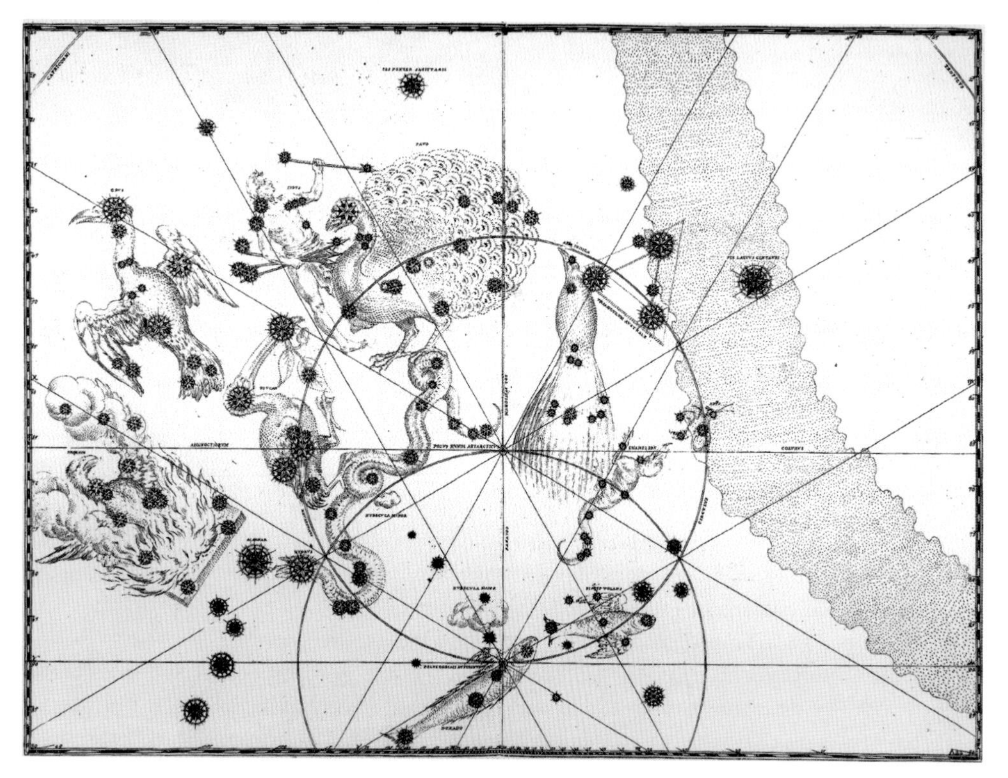

Fig. 1.11 **Bayer's southern hemisphere** The "modern constellations" begin to show up on Bayer's map of the southern hemisphere, where we see Phoenix, Hydrus, Chameleon, and other animals. (Johannes Bayer, *Uranometria*, 1603, courtesy of the Rare Book and Special Collections Library, University of Illinois.)

with the Greek letters, he added lower-case Roman letters, then upper case letters, stopping in the extreme at "Q." Most Roman letters were dropped and ring loudly today in only a few instances, such as "G Scorpii" and in the most famous case "h Persei," one of the clusters of stars [9.2] that make the Double Cluster in Perseus (the other "Chi Persei").

Over a century later, a better system was initiated by John Flamsteed (1646–1719), the first of England's Astronomers Royal, who measured stellar positions telescopically with unprecedented accuracy (which were desperately needed for navigation). His catalogue was the base for the *Atlas Coelestis*, first published in 1729, ten years after his death (Fig. 1.10). Around the time of the publication of the third edition in 1781, the French astronomer Joseph de Lalande numbered the stars within each of the constellations from west to east, in order of right ascension [2.3.3] (though there is some argument that Newton and Halley did it first). Though Flamsteed never knew of them, these Flamsteed numbers honor the great astronomer to this day.

In Bayer's time, nor even in Flamsteed's, there were no formal constellation boundaries. Constellations were instead located by artistically impressive drawings of the mythological figures. Stars that fell outside of them might receive Greek letters too, even if now they are included in another constellation. Thus Alpha Canis Majoris (Sirius) was also listed by Bayer as Pi Leporis, and Rigel in Orion as Xi Leporis, names by which they are *never* now known. Johannes Bode, in his *Uranographia*, solved the whole problem of where the stars belong simply by drawing dotted boundary lines. Such lines were curved and fluid and depended on the author until the IAU set the rectangular borders we have today. Since the Bayer letters and Flamsteed numbers were assigned before the final boundaries were drawn, however, an unfortunate few stars are exiled to the wrong constellations: 25 Camelopardalis is in Auriga, 77, 78, and 79 Draconis are in Cepheus, and 34 Vulpeculae is in Pegasus.

Two bright stars that link constellation pairs and are important to both figures merit special attention. Beta Tauri, the Bull's northern horn, is also Gamma Aurigae, and the star that anchors the southern end of Andromeda, Alpha Andromedae, makes the northern corner of the Great Square of Pegasus and as such is Delta Pegasi. The IAU boundaries place the stars in Taurus and Andromeda, so in a sense "Gamma Aur" and "Delta Peg" no longer exist. Two other ancient constellations contain additional anomalies. Bayer assigned Greek letters to Argo. When the ship was broken up, the letters went with the parts, Alpha and Beta to Carina, Gamma and Delta to Vela, Zeta to Puppis, and so on. There are therefore no alpha stars in either Vela or Puppis, nor zetas in Carina or Vela. The Greek letters were also distributed among both parts of Serpens: Alpha through Eta Serpentis are on the west side (Serpens Cauda), while Zeta and Eta Serpentis lie to the east, in Serpens Caput.

Tycho could not see very far into the southern sky. From Denmark, stars below declinations 35° or so south [2.3.2] are perpetually below the horizon (Fig. 1.11). Even Bayer was limited to declinations north of about −40°. He therefore had to rely on reports and measures by others, including explorers to the southern hemisphere who were hardly accurate observers. Thus Alpha and Beta Sagittarii are erroneously given alphabetic first and second rank and presented as such on Bayer's chart, when in fact they are far down the brightness list. His charts for both Argo and Centaurus contain large errors, and Crux is hopelessly misplaced. Nor were the numbers of stars on his southern map as great as they are in the north. Other astronomers, particularly Lacaille, later assigned additional Greek letters. (Others re-assigned letters in both the northern and southern hemispheres, leading to occasional confusion.)

Flamsteed too was limited: from Greenwich, no star below −36° declination can be seen. Indeed, Flamsteed's atlas stops at −30°, below which there are no Flamsteed numbers. Similar numbers were assigned by Bode, but his maps were so cluttered with now-defunct constellations that the numbers could hardly prevail. Bode's numbering system survives only through the names of a few special objects, particularly the giant nebula in the Large Magellanic Cloud known both as the Tarantula Nebula and 30 Doradus [11.1.3; Fig. 4.11], and the great southern globular cluster 47 Tucanae [9.4.1].

1.4 Catalogue names

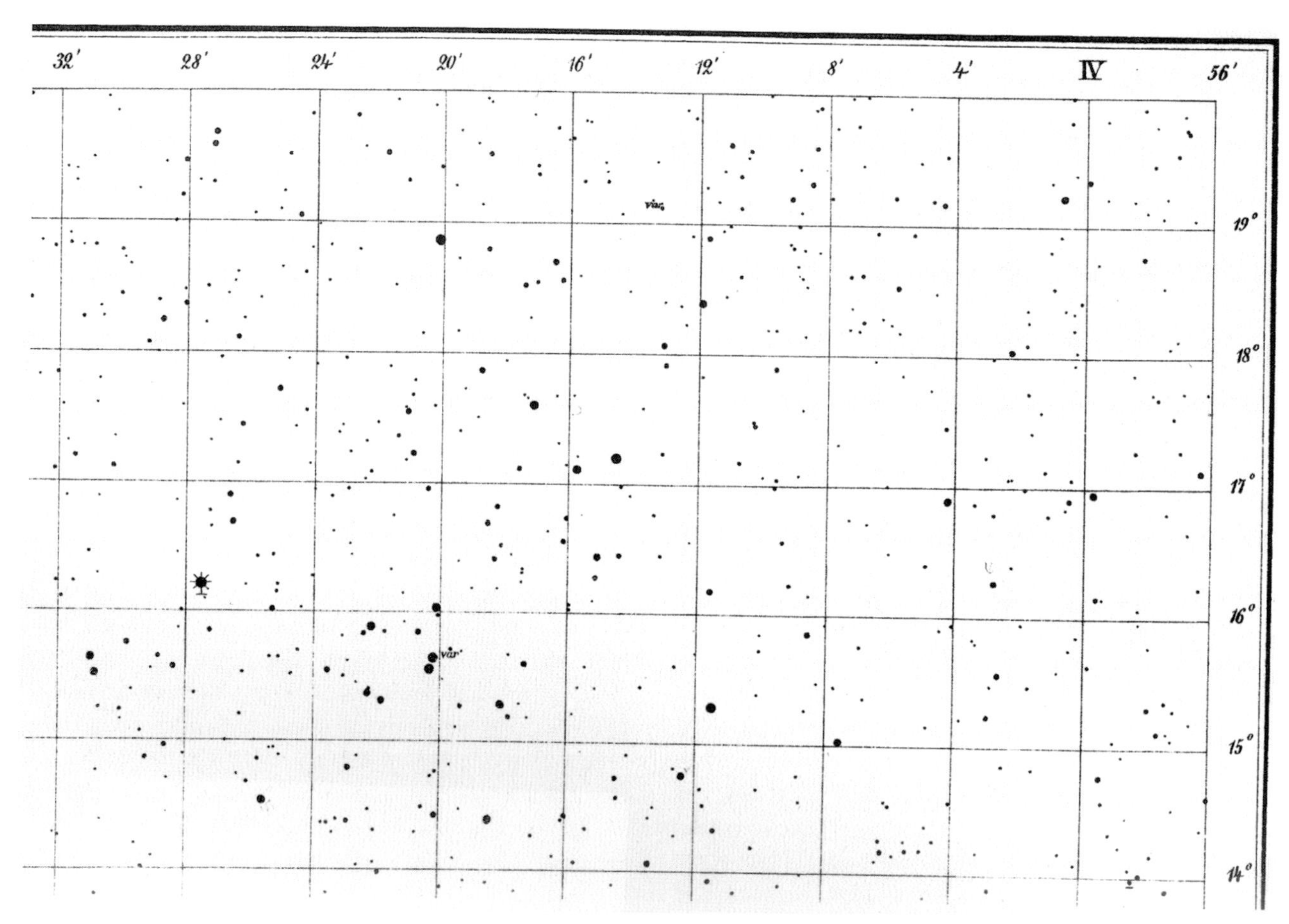

Fig. 1.12 **The Bonner Durchmusterung** A page from the "Bonn Survey" shows the Hyades star cluster and the bright foreground star, Aldebaran. The hand-plotted atlas has a one-degree grid. The stars are numbered from west to east within each declination band. The center-left portion of the atlas appears in Fig. 1.13. (Argelander F. W. A. (1899). *Atlas des Nördlichen Gestirnten Himmels*, Bonn: A. Marcus and E. Weber's Verlag.)

A new era of star names dawned in the mid nineteenth century with the move toward much deeper telescopic catalogues by the German astronomer Friedrich Argelander (1799–1875) of the Bonn Observatory. With a relatively small telescope, he and his assistants began the visual measurement of the coordinates of what would eventually be all the stars of the sky brighter than around 10th magnitude [3.1], the positions generally correct to within a few seconds of arc. This monumental celestial survey was published in several phases that were eventually combined into two separate works, one for the northern hemisphere (or for the stars visible therefrom), the other for the southern.

Each set (or subset) consists of a catalogue that gives both magnitude and position plus a folio-sized atlas on which the stars were hand-plotted on large charts. The first subset (containing 325 000 stars and published in 1859) goes from the north celestial pole to $-2°$ declination, while the next (adding 135 000 stars and produced 27 years later) extends the work to $-23°$. Working from Cordoba, Argentina, astronomers then took up the challenge of the southern hemisphere, making the measurements and publishing the results of 614 000 stars in stages from $-23°$ to the south celestial pole. This vast undertaking, which altogether contains just over a million stars, was finally completed in 1929. The northern portion down to $-2°$ is called the *Bonner Durchmusterung*, "Bonn Survey," or just the BD (Fig. 1.12). The extension from $-2°$ to $-23°$ is the *Südliche Bonner Durchmusterung*, the "Southern Bonn Survey," or "SBD." The two were later combined into one, and are known together as simply the BD. The part from $-23°$ to the pole is the *Cordoba Durchmusterung*, or the CD. The original BD is for the equinox for 1855 (the year for which the coordinates are valid relative to precession), the CD, 1875 [2.4].

The charts have no constellations, neither fanciful figures nor boundaries, no markings of any kind; just huge numbers of stars plotted by position. Argelander divided his catalogue – and atlas – into one-degree-wide declination strips, and then serially numbered the stars in each strip beginning at 0 hours right ascension. Arcturus is thereby BD+19° 2777, the 2777th numbered star between +19° and +20°, while Sirius is BD−16° 1591, the 1591st star between −16° and −17°. Below −23°, the stars were given CD numbers; Achernar is also CD−57° 334. The positions of the southern stars between −18° declination and the south celestial pole were also measured photographically in the *Cape Photographic Durchmusterung*, published in 1896. The catalogue's 455 000 stars (75 percent the number in the CD even though covering more sky) are named similarly to those of the BD and CD (and have the same equinox of 1875), but are prefixed CPD; Alpha Corona Australis, for example, is both CD−38°13350 and CPD −38° 7723. (The running numbers of the CD and CPD track each other within a declination strip and are sometimes even identical.) For simplicity, the degree symbols in these catalogues are now usually dropped, Arcturus becoming just BD+19 2777.

Though the BD and CD (and CPD) prefixes are commonly combined into DM for "Durchmusterung" (Arcturus is thus also DM+19 2777), the two really are separate. A larger telescope was used for the southern hemisphere (from −2° down), so more stars are included. Moreover, BD and CD are set for slightly different equinoxes, and as a result the original positions do not quite match. (Obviously, they can be converted to any common year one chooses). Precession has also taken a bit of a toll on the names, which always refer to the original adopted equinox. Over the past 150 years, precession has shifted many stars out of their original declination bands, yielding inconsistent positions. For example, fourth magnitude Alpha Cancri, BD+12 1948, has a current (equinox 2000) declination of +11° 51′. Precession also shifts number counts within declination bands. Despite that flaw, the BD-CD-DM names quickly achieved currency, and are still in frequent use, especially when other common names are not available.

Four other catalogue naming systems (three of which avoid declination bands) merit close attention. Around the turn of the nineteenth century into the twentieth, while German astronomers were occupied with the BD, Harvard astronomers were busy with the classification of stars into the familiar classical spectral sequence OBAFGKM [6.2]. By 1924, Annie Cannon (working with the director, E.C. Pickering) had classified 225 300 stars. The work was published in *Harvard Annals* volumes 91 through 99 as the *Henry Draper Catalogue*, honoring the astronomer-physician whose widow helped finance the spectroscopic survey. Each volume of the catalogue contains a range of right ascensions, with the stars numbered sequentially west to east for the equinox of the year 1900 (the coordinates taken from a variety of sources, the BD, CPD, etc.). The names are prefixed HD, and thus Arcturus, Alpha Boötis, BD+19 2777, is also HD 124897 (the numbers never separated by commas). Another 133 782 fainter stars were added in the *Henry Draper Extension* of 1948. These pick up where the original HD numbers left off. But since they begin all over again at 0 hours right ascension, they are usually tagged as HDE to separate them from coordinates of the HD. The HD catalogue goes about as faint as the BD/CD catalogues, but contains fewer than half the stars, as neither catalogue is complete to any magnitude limit, and there is significantly incomplete overlap. Nevertheless, in spite of encroachment by other systems, the HD numbers are still usually the default in the naming of stars fainter than those visible to the naked eye, and indeed are even used for many of those that are.

Competing with the use of proper names, Greek letter names, Flamsteed numbers, even HD numbers, are the numbers of the *Bright Star Catalogue*, or the BSC. The BSC, also known as the Yale *Bright Star Catalogue*, was derived from an original produced at Harvard, hence the usual HR prefix for the 9110 stars in the catalogue, though BS is occasionally used as well. The catalogue, of which the fourth revised 1982 edition is most extant, contains an immense amount of information on stars brighter than magnitude 6.5, the canonical limit of human vision (for a 6-year-old child under a clear desert sky 1000 kilometers from any population center). Compiled by Dorrit Hoffleit with the collaboration of Carlos Jaschek, the 1983 catalogue numbers stars according to right ascension for the equinox of 1900 without regard to declination. Included are equatorial coordinates for the years 1900 and 2000, ecliptic coordinates [2.3.4], apparent

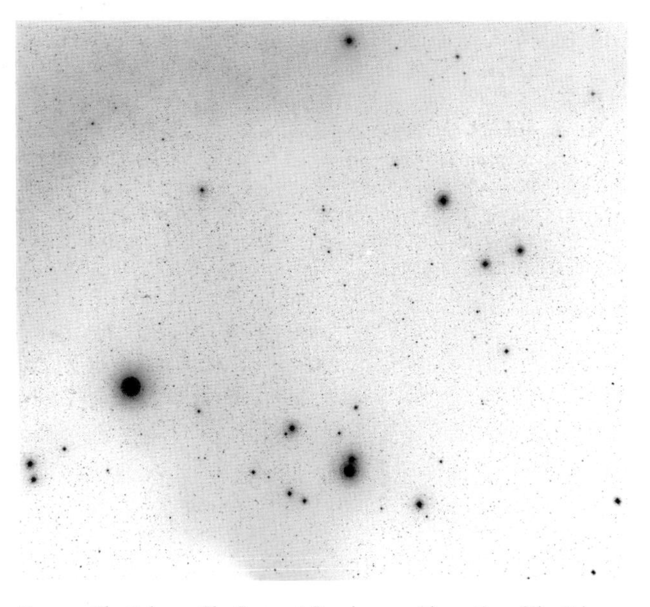

Fig. 1.13 **The Palomar Sky Survey** A five-degree-wide section of the Palomar photographic sky survey shows Aldebaran and the major stars of the Hyades cluster (see Fig. 1.12). In this blue-light negative photographic image we can see stars eight magnitudes fainter than those in the BD. (Courtesy of Palomar Observatory/California Institute of Technology.)

Proper name	Drawn from a variety of sources and languages, a few Greek and Latin, but most are corrupt, reduced, and mistranslated Arabic. Examples: Procyon, Bellatrix, Deneb Algedi.[1]
Greek letter	Applied to (mostly) bright stars within a constellation in some rough accord with brightness, but more often than not by position. Takes Latin genitive of constellation name (or its abbreviation). Examples: Alpha Canis Minoris (α CMi), Gamma Orionis (γ Ori), Delta Capricorni (δ Cap). Invented by Bayer; extended to the southern hemisphere by others.
Flamsteed number	Applied to brighter naked-eye stars from west to east within a constellation. Takes Latin genitive of constellation name. Examples: 10 Canis Minoris, 24 Orionis, 49 Capricorni. Not used in the deep southern hemisphere.
HR number	"Harvard Revised," from the Yale *Bright Star Catalogue*. Numbers all naked-eye stars in order of right ascension for equinox 1900. Examples: HR 2943, HR 1790, HR 8322.
BD/CD number	*Bonner Durchmusterung* north of −23° declination, *Cordoba Durchmusterung* south of −23°. Numbers stars from 0 hours right ascension within 1°-wide declination bands for equinox 1855 (BD), equinox 1875 (CD). Examples: BD+05 1739, BD+06 919, BD−16 5943.
HD number	*Henry Draper Catalogue* of spectral classes. Numbers stars from 0 hours right ascension, equinox 1900. Examples: HD 61421, HD 35468, HD 207098. Extension after HD 225300 called HDE.
SAO number	*Smithsonian Astrophysical Observatory*. Numbers stars serially within 10°-wide declination strips from 0 hours right ascension (equinox 1950), each strip picking up the number from the last. Examples: SAO 115756, SAO 112740, SAO 164644.
HIC = HIP number	*Hipparcos Input Catalogue*. Numbers stars from 0 hours right ascension (equinox 1991.25). Same as *Hipparchus Catalogue* number. Examples: HIP 37279, HIP 25336, HIP 107556.
TYC number	*Tycho Catalogue* from Hipparcos. Numbers over a million stars numbered from right ascension 0 hours (equinox 1991.25).
Guide Star number	Numbering system for 15 million stars from sixth magnitude and fainter used to guide the Hubble Space Telescope.
Messier number (M)	Catalogue of non-stellar objects by Charles Messier and Pierre Méchain, go to M110.
NGC number	*New General Catalogue* of non-stellar objects, by J.L.E. Dreyer, which adds discoveries to the Herschel's *General Catalogue*, circuits from right ascension 0 hours (equinox 1860).
IC number	*Index Catalogues* of non-stellar objects, two of them, each separately circuiting from right ascension 0 hours (equinox 1860).

[1] The same three stars are listed as examples in the following entries.

Table 1.3 **Names in a nutshell**

visual magnitudes [3.1.1], $B-V$ and $U-B$ colors ($R-I$ when known) [3.4], spectral classes [6.2, 6.3], proper motions [5.2.1], pre-Hipparcos parallaxes [4.2], rotational velocities [7.2.5], double star information [Chapter 8], and detailed notes on individual stars. An extension (without HR numbers) is available to magnitude 7, as is the fifth revised edition in preliminary form on the Web. HR numbers are commonly used (competing with HD numbers) when neither Greek letter names nor Flamsteed numbers are available. In 1900 coordinates, HR 9110 was last in line. By the year 2000, precession had caused HR 9076 to end the long right-ascension march and HR 9077 to start it.

In 1966, the Smithsonian Astrophysical Observatory at Harvard University (the combined group called the Harvard-Smithsonian Center for Astrophysics) produced a comprehensive star catalogue that combined a variety of earlier catalogues, restricting itself to stars that had proper motions (angular movements across the plane of the sky). The 258 997 stars are counted in order of right ascension in 10-degree-wide bands, the numbers in the zone from 80° to 90° continuing with those from 70° to 80° and so on, with no indication of declination zone. Like the BD, the SAO stars slowly shift with precession into neighboring declination bands. While the system was fairly popular in the naming of stars in the 1960s and 1970s, it is now rarely used.

In 1989, the European Space Agency launched the astrometric satellite Hipparcos [4.2.3], which was dedicated to the measurement of ultra-precise star positions, motions, and in particular parallaxes, the purpose being to expand our knowledge of the distances to the stars. To effect the survey, the stars had to be selected in advance and entered into the *Hipparcos Input Catalogue*, or HIC, which contains a little over 118 000 stars and a variety of data derived from ground-based observations. Each star has a running HIC number that begins at 0 hours right ascension for the equinox of 1991.25 (but commonly seen precessed to the year 2000). The results of the mission, which concluded with publication in 1997 and contains the same stars, were deposited into the *Hipparcos Catalogue*, where the star names are prefixed by HIP. The HIP and HIC numbers are therefore identical; the redundancy is needed to separate the data in the catalogues. The HIP/HIC includes almost all stars down to magnitude 7.3, many to magnitude 9 (depending on stellar positions and properties), and with large gaps extends as faint as magnitude 12.

Three more catalogues contain vastly more stars. The *Hipparcos* guidance system observed great numbers with lower precision. The results, placed in the *Tycho Catalogue*, include over a million, and incorporate 97 percent of the *Hipparcos Catalogue*. While it does not go as faint as the *Hipparcos Catalogue*, it is more complete to around 10th magnitude. *Tycho Catalogue* stars are designated TYC followed by three separated sets of numbers: one denoting a region of the sky, the next denoting a number within the region, the last a specific "component number." Betelgeuse, Alpha Orionis, which is also HR 2061, HD 39801, SAO 113271, HIP 27989 = HIC 27989, is also TYC 129 1873 1. The *Tycho-2 Catalogue*, the result of advanced reduction techniques applied to the same set of data, contains 2.5 million stars.

Finally, the *Hubble Guide Star Catalogue* (rather obviously needed to guide, or direct, the Hubble Space Telescope) was compiled from positions measured from the *Palomar Sky Survey* (Fig. 1.13) and a later survey undertaken with the United Kingdom Schmidt Telescope. Because of the sensitivity of Hubble's instruments, bright stars are unsuitable as pointing guides. The catalogue, which contains 19 million stars and other objects, therefore begins near the limit of naked-eye vision (sixth magnitude) and ranges to 15th. Prefixed GSC, the catalogue numbers are (rather like those of the later *Tycho Catalogue*) presented in two groups, one giving a region, the other a running number. Neither Sirius, Arcturus, Betelgeuse, none of the stars with Flamsteed numbers, and few with Greek letters have GSC numbers (in the southern hemisphere, a few stars below naked-eye vision were oddly given them). As a random example, HD 123456, a modest ninth magnitude star in Hydra (and also known as CD-25 10180, CPD −25° 5348, HIC 69081, and TYC 6738 988 1) is GSC 06738-00988.

Star-naming schemes are summarized in Table 1.3. There are yet more names, but they mostly involve specialty lists, some stars having more than 50 identifiers. Indeed, the process is never ending as astronomers reach ever deeper into the cosmos, toward and beyond the recording of hundreds of millions of celestial objects.

1.5 Non-stellar objects

Neglecting the bodies of the Solar System, nearly everything visible in the sky with the unaided eye is a star, or at least a cluster of stars. With the telescope, the scenery changes. Now we can view various fuzzy objects, once known generically as "nebulae" (Latin for "clouds"). In our own time, these fuzzballs have been deciphered as compact clusters, clouds of interstellar gas and dust, and various kinds of galaxies. In the early years of telescopic study no one knew the difference.

To help distinguish these from the equally-fuzzy (but moving) comets that continued to make both the telescopic and naked-eye scene, Charles Messier (1730–1817) compiled a list of "permanent" diffuse bodies that might fool the eye into thinking a new comet had been discovered. Messier's original catalogue, fully published in 1781, contained 103 "non-stellar objects" (so-called even if, like star clusters, they are themselves made of stars). Added to by Messier's colleague Pierre Méchain, the catalogue now numbers 110 objects (prefixed simply with M). Though nearly all are telescopic, a few – the Andromeda Nebula (M31), the Pleiades (M45), the Beehive cluster (M44) – are seen with the unaided eye. The catalogue, its objects commonly referred to by their M-numbers, is a treasury for both the beginning and advanced amateur (Appendix 1).

With improved telescopes, the discovery of non-stellar objects continued at an ever-accelerating pace. The individual champion was William Herschel (1738–1822), who, with his sister Caroline (1750–1848), discovered over 2500 non-stellar objects in the skies visible from England. Herschel's son John (1792–1871) did the same for the southern hemisphere. Adding his (and others') discoveries to those of his father's and his aunt's, John produced the first really comprehensive listing of such objects, a set of more than 5000, in the *General Catalogue of Nebulae* (the GC) of 1847, which also includes star clusters, not just what we now refer to as "nebulae."

As discoveries continued unabated, the GC rapidly grew out of date. In 1888, J.L.E. Dreyer (1852–1926) of the Armagh Observatory added myriad new ones to create the *New General Catalogue of Nebulae and Star Clusters*, the NGC. Still a foundation of nomenclature, the NGC contains 7840 objects, numbered around the sky in order of the right ascensions of the equinox of 1860. In 1895, Dreyer added the first of two *Index Catalogues*, the IC (or IC I), and in 1908 the second one, the IC II. The IC I begins at 0 hours (also for the year 1860), and wraps an additional 1529 objects around the sky. The IC II begins at object 1530 and runs to IC II 5386, but begins over again at 0 hours (again using 1860 coordinates). In the single *Index Catalogue*, the distinction between the two parts is important only if one wishes to know which circuit of the sky is involved; when it is presented whole in order of right ascension, the numbers of the two sets intermingle. Together, the NGC/IC catalogue includes 13 226 non-stellar objects. The Andromeda Galaxy, M31, is also NGC 204, while a well-known planetary nebula [13.4] in Lepus is referred to as IC 418.

Rapidly advancing telescopes and techniques have far-outstripped the ability, even the need, for additional comprehensive catalogues. Planetary nebulae, open clusters, galaxies all have their specialty listings. The stellar and non-stellar catalogues, however, provide a fine way to begin to know the sky on an intimate basis, wherein all the brighter bodies have personal names, sometimes a great many of them that begin with the ancient constellations and run to the systems derived from today's extraordinary technologies.

CHAPTER TWO

Location

To study the stars, we must know where they are, and simple placement within their constellations [1.1, 1.2] is nowhere near good enough. Instead, we need their three-dimensional locations, two on the apparent sphere of the sky, the other the distance. Only then can we make sense of stellar origins, patterns of evolution, deaths, and comings and goings about the Galaxy. The first two positional measures, different from the third, involve placement within some kind of defined coordinate system, and have been assessed for thousands of years. The third [Chapter 4], though intimately entwined with the other two, has traditionally needed a different and higher technology, one accompanied by a stern measure of astrophysics.

Though astronomers will never be satisfied with the precision of measurement, the problem of position has in principle been solved. We can now find coordinate locations to within a thousandth of a second of arc (0.000 000 3 degree) and reasonably accurate distances for all the brighter stars within about 300 parsecs (as well as good ones for selected stars and other objects at much greater distances). Start with simple location on the sky.

2.1 Earth

To find your celestial way, first find it on the ground. Landmarks to navigate are fine when available, but they are often not, especially at sea. Our "landmark" then becomes a coordinate grid, and we position ourselves within it by external landmarks: the stars. Though as a result of its rotation the Earth is not quite a sphere, assume it to be one. On any sphere, we can draw an infinite number of *great circles*, those circles whose centers are coincident with the center of the sphere. Draw a line through the center of any great circle perpendicular to its plane, and it pokes through the sphere at two opposed *poles*, each of which are 90° from the circle itself. The Earth's rotation axis defines two natural poles (the *north pole* in the Arctic Ocean and the *south pole* in Antarctica) and a great circle that lies between them, the *equator*, which can be used as a formal *fundamental great circle* (the "FGC") of a superimposed coordinate system (Fig. 2.1).

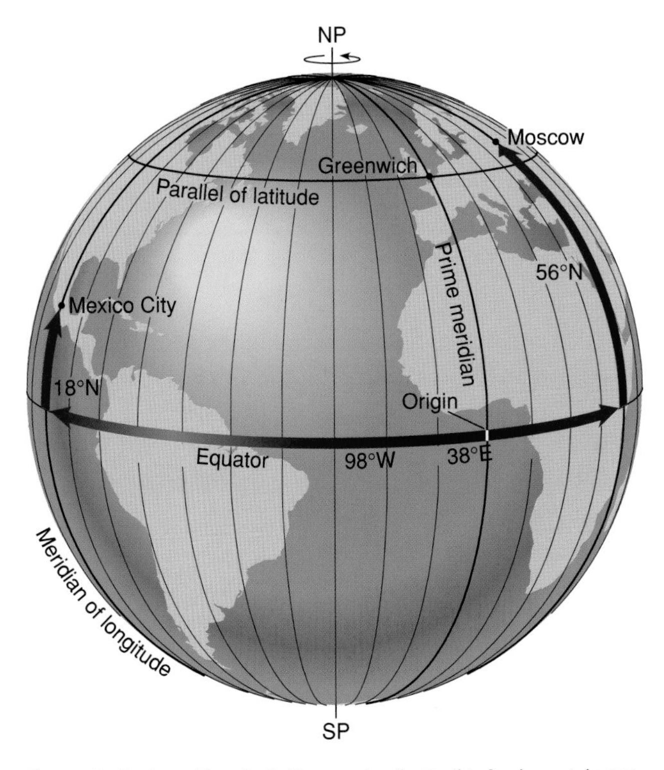

Fig. 2.1 **Latitude and longitude** The equator, the Earth's fundamental great circle, is defined by the rotation poles that lie perpendicular to it. The longitude of a point on Earth is measured along the equator east and west of an arbitrary prime meridian, while its latitude is measured north and south of the equator along the meridian passed through that point. (Kaler, J.B. (1994). *Astronomy!* New York: HarperCollins, reprinted by permission of Pearson Education, Inc.)

Each great circle defines an infinite number of *secondary circles* that run through the poles perpendicular to the chosen fundamental great circle. On Earth we halve them to semicircles, and call them *meridians*. For cultural and political reasons, the meridian that passes through an observatory in Greenwich, England – the *prime meridian* – is chosen as special. The prime meridian cuts across the equator in the Atlantic Ocean off the west coast of Africa.

Move to Denver, Colorado, USA. Draw a meridian from the north pole through Denver to the south pole. See where it crosses the equator. Denver's *longitude* (λ) is the angle (rather the arc) measured along the equator from the prime meridian to Denver's meridian. By convention, longitudes are measured east and west of the prime meridian up to 180°, where the two directions meet. Then measure the angle (or arc) along Denver's meridian from the equator up to the city and you have Denver's *latitude* (ϕ). Latitudes are given as north (N) or south (S) of the equator (respectively plus or minus). The terrestrial coordinates of Denver are 105°W, 40°N (or +40°). Those of Adelaide, Australia, are 138°E, 35°S (−35°). Any coordinates can be refined to subdivisions of a degree. Each degree contains 60 minutes of arc (written 60′), each minute of arc contains 60 seconds of arc (60″). The coordinates of the University of Chicago's historic Yerkes Observatory are 88°33′24″ W, 42°34′12″ N.

Directions north and south are travelled along meridians of longitude, while east and west direct us along lines of constant latitude (*parallels* of latitude). The latitudes of the north and south rotation poles are 90°N and 90°S. The poles have no longitudes, as they are the origins of all meridians; that is, from the north pole all directions are south, and vice versa.

2.2 The celestial sphere

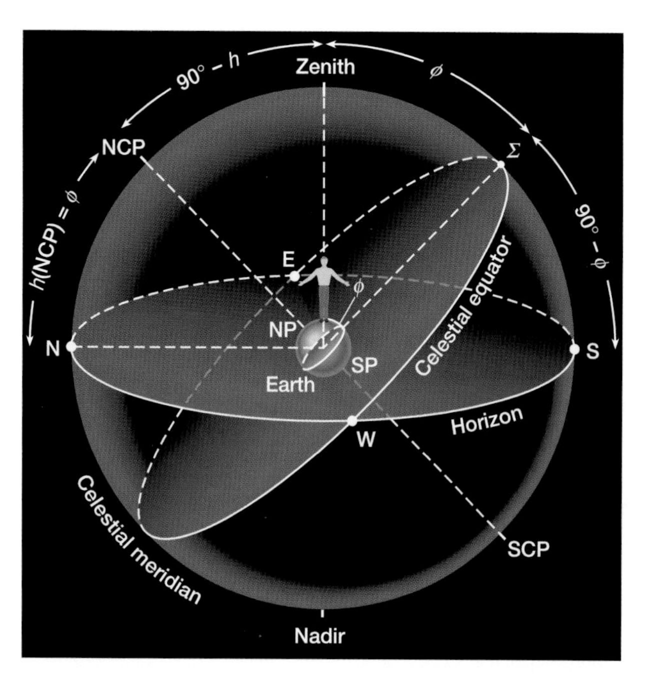

Fig. 2.2 **The celestial sphere** The celestial sphere encloses the Earth and the observer. The zenith is above the observer's head, the nadir opposite. The celestial poles and celestial equator are the extensions of their terrestrial analogues. The horizon divides the sky into visible and invisible hemispheres, the celestial equator into northern and southern hemispheres, the celestial meridian (defined as the circle through the celestial poles and the zenith) into eastern and western hemispheres. The observer, at a latitude of 45°N, appears to stand atop the Earth. As a result, the north celestial pole (NCP), defined by the Earth's rotation axis, has an altitude of 45°. The intersections of the meridian and equator with the horizon define the compass points, while that between the equator and meridian defines the equator point, Σ. (Kaler, J.B. (1994). *Astronomy!* New York: HarperCollins, reprinted by permission of Pearson Education, Inc.)

Fig. 2.3 **Polar star trails** As the Earth rotates, stars trace out daily paths around the south celestial pole, which is the point of apparent zero rotation. Because the latitude of the observatory in the background is 31°S, the celestial pole is 31° above the horizon. (Courtesy of Anglo-Australian Observatory, David Malin Images.)

Set the Earth into its surroundings. Assume that the apparent sky is a bowl over your head and that the Earth is enclosed within it (Fig. 2.2). This bowl, the *celestial sphere*, is infinitely large. The Earth is a mere speck on its scale, such that there are no parallax effects as we move around. In the daytime, the celestial sphere is blue and holds the Sun and perhaps the Moon on its apparent curved surface; at night it is black and speckled with stars, which in our fiction are all at the same distance. That is, we look at the sky as it appears rather than as it actually is, and harken back to the time when most people really believed (as indeed some still do) that the Earth is at the center of the Universe.

For the celestial sphere to be any use, we must define points and circles on it relative to us on Earth. Above your head, directly away from the center of the Earth, is your *zenith*. A carpenter's plumb bob, drawn to the terrestrial center by gravity, nicely points to it. Directly opposite the zenith, beyond the Earth's center, lies your *nadir*. Since the Earth is spherical (the shape known since ancient times), someone next to you will be slightly around the terrestrial curvature, and will have a different zenith and nadir: these points are therefore personal and belong to you only. Next, extend a plane outward in all directions, one tangent to the Earth at your feet. It will slice the celestial sphere at the *horizon*. Stars on the sphere above the horizon are all visible, whereas those below it are not. The visible horizon at sea or in the flatland prairie is a reasonable approximation to this true, or *astronomical*, horizon.

In your mind, walk to a latitude of 45°N, halfway between the equator and the north pole (see Fig. 2.2). You are pulled to the Earth's center by gravity, and stand perpendicular to the terrestrial surface, with the zenith directly above you. The Earth's axis is now set at an angle of 45° to this perpendicular (that is, to the line between the zenith and nadir) as well as to the plane of the horizon. Extend the Earth's axis in both directions, and it intersects the celestial sphere at the *north* and *south celestial poles* (the NCP and SCP), heavenly analogues to the Earth's poles. Extend the plane of the Earth's equator outward in all directions, and it cuts the celestial sphere in half at the *celestial equator*. The Earth rotates about its axis, identified by the unmoving north and south terrestrial poles. The motion is so smooth that you do not feel it. Instead, you see the reflexive motion, as the sky appears to move the opposite way about the north and south celestial poles. The NCP and SCP can be identified by the observer as the points of zero celestial rotation, the only points in the sky that seem stationary (Fig. 2.3).

Now draw a great circle through the zenith, the NCP, the nadir, and the SCP, to locate your *celestial meridian*, which divides the sky into equal eastern and western hemispheres. Each of the three great sky-circles intersects each of the others at two points. The celestial meridian and the horizon meet at *north* and *south*, the celestial equator and horizon at *east* and *west*, thus defining the four "compass points," which are points not on the Earth, but on the sky. Now you have direction. Face the north celestial pole and walk along a meridian of longitude, and you will eventually arrive at the north terrestrial pole. The equator and meridian meet above the horizon at the *equator point* (or *sigma point*, Sigma, Σ). The sky is set in place. Now to use it.

2.3 Coordinates on the celestial sphere

Millions of stars are on display, and we must have some sort of accurate way of sorting them out. Moreover, to study their characteristics – for purposes of stellar astrophysics – we must be able to locate them quickly with telescopes. Coordinates in the sky are found the same way as on Earth. First identify a fundamental great circle, which defines the poles (or vice versa) and the secondary circles, and then define a zero point on the fundamental. Along the way, give names to the circles, poles, and to the coordinates themselves, and also define directions of measurement. The coordinates of a point on the sky are always found by passing the secondary great semicircle from one pole through the point to the other pole. Unlike Earth, however, a number of different systems are available, each having its own unique use. The concepts are the same as for Earth; only the names change.

2.3.1 Horizon coordinates

The celestial sphere's most obvious fundamental great circle is the horizon, whose poles are the zenith and nadir. The north point is usually used as "zero." Locate a star and pass the secondary circle (or *vertical circle*) through it and see where it hits the horizon (Fig. 2.4). The star's angular elevation above the horizon along the vertical circle – the analogue to latitude – is the star's *altitude*, h. The altitude of the zenith is $+90°$. Below the horizon, altitudes are negative, so that of the nadir is $-90°$. The complementary angle, from the zenith down, is the *zenith distance*, z, which must be $90° - h$.

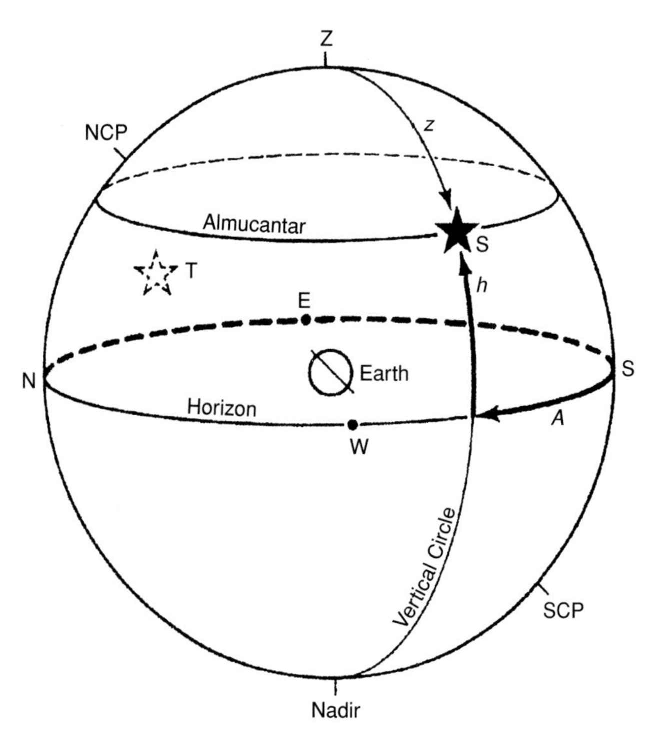

Fig. 2.4 **Horizon coordinates** The altitude of a star is measured vertically up or down from the horizon, while the azimuth is measured along the horizon from north through east. (Kaler, J.B. (1996). *The Ever-Changing Sky*, Cambridge: Cambridge University Press.)

This system's analogue to terrestrial longitude, the star's *azimuth* (A), is the angle between north and the intersection of the vertical circle and the horizon, measured in the clockwise direction as seen looking down from the zenith (that is, passing from north through east). Face the star. The azimuth (or "heading") is the angle between north and the direction you face. The azimuths of the compass points north, south, east, and west are respectively 0°, 90°, 180°, 270°, north returning again as 360°, at which point we start over. Since the zenith and nadir are the poles of the horizon, neither has an azimuth.

Horizon coordinates provide the first lesson in celestial navigation. Return to 45°N latitude. Since the axis of the Earth extends upward at 45° from the horizon, the altitude of the north celestial pole must also be 45°. Now go to the Earth's north pole, $\phi = 90°$N. The rotation axis now extends through you to the zenith, hence h (NCP) $= 90° = \phi$. At the equator, the Earth's axis is parallel to the horizon, and h (NCP) $= 0° = \phi$. That is *the altitude of the pole equals the observer's latitude* (see Fig. 2.2). If you are south of the equator, the NCP is below the horizon (with a negative altitude), and the SCP above it. Find the pole, measure its altitude, and you know at least part of where you are. Within a degree, the NCP is marked by Polaris [Fig. 10.4], the North Star, while the SCP is tagged with the much dimmer Sigma Octantis ("Polaris Australis"). The effect of latitude on the sky is dramatically visible by comparing Fig. 2.5 with Fig. 2.3.

While horizon coordinates are necessary for navigation and for setting the directions of certain kinds of telescopes, they are useless as permanent stellar locators, since as stars wheel around the sky on their individual paths, altitude and azimuth continuously change in complicated fashion. We have to find something else, a coordinate system that is fixed, not to the observer, but to the stars, one that rotates around the sky along with them.

2.3.2 Equatorial coordinates I

There are several ways to establish fixed celestial coordinates. Since the stars appear to rotate around the sky parallel to the celestial equator, the equator again becomes a

Fig. 2.5 **Effect of latitude** From two degrees south of the equator, the south celestial pole lies much closer to the horizon than it does from 31°S in Australia (Fig. 2.3). (Courtesy of Gregory Dimijian.)

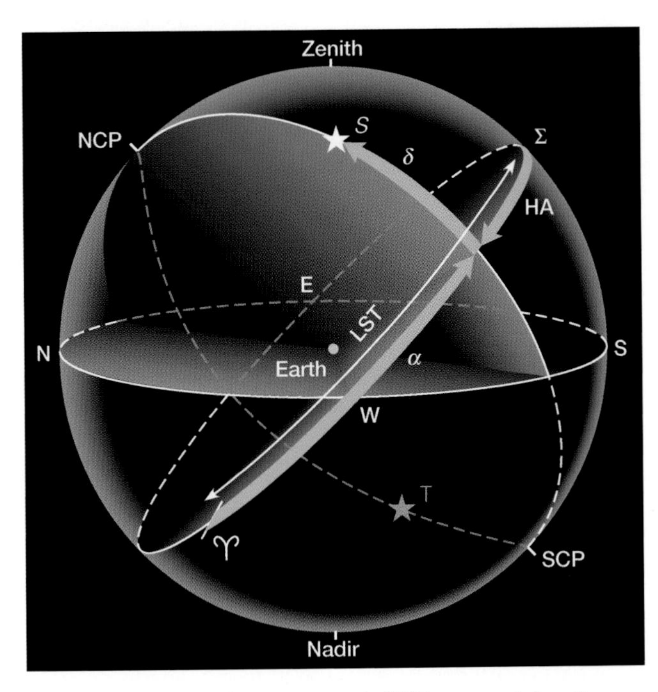

Fig. 2.6 **Equatorial coordinates** Hour angle (HA) is measured along the celestial equator to the west from the equator point (Σ) to the foot of the hour circle that passes through the star. Right ascension (α) is measured opposite (to the east), from the Vernal Equinox (♈) to the foot of the hour circle. Declination (δ) is measured along the hour circle to the star, north or south of the equator. LST (local sidereal time) is the sum of HA and α. (Kaler, J.B. (1994). *Astronomy!* New York: HarperCollins, reprinted by permission of Pearson Education, Inc.)

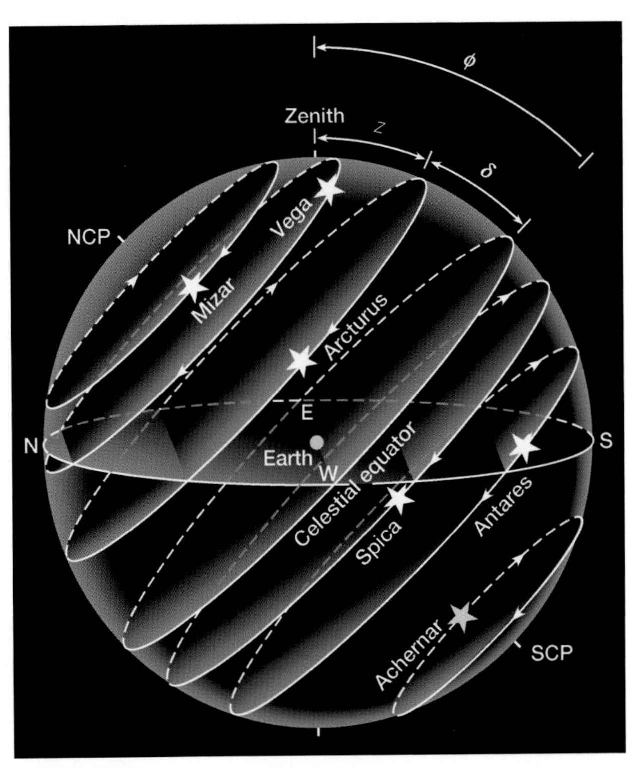

Fig. 2.7 **Latitude and circumpolar stars** For a star south of the celestial equator, as seen by an observer in the northern hemisphere, latitude (ϕ) equals the sum of the zenith distance (z) and the declination (δ). For a star north of the zenith, $\phi = \delta - z$. The daily paths of stars are parallel to the celestial equator. From 45°N latitude, Vega, Arcturus, Spica, and Antares all rise and set, though at different points on the horizon. Mizar, however, is circumpolar, while Achernar never rises. (Kaler, J.B. (1994). *Astronomy!* New York: HarperCollins, reprinted by permission of Pearson Education, Inc.)

natural fundamental great circle for a coordinate system. Astronomers use two complementary systems of *equatorial coordinates*, one that is half-fixed to the observer, the other completely fixed to the stars. Both are necessary to find our way, not only on our planet but across the celestial globe (Fig. 2.6).

In equatorial coordinates, meridians of longitude become *hour circles*, while latitude becomes *declination*, the latter common to both equatorial systems. To measure the declination of a star, proceed as on Earth. Pass a great semicircle – the hour circle – from the north celestial pole through the star to the south celestial pole. Then measure the declination along the hour circle in degrees north or south of the equator. The declination of the celestial equator is 0°, of the north celestial pole 90°N (+90°), of the south celestial pole 90°S (−90°).

Declination affords another way of finding latitude (Fig. 2.7). Since the angle between you and the Earth's equator is the latitude, the angle between the celestial equator and your zenith also equals latitude. This angle is both the zenith distance of the equator point (Σ) and the declination of the zenith; the two angles are just measured in opposite directions. Or, $\phi = z\,(\Sigma) = \delta$ (zenith). Find a star that passes overhead, and your latitude is the star's declination, which is available from a star catalogue. You do not even need a zenith star. In Fig. 2.7, note that the zenith distance of any star (∗) crossing (*transiting* or *culminating*) the meridian to the south of the zenith equals its declination plus its zenith distance, or

$$\phi = \delta(*) + z\,(*).$$

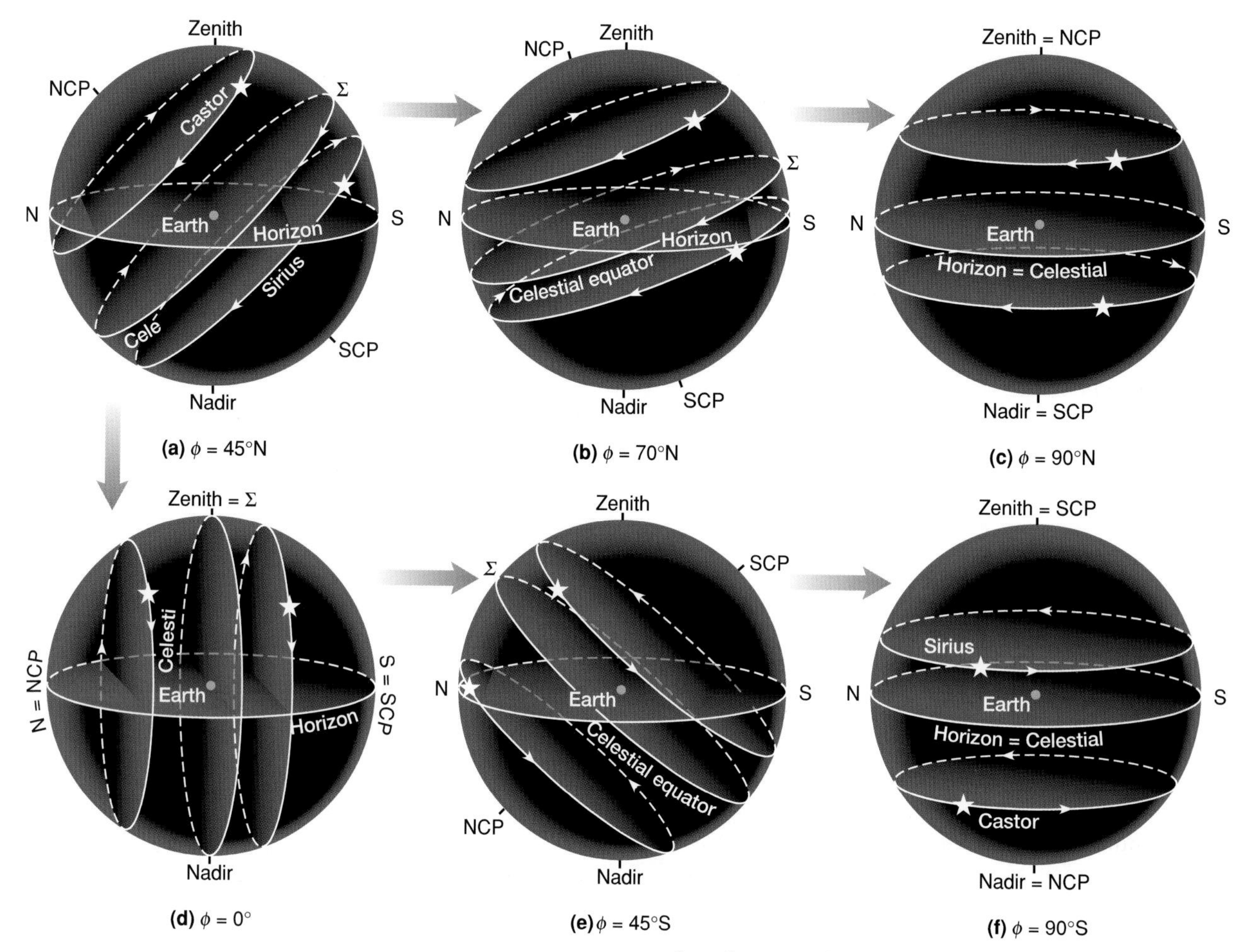

Fig. 2.8 **Effects of latitude** As the observer travels north (a-b-c), the north celestial pole rises, the celestial equator flattens out to the horizon, and more circumpolar stars are on view. At the Earth's north pole, the stars travel parallel to the horizon, the entire northern hemisphere is circumpolar, and none of the southern sky can be seen. Travelling southward (a-d-e-f), upon crossing the Earth's equator the observer sees the north celestial pole set and the south celestial pole rise. When the observer reaches the Earth's south pole, the south celestial pole is at the zenith and the entire southern hemisphere is perpetually on view. (Kaler, J.B. (1994). *Astronomy!* New York: HarperCollins, reprinted by permission of Pearson Education, Inc.)

If the star transits to the north of the zenith, then $90° - \phi = z + 90° - \delta$, or

$$\phi = \delta(*) - z(*).$$

Declination also relates directly to what stars can be seen from any place on Earth and where they will rise and set. As the Earth rotates around its axis and the sky appears to rotate about the celestial poles, the stars seem to be carried along on *daily paths* that are parallel to the celestial equator (Fig. 2.7). As we watch, stars rise above the eastern horizon and drop below the western. The stars progressively north of the celestial equator (toward more positive declinations) rise more and more north of east, and set more and more north of west, until they do not set at all, but are perpetually visible – *circumpolar* – as they swing beneath the pole (see Fig. 2.3). Alternatively, the stars progressively south of the celestial equator (toward negative declinations) rise more and more south of east, setting more and more south of west, until they do not rise at all, but are perpetually invisible, never reaching the horizon even at their highest arcs. From 45°N latitude, the Big Dipper is never lost to sight, while the Southern Cross never appears. From 45°S we see the reverse.

The appearance of the celestial rotation and the amount of the sky that is circumpolar change dramatically with latitude (Fig. 2.8). The farther north you go, the flatter the angles the daily paths make with the horizon, the more circumpolar stars you can see, and the more stars that never rise. At the north pole the entire northern celestial hemisphere is circumpolar, while the southern is perpetually invisible. Here, no stars ever rise or set. Toward the Earth's equator, the rising and setting angles steepen, until at zero degrees latitude daily paths are perpendicular to the horizon. Here, in the middle of the rotating world, the celestial poles are on the horizon. Thus there are no circumpolar stars, and all stars (ignoring daylight) are visible. Since the altitude (h) of the north celestial pole equals the latitude ϕ, the declination of the north point of the horizon is $90° - \phi$ and that of the south point is $-(90° - \phi)$. Therefore, for any latitude, any star with a declination greater than $90° - \phi$ must be circumpolar, while any with δ less than $-(90° - \phi)$ is permanently out of sight.

To define a true celestial analogue to longitude, use the equator point Σ as the zero point (see Fig. 2.6). The *hour angle* (HA) of a star is measured along the equator to the west of the equator point all the way around the sky and back through 360°. From the northern hemisphere, the hour angle of the meridian above the pole (or that of the equator point and the south point of the horizon as well) is 0°, while that of the west, north, and east points of the horizon are respectively 90°, 180°, and 270°. From the southern hemisphere, the hour angles of north and south are reversed, while those of east and west remain the same. As longitude has no meaning at the poles, neither does hour angle, as the hour circles all converge at these points. Unlike longitude, the default measures hour angles to the west, so no indication of direction is needed. When measured oppositely, to the east (which is acceptable), "east," "E," or a minus sign is appended to the hour angle.

While measurement of hour angle in degrees is never wrong, astronomers more commonly use time units. The 360° of a circle, which comes from the number of days in the year (360 being the closest widely divisible number to 365), is ancient and traditional, but hardly mandated. You can as easily divide the circle into 24 units, and from analogy with a clock, call them *hours*, h. There are therefore $360/24 = 15°$ per hour. The northern-hemisphere hour angles of south, west, north, and east are therefore 0^h, 6^h, 12^h, and 18^h.

Dealing with time units is at first confusing and rather an astronomical "rite of passage." Just as a degree is subdivided into 60 minutes of arc and a minute into 60 seconds of arc, so an hour is subdivided into 60 minutes of "time," and the minute into 60 seconds of time. However, the two angular units for which the word "minute" is used are different (as are those for "second"). The difference is expressed symbolically, whereby minutes and seconds of arc are expressed by prime and double prime ($'$ and $''$), and minutes and seconds of time units by superscripted "m" and "s." Given 15° per hour,

$$1 \text{ hour} = 15° = 60^m$$
$$1° = 4^m = 60'$$
$$1^m = 15' = 60^s$$
$$1' = 4^s = 60''$$
$$1^s = 15''$$
$$1'' = 1/15^s = 0.0667^s.$$

The reason for the unit is obvious. Since the sky turns a full cycle in 24 hours of actual time, the hour angle of a celestial body just keeps pace with the clock. Once we know the hour angle at any particular time, we can easily determine what it will be at any time in the future. Declination and hour angle are useful in setting telescopes to directions that change as a result of the rotation of the Earth. However, while declination remains fixed, hour angle is – like altitude and azimuth – still useless for a permanent recording of stellar position.

2.3.3 Equatorial coordinates II

A system that relates not to the observer but to the stars alone can be defined by using a zero point on the celestial

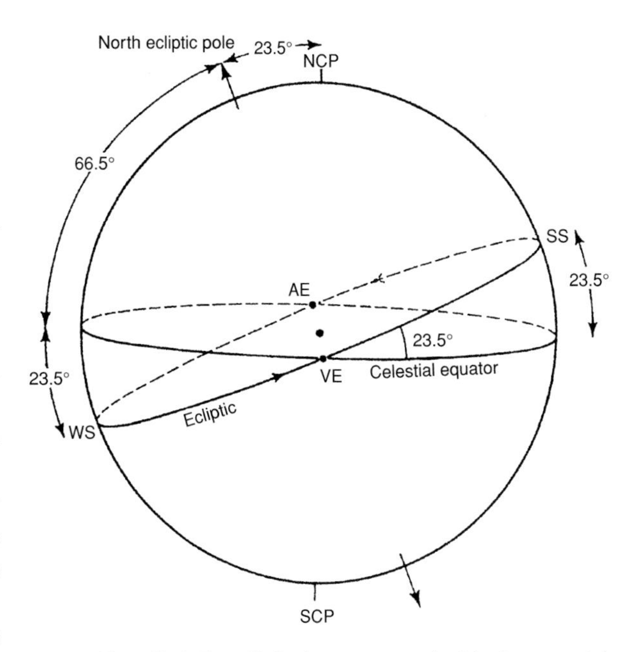

Fig. 2.9 **The ecliptic** The ecliptic, the apparent path of the Sun around the Earth, intersects the celestial equator at an angle of 23½°. While travelling north, the Sun crosses the equator at the Vernal Equinox (VE) on March 20, and while travelling south crosses again at the autumnal equinox (AE) on September 23. The Sun reaches its highest declination of 23½°N at the Summer Solstice, which it crosses on June 21, while its lowest point is defined by the Winter Solstice at 23½°S. Celestial (ecliptic) longitude is measured along the celestial equator from the Vernal Equinox, celestial (ecliptic) latitude perpendicular to it toward the ecliptic poles. (Kaler, J.B. (1996). *The Ever-Changing Sky*, Cambridge: Cambridge University Press.)

equator that turns along with them. As we use Greenwich on Earth to give a prime meridian, we could use a bright star in the sky, and define a zero point as the intersection of the hour circle through the star and the celestial equator. Unfortunately, the star is moving relative to its neighbors, that is, it has an angular motion (its "proper motion") across the sky that is independent of any movement of the Earth [5.2.1]. If we use a star for a zero point, then (as we watch the coordinates change with time) we find all other angular motions relative to the fundamental star, which vitiates any studies of the dynamics of the Galaxy. We must instead use a zero point that is, at least in principle if not entirely in practice, independent of the stars.

As the Earth orbits the Sun, the Sun seems to go around the Earth along a fixed great circle, the *ecliptic* (Fig. 2.9). The ecliptic defines the plane of the Earth's orbit, and also defines the Zodiac [1.1.1], the set of constellations through which it passes. The perpendicular to the ecliptic plane defines the *north* and *south ecliptic poles*, which are respectively found in the constellations Draco and Dorado. By accident of history, the Earth's rotation axis is inclined to the ecliptic's perpendicular by an angle close to 23½° (23°26′21″). Since the celestial equator and ecliptic are each 90° from their respective poles, they must intersect each other at the same angle, which is then called the *obliquity of the ecliptic*, ϵ.

Position	Right Ascension α (2000)	Declination δ (2000)
Vernal Equinox	0^h	0°
Summer Solstice	6^h	23°26′21.4″[a]
Autumnal Equinox	12^h	0°
Winter Solstice	18^h	−23°26′21.4″
North ecliptic pole	18^h	66°33′38.6″
South ecliptic pole	6^h	−66°33′38.6″
North galactic pole[b]	$12^h51^m26.3^s$	+27°07′41.7″
South galactic pole[b]	$00^h51^m26.3^s$	−27°07′41.7″
Galactic center[b]	$17^h45^m37.2^s$	−28°56′10.2″
Galactic-celestial equator intersection[b]	$18^h51^m28^s$	0°

[a] Also the obliquity of the ecliptic.
[b] Modern Galactic longitude, l^{II}, b^{II}, for which the Galactic center is the zero point.

Table 2.1 **Equatorial coordinates of major celestial points (J 2000)**

The Sun moves along the ecliptic at a rate just short of a degree per day, and crosses the celestial equator twice, on its way north about March 20, and south about September 23. If the Sun is on the celestial equator, it must rise exactly east and set exactly west, so that days and nights are of equal length, hence the term equinoxes for the two, *Vernal Equinox* at the northward crossing in Pisces, *Autumnal Equinox* at the southward crossing in Virgo. The point at which the Sun is farthest north, at a declination of 23½°N, is the *Summer Solstice* in traditional Gemini (but actually just across the artificial border in Taurus), which it passes about June 21; the most southerly point (23½°S) is the *Winter Solstice* in Sagittarius, which the Sun hits about December 22.

The solar crossings of these four points define the beginnings of the northern-hemisphere seasons (and were obviously named by northerners). At the Summer Solstice, the Sun is most nearly overhead at mid northern latitudes and provides the most heat (as the Sun's rays are spread out over the least area), while at the Winter Solstice its maximum altitude is as low as possible, and it thereby gives us the least heat. The reverse is true in the mid southern hemisphere.

The Vernal Equinox, which (except for precession [2.4]) is fixed to the stars, is an obvious zero point upon which to base a coordinate system. The equinox defines a celestial "prime meridian," the *equinoctial colure*, a great circle that passes through both equinoxes and both celestial poles. To declination, which remains the same as before, now add *right ascension*, α, sometimes RA (see Fig. 2.6). Again pass an hour circle through a star. Right ascension is measured counterclockwise (eastward) along the celestial equator from the Vernal Equinox to the intersection of the equator and the star's hour circle. As the sky appears to rotate around us, the Vernal Equinox and the star move together and maintain the same angle between them. Right ascension and declination are thus fixed and can be recorded in a catalogue of stars. Like hour angle, right ascension is traditionally and conveniently measured in time units, though degrees are not wrong. Table 2.1 gives the equatorial coordinates of the equinoxes, solstices, and other relevant points.

Hour angle (fixed to the observer) and right ascension (fixed to the stars) are related through *time*. Time is measured through hour angle. There are two broad kinds, *solar time* and *sidereal time*. Solar time is defined as the hour angle of the Sun plus 12 hours (so that the day starts at midnight). Sidereal time is similarly defined as the hour angle of the Vernal Equinox (without the 12-hour addendum). Because the Sun is continually moving eastward relative to the Vernal Equinox by just under a degree per day counter to its movement along its daily path, it takes the Sun longer to make a full turn of the sky than does the equinox. The day defined by the Sun (the *solar day*) is therefore just under four minutes ($3^m\ 56.56^s$) longer than the *sidereal day* that is defined by the apparent rotation of the stars.

Each day is broken into 24 hours, each hour into 60 minutes, and so on. Solar hours, minutes, and seconds are therefore proportionately longer than their sidereal counterparts. "Mean solar time" (adjusted for the eccentricity of Earth's orbit and the obliquity of the ecliptic, which make raw solar time somewhat variable) is kept by a solar clock, sidereal time by a sidereal clock that runs not quite four minutes fast per day relative to the solar clock. The two kinds of time are connected through the date. On the day the Sun crosses the Autumnal Equinox, the Sun is opposite the Vernal Equinox. If the Sun is crossing the celestial meridian at noon and has an hour angle of 0 hours, the Vernal Equinox has an hour angle of 12 hours. The solar time ($0 + 12 = 12$ hours) therefore equals the sidereal time (12 hours). For every day thereafter, since the sidereal clock gains not quite

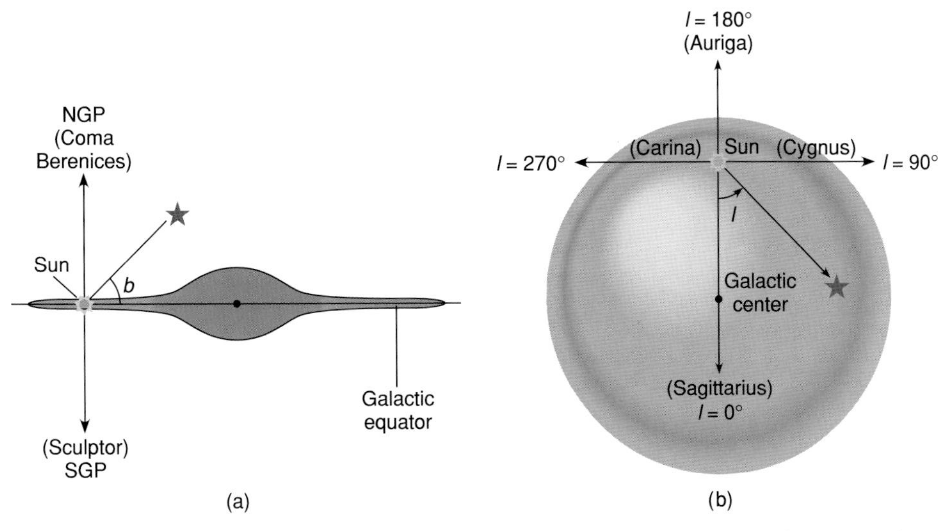

Fig. 2.10 **Galactic coordinates** The Sun is embedded in the disk of our flat Galaxy. On the left, the Galactic latitude (*b*) of a star is measured upward or downward perpendicular to the Galactic equator toward the Galactic poles. On the right, Galactic longitude (*l*) is measured along the Galactic equator counterclockwise (as viewed from the north Galactic pole) from the Galaxy's center. (Kaler, J.B. (1994). *Astronomy!* New York: HarperCollins, reprinted by permission of Pearson Education, Inc.)

4 minutes on the solar clock, it picks up a whole day over the course of the year (the divergence is, of course, continuous). The year contains 365.2422 . . . solar days, and therefore 366.2422 . . . sidereal days. All observatories must have both solar and sidereal clocks to access the sky.

The hour angle of the Vernal Equinox and the right ascension of the equator point are the same, just measured in opposite directions. Therefore, sidereal time (defined as the hour angle of the Vernal Equinox) must also be the instantaneous right ascension of the meridian (or the equator point). From Fig. 2.6, note that the sum of the hour angle and right ascension of any star must equal the sidereal time. Since for any celestial body, sidereal time $= \alpha + \mathrm{HA}$, $\mathrm{HA} =$ sidereal time $- \alpha$. You can then find a star in the sky by looking up its declination and right ascension and calculating its hour angle from your sidereal clock. Telescopes at professional observatories (as well as high-end amateur telescopes) have built-in clocks. The operator need only enter the coordinates on a keyboard, and the computer does the rest, guiding the instrument to its proper place.

2.3.4 Ecliptic coordinates

Since you can establish a coordinate system on any great circle, why not on the ecliptic as well? Indeed, because of the cultural and spiritual importance of the once-mysterious seven moving bodies of the sky (Sun, Moon, and five bright planets), all of which closely follow the ecliptic path, *ecliptic coordinates* were at one time much more important than equatorial coordinates. Their significance lies in the system's very name, as ecliptic coordinates are equally well known as the original *celestial coordinates*.

The celestial/ecliptic system works much the same way as do right ascension and declination. The poles of the coordinate system are the north and south ecliptic poles defined by the perpendicular to the ecliptic. The north ecliptic pole lies at $\alpha = 18^{\mathrm{h}}$, $\delta = 66\frac{1}{2}$°N (the complement to the obliquity), the south ecliptic pole at $\alpha = 6^{\mathrm{h}}$, $\delta = 66\frac{1}{2}$°S (see Table 2.1). The zero point is the Vernal Equinox. Pass a great semicircle from the north ecliptic pole through a star to the south ecliptic pole. *Celestial longitude* (or *ecliptic longitude*), λ, is the arc along the ecliptic to the east (counterclockwise) from the Vernal Equinox to intersection of that great semicircle and the ecliptic. *Celestial latitude (ecliptic latitude)*, β, is the angle measured along the great semicircle north or south of the ecliptic to the star. Both sets of coordinates can be seen in the old star maps in Figs. 1.9, 1.10, and 1.11.

Celestial coordinates are no longer commonly seen, having been mostly supplanted by equatorial coordinates. They are still used in planetary studies, however, and have great advantage for solar position, as the celestial latitude of the Sun is by definition always 0°, and the solar celestial longitude increases more or less uniformly with time (the eccentricity of the Earth's orbit causing the Sun to move a small angle ahead or behind the average).

2.3.5 Galactic coordinates

While celestial coordinates are rarely used in stellar astronomy, Galactic coordinates surely are. After all, stars are among the most defining features of the Galaxy [Chapter 5], and we have to know where they are in a Galactic context, one entirely divorced from the rotating or revolving Earth. Since the Galaxy's disk is its primary organizing feature, and since the Milky Way is the main manifestation of the disk, the Milky Way's centerline provides an obvious start in obtaining a fundamental great circle, a *Galactic equator*. A line perpendicular to the Galactic equator defines the *north* and *south Galactic poles*

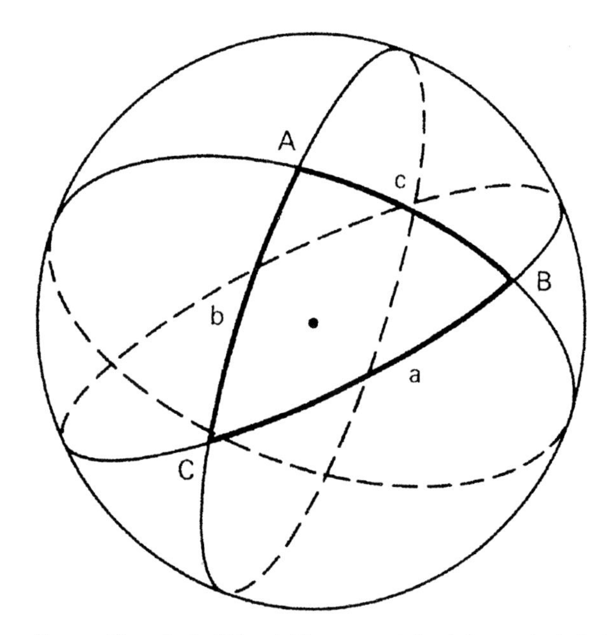

Fig. 2.11 **The spherical triangle** Three great circles define a spherical triangle with angles A, B, and C and opposing sides a, b, and c. All are measured in angular units and are connected via a variety of formulae. (Kaler, J.B. (1996). *The Ever-Changing Sky*, Cambridge: Cambridge University Press.)

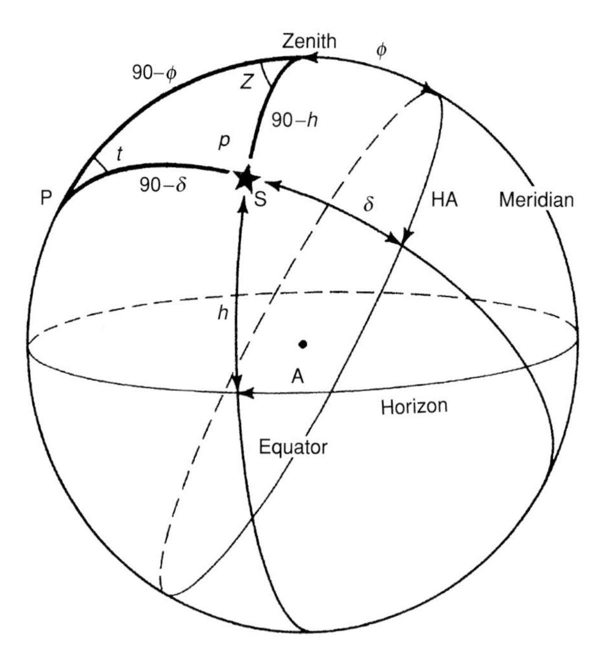

Fig. 2.12 **The astronomical triangle** The astronomical triangle is defined by the great circles that connect the star, the zenith, and the visible celestial pole. The angles are the hour angle (t), the "zenith angle" (Z, which is here 360° – azimuth), and the parallactic angle (p). The sides are 90° – latitude (ϕ), 90°–declination (δ), and 90° – altitude (h). The triangle allows conversion between altitude and azimuth, and hour angle and declination. (Kaler, J.B. (1996). *The Ever-Changing Sky*, Cambridge: Cambridge University Press.)

Box 2.1 Coordinate conversions

Stars have coordinates in all the different systems presented here, and it is necessary to be able to convert among them. Conversion is done through the relatively simple formulas of spherical trigonometry. Draw three great circles on a sphere. Each intersects the other two and forms a *spherical triangle* made of the truncated arcs (Fig. 2.11). All triangles have three sides and three angles. On a plane, the sides are measured in physical units, in feet, kilometers, etc. The sides of a spherical triangle, however, are measured in degrees of arc, so there are really six angles to consider rather than three. Laws connecting the six can be derived from the standard rules of plane trigonometry. If the spherical triangle is described by sides a, b, c, and opposing angles A, B, C, the *law of sines* states:

$$\sin A/\sin a = \sin B/\sin b = \sin C/\sin c,$$

while the *law of cosines* is:

$$\cos a = \cos b \cos c + \sin b \sin c \cos A,$$

with similar formulae for the other sides and angles.

An important application of these rules is the conversion of hour angle and declination into altitude and azimuth, which connects star positions with locations relative to the observer. It allows celestial navigation, and also tells how to point telescopes whose rotation axes are set up on the horizon system. The conversion involves the solution of the *astronomical triangle* (Fig. 2.12), at whose apices lie the zenith, the star, and the visible pole. The three arcs of the astronomical triangle consist of the arc on the celestial meridian that connects the visible celestial pole with the zenith ($90^\circ - \phi$), the arc along the vertical circle that connects the zenith with the star ($90^\circ - h$ = zenith distance), and that along the hour circle that connects the star and the pole ($90^\circ - \delta$). The three opposing angles are: at the pole, the hour angle HA (or 360° – HA) = t; at the zenith, the azimuth A (or $360^\circ - A$); and at the star, the *parallactic angle*, p, which is rarely used and which we can forget.

Applying the law of cosines to Fig. 2.12, $\cos(90^\circ - h) = \cos(90^\circ - \delta)\cos(90^\circ - \phi) + \sin(90^\circ - \delta)\sin(90^\circ - \phi)\cos t$, or (from the standard conversions of trigonometry),

$$\sin h = \sin\delta \sin\phi + \cos\delta \cos\phi \cos t.$$

Applying the law of sines (where in Fig. 2.12, Z = 360° – azimuth),

$$\sin Z / \sin(90^\circ - \delta) = \sin t/\sin(90^\circ - h),$$

or

$$\sin Z = \cos\delta \sin t/\cos h.$$

In the other direction,

$$\sin\delta = \sin\phi \sin h + \cos\phi \cos h \cos Z,$$

$$\sin t = \cos h \sin Z / \cos\delta.$$

One must only be careful to pick the correct quadrant of the sphere. Similar applications yield conversions between equatorial and celestial coordinates, between equatorial and Galactic coordinates, and so on.

Name	Terrestrial	Horizon	Equatorial I	Equatorial II	Celestial (Ecliptic)	Galactic I (old)	Galactic II (new)
FGC	terrestrial equator	horizon	celestial equator	celestial equator	ecliptic	Galactic equator	Galactic equator
SGC	meridians	vertical circles	hour circles	hour circles	. . .	. . .	. . .
Prime SGC	prime meridian	celestial meridian	celestial meridian	equinoctial colure[a]	. . .	. . .	. . .
Zero point	intersection of equator and prime meridian	north point of horizon	sigma point	Vernal Equinox	Vernal Equinox	intersection of Galactic and celestial equators	center of Galaxy
FGC coordinate	longitude, λ	azimuth, A (or Z)[b]	hour angle, HA	right ascension, α	celestial longitude, λ	Galactic longitude, l^{I}	Galactic longitude, l^{II}
Units	degrees/hours	degrees	hours/degrees	hours/degrees	degrees	degrees	degrees
Direction	east and west to 180°	east	west (clockwise), or east[c]	east (counter-clockwise)	east (counter-clockwise)	north	north
SGC coordinate[d]	latitude, ϕ	altitude, h	declination, δ	declination, δ	celestial latitude, β	Galactic latitude, b^{I}	Galactic latitude, b^{II}
Pole	north, south pole	zenith, nadir	north, south celestial poles	north, south celestial poles	north, south ecliptic poles	north, south Galactic poles	north, south Galactic poles

[a] The hour circle through the solstices is called the solstitial colure.
[b] Zenith angle, Z, is measured in both directions from north to 180°.
[c] Hour angles measured east are negative.
[d] Small circles for which this coordinate is constant are called *parallels*; parallels of altitude are called *almucantars*.

Table 2.2 **Summary of coordinate systems**

(the NGP and SGP). *Galactic longitude*, l, is measured along the equator, *Galactic latitude*, b, perpendicular to it in both directions toward the poles (Fig. 2.10).

However, neither the optical Milky Way nor the nearby stars work well to establish the Galactic equator. Interstellar dust hides the more distant stars, preventing us from seeing the whole Galaxy, while the dust's uneven distribution distorts the local scene. Moreover, different kinds of stars have different Galactic distributions, and any stellar selection will necessarily skew the coordinate system. Much better is the use of radio observations, which can punch through the dust, allowing access to the entire Galaxy. The best of these is radiation from the 21-centimeter line of interstellar neutral atomic hydrogen [5.3.3; Box 11.2], which forms a very thin disk down the middle of the optical Milky Way. A best fit to the 21-cm data that had been acquired by the late 1950s defines the Galactic equator, and locates the NGP at right ascension $12^h46.6^m$, declination $+27°40'$, for the equinox of 1900. Precession has since moved the poles to the positions listed in Table 2.1. The Galactic equator crosses its celestial counterpart in Aquila at a 63° angle at a right ascension of 18^h51^m, and by coincidence crosses the solstitial colure (the great circle that connects the celestial poles and the solstices) near the Summer and Winter Solstices. The NGP lies in Coma Berenices, the SGP in Sculptor. In these directions we find the fewest stars, and because of the lack of dust obscuration, the greatest number of external galaxies.

The next step is to define a zero point. In the early days of radio astronomy, before anyone knew the richness of the radio sky, astronomers gave Roman letters to the first radio sources found within a constellation. Since these are obviously the brightest and most famed, the system, which has long-since been supplanted by complex cataloguing, is still in informal use. "Cassiopeia A," for example, is now recognized as the most recent known supernova remnant [14.3.4]. Deep within the dark dust clouds in the thickest portion of the Milky Way lies radio-brilliant Sagittarius A. Observations with very long baseline interferometers [2.6.2] reveal a near-point source of radiation within Sagittarius A, now referred to as Sagittarius A*, which is the true center of the Galaxy and seems to be a supermassive black hole [14.3.3] 8000 parsecs away. (The Galactic center is totally obscured in the optical spectrum by over 30 magnitudes of dust extinction [3.5], and is accessible only in the radio and infrared.) Sagittarius A, which lies on the Galactic equator, is thus the natural zero point. Galactic longitude is measured counterclockwise (looking down from the NGP) from the zero point toward the north, toward Aquila and Cygnus. In Figure 1.1 (and frontispiece), longitude starts at 0° at the center of the picture and goes to 180° at the left-hand edge (the anticenter: see Appendix 5), then continues from the right-hand edge back to the center.

This system of Galactic coordinates replaced an earlier system used prior to 1960. The north pole of the original system was defined by the problematic distribution of stars in the Milky Way at $\alpha = 12^h40^m$, $\delta = +28°$ (equinox 1900), 4.6 minutes of time west and 20 minutes of arc north of the modern pole. The older system also used as a zero point the intersection of the Galactic equator with the celestial equator at $\alpha = 18^h40^m$ (1900). During the transition period in the 1960s, the earlier system was known as system I, the newer as system II, yielding coordinates respectively called l^I and b^I and l^{II} and b^{II}. The newer system is now in exclusive use and the superscripts are dropped, but readers of older literature need to be aware of the original version and particularly of the difference in Galactic longitude between the two. All the various coordinate systems, which are easily converted one to another (Box 2.1), are summarized in Table 2.2.

2.4 Precession

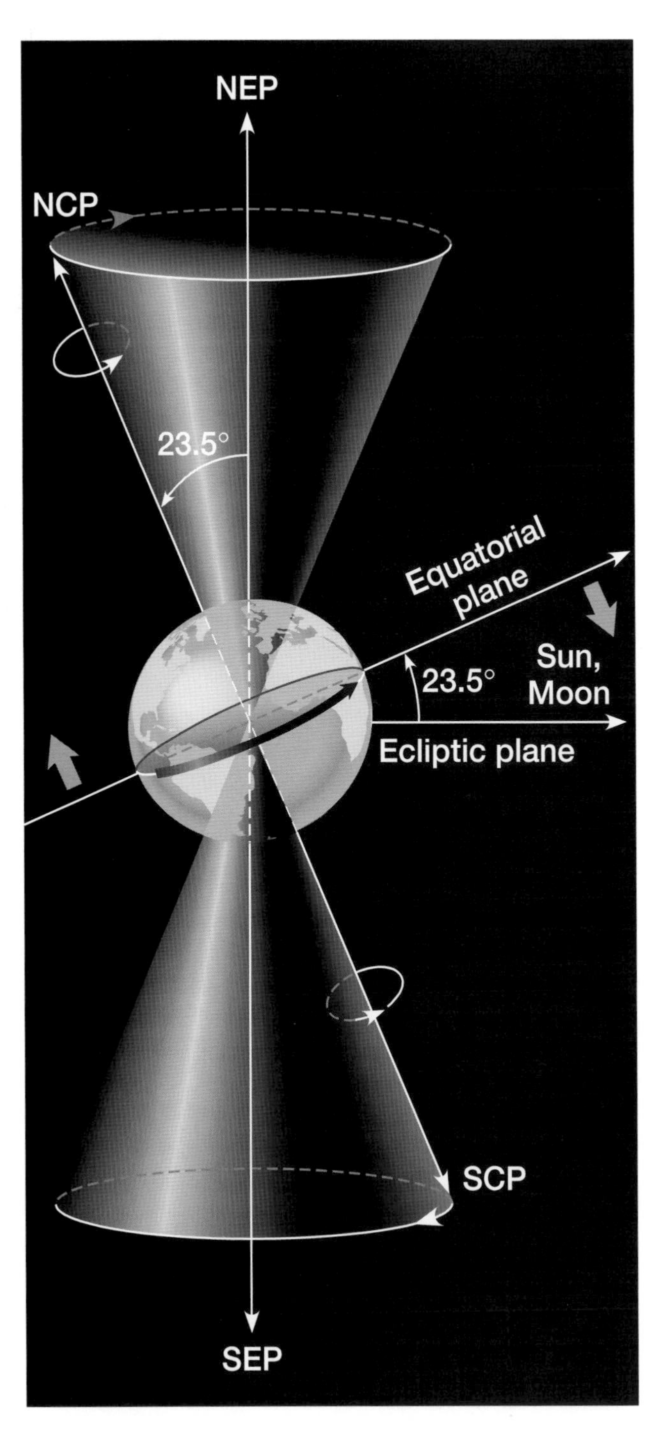

Fig. 2.13 **Precession** The gravitational pull of the Moon and Sun on the Earth's equatorial bulge causes the Earth's axis to precess over a 26 000-year period, defining a cone with an opening angle of twice $23\frac{1}{2}^{\circ}$. (Kaler, J.B. (1994). *Astronomy!* New York: HarperCollins, reprinted by permission of Pearson Education, Inc.)

The student of astronomy quickly learns that nothing is still. Equatorial coordinates, right ascension and declination, are "fixed" only in the sense that they change vastly more slowly than do hour angle, altitude, and azimuth. But change they do. The chief culprit is the precession of the Earth's axis, though many other effects factor in as well. As the Earth spins, it throws itself out a bit at the equator, rendering our non-spherical planet 40 kilometers larger across its equator than through its poles. The Moon and Sun, however, do not lie in the equatorial plane, but in the ecliptic plane. They therefore exert a gravitational force on the equatorial bulge and a resulting torque on the Earth that causes its rotational axis to *precess*, or to rotate around the axis of orbital revolution (Fig. 2.13). The action is similar to the wobbling of a spinning top. The Earth's axis maintains a more or less constant obliquity while it wobbles about the orbital axis. The result is a continuous change in equatorial and ecliptic coordinates.

The gravitational torque is not large, hence the period of precession is long, 25 800 years. Hardly obscure, however, precession was discovered with the naked eye around 150 BC by the Greek astronomer Hipparchus. The most obvious feature of precession is the change in the direction among the stars of the celestial poles. Over the course of 26 000 years, the north celestial pole will make a circle $23\frac{1}{2}^{\circ}$ in diameter about the north ecliptic pole, as will the south celestial pole about the south ecliptic pole (Fig. 2.14). At the moment, the NCP is within $\frac{3}{4}$ degree of Polaris, making it a fine "pole star." The SCP is within a similar distance of much fainter Sigma Octantis. As the north rotation pole moves, it will approach Polaris over the next century, and then pull away. Around AD 7500, Alpha Cephei will be a modest north pole star, and around AD 14 000 we get Vega! Near the time of the building of the Great Pyramid at Khufu, Thuban (Alpha Draconis) was near the pole. Over the next 10 000 years, the SCP will make a pass through the rich star fields of Argo. Note, however, that precession does not cause any changes in terrestrial latitude. If you live at 30°N, the NCP will always have an altitude of 30°. Since you are on the moving – here precessing – body, it will appear that the stars are moving past the pole rather than that the pole is moving against the stars.

Polaris is now near the NCP, close to 90° north declination. In half the precessional period, the stars will shift such that the NCP will be on the other side of the north ecliptic pole, and Polaris will be 47° (double the obliquity) from the NCP, giving it a declination of +43°. Such must be true of all stars. As the axis precesses, their declinations change through a range of 47° over 26 000 years. In 13 000 or so years,

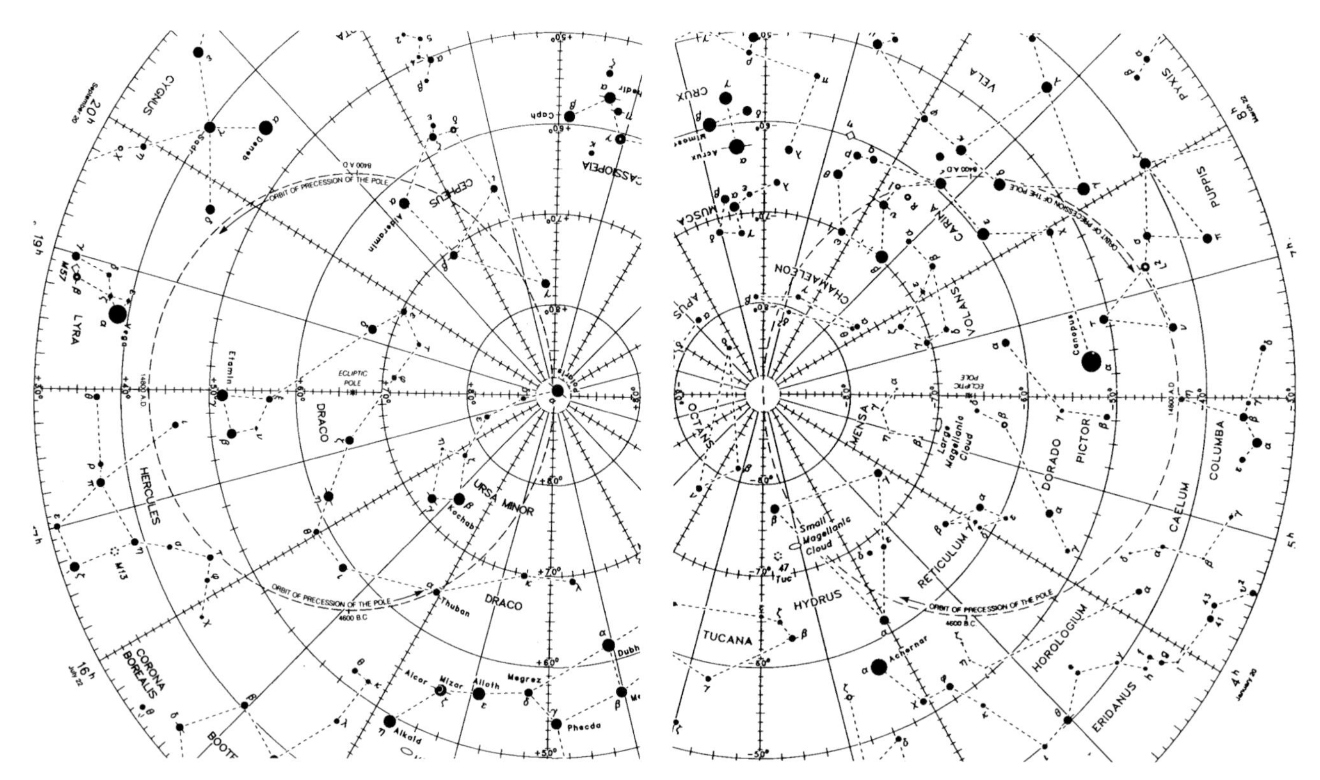

Fig. 2.14 **Polar precession** At left, the north celestial pole takes a 26 000-year tour of the northern sky, while on the right the south celestial pole travels the southern sky. (Epoch 2000 SC2 and SC3 Constellation Charts, courtesy of Sky Publishing Corporation.)

"Polaris" (quite misnamed) will be passing nearly overhead at New York or Paris.

As the axis wobbles, so must the equatorial plane. The result is that the intersections of the equator and the ecliptic – the equinoxes – move westward against the background stars, each taking 26 000 years to move through all the constellations of the Zodiac, the solstices following along always between them. Astrologers fix the signs of the Zodiac [1.1.1] to the Vernal Equinox. Some 2500 years ago, that point was in Aries. Still astronomically called "the first point of Aries," its symbol is a ram's horn, even though about 100 BC the equinox passed the current formal boundary line into Pisces. In the year 2700 or so, it will move into Aquarius, then in AD 4400 into Capricornus, and so on.

Again recall that we are on the moving body. As a result, we see the reflexive motion, such that a long-lived terrestrial observer would see the zodiacal constellations move counterclockwise (as viewed from the north), to the east past the equinoxes and solstices as if on a moving belt. In 13 000 years, the constellations that hold the equinoxes will have switched positions, vernal then in Virgo, autumnal in Pisces. More to the point, so will have the solstices. In 13 000 years, Gemini will hold the Winter Solstice, while Sagittarius will hold the summer. From the point of view of a northern-hemisphere observer, Gemini will then appear low in summer, declination -23° or so, while Sagittarius will appear high in winter near $+23^\circ$ (and vice versa in the southern hemisphere), the declinations of all stars again ranging across 47°. The Southern Cross was once visible from central North America, and in 13 000 years, Canis Major and Sirius will be lost from sight.

If the Vernal Equinox moves all the way around the ecliptic westerly against the stars, right ascensions will generally increase with time and must change through a full 360°, or 24 hours, over the precessional cycle (though near the ecliptic poles, they can decrease for a time). Dividing the number of seconds of arc in a circle by 25 800 years gives an average backward drift of 50.3 seconds of arc per year, a degree in a human lifetime. Because of the angle the equator makes with the ecliptic, the increase in right ascension of a star exactly at the equinox is $46.1'' = 3.1^s$ per year, a virtual barn door compared to our second-of-arc ability to point a telescope.

Because equatorial coordinates continuously change, star catalogues must specify the date, or *equinox* (sometimes called the *epoch*) for which they are instantaneously valid. The *Bonner Durchmusterung*, for example, is set for an equinox of 1855, while modern star catalogues use the year 2000. In 50 years we will be using coordinates for 2050. Just specifying a year, though, is not sufficient. The coordinates for 1950 are pegged at 1950.0, defined by the beginning of the "Besselian year," the moment at which the Sun passed celestial longitude 280° (which falls close to the beginning of the calendar year).

They are now referred to as B1950 coordinates. In 1983, the International Astronomical Union slightly changed the definition and pinned the moment of 2000.0 at January 1, 2000, 12 hours Terrestrial Time (close to January 1.5 Greenwich Mean Time, but corrected for the effects of Einstein's relativity). These are known as J2000 coordinates. Both B1950 and J2000 positions are used in the naming of celestial objects by coordinate numbers.

To find a star in the sky, one needs to look up the coordinates for a given equinox and then "precess them" with a coordinate transformation. Here is where celestial or ecliptic coordinates really shine. Since it is the equator that is moving and not the ecliptic, celestial latitudes are unaffected and celestial longitudes simply increase steadily at the rate of 50.3 seconds of arc per year. Galactic coordinates are entirely unaffected.

2.5 Nutation, aberration, and the rest of it

In a search for parallax [4.2] using the star Gamma Draconis, which passes nearly overhead at London, in 1748, James Bradley announced the discovery of *nutation*. The progress of precession clearly depends on the gravitational torques of the moment, which are not constant. The lunar orbital plane is tilted by just over 5° relative to the ecliptic plane, and as a result, the sum of the forces of the Moon and Sun vary over the course of the month. Since the Moon has by far the larger effect in producing precession, the Earth's rotation poles really revolve around the axis of the tilted lunar orbit. However, the lunar orbit also precesses over an 18.6-year period (called the "regression of the nodes," where the nodes are the points where the lunar orbit crosses the ecliptic). As a result, the lunar orbital plane averages out to the ecliptic plane, but since the Earth is following the Moon, the nodal regression induces an 18.6-year wobble on the precessional wobble with a size of 17 seconds of arc in the direction of precession and 9 seconds of arc perpendicular to it. This small periodic change continuously factors into 18.6-year changes in right ascension and declination.

Moreover, the Earth's orbit is elliptical, and so (to a rough approximation) is the Moon's. As a consequence, the gravitational forces change with solar and lunar distance, which add nutational effects with periods of a year and a lunar month. Large deviations from lunar orbital ellipticity add more. Nutation has over 100 more wobbles associated with it. Then add the fact that the obliquity of the axis changes periodically with time over nearly three degrees, and we see that even the average circle of precession does not quite close on itself.

Twenty years before Bradley found nutation, he discovered the aberration of starlight. On a windless day, walk into falling rain and it appears to fall into your face. The faster you walk, the more the rain seems to come from the forward direction. So it is with light. As the Earth orbits, the stars all appear to be shifted in the direction of motion. The degree of shift depends on the speed of the Earth (along the direction perpendicular to the direction to the star) relative to the speed of light. If you are moving right at the star (one on the ecliptic 90° to the west of the Sun) there is no shift at all. If you are moving perpendicular to the direction to the star (one at either of the ecliptic poles, or 180° from the Sun), the shift is maximized at 20.5 seconds of arc. Even a tiny shift caused by the rotation of the Earth, which at the meridian shifts equatorial stars by 0.19 seconds of arc to the east, must be considered. All these movements factor into small coordinate changes.

In addition, the Earth is gradually slowing down as a result of tides raised by the Moon and Sun. Moreover, there are a variety of irregular movements that can temporarily slow the Earth yet more, or even speed it up. To derive final positions of the stars we have to find corrections to the observers' clocks through continuous measures of time.

Given these various problems, astronomers adopt three kinds of coordinates. The *apparent place* is that actually observed, the *true place* smooths aberration, and the *mean place* for a specific date gives the coordinates with both aberration and nutation averaged out.

2.6 Astrometry

Positional astronomers practice *astrometry*, the "measure of the stars," to find exactly where the stars are. Though the principles are not difficult, the practice requires great expertise and time. Nevertheless, astrometrists can now obtain positions of celestial objects with accuracies of better than a thousandth of a second of arc (a *millisecond*, *milliarcsecond*, or mas).

The problem is one of defining the fundamental directions and then in measuring the positions of the stars and other objects with reference to those fundamental directions. There are three overlapping systems: a traditional optical system as observed from the Earth's surface, one based on astronomical radio sources as observed from Earth, and an optical system as observed from space. All are calibrated against the others, and one – the radio system – is picked as primary.

2.6.1 Traditional positional astronomy

The declination of a star is found from its altitude at upper culmination. For a star south of the zenith, latitude ϕ equals declination δ + zenith distance z, [2.3.2; Fig. 2.7] and since altitude $z = 90^\circ - h$,

$$\delta = \phi - z = \phi + h - 90^\circ.$$

Since for a star north of the zenith, $\phi = \delta - z$,

$$\delta = 90^\circ + \phi - h.$$

Set up a telescope so that it rotates only on one axis and points only around the great circle of the celestial meridian. Fit it with a graduated circle rather like a giant protractor, and calibrate it to read 90° at the zenith, 0° at the horizon. Center a crosshair in the eyepiece and watch the Earth's rotation send a star through the field of view. Adjust the telescope so that the star hits the center of the crosshair, read off the altitude, and calculate the declination. The necessary latitude can be found with the same telescope. Pick a circumpolar star and measure its altitude at both upper and lower meridian transit. Take the average, and you have the altitude of the center of the daily path, which must be the altitude of the pole, and therefore the latitude.

Many are the problems. You must obtain excellent machine work to build the telescope and the graduated circle, and must precisely align and level the instrument to find the zenith. Then you must attend to atmospheric refraction. As a ray of starlight enters the air from the near-vacuum of space, it bends toward the perpendicular. As a result, all stars – and the celestial pole as well – are lofted above their correct positions by an amount that itself depends on the altitude. At an altitude of 45°, stars are raised upward by a minute of arc, and at the horizon by thirty times that. At the subsecond of arc level, the effect depends critically upon atmospheric pressure, temperature, water vapor content, and the wavelength of observation. Moreover, variable refraction produced by turbulence in the Earth's atmosphere causes the images of the stars to jump around, preventing you from pinning down their exact locations. As a result, not only are stellar declinations subject to inevitable errors, but so are latitudes.

Sidereal time is the instantaneous right ascension of the celestial meridian. To measure the right ascension of your star, note the time from the sidereal clock at the moment the star passes the vertical crosshair. But you get the sidereal time from the right ascensions of stars that cross the meridian! The circularity is resolved by finding the Vernal Equinox through daytime measures of the declination of the center of the Sun. The point at which the solar declination is zero identifies one of the equinoxes, which can be fixed to the stellar background through daylight or twilight observations of planets and stars. Over a long period of observation, a knowledge of sidereal time and right ascensions is obtained through successive improvements in the data. Error from refraction still enters into right ascensions via solar declination measures.

Next you must contend with precession, nutation, aberration, the Earth-rotation irregularities. All these matters are time-dependent. Say you have 10 000 stars in your observing program and can observe 10 a night. Assuming clear weather, it will take you 1000 nights, or nearly 3 years. Each night you are observing relative to an equinox shifted from the night before, and during the day must calculate the amount of precession, nutation, and aberration to reduce your data to the equinox of your first night of observation, or better to some standard equinox, so that all stars will be on the same system of coordinates.

That done, you discover that you have not yet contended with proper motion [5.2.1] or parallax [4.2.1]. Stars move with respect to one another, and therefore with respect to the coordinate frame defined for a particular equinox. From one observation, you cannot tell how the star is moving; by the time you determine the position of star number 10 000, star 1 and all the others you have measured on previous nights are somewhere else. You want a snapshot of the sky, but because of the time it takes to make the observations, you cannot get it. The only way to find the proper motions of your stars is to go back and make the measures all over again, and after another three years solve for both the position and the proper motion of each star. Many of your stars are also nearby, so you also have to take the distance-dependent wobbles induced by parallax into account. Measurement of absolute parallax requires additional cycles of observation.

However, how do you know the rates of precession and nutation in the first place? The only way is to measure the changes in right ascension and declination with time. But to get coordinates of stars, you need to be able first to correct them for precession, so you are again in the fix of circular reasoning. You do it by making a massive simultaneous solution of the right ascensions and declinations of the stars in which you adjust the rates of precession and nutation such that the proper motions that you determine from your two sets of observations are directionally random. If for physical reasons they are not (which can well be the case), then you have made errors in your precessional constants. To improve the precision of your results, you do it all over again with another three years of observation.

The results of intense work from observatories all over the globe led to the astronomer's "fundamental catalogue" of 1535 stars, the "FK" from the German "Fundamental Katalog," which was compiled at Heidelberg. We are now in the fifth generation, the FK5. Which is where it stops as a result of a great explosion of technology: there will be no FK6.

2.6.2 Radio positions

The ability of a telescope to see, or resolve, fine detail depends on the aperture of the instrument and upon the wavelength of observation. Through a telescope with a diameter of 10 meters you can in principle see twice as much detail as you can with a 5-meter telescope. The effect involves the interference of light waves that are refracted or reflected by the instrument's lens or mirror. Long waves interfere over a larger scale. As a result, radio telescopes are inherently less discriminating than are optical telescopes. To achieve the same resolution of a source as an optical telescope, a radio telescope must be made much larger. Radio waves are so long, however, that a single huge radio telescope can never compensate. We can, however, use two separate radio telescopes, link them, and make simultaneous observations with both. In practice, we observe the interference of the radio waves that arrive at the two telescopes. When we do, we achieve the resolution of a single telescope equal in size to the separation between the two. Three or more telescopes greatly improve the view. By using extremely accurate atomic clocks to synchronize the observations made with individual telescopes, astronomers can obtain the effect of a telescope up almost to the size of Earth: a *very long baseline interferometer*, or VLBI. Such instruments have resolving powers of under a milliarcsecond, far better than optical telescopes.

Stars have appreciable proper motions because they are close to us. Many of the sky's radio sources, however, are not local, but are the ultradistant nuclei of galaxies (quasars) at distances of billions of light years. They are not expected to have any observable proper motions, and can therefore be taken as points in a fixed reference frame. Observations of the relative positions of these distant sources allow the determination of the rotation of the Earth, and of sidereal time, to great accuracy. Modern time is told by their positions. These radio sources also allow the establishment of a celestial reference frame that is divorced from our irregularly rotating planet, yet another displacement from our provincial viewpoint as "center of the Universe."

The axes – the framework – of the *International Celestial Reference System* (the ICRS) were initially aligned as closely as possible with the FK5, using sources common to both the radio and optical spectral regions and a variety of calibrations. The complete set of optical and radio data then constitute the *International Celestial Reference Frame* (the ICRF). The coordinates of stars no longer depend on the celestial equator and ecliptic, but are as much as possible absolute. The positions of stars as viewed on Earth are calculated from the ICRF on the basis of precessional theory, which from optical observations is continuously upgraded. The whole system will be one of coordination between the absolute radio framework and the optical that is used for us to look out into the Universe.

The best optical positions, with errors of the order of a milliarcsecond, are those derived from the Hipparcos parallax mission [4.2.3]. The observations, however, were restricted to brighter stars and there is little overlap with radio positions used to define the ICRS. The "natural" Hipparcos reference frame was subsequently tied to the ICRS through a variety of separate and linking observations. These were accomplished through radio observations of radio-bright Hipparcos stars and ground-based and Hubble Space Telescope observations that could tie Hipparcos stars to distant radio-bright quasars. The Hipparcos catalogue is thereby tied to the radio frame to within about 10mas.

As the radio data improve, astronomers are able to weed out those sources that might provide misleading results. Improved positional satellites are on the way as well. Before long, we hope not to be observing positions at the millisecond level, but at the microsecond level, allowing us to see and know the Universe as never before. And it all started by simply looking up at the night sky.

CHAPTER THREE

Magnitudes

Fig. 3.1 **Countless stars** The star cloud in the thick of the Milky Way in Sagittarius embraces vast numbers of stars of different brightnesses and colors. (Courtesy of Hubble Heritage Team – AURA/STScI/NASA.)

The first impression of a dark nighttime sky is that of myriad stars, typically some two thousand of them. Brighter ones immediately draw the eye to themselves and to the patterns they seem to form. Yet the awesome power of the sky is perhaps better revealed through the sheer quantity of stars, the number of fainter ones overwhelming that of the brighter, until the dimmest ones that spatter the background simply fade into the void. Through even a small telescope, millions are available, through a big one, billions (Fig. 3.1).

We organize the stars in different ways: first by constellation pattern [1.1], second by coordinate position [2.3]. But we can also organize by individual properties: color, the appearance of the stars' spectra [6.2], and by apparent brightness. Why are some stars bright, why are others so faint? Is the reason – argued historically – that stars have different intrinsic luminosities [7.2.1], that they are at all different distances [Chapter 4], or perhaps because of a combination of the two? The only way we will know is to measure, first by quantifying apparent stellar brightness, and then by determining distances. With the data in hand, some stars are found to be millions of times more luminous than the Sun, while others are a million or more times less luminous. We can then progress toward learning what the stars really are. And the journey starts by finding a way of scaling stellar brightnesses to one another, by attaching numbers that tell both how bright stars appear to be and how bright they really are.

3.1 Apparent magnitudes

If we were today to begin to measure the *apparent* brightnesses of stars, that is, how bright the stars appear to be, we would likely use a unit that relates to the physics lab – to the amount of energy that passes through a unit area per second as viewed from above the Earth's absorbing atmosphere: ergs per square centimeter per second (ergs $cm^{-2}\,s^{-1}$), or more standard joules per square meter per second (j $m^{-2}\,s^{-1}$) = watts per square meter. But we are not doing it today. Instead, the process began more than 2000 years ago, long before there were any such physical measures.

To organize stellar brightness, the ancient Greek astronomer Hipparchus divided the night's stars into six groups, now called "magnitudes." The top twenty or so stars (Betelgeuse, Rigel, Antares, Arcturus) fell into the brightest group of "first magnitude," while the faintest that could be seen with good eyes were of the "sixth magnitude" (Fig. 3.2). The stars of the Big Dipper (Britain's Plough) are mostly second magnitude, and those in the Little Dipper range from second to fifth. The system begun in ancient Greece thrives today in highly quantitative form, its magnitudes measured across the spectrum, corrected for distance, and even for the deleterious effects of interstellar dust.

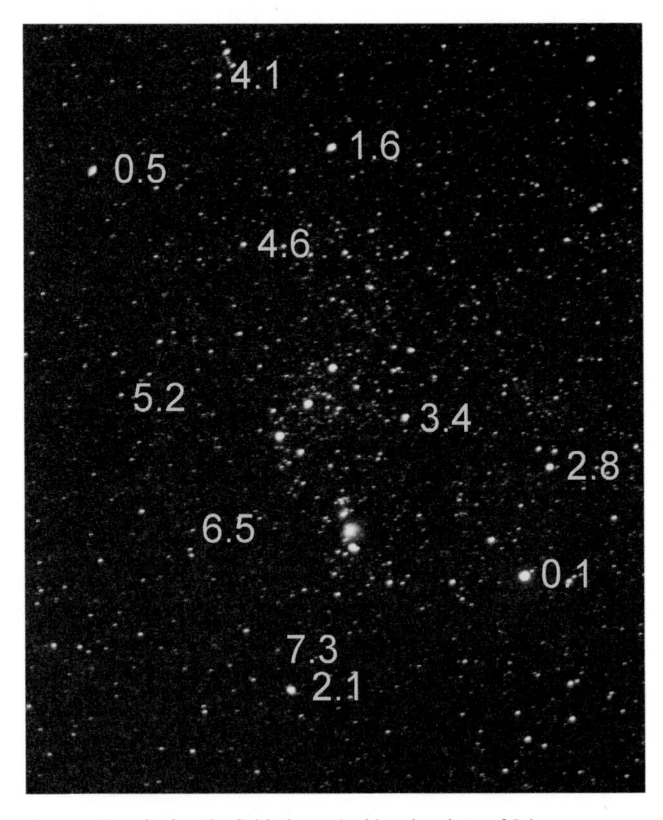

Fig. 3.2 **Magnitudes** The field of stars in this color photo of Orion appears very much as if viewed with the human eye. At center is the prominent Belt, from which dangles the Sword, the central star of which holds the reddish Orion Nebula. Various stars are labelled to the immediate right with their magnitudes. Betelgeuse is at upper left, Rigel at lower right. (J.B. Kaler.)

3.1.1 Apparent visual magnitudes

While the eye is an amazing instantaneous recording device, it does not store data. It is therefore not possible to convert a visual impression directly to a number. Even if one could, it would mean little, as such impressions of apparent brightness are not linear. While one star may seem twice as bright as another, it really may be many more times brighter. The eye compresses brightnesses in order that it can cover its astounding range, just as the ear does for sound.

Nevertheless, the eye can still be used indirectly. The apparent brightness of a star seen through the telescope depends on the amount of light gathered by the lens or mirror, which in turn is proportional to the square of the lens/mirror diameter. We could use a mask to diminish the telescope aperture to find the minimum size needed to see the star. The minimum light gathering area is thus inversely proportional to apparent stellar intensity. We could also insert an opaque filter between the eye and the star and note the degree of opacity required to make the star disappear.

Such experiments in the nineteenth century made a case that the visual response to light was more or less logarithmic. The British astronomer Norman Pogson (1829–1891) then suggested a more formal relation, that the difference in magnitude between two stars was 2.5 times the logarithm of their brightness (intensity) ratio as would be measured in physical units of watts. If we have two stars whose apparent visual intensities are actually i_1 and i_2, the magnitude (given by lower case m) difference

$$m_2 - m_1 = 2.5 \log i_1/i_2.$$

Say star 1 radiates a visual intensity that is 100 times brighter than star 2, that is, $i_1/i_2 = 100$. The logarithm of 100 = 2 (another way of saying that $10^2 = 100$). Consequently, $m_2 - m_1 = 2.5 \times 2 = 5$. Note that m_2 is a larger number than m_1. The brighter star has the lower magnitude number, as suggested by Hipparchus. In the scale of magnitudes, a difference of exactly 5 magnitudes between stars corresponds to a factor in apparent intensity of radiation of 100, that is, it takes 100 sixth magnitude stars to equal the apparent brightness of one first magnitude star. One magnitude difference would thus correspond to the fifth root of 100, or

2.512 . . . To check, substitute the rounded-off 2.512 in the equation above, and see that

$$m_2 - m_1 = 2.5 \log 2.512 = 2.5 \times 0.400 = 1.$$

Such magnitudes are now called *apparent visual magnitudes* because they are determined by the human eye.

This so-familiar magnitude system is relative, and in comparing one star with another, relates magnitude *differences* with intensity *ratios*. In the nineteenth century, the measure of quantitative visual magnitudes attained precisions of a tenth of a unit or so. With decimalization, generic "second magnitude" spans the scale between 1.5 and 2.5, third between 2.5 and 3.5, and so on. The next step is calibration on some kind of a fixed scale. Early work in the 1800s adopted Polaris, which is circumpolar [2.3.2] from the northern hemisphere, to be the fundamental comparison star, and set it to apparent visual magnitude 2.00. Similar systems used an average of fainter fifth magnitude circumpolar stars, but the result was much the same.

Upon absolute calibration, the set of "first magnitude" stars is seen to span a large range that must include negative magnitudes. Several stars – Capella, Vega, Arcturus, and Alpha Centauri – fall between −0.5 and +0.5, and are thus magnitude zero, while both Canopus and Sirius are of the "minus first magnitude" (Sirius nearly −2). Continuing on, Venus at her brightest reaches nearly to −5, the full Moon to −12, and the Sun to −27.

3.1.2 Apparent photographic magnitudes

The dawn of July 17, 1850, witnessed an astronomical revolution, the first permanently recorded stellar image, a photograph of zero magnitude Vega. With ever-more sensitive photographic plates, astronomers could permanently record the sky and take their work into the daylight for analysis. Moreover, whole fields of stars could be observed at one time (Fig. 3.3).

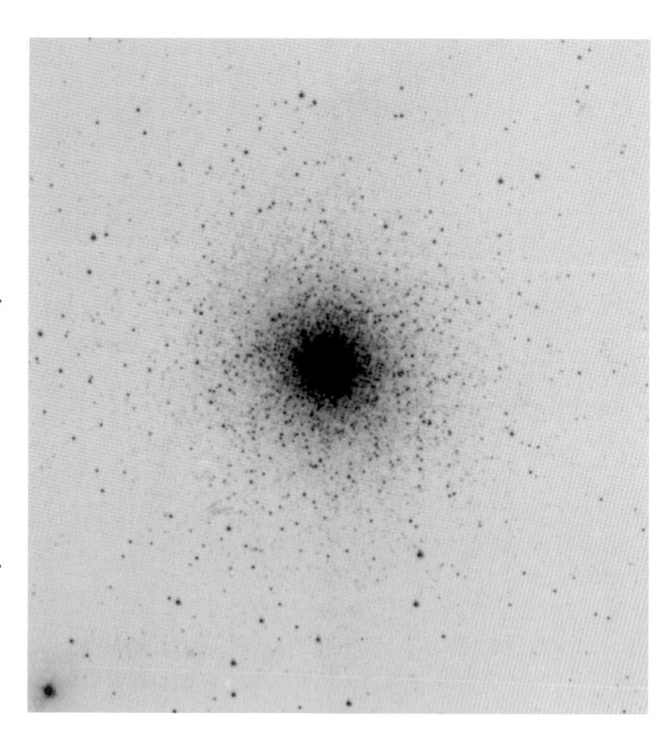

Fig. 3.3 **The photographic plate** This photograph of the globular cluster Messier 13 [9.4] is presented as a negative in which black and white are reversed. The original photographic negative is always used for research. (Courtesy of University of Illinois Prairie Observatory.)

Unlike the "human-eye" color view seen in Fig. 3.2, the early photographic plates (as well as nearly all those ever used for astronomical research) recorded only in black and white. A photographic plate consists of a light-sensitive emulsion of silver halide grains set into gelatin and spread on a glass plate (or film). When struck by light and then chemically treated in a developing bath, the grains turn black. The image of a star is therefore a collection of tiny blackened dots that are densely clumped in the middle and fade out to the edge. Unlike the eye, which sees in real time, photographic plates are integrating devices. The more light that strikes the plate, the more blackened grains there are. The longer the exposure, the darker the images, and the fainter the stars that can be detected. Photography thereby opened up the depths of the Universe.

The density of blackened grains near the center of a stellar image is in some way proportional to the apparent brightness of the star multiplied by the exposure time. However, the photographic process is very inefficient (taking 50 or more photons to darken a grain), highly non-linear, and non-reciprocal (that is, brightness and exposure time do not evenly trade off). Photographic emulsions also have a bottom rung, a low-light threshold below which no star can be recorded no matter what the exposure time, and a top rung in which the images become saturated, or as dark as possible. Moreover, the relations between intensity and blackening can change from one batch of plates to the next. As a result, each photograph must be calibrated with a device that superimposes images of known brightness on the same plate that was used to take a picture of the sky.

Other than image persistence (the ability of photography to integrate and retain the image), the most profound distinction between visual and photographic observations is their different responses to different colors. The eye sees best in the yellow–green part of the spectrum, around 5000 angstroms [Box 7.1]. Untreated photographic emulsions, however, work best under the action of more energetic photons with shorter wavelengths, blue and violet. (It is for this reason that spectral classes are traditionally determined by blue–violet spectra [6.2]. Modern films and plates are much better at longer wavelengths. "Panchromatic" films do well over the whole visual range so as to replicate the view seen with the human eye.)

As a result, a black-and-white photographic plate "sees" the sky and stars differently than does the eye. Hot (O, B, and A) stars [6.2], which emit abundant amounts of blue light, appear relatively brighter photographically than they do to the

Fig. 3.4 **Photographic magnitudes** Betelgeuse, the reddish first-magnitude upper left-hand star of Orion, appears as only third magnitude in this black-and-white professional photograph. Compare with the color photo in Fig. 3.2, in which the star appears to be much brighter. (Courtesy of Palomar Observatory/California Institute of Technology.)

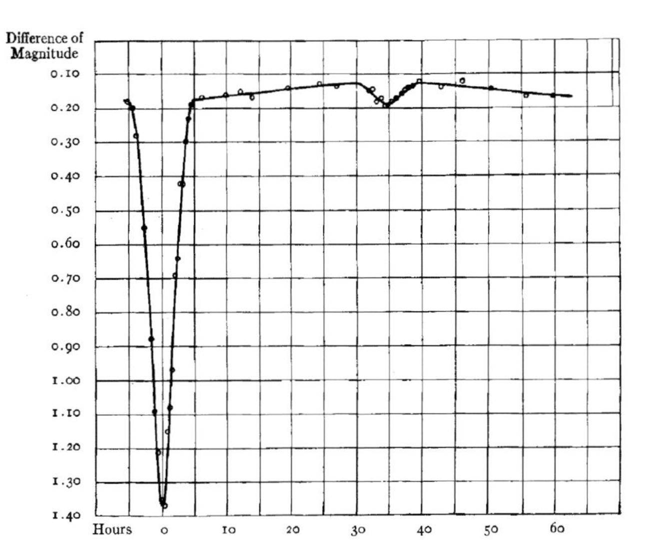

Fig. 3.5 **The first photoelectric light curve** The eclipses of Algol were accurately recorded photoelectrically around 1910. (Stebbins, J. (1910). *Astrophysical Journal*, **32**, 185.)

eye, whereas red stars, in which the blue-light component is close to absent, appear photographically dim (Fig. 3.4). A system of *photographic magnitudes* is therefore quite different from that of visual magnitudes, the former emphasizing blue stars, the latter, yellow to red ones. However, rather than being a disability, the difference between photographic and visual magnitudes became profoundly important to astronomy, as it led to the concept of the "color index," CI, which is the algebraic difference between the two [3.4]. Color index parallels temperature, and therefore substitutes for spectral class and temperature.

Like visual magnitudes, photographic magnitude determinations are relative, involving the comparison of the apparent brightness of one star with another. The final photographic magnitudes are derived by scaling them to visual magnitudes for white stars (that is, stars of "no color") of class A0 [6.2.1]. Among the original standards was the "north polar sequence" (NPS), which was originally a set of white fifth magnitude stars immediately surrounding the north celestial pole that is always visible from most northern hemisphere observatories. Later expanded to both brighter and fainter stars, the NPS magnitudes were also scaled to the visual magnitude of Polaris (corrected for color) and the magnitudes derived for various visual star catalogues. By using a combination of treated, more-visually sensitive emulsions, together with yellow–green filters that could match the human eye, photography could also be used to record and measure visual magnitudes, or *photovisual magnitudes*, so that the eye was not needed at all.

3.1.3 Photoelectric magnitudes

Problems with the intensity calibration of photographic plates, as well as their extreme and inconsistent non-linearity, render photographic magnitudes inherently uncertain. Moreover, the poor efficiencies of the emulsions placed limits on the faintest stars that could be observed. Not all that long after photography was introduced, the astronomical world saw the first glimmer of revolution in the introduction of the electronic detector. Around 1910, the American astronomer Joel Stebbins (1878–1966), then of the University of Illinois, used a selenium detector to construct a light curve for the eclipsing binary Algol [8.4] in which the magnitudes were accurate to nearly a hundredth of a division, ten times better than photography could achieve (Fig. 3.5).

Early photoelectric astronomy struggled with both low sensitivity and with the restriction of being able to measure the magnitude of only one star at a time. The first problem was broken in the 1940s with the introduction of the photomultiplier tube, epitomized by the venerable RCA 1P21, which finally opened the door wide to photoelectric techniques. In such a device, light hitting a cathode kicks off electrons that are accelerated to another cathode that kicks off more electrons, and so on. The result is a huge cascade of electrons onto the final receiving anode, allowing a small signal of light to produce an easily measurable electric current. The second problem was resolved with the introduction of the "charge-coupled device" (the CCD), which can electronically record

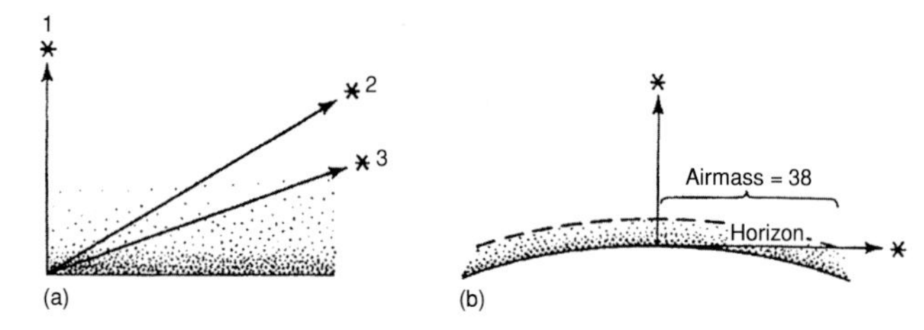

Fig. 3.6 **Path lengths in the Earth's atmosphere** Absorption by the Earth's atmosphere makes stars at high zenith distances [2.3.1] appear fainter than when they are overhead. In a simple approximation at left (a), the air lies in a plane above a flat landscape. At a zenith distance of 60° (Star 3), we look through twice as much air as we do overhead, while at a zenith distance of 81° (Star 3), we look through three times as much. Since the atmosphere is really a curved sheet (b), the approximation breaks down for higher angles. At the horizon, we look through 38 times more air (*M*, the airmass) than we do when looking overhead. (Kaler, J.B. (1996). *The Ever-Changing Sky*, Cambridge: Cambridge University Press.)

whole fields of stars, and is to the photomultiplier as photography was to the eye. With the use of a computer, the magnitudes of hundreds, even thousands, of stars, can be measured nearly simultaneously.

Photographic and photovisual magnitudes are replicated photoelectrically by the use of filters that match the color response of the photographic plate and the human eye, allowing a continuation of the original concept of color index. But the ease of photoelectric observations and the wider color responses of electronic measuring devices opened the whole magnitude system to a much broader range of wavelengths. The most-used, now-classic system, developed by the American astronomers H.L. Johnson and W. W. Morgan in the early 1950s, uses three colors: ultraviolet (U), blue (B), and yellow–green, or visual (V) [Box 7.1]. Stellar magnitudes go by the same names: apparent visual magnitudes are called "*V*", what were once "photographic magnitudes" are

Box 3.1 Measurement of magnitude

We measure magnitudes by comparing the response of a detector to the "program star" (the one to be measured) with its response to a standard star, one whose magnitude is already known. But there is far more to it. Anything that affects the apparent brightnesses of the two stars must be accounted for. Stellar magnitudes are appropriate to what we would measure with a perfectly efficient detector from outside the Earth's atmosphere. Though the air is partially transparent, it still absorbs light, made obvious by the observation that stars appear dimmer toward the horizon than they do overhead. The corollary is that we observe far fewer stars near the horizon than we do near the zenith.

The air is not on a uniform inverted bowl, but is a layer parallel to the ground that is much thicker near the horizon than overhead (Fig. 3.6). At higher altitudes (lower zenith distances) [2.3.1] it behaves like a flat sheet such that the thickness compared to the "unit atmosphere" (that directly overhead) scales as the inverse of the cosine (that is, as the secant) of the zenith distance. Since the atmosphere is really curved, at lower altitudes the formula becomes more complex, and the thickness increases at values less than the secant of z. Nevertheless, at $z = 90°$ we still look through 38 times more air (the *airmass*) than we do at the zenith (Figs. 3.6 and 3.7).

Moreover, the opacity (the inverse of the transparency) of the air is a strong function of the color, or wavelength, of the light. Air is very clear at the longest visual wavelengths, in the red, and also in the near-infrared. The opacity of the atmosphere rapidly decreases toward the blue and violet, and below 3000Å becomes infinite, that is, the air becomes totally opaque, a good thing as the air thus blocks harsh – and deadly – ultraviolet radiation from the Sun (Fig. 3.8).

The intensity of a beam of light moving through a medium decays exponentially. If the intensity of starlight as it hits the Earth's atmosphere is $I_*(\lambda)$ and the degree of extinction at the zenith is f_λ (do not confuse with the interstellar extinction function [3.5.1]), the intensity of the starlight as it hits the Earth's surface is $I_{obs} = I_* \times 10^{-f_\lambda M}$, where M is the airmass. Since magnitudes are logarithmic and $m_* - m_{obs} = 2.5 \log I_{obs}/I_*$, $m_* - m_{obs} = 2.5 \log 10^{-f_\lambda M}) = -2.5f_\lambda M$, or $m_* = m_{obs} - 2.5f_\lambda M$. Figure 3.8 plots $2.5f_\lambda$. The extinction at any wavelength and zenith distance is 2.5 times the product of the values found from Figs. 3.7 and 3.8.

The telescope optics also absorb and scatter radiation, making the star dimmer than it would appear if the optical system were 100 percent perfect. However, since magnitudes are determined by comparing the photometer's response of one star with that of another, the effect cancels out. You could cancel the effect of the Earth's atmosphere by comparing the program star with a known standard star at the same zenith distance. Though it is wise to attempt to do so, the match can never be perfect, and is sometimes far from it. Since the opacity function f_λ can change from one night to the next (for that matter even over a few hours), you observe a particular standard star at several zenith distances to derive it afresh (or use several stars at different zenith distances).

With a single-channel photometer, the program and standard stars are measured with the same detecting surface, one after the other. However, a CCD acts like an ultrasensitive – though linear – photographic plate on which standard and program stars are recorded simultaneously at different positions that will have slightly different sensitivities. The CCD therefore requires additional calibration, the observation of a uniformly illuminated surface, that smooths out the variations.

Now you have magnitudes on the "observatory system." But other observatories use telescopes and filters with somewhat different wavelength response magnitudes than yours. Observatories are also at different altitudes and under different local weather conditions, producing differences in the transmission of the Earth's atmosphere from one place to the next. The final step in the "reduction" of the observations to useful numbers is a complex "transformation" to the standard defined system, effected by observing numerous standard stars on your own system and by defining the mathematical constants that can be applied to standard transformation equations. Now you are ready to publish your work.

called "B," ultraviolet magnitudes "U." Huge numbers of stars have had their magnitudes measured in what evolved into the carefully-crafted *Johnson UBV* photometric system, which even specified the altitude of the observatory to allow for the proper coloration of stars produced by the Earth's atmosphere (Box 3.1).

V magnitudes were broadly scaled to the earlier photovisual system. In the simplest sense, they are calibrated to Vega, for which $V = 0.03$, but in the broader sense they are scaled to the self-consistent set of all the *V* magnitudes found in various catalogues. As was the case for the original photographic magnitudes, *U*, *B*, and *V* are set equal for white A0 stars [6.2.1], effectively again for Vega. Johnson *UBV* photometry uses wide, or broad-band, filters that cover a rather large wavelength domain of several hundred angstroms, allowing a great deal of light to enter the photometer and allowing the astronomer to observe fainter stars. A disadvantage of the system is that fine-scale information in the spectrum is lost. Although no longer done, *UBV* magnitudes can be measured photographically with the use of appropriate filters, just as they are now with wide-angle CCDs.

The range of visual magnitudes (and those of other colors) is astonishing. At the bright end, Sirius has an apparent visual magnitude of nearly −2. At the faint end of the observational scale, we can detect stars with the Hubble Space Telescope to magnitude 30, a factor of some 6 trillion in brightness. Faint stars make up for their dimness, however, by their sheer numbers (Box 3.2).

Box 3.2 How many stars?

How many stars can be seen? Table 3.1 shows the number of stars that are brighter than a given magnitude. Doubles and multiples are counted as single units. Exact numbers are not possible because of variability. From magnitude 3.0 upward, the counts are rounded off. The numbers start out small at low magnitudes (the brightest stars) and increase very rapidly as the stars become fainter. Brighter than magnitude 0.0 there are but four stars: Sirius, Canopus, Alpha Centauri, and Arcturus, while brighter than 1.0 there are 14 of them. Seven more are spread between 1.0 and 1.5 to make 21 first magnitude stars (see Appendix 1).

Each magnitude is 2.5 times fainter than the next brighter one. Assuming all stars are the same, a decrease in brightness of a factor of 2.5 would require an increase in distance of $\sqrt{2.5} = 1.58$. Since the volume of a sphere is proportional to the radius cubed, each lower magnitude number incorporates a factor of $1.58^3 = 3.95$ times more stars. The actual star counts increase at a much lower pace because: (1) the distribution of stars is in the form of a flat disk (the Milky Way) and is far from spherically symmetric; and (2) distant stars are dimmed by interstellar dust. That stars are hardly all the same brightness actually raises the factor, as each magnitude number incorporates more and more intrinsically fainter stars that are nearby but lie farther down the main sequence [6.3.1].

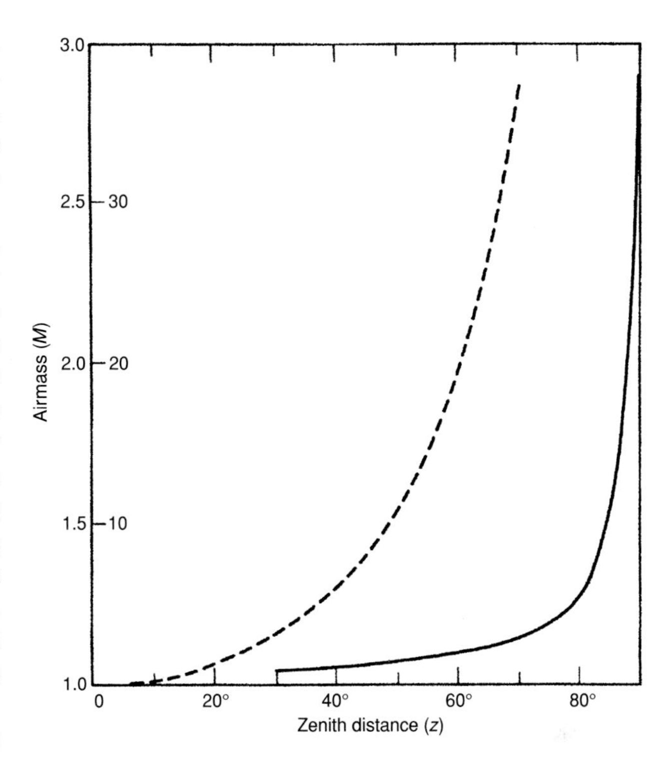

Fig. 3.7 **Airmass** Airmass (*M*) climbs quickly with increasing zenith distance (*z*). The dashed curve is read on the outside scale, the solid curve on the inside scale. (Kaler, J.B. (1996). *The Ever-Changing Sky*, Cambridge: Cambridge University Press.)

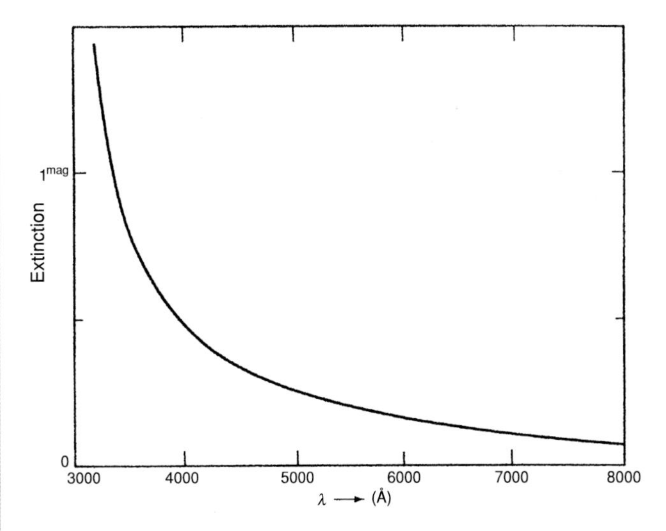

Fig. 3.8 **Atmospheric extinction** The atmospheric extinction of radiation of starlight for a star at the zenith climbs very quickly with decreasing wavelength. Increasing the zenith distance greatly enhances the effect (see Fig. 3.7). (Kaler, J.B. (1996). *The Ever-Changing Sky*, Cambridge: Cambridge University Press.)

Any of these magnitudes can be related to brightness in true physical units by comparing the *V* magnitude of a bright star with that of an Earth-based blackbody [7.1.2] at a precisely-known temperature and luminosity, most commonly a metal at its melting point. The measurements are compromised mostly by the difficulty of evaluating the

Visual magnitude	Number of stars brighter than	Visual magnitude	Number of stars brighter than
0.0	4	13.0	4.1 million
1.0	14	14.0	9.7 million
2.0	48	15.0	22 million
3.0	180	16.0	46 million
4.0	500	17.0	92 million
5.0	1600	18.0	180 million
6.0	4900[b]	19.0	330 million
7.0	14 300	20.0	590 million
8.0	41 300	21.0	1.1 billion
9.0	116 000	22.0	1.7 billion
10.0	335 000	23.0	2.6 billion
11.0	903 000	24.0	3.6 billion
12.0	1.5 million	25.0	4.8 billion

[a] From *Astrophysical Quantities*, C.W. Allen, 3rd ed., Athlone, London, 1973; *Allen's Astrophysical Quantities*, 4th ed., A.N. Cox, ed., AIP Press, Springer, New York, 2000; *Observer's Handbook*, Royal Astronomical Society of Canada, University of Toronto Press, 2001. The numbers in the far right-hand column are based upon statistical models.
[b] The *Bright Star Catalogue*, D. Hoffleit and C. Jaschek, 4th Ref. Ed., Yale University Observatory, 1982, gives 9110 stars brighter than magnitude 6.5, the canonical naked-eye limit.

Table 3.1 **Star counts**[a]

degree of atmospheric absorption, as the light from the blackbody takes a very different path to the telescope than that taken by starlight. For $V = 0.00$, the flux of radiation outside the Earth's atmosphere (defined as the power flowing through a unit area per unit time per angstrom) at 5480Å (548nm) is 3.64×10^{-9} ergs^{-1} s^{-1} cm^{-2} Å^{-1} = 3.64×10^{-7} watts m^{-2} nm^{-1}. Magnitudes at all the other colors are similarly calibrated.

The properties of the brightest 51 stars as measured in V, which includes those to the brightness of Polaris (the original calibration star, magnitude 2.02), are given in Appendix 2.

3.1.4 Red and infrared extension

Observations made only in the blue and visual ignore all the information that a star can provide from its red and infrared radiation [Box 7.1]. However, until near the end of its run, photography was hopelessly bad at observing at longer wavelengths. The domain really belongs to the electronic age. The window was opened by G.E. Kron and J.L. Smith in 1950 with the introduction of broadband red (R) and infrared (I) magnitudes centered approximately at 6800 and 8300 angstroms. Like any new system, it evolved to meet the needs. Johnson redefined R and I, and then the South-African astronomer A.W.J. Cousins re-redefined it. The current Johnson–Cousins *UBVRI* system now covers the spectrum from 3670Å at U to 7970Å at I (Table 3.2).

If R and I are useful, why not extend the system farther into the infrared? While in the visual and near-infrared spectrum the air becomes more transparent with increasing wavelength, in the deeper infrared it becomes more opaque as a result of absorption of light by water vapor, carbon dioxide, and other molecules. Fortunately, there are a few transparent windows between the absorption bands through which we can look to the outside of the atmosphere. Johnson took advantage of these by extending the system to include J, K, L, and M magnitudes that range from 1.22 microns (or micrometers, 12 200Å) to 4.75 microns. These were redefined by I.S. Glass, who added an H magnitude at 1.63 microns. All are calibrated such that the individual magnitudes are the same for white stars of spectral class A0. The "final" Johnson–Cousins–Glass system is given in Table 3.2.

The *JHKLM* infrared magnitudes are not "stable," in the sense that every observatory seems to run on a somewhat different system. Transformations from one system to another are possible, but always fraught with errors. Some 50 years after they were invented, however, we finally see some real standardization.

3.1.5 And many more

Magnitude systems are the rabbits of astronomy – they multiply almost as fast as you can count them. Astronomers continually invent new ones that either fit well with new observational systems, or can provide additional information about stars that older systems cannot. The widespread intermediate-band Strömgren "*ubvy*" system is a good example, its four "colors" centered at 3490 (u), 4110 (b), 4670 (v), and 5470Å (y) (see Table 3.2). They are set up so that

Standard[b] (Broad-band)		Strömgren (Intermediate)		Sloan (Broad)	
U	3670Å	*u*	3490Å	*u′*	3500Å
B	4360	*b*	4110	*g′*	4800
V	5450	*v*	4670	*r′*	6250
R	6380	*y*	5470	*i′*	7700
I	7970			*z′*	9100
J	1.22 microns				
H	1.63				
K	2.19				
L	3.45				
M	4.75				

[a] The actual central wavelengths depend on the color of the star.
[b] Johnson–Cousins–Glass.

Table 3.2 **Three magnitude systems**[a]

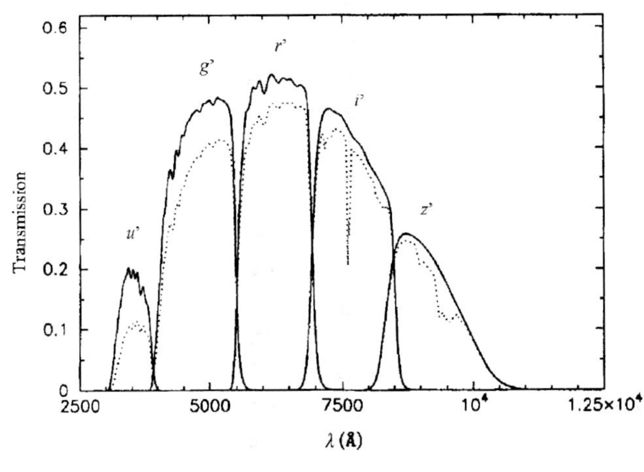

Fig. 3.9 **Transmission curves** The solid curves plot the transmission of each of the *u′g′r′i′z′* filters used for the Sloan Survey as a function of wavelength. Atmospheric extinction affects the energy distribution, or color, of a star and therefore also the effective transmission wavelength. The dotted curves show the slight changes in the effective transmissions for a star seen though an air mass of 1.2. (Fukugita, M. *et al.* (1996). *Astronomical Journal*, **111**, 1748.)

color differences provide information on both temperature and metallicity. The "y" magnitude is close to Johnson *V* and can easily be transformed into it. Various color indices (defined below) give information on stellar characteristics. The ubvy magnitudes are usually coupled to the ratio of the star's brightness at the H-beta line [6.1] as observed with wide and narrow filters, which provides information on the broadening of the absorption line and thus on surface gravity [7.2.4].

Variations on the *UBV* theme are found in Geneva and Walraven photometry. Astronomers at Canada's David Dunlap Observatory came up with six-color DDO photometry that stretches from 3490Å to 4890Å. The Hubble Space Telescope uses its own multicolor system from filters that have to serve a variety of masters. The Hipparcos satellite [4.2.3] produced a very broad-band H_p magnitude (not to be confused with the infrared H magnitude). The Tycho portion of the mission [1.4] contained photometers that allowed observation of B and *V* magnitudes (called B_T and V_T) that were transformed to Johnson B and *V*.

Among the most important of the other magnitude systems is that used for the Sloan Digital Sky Survey (the electronic version of the photographic Palomar Sky Survey [1.4]), u′g′r′i′z′, whose filters overlap and run from 3000 to 10 000Å (Fig. 3.9 and Table 3.2). The primes on the letters separate it from its distant parent, the uvgr (ultraviolet, violet, green, red) system established in 1976 by T.X. Thuan and J.E. Gunn. The idea is to span the entire optical–infrared window to the long-wave limit of the CCD detector. Nothing in it comes close to standard *V* magnitudes, and it stands quite on its own.

One wonders if astronomers will ever settle on a truly universal system. Indeed, given the specific problems they work on, a universal system probably would not work. So much effort has been put into the magnitude system that it will – to the dismay of physicists – probably last for as long as there are astronomers to use it. Hipparchus lives.

3.2 Absolute magnitudes

Apparent magnitudes give the view from Earth. By themselves, they tell us nothing about what we are really interested in: the true luminosity of the star, which is ultimately to be measured in watts. Only then can we relate stellar luminosity [7.2.1] and spectral class [6.2] to the predictions made by theories of stellar structure and evolution [12.2.1; Chapters 13 and 14]. Since magnitudes are endemic to the astronomical trade, we continue their use in the expression of luminosity.

In 1922, the International Astronomical Union adopted the parsec as the standard stellar distance unit (equal to 206265AU and 3.2616 . . . light years) and the standard distance for comparing stars as 10 parsecs. The *absolute magnitude* (M) of a star is thus defined as what the apparent magnitude (m) would be at that distance. Apparent and absolute magnitudes are related through the *magnitude equation* (Box 3.3),

$$M = m + 5 - 5 \log d \,(\mathrm{pc}).$$

If $d = 10$pc, the standard distance, $M = m + 5 - 5\log 10$. But $\log 10 = 1$ (that is, $10^1 = 10$), so $M = m + 5 - 5 = m$, as expected. Now, say $m = 10$ and $d = 100$pc. Then $M = m + 5 - 5\log 100$. The log of 100 is 2 ($10^2 = 100$), so $M = 10 + 5 - 10 = 5$.

It does not matter which kind of apparent magnitude we use so long as we are consistent. Apparent visual magnitudes give absolute visual magnitudes, apparent blue magnitudes absolute blue magnitudes; apparent *V* magnitudes transform into M_V, B transforms to M_B, and so on. Try it on the Sun. The apparent *V* magnitude of the Sun is -26.75. The mean solar distance of $1\mathrm{AU} = 1/206\,2\,65\ \mathrm{pc} = 4.848 \times 10^{-6}$ pc so $M_V(\mathrm{Sun}) = -26.75 + 5 - 5 \log\,(4.846 \times 10^{-6}) = -26.75 + 5 - 5(-5.31) = 4.83$. Push our brilliant Sun to a mere 10pc and it glows just a bit more brightly than Eta Ursae Minoris, the faintest star of the Little Dipper.

The range of absolute magnitudes is as amazing as the range of apparent magnitudes. The most luminous stars have M_V up to -10. The faintest stars of the standard main sequence [6.3] hover around $+21$, over a trillion times less luminous than the brightest supergiants. You can stretch the range much more. Supernovae can reach -19, whereas the coolest class T brown dwarfs [6.2.1, 12.4.2, 13.1] are so dim that at visual wavelengths they cannot be seen at all.

Box 3.3 The magnitude equation

As the radiation from a light bulb, a star, or other source, moves outward into space, it spreads out over a sphere of increasing radius R. The surface area of a sphere is $4\pi R^2$. As a result, the amount of energy passing per unit time through a unit area of a sphere surrounding the source is diluted by a factor of $1/R^2$. The apparent brightness of a light source, which is the amount of radiation received per unit time per unit area (watts per square meter, for example), therefore depends on the inverse square of the distance from the source.

Site a star with apparent brightness i_1 at a distance d_1 parsecs. Move the star to distance d_2, where it will appear with a brightness of i_2. Then

$$i_1/i_2 = (d_2/d_1)^2.$$

Magnitudes are defined from intensities by the relation

$$m_2 - m_1 = 2.5 \log(i_1/i_2).$$

From the previous equation,

$$m_2 - m_1 = 2.5 \log(d_2/d_1)^2 = 5 \log d_2 - 5 \log d_1.$$

Now take d_1 as the true distance of the star, m_1 as its apparent magnitude (m), and d_2 as the standard distance of 10pc, whereupon m_2 becomes the absolute magnitude M. The last equation then becomes

$$M - m = 5 \log 10 - 5 \log d,$$

whereupon

$$M = m + 5 - 5 \log d.$$

3.3 Bolometric magnitudes

Visual, photographic, *UBVRI*, and all the rest of the magnitudes measure only the amount of starlight that comes through a particular filter. By themselves, they say little about how much total radiation a star produces. Take an extreme case, a red giant star of 2000K encased in a dust shroud [13.3.2]. It can be brilliant in the infrared, yet radiate nearly nothing at visual wavelengths. We have to find a way of correcting any individual magnitude so that it reflects the totality of radiation, which is given by the **bolometric magnitude**. (The term comes from "bolometer," an instrument designed to detect heat from a radiating source.)

To measure an ideal bolometric magnitude directly would require a space observatory and a detector that could span the entire spectrum of radiation, which is at best impractical (not to mention expensive). Instead, we apply to an observed magnitude a **bolometric correction** (BC) that is a function of the star's temperature. Such corrections can be theoretically calculated. Or we can make special efforts to measure the flux of radiation in whatever wavelength bands are available to us for a variety of stars, relate such measures to the visual flux, and "connect the dots" to link all the observations together into some kind of whole. Great effort has been spent on determining bolometric corrections. Not all observers (or theoreticians) agree on the values, especially at or near the extremes of temperature, where the large bulk of the starlight is emitted; in the far ultraviolet or infrared corrections can be large and uncertain.

While bolometric magnitudes can be either apparent or absolute, they are nearly always considered absolute, which is the default. And while a correction can be applied to a magnitude measured at any specific wavelength, the most common rule (and again the default) is to apply it to the visual magnitude (unless the star is too cool to have one). The absolute bolometric magnitude of a star (M_{bol}) is defined as

$$M_{bol} = M_{vis} + BC.$$

Like observed magnitudes, bolometric corrections are geared to the human eye. Experiments and calculations in the early part of the twentieth century showed that the eye sees blackbody radiation [7.1.2] most efficiently at a temperature of 6500 kelvin, which corresponds to a star of spectral class F5. The BC for a star of 6500K was then taken as 0.00. The implication is that at 6500K, the eye responds to the total radiation of a star, which is far from true. The zero point merely provides the origin for a relative scale.

Since the Sun (5780K) is cooler than the 6500K standard for the zero point, it will radiate somewhat more strongly in the infrared than will an F5 star. Bolometrically, the Sun will have to be brightened a bit relative to the F5 star, and thus has a bolometric correction of −0.08 (lower magnitude numbers meaning brighter stars), rendering its bolometric magnitude 4.75. Early measures of bolometric corrections were clearly crude. Improvements in observation and theory allowed ever-better values to be obtained. The standard, however, was over the course of time moved to the Sun, with its BC = −0.08 (BC = −0.07 in some systems). Improved measures then forced the corrections for stars between about 6500K and 8500K to be positive instead of negative. The implication is that stars with classes between A1 and F5 are fainter bolometrically than they are visually, which makes no physical sense. The reason is that bolometric corrections are all relative to the solar correction, and that the solar "zero point" (actually that the solar BC = −0.08) is arbitrary. Bolometric corrections are graphed in Fig. 3.10.

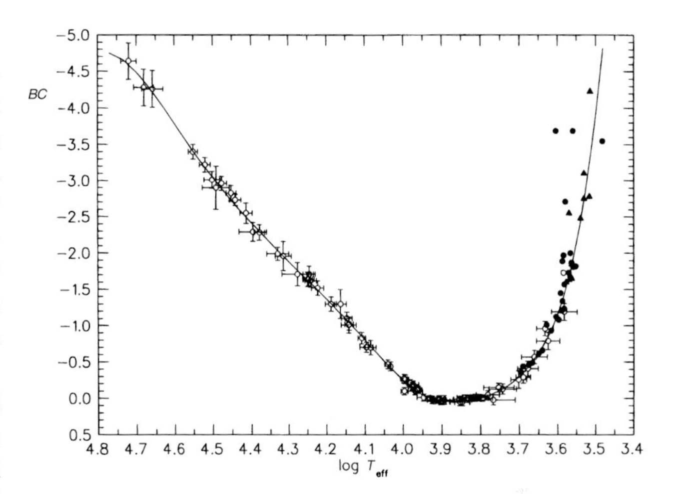

Fig. 3.10 **Bolometric corrections** Bolometric corrections (*BC*) are plotted against the logarithm of stellar temperature. Curves from other authors may differ considerably. (Flower, P. J. (1996). *Astrophysical Journal*, **469**, 355.)

The bolometric magnitude immediately allows the calculation of the luminosity of the star (∗) in solar (⊙) units. Since

$$M_{bol}(\odot) - M_{bol}(*) = 2.5 \log L_*/L_\odot,$$
$$\log (L_*/L_\odot) = 0.4\,[M_{bol}(\odot) - M_{bol}(*)]$$

and thus

$$L_*/L_\odot = 10^{0.4[M_{bol}(\odot) - M_{bol}(*)]}.$$

Since we know that $L_\odot = 3.845 \times 10^{26}$ watts [12.1.1], we can know the absolute radiative power of any star. A star of absolute bolometric magnitude 0.00 has a luminosity of 3.05×10^{28} watts (3.845×10^{35} ergs per second).

3.4 Color

That stars have color comes as no surprise to anyone who has looked at the nighttime sky. Orion (Fig. 3.1) is famed for red Betelgeuse and for the blue stars of the Belt. But while in the literature star colors are vividly described, in reality they are quite subtle, Betelgeuse appearing more as orange, even yellow–orange, and the Belt stars as white with perhaps just a tinge of blue. Star colors can be the source of vigorous argument. At the end, such debate is futile, as different eyes see colors differently and contrast effects can powerfully enhance them. Seen through a telescope against a blue twilight sky, Betelgeuse becomes a vivid red, its color seeming to fade as the night turns to black. The closeness of the class B and K stars of binary Albireo magically transforms them to brilliant blue and orange. The descriptions of double star colors in nineteenth-century astronomical literature are charming to read: "dusky blue," "pale gray," "rose," and so on. The colors of stars of the same classes seen singly are more washed out, since what we see as "color" is a mix of radiation that tends to white.

One of the two exceptions to the subtlety of color is the set of carbon stars [6.4.1], whose molecular absorptions pull so much shorter-wavelength radiation from the flow of light that they indeed appear ruby red; among the most famed is "Hind's Crimson Star" (R Leporis). The other exception consists of those stars highly reddened by interstellar dust [3.5.1], few of which (like the carbon stars) are visible to the naked eye.

To a stellar astronomer, "color" means something quite different from the visual impression. "Red" refers to "longer wavelengths," "blue" to "shorter wavelengths." Cool stars put out most of their radiation in the red and infrared and, as a result, they are called "red stars," even though the colors seen by eye might be no more than a pale rose. We therefore speak of "red giants" and "blue supergiants" even when the latter seem almost white.

Color, in the sense of the distribution of light within the spectrum, that is, of where the bulk of the radiation actually lies, is quantifiable by comparing magnitudes measured in different wavelength bands. The grandparent of a plethora of systems was the original *color index*, the difference between photographic and visual (later photovisual) magnitudes, or

$$CI = m_{ptg} - m_{vis}.$$

(In older literature, the color index is also given as "$P - V$", where P is the photographic magnitude and V the photovisual.) With the advent of multicolor systems (specifically, Johnson *UBV*), the color index transmuted to "B–V," which

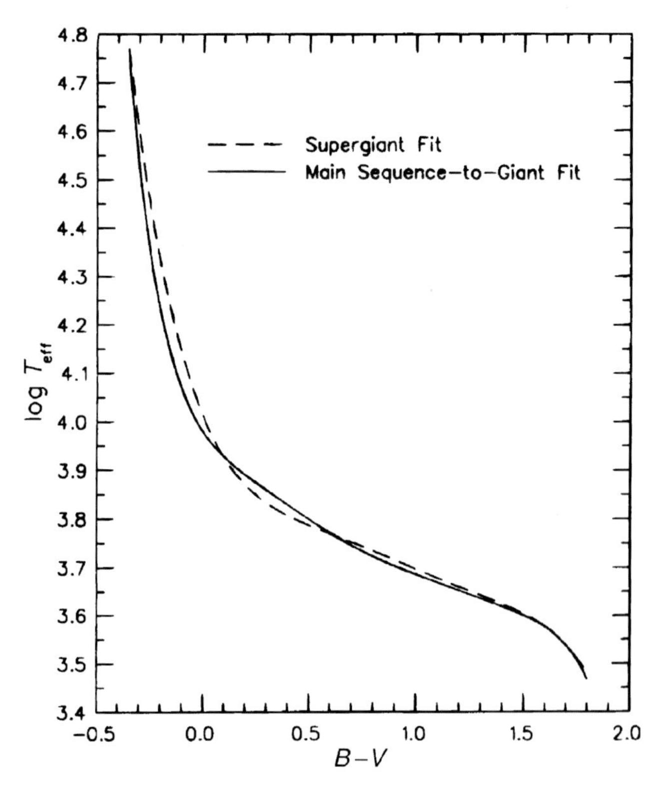

Fig. 3.11 **Temperature and color** As the $B - V$ colors of stars increase from a low of −0.4 to nearly 2.0, stellar temperatures (expressed through their logarithms) fall from near 60 000K to 3150K. Color thereby serves as a means of estimating temperature. (Flower, P. J. (1996). *Astrophysical Journal*, **469**, 355.)

is (a tautology if ever there was one) equal to the magnitude difference $B - V$. Given that the classic Johnson system has three magnitudes, "U–B" $= U - B$ serves as a second color index. The default for the term "color index," however, is always $B - V$.

Since U, B, and V are all calibrated to be the same for an A0 star (specifically Vega), $B - V = U - B = 0$ at that class. Hotter stars of classes O and B will be brighter in U and B than they will be at V, and will therefore have lower magnitude numbers (Figs. 3.11, 3.12). As a result, $U - B$ and $B - V$ become progressively more negative as temperature increases, the spectral class becomes "earlier" (hotter) [6.2.1], and the color bluer. Since stars "later" (cooler) than those at class A0 are brighter at longer wavelengths, their $B - V$ colors become progressively more positive. The color index therefore provides a fine way of estimating the spectral class, hence temperature. Indeed, color index stands in well for class, and is commonly used in a version of the HR diagram, known as a *color–magnitude*

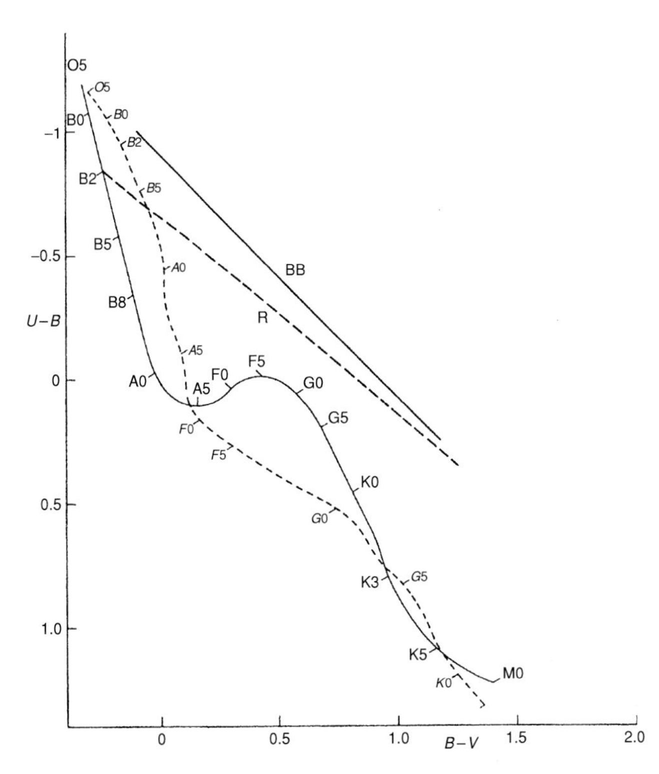

Fig. 3.12 **The color–color diagram** In a classical "color–color" plot, $U - B$ is graphed against $B - V$. Unreddened stars fall on the wavy lines (solid, dwarfs; dashed, supergiants), below that calculated for a blackbody (labelled "BB"). Reddening of starlight by interstellar dust (see Section 3.5) causes the star to fall on a sample locus presented for a B2 star (labelled "R") that extends down and to the right. The degree of reddening can be found by tracing the star's colors back to the proper position. (Kaler, J.B. (1989). *Stars and their Spectra*, Cambridge: Cambridge University Press.)

diagram, whose lower axis can use any color, not just $B - V$ [6.3.3].

The negative range of $B - V$ is rather small. Hotter stars shift the peaks of their energy distributions progressively farther into the blue, violet, and ultraviolet parts of the spectrum. As they do, the slope of the energy distribution from long to shorter visual wavelengths (energy flux plotted against wavelength) approaches an upward-rising, but constant, curve. Between 4000Å and 8000Å the spectrum of a 100 000K star is effectively the same as that of a 50 000K star. As a result, for hot stars, $B - V$ does not decrease much with temperature, and asymptotically approaches a limiting value of −0.43. Consequently, for hot stars, color index does not provide a very accurate thermometer. The cooler end of stellar life is much more receptive to color, $B - V$ climbing quickly with temperature.

Since the flux of ultraviolet radiation is more sensitive to high temperatures, the negative range in $U - B$ is much larger. At the high temperature limit, $U - B$ reaches to −1.2. The relation between the two colors is shown in the color–color plot of Fig. 3.12. Here, $U - B$ is graphed against $B - V$ in typical perverse astronomical fashion, in which $U - B$ becomes more positive downward and $B - V$ more positive outward. The relation between the colors for a perfect radiator (the blackbody [7.1.2]) is indicated by the upper straight line. As expected, the blackbody's $U - B$ and $B - V$ both steadily increase in lock step with decreasing temperature (advancing spectral class). Stars, however, fall (with considerable scatter) along the lower curves, clearly showing that they deviate more and more from blackbodies as temperature declines, the deviation being the result of an increasing number of absorption lines, bands, and absorption continua [Chapter 6] that redistribute radiation within the stellar spectra.

Moreover, the $U - B$ colors for dwarfs like the Sun [6.3.1] do not monotonically increase with dropping temperature, but between about class A5 and F5 go in the other direction and actually reverse and decrease, producing the well-known "hook" in the middle of the graph (solid curve). The culprit is the "Balmer jump" produced by the onset of the Balmer continuous absorption at 3646Å [7.1.7]. The U filter is to the short-wave side of the Balmer jump, while the B filter is to the long-wave side. U magnitudes are therefore strongly affected by the absorption of outflowing starlight by the Balmer continuum, which peaks in strength at about class A0 and then, toward lower temperatures, decreases to the point that the $U - B$ color actually becomes "bluer" rather than "redder." The Balmer jump is less prominent in lower-pressure giants (short-dashed curve), so that only a bend, but not the hook, appears.

Color indices in other magnitude systems have similar uses, $g' - r'$ a substitute for $B - V$. Color–color plots in which the short-wave filter is longward of the Balmer jump do not show the hook. Photometry at more wavelengths, which produces more colors, extends the information that can be gleaned from the spectrum. A fine example is Strömgren four color intermediate-band "uvby" photometry (see Table 3.2). Four colors allow for three color indices: $b - y$, $v - b$, and $u - v$. The first of these, $b - y$, is similar to Johnson $B - V$ and can easily be transformed into it. The system also uses two color differences that are affected by both the Balmer jump (which is sensitive to temperature and surface gravity) and metallic lines in the spectrum:

$$c_1 = (u - v) - (v - b),$$
$$m_1 = (v - b) - (b - y).$$

The c_1 index allows for a measure of gravity (hence density) in hot stars, and temperature in cooler stars, as does the H-beta index, the ratio of the light passing through the wide and narrow filters at the H-beta absorption line. The m_1 index, on the other hand, gives the darkening of the spectrum in violet and blue light, where there are hosts of metallic absorption lines, and is thereby a measure of the metallic content, or gross chemical composition, of the stellar atmosphere [Chapter 6].

3.5 Interstellar extinction

A chaotic mixture of gas and dust, the *interstellar medium*, permeates the spaces between the stars [11.1]. In its simplest approximation, it is composed of denser, cooler interstellar clouds set into a medium of much warmer – sometimes hot – and more tenuous material. Mixed into all of it are grains of carbon and silicate dust that constitute roughly one percent of the mass. The thicker dusty clouds are easily seen as dark splotches against the Milky Way, conglomerations of them making much of the Milky Way's visible structure and beauty [11.1.1]. The dust also makes a mess out of the observations of stars.

In spite of the obvious dark clouds of the Milky Way, some so thick and black that the Incas made constellations out of them, astronomers long believed that the space between the stars was clear (perhaps a product of wishful thinking). The devastating truth was revealed in 1930 when the American astronomer R.J. Trumpler (1886–1956) calculated distances of open clusters [9.2] from the main sequences of their HR diagrams [4.4.2]. From the clusters' distances and angular diameters, Trumpler calculated their physical diameters in parsecs or light years. He found that the farther the cluster, the larger it seemed. With smaller and smaller clusters converging on it, the Earth seemed to be in a "special place," a sure sign that something was wrong. Trumpler was progressively overestimating the clusters' distances, the error increasing steadily with increasing distance. The conclusion was that something in space was absorbing starlight and making the stars seem too faint and too far away. It had to be the same dust that made the obvious dark clouds.

3.5.1 Stars

Interstellar dust is as fine as smoke, the particles typically under a micron across. Outside of the obvious clouds, within the so-called clear pathways, there is only about one grain per cubic meter. Yet the lines of sight to the stars are so great that all distant stars within the Milky Way are at least somewhat dimmed. The dimming is accomplished by two different phenomena. The grains can absorb stellar photons, which heat the grains, or they can simply scatter (or "reflect") them. In either case, a beam of stellar photons heading toward us and our telescopes is diminished. The two effects are usually lumped together under the rubric of *interstellar extinction*.

The efficiency of interstellar extinction depends on the size of a light wave relative to the size of the grain. Longer waves pass the small grains by without "seeing" them. Shorter waves, more comparable to the grain sizes, cannot. The exact relation between grain size and extinction efficiency depends on the distribution of grain size and to some extent on their chemical compositions. On the whole, the degree of extinction in the optical part of the spectrum is roughly inversely proportional to wavelength (λ), or proportional to $1/\lambda$. B magnitudes are therefore affected more than V. The effect drops quickly: R and I magnitudes are not affected much, and from J longward hardly at all. Radio waves are totally unaffected.

Since interstellar dust makes a star fainter and thereby increases its magnitude number, the extinction in magnitudes (A) must be subtracted from the observed apparent magnitude. The magnitude equation thus becomes

$$M = m + 5 - 5 \log d\,(\mathrm{pc}) - A,$$

where A is appropriate to the observed wavelength (that is, the kind of magnitude). The default is to measure the absorption at V, A_V, the magnitude equation then becoming

$$M_V = V + 5 - 5 \log d\,(\mathrm{pc}) - A_V.$$

The dust is confined almost entirely to a thin plane in the disk of our Galaxy (Fig. 1.1), and is thicker toward the Galaxy's center [5.1, 11.1] (also see Fig. 3.13). The dimming by interstellar extinction then depends on both Galactic latitude and longitude [2.3.5]. When we look perpendicular to the disk, in the direction of the north and south Galactic poles, the dust has little impact. For high latitudes (b), the degree of interstellar extinction behaves just like absorption of starlight in the Earth's atmosphere (see Box 3.1). At the Galaxy's "zenith," at a Galactic latitude of 90°, for a star farther away than a few hundred parsecs we see a minimum extinction A_V (min) of about 0.2 magnitudes. As the latitude decreases, we look through more and more of the disk. Above about $b = 15°$ (where we treat negative latitude toward the south Galactic pole as positive), the thickness through the plane and the extinction increase as $1/(\sin b) = \text{cosecant } b$, that is,

$$A_V = 0.2/\sin b = 0.2 \csc b,$$

where "b" is the absolute value (i.e. always considered positive).

As our line of sight approaches the Galactic plane ($b = 0°$), the extinction becomes increasingly greater and more dependent on distance. As long as we do not go too far away, at low latitudes starlight is dimmed by about 1 magnitude per kiloparsec (1000 parsecs) a path length of 3×10^{19} meters, which is sufficient for even thinly spread dust to make a difference. Most of the stars that make the constellations are near enough, however, that interstellar extinction is of little

Fig. 3.13 **Dust in an edge-on galaxy** Obscuring dust in the edge-on spiral galaxy NGC 4565 is confined to a narrow lane that runs right down the middle. (Courtesy of Loke Kun Tan, Starryscapes.)

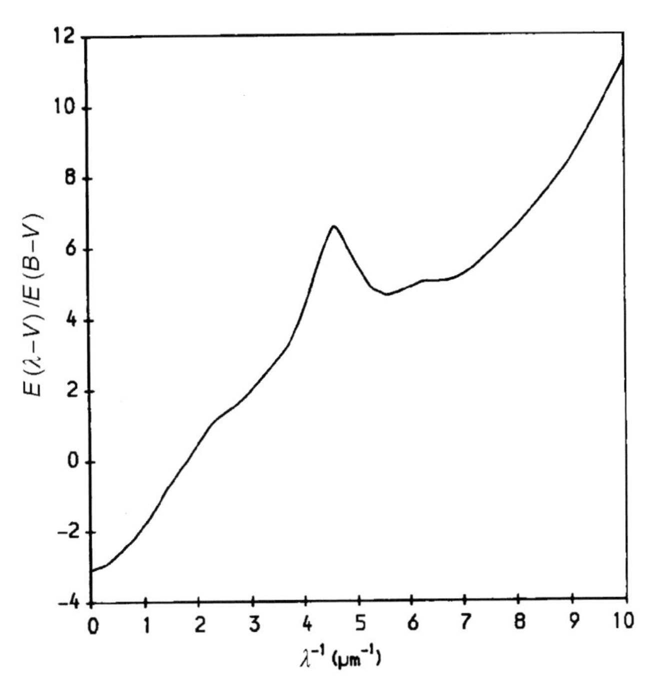

Fig. 3.14 **Interstellar extinction** The degree of interstellar extinction at different wavelengths (λ), f(λ), is expressed through color excess $E_{(\lambda-V)}$ relative to $E_{(B-V)}$. (The curve is traditionally plotted according to the inverse of the wavelength $1/\lambda$, where λ is in microns, μm, or nanometers.) As wavelength drops, the extinction curve steadily climbs, with a sharp peak caused by carbon superimposed near 2150Å ($\lambda^{-1} = 4.6$). (Adapted from Savage, B. and Mathis, J. (1979). *Annual Review of Astronomy and Astrophysics*, **17**.)

consequence. Off in the distance, though, the thick Galactic dust makes it impossible to see external galaxies that lie in the plane, and creates the so-called "zone of avoidance" that so confused earlier generations of astronomers. At optical wavelengths, galaxies are always seen best at high Galactic latitudes.

General rules do not apply well, though. The magnitude-per-kiloparsec dictum breaks down quickly as our sight penetrates toward the Galaxy's center in Sagittarius, which is dimmed by over 30 magnitudes and is effectively invisible at optical wavelengths (less than one photon in a trillion getting through). Worse, the dust is terribly patchy. Just look again at the Milky Way, either outdoors or in Fig. 1.1. We must therefore take each star as a special case and measure its specific extinction. Fortunately, since extinction efficiency increases with decreasing wavelength, the dust selectively extracts the blue component of starlight and reddens it. The greater the extinction, the greater the *interstellar reddening*. Extinction can indeed be so great that blue–white class B stars can really appear red. $B - V$ can substitute for spectral class and temperature only if the color is first corrected.

In principle, the corrections are easy to make. All we need do is to find the relation between reddening and extinction, measure a star's reddening, and calculate its extinction. The classic measure of interstellar reddening is the *color excess*, E_{B-V}, which is the difference between observed color index, $(B - V)_{obs}$, and the color the star *should* have on the basis of its spectral class or temperature, $(B - V)_{true}$ (Figs. 3.11 and 3.12), or

$$E_{B-V} = (B - V)_{obs} - (B - V)_{true}.$$

Since $B - V$ increases as stars become redder, E_{B-V}, also called the *selective extinction*, is always positive.

The *total extinction*, A_V, is related to E_{B-V} through a factor R such that

$$A_V = R\, E_{B-V}.$$

All we need to solve the whole problem is to derive R for a few stars and we are home free. To do that, astronomers carefully observe the dust's actual degree of extinction – the "extinction curve" – as a function of wavelength (Fig. 3.14); the curve is also known as the "reddening function," f(λ) (do not confuse with the atmospheric extinction function [Box 3.1]). A_V proportional to $1/\lambda$ is but an approximation in the optical. Extrapolating the curve to zero extinction at long infrared wavelengths provides the needed ratio between A_V and E_{B-V}, $R = A_V/E_{B-V}$, the famed *ratio of total to selective extinction*. The long-taken standard value is $R = 3$ (usually refined to 3.1 or 3.2). To find the extinction to any star, observe its spectral class, which is related to $B - V$ through the observation of nearby unreddened stars. With true $B - V$

known, the observed $B - V$ leads to E_{B-V} and thus to A_V. Once we observe enough stars, we can make dust maps that give the dust distribution as a function of Galactic coördinates and distance. However, the patchiness of the dust can exist on such a fine scale that it is always best to make individual determinations of extinction for any star of interest.

It seems as if the problem is solved. However, R depends critically on the nature of the dust. In some directions, "anomalous dust" can raise R significantly, to as high as 6. We can identify some regions with odd values of R, but we cannot a priori tell exactly what the value may be. As a result, the calculation of A_V is still fraught with difficulty.

Reddening evaluations can be extended to other wavelengths. As wavelengths decrease in the ultraviolet, the extinction broadly bulges near a wavelength of 2150Å as a result of absorption of starlight by carbon grains. The effect creates something of an ultrabroad spectrum line observable by orbiting satellites (see Fig. 3.14). The size of the "2150 bump" is related to A_V, which can therefore be found as long as there is enough ultraviolet light from the star to highlight it.

Extinctions can also be evaluated through color–color plots. The reddening that accompanies interstellar absorption causes both $U - B$ and $B - V$ to increase, not only in different ways, but also in ways quite different from the color changes caused by stellar temperature. Figure 3.12 contains a "reddening line" marked "R" for a hot class B2 star. Such reddening lines are parallel to each other. So one need only enter the colors of a star onto Fig. 3.12 and then trace backward along a reddening line to obtain both the color excess in $B - V$ and the true spectral class. A major problem is caused by the hook in the stellar distribution. For dwarf stars between about B5 and K5, which incorporates most of the main sequence, the traceback is ambiguous. A dwarf with $B - V = 1.3$ and $U - B = 0.5$ could be class K0, F0, or A0 depending on the degree of reddening. Supergiants have no such color–color hook, so they present no such problem. The matter is avoided in clusters in which stars with a range of classes are present [4.4.2]. If all the stars are plotted, the hook will be evident in the reddened starlight. A backward fit to the unreddened color–color curve along the reddening lines then allows for an accurate determination of the reddening.

3.5.2 Nebulae

Planetary and diffuse nebulae [11.1.3, 13.4], illuminated by embedded or nearby stars, provide their own excellent ways of evaluating interstellar reddening, hence extinction, both for the nebulae themselves and for their associated stars. Since various emission line intensity ratios are precisely known from atomic theory, any deviation from them can be ascribed to selective absorption by interstellar dust, which reduces the flux (intensity at the Earth, F) from the shorter-wavelength emission line more than it does that from the longer-wavelength line [13.4.1]. The nebular standard is the extinction of the nebula's hydrogen Hβ flux as seen at the Earth, C, where

$$C = \log[F_{true}(H\beta)/F_{observed}(H\beta)].$$

To evaluate C, scale the extinction curve $f(\lambda)$ in Fig. 3.14 to unity (1) at Hβ and zero at infinitely long wavelengths. The true-to-observed ratio for an emission line at any other wavelength λ will then be

$$F_{true}/F_{observed} = 10^{Cf(\lambda)}, \text{ or } \log F_{true} = \log F_{observed} + Cf(\lambda).$$

The most straightforward way of evaluating C is to measure the radio flux (radio intensity at a given wavelength) and use theory to calculate the value of the true Hβ flux. Since the radio flux is not affected by interstellar dust, the difference between the logarithms of the expected and observed Hβ flux immediately gives C.

We can also use the ratios of optical emission line fluxes known from theory. If $F_{\lambda 1}/F_{\lambda 2}$ is the ratio of a pair of emission lines, where the wavelength $\lambda 1$ is greater than $\lambda 2$,

$$\log[F_{\lambda 1}/F_{\lambda 2}]_{true} = \log[F_{\lambda 1}/F_{\lambda 2}]_{observed} + C[f(\lambda 1) - f(\lambda 2)],$$

which is again easily solved for C. All we need is an accurate extinction function. Many ratios of emission line fluxes are available. Relative hydrogen line fluxes are known from theory. That of the Hα to Hβ lines is a favorite to use in deriving extinction. Helium lines can be used as well, as can the ratios of a great number of forbidden lines [13.4.1] whose flux ratios are also fixed by atomic theory. If the extinction function is in good shape, all the derived values will be the same.

Nebular astronomers use the Hβ extinction, while stellar astronomers use A_V (the extinction in magnitudes at V, centered at 5450Å), which is commonly derived from the selective extinction E_{B-V}. From the extinction function, if $C = \log F_{true}/F_{observed} = 1$ at Hß, the analogous extinction at 5450Å is 0.87. To convert to magnitudes, multiply the logarithmic ratio above by 2.5. A_V is then $(2.5 \times 0.87)C$ or 2.18C. Since $A_V = 3.2\,E_{B-V}$, $E_{B-V} = (2.18/3.2)C = 0.68C$.

Of all the problems in astronomy, the issue of the correct evaluation of interstellar extinction is among the most critical. And all of these calculations can be confused if the extinction function is different from that expected. And it commonly is, leading to a variety of discrepancies and frustrations that are yet to be fully resolved.

CHAPTER FOUR

Distances

Fig. 4.1 **Stellar illusion** Though stars all appear to be on a flat plane, in reality they are at different distances. Bright stars (like Delta Scorpii at the top) may be very far, while faint ones may be quite nearby. The challenge is to construct a three-dimensional picture of our surroundings. (Courtesy of Lok Kun Tan, Starryscapes.)

Little in astronomy is as important as distance, as it opens the door to masses, luminosities, and stellar life cycles. Stellar distances in turn lead to the distances to nearby, then very distant galaxies, allowing us to learn about the nature and origin of the Universe (Fig. 4.1). And little has been harder to obtain. The first stellar distance was announced only in the mid nineteenth century, and for all our modern sophistication and successes, we are far from the end of the quest. Many kinds of stars and stellar groupings still have no decent measures. Even for those that do, we are never satisfied that the errors are small enough. As a result, astronomers will go to almost any length to get distances, and in the process have devised a mind-bending mix of stratagems. The subject is a lively one, with advances being made at such a great pace that within a few decades most of the big problems should be solved. The astronomers of the future will perhaps look back on us with some amusement at our tribulations as we struggle to learn how far away things are.

4.1 The cosmic distance ladder

Many are the distance methods. There is, however, no single one that can take us all the way from Earth to the farthest parts of the Universe. Instead we use one or two schemes for nearby stars to calibrate others for more distant stars, that are in turn used to calibrate still others, until we reach the visible horizon of space. The first stellar distances were measured by triangulation, or parallax [4.2.1]. Over most of the twentieth century, parallaxes were not quite good enough to get us started on the trip. Instead, we relied more on the distance to the Hyades cluster [9.2.2] found by the "moving cluster method" [4.3.1], through which we could establish the Hyades HR diagram [6.3.1]. Main sequence fitting [4.4.2] to more distant and younger clusters like the Pleiades and the Double Cluster, which contain the upper main sequence missing among the Hyades stars [13.1], allowed the calibration of the entire HR diagram. Hipparcos distances [4.2.3] voided much of the need for the traditional moving cluster method, and directly provide enough excellent parallaxes and absolute magnitudes [3.2] to calibrate the HR diagram, while also improving the application of main sequence fitting for distant star clusters.

Through various applications of spectroscopic parallaxes and astrophysical theory, we can find the distances to the Magellanic Clouds [5.1; see Fig. 4.11], which are rich in Cepheid variables, allowing the calibration of their period–luminosity relation [4.4.3]. The supergiant Cepheids are so bright that their light curves can be followed in galaxies [5.1] over 20 megaparsecs away [see Fig. 4.13], which includes enough of them to calibrate the light curves of Type Ia (white dwarf) supernovae [14.2.5], which in turn are visible in galaxies billions of light years off (the light year is the distance a beam of light will travel in one year; see Box 4.2). As a result, we can analyze the details of the expansion of the Universe. Along the way are numerous other methods, some fit for only particular stars or kinds of stars, others that work their way into the distance ladder. And it all starts right here, at home.

4.2 Parallax

Begin with Earth. The Earth orbits the Sun at an average distance (the semimajor axis of its slightly elliptical orbit) of 1.496×10^8km, a fundamental unit appropriately called the *Astronomical Unit*, or AU (Box 4.1). As the Earth orbits the Sun, stars must appear to move in the opposite direction, to swing back and forth in the same way your outstretched finger appears to move as you look at it with one eye and then the other. Such stellar *parallaxes* proved to be devilishly difficult to find. On the road to discovery, astronomers discovered orbiting double stars [8.2.3], the aberration of starlight, and nutation [2.5]. Not until 1838 did Friedrich Bessell succeed with the measurement of the parallax, hence distance, of 61 Cygni (Fig. 4.3).

4.2.1 Parallax defined

Focus on a nearby star that lies in the ecliptic plane (Fig. 4.4). As the Earth orbits the Sun, the star will appear to change its position against a background of other stars so distant that they are effectively unaffected by terrestrial movement. The formal parallax of the star, p or π (expressed in seconds of arc), is taken as half the total shift angle. From the point of view of the star, p is the angle subtended by the Earth's orbital radius (rather, the semimajor axis of the orbital ellipse), the Astronomical Unit, or AU. Observation of another star twice as far away shows the parallax to be smaller, only half that of the first star; that is, distance and parallax are inversely proportional to each other. The distance to a star is therefore defined as $d = 1/p''$, where the distance unit is called the *parsec*. If p is $1''$, d is 1 parsec, if $p = 0.01''$, $d = 100$pc. One parsec is also the distance at which the semimajor axis of the Earth's orbit (the AU) would subtend an angle of one second of arc, and is (rounding figures) 206 265AU = 3.2616 light years = 3.086×10^{13}km long (Box 4.2).

No star is as close as one parsec. The nearest naked eye star, the binary Alpha Centauri, has a parallax of $0.7421''$, rendering it 1.35pc = 4.40 light years away. The nearest star of all, Alpha Centauri's distant companion Proxima Centauri, has a marginally larger parallax of $0.7723''$ and a distance of 1.29pc. The closest star in the northern hemisphere, quickly moving Barnard's Star ($p = 0.549''$), is 1.82pc away. Bessell's 61 Cygni ($p = 0.286''$), recognized as close and therefore

Box 4.1 The Astronomical Unit

At home we measure distances by tape measure. Over long spans the means are indirect. The first step is the measure of the Earth's radius, R. The classic method goes back to Greece in the third century BC, and Eratosthenes of Cyrene. On the first day of summer in Syene (in upper Egypt), the Sun appeared directly overhead. In Alexandria, however, it appeared to the south of the zenith by 1/50th of a circle (7°). The Earth's circumference must be in proportion to the offset; that is, the circumference must be 50 times the distance between Alexandria and Syene, which was short enough that it could be paced off. The terrestrial radius is the circumference divided by 2π (where $\pi = 3.14159\ldots$). From what we know of Greek distance units, Eratosthenes determined the circumference to within 15 percent. Now we gauge the Earth's equatorial radius to be 6378.136km (accurate to better than one meter).

With this terrestrial scale, we extend distance measure into the Solar System to determine the length of the Astronomical Unit. In Fig. 4.2a, observe body B at distance d from two points on Earth E_1 and E_2. The sight lines from each point form an angle called the *parallax*, p. Since E_2 is on a line to the center of the Earth, C, p is a special case, the *horizontal parallax*. B, C, and E_1 make a right-angled triangle. In the triangle, $\sin p = R/d$, so the distance $d = R/\sin p$. In practice, observations are made from two arbitrary points and converted to horizontal parallax after the fact.

Long thin triangles like that in the figure have simple solutions. The "natural" unit of angular measure is the *radian*. In Fig. 4.2b, two lines form angle X and define an arc of length a in a surrounding circle of radius r. In radian measure, $X_{rad} = a/r$. Since the circumference of the circle is $2\pi r$, the whole circle of 360° defines an angle $2\pi r/r = 2\pi$ radians. Each radian therefore contains $360°/2\pi = 57.2958\ldots$ degrees, $3437.75\ldots$ minutes of arc, and $206\,264.8\ldots$ seconds of arc, the latter usually rounded to $206\,265''\text{radian}^{-1}$. Consequently, $X'' = 206\,265\, X_{rad}$. Because astronomical distances are so large, the radius of the Earth in Fig. 4.2a is an excellent approximation to the arc of a circle drawn around the body B. Therefore $p_{rad} = R/d$ and $d = R/p_{rad} = 206\,265\, R/p''$, thus avoiding the trigonometric functions.

The horizontal parallax of the Sun is small (only $8.9''$) and difficult to observe directly. Instead, since distances to all solar system bodies are known in AU from observations of planetary orbits and the well-known laws of their collective motion, we can derive an accurate length for the AU by observing the distance of a much closer body. Historically, astronomers observed the parallaxes of close-passing asteroids. Now we use the distance of Venus as found by radar and the round-trip flight time of a radio signal to get the length of the AU to within the accuracy of the speed of light: $1.495\,978\,707\ldots \times 10^8$km.

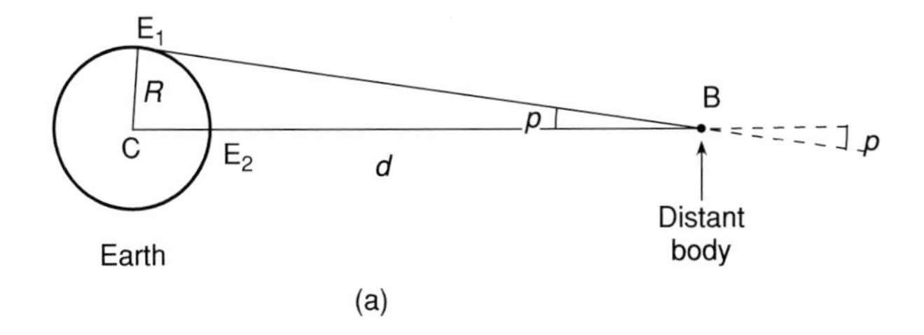

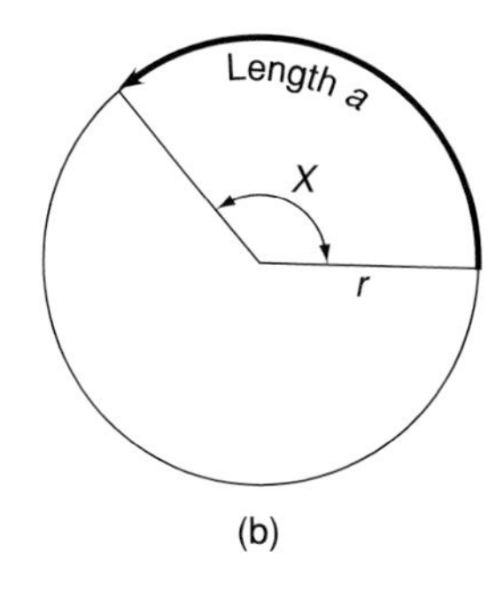

Fig. 4.2 **Horizontal parallax and radian measure** At the top (a), the horizontal parallax (*p*) of a celestial body B (at distance *d*) is the angle through which the position seems to change as viewed from two ends of the radius of the Earth. It is also the angle subtended by the radius of the Earth as seen from the body. At bottom (b), the angle *X*, as measured in radians, is defined as length of the arc *a* divided by radius *r*.

measureable because of its large proper motion [5.2.1], falls at 3.49pc and is effectively tied with Procyon, the closest first magnitude star of the northern hemisphere. The stars within 4 parsecs (13.1 light years) and their properties are listed in Appendix 3. One *kiloparsec* (kpc) is 1000 pc, one *megaparsec* (Mpc) a million pc.

4.2.2 Measurement and complications

The orbiting Earth causes stars exactly on the Earth's orbital (ecliptic) plane to appear to move back and forth in a straight line, while those at the ecliptic poles will go round in circles in which the parallaxes are the angular radii. In between, the stars trace out parallactic ellipses, for which the astronomer measures the semimajor axes. Many are the complications. Because all stars also have proper motions [5.2.1], the parallax effect will appear as a wobble in the proper motion vector (Fig. 4.5). The two movements must be separated, which requires observations over several years. Stars in binary systems also have orbital motions that must be taken into account.

To derive a parallax with a ground-based telescope, the astronomer takes an image of a fairly small field of view and measures the parallactic shift of a nearby program star (the one being studied) relative to a few reference stars that, at first,

Fig. 4.3 **The first parallax star** Cygnus, the Swan, flies from upper left to lower right, Deneb at its tail, Albireo at its head, the wings outstretched from upper right to lower left. Fifth magnitude 61 Cygni (a double star) was the first parallax star. Brightness is no indicator of distance. Supergiant Deneb is 225 times farther away than 61 Cyg. (J.B. Kaler.)

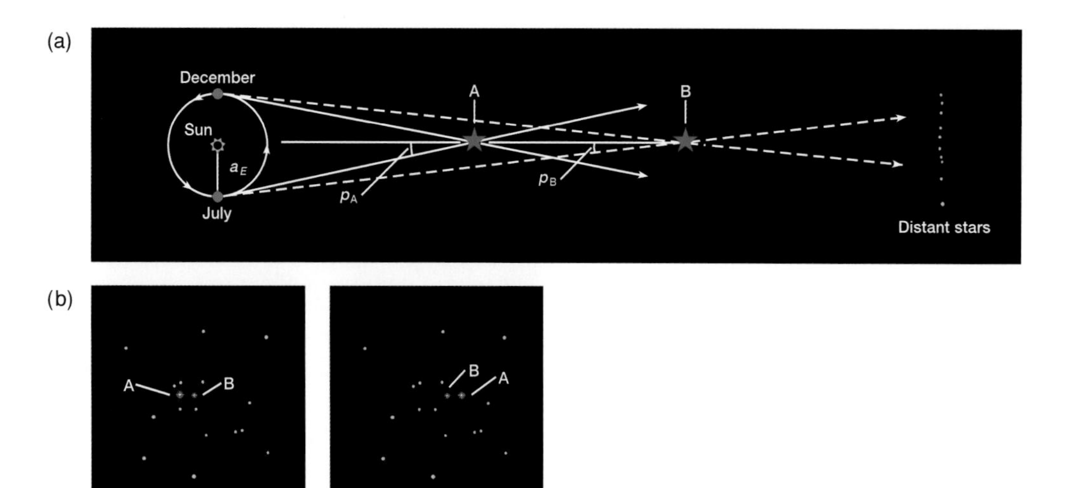

Fig. 4.4 **Stellar parallax** As the Earth orbits the Sun, nearby stars appear to shift back and forth relative to those at great distances. In (a), star A appears to shift through an angle of $2p_A$, where the parallax p_A is the angle subtended by the Earth's orbital radius as seen from the star. Star B is twice as far away, and therefore has a parallax p_B half of p_A. The lower two panels show simulated images of the observed shifts of the two stars as seen against the background, star B shifting through half the angle displayed by star A. (Kaler, J.B. (1994). *Astronomy!* New York: HarperCollins, reprinted by permission of Pearson Education, Inc.)

are assumed to be so far away that they have no appreciable parallaxes themselves. In fact, the reference stars all have finite distances and parallaxes too, however small. Since the reference stars shift in the same direction as the program star, the program star's derived *relative parallax* will be less than the true absolute value that would be measured against an infinitely distant background. The star will therefore be measured to be too far from Earth. If we know the average distance of the reference stars, their average parallax can be added to the program star's measured parallax, allowing it to be corrected

Box 4.2 Scaling the parsec

In Fig. 4.4, the parallax in radians p_{rad} equals the radius of the Earth's orbit a_E divided by the distance d, where both are measured in Astronomical Units, or $p_{rad} = 1/d_{AU}$. Since one radian equals 206 265 seconds of arc, $p''/206\,265 = 1/d_{AU}$, or $p'' = 206\,265/d_{AU}$, which by definition equals $1/d_{pc}$. Therefore, $d_{AU} = 206\,265 d_{pc}$. That is, there are 206 265AU in one parsec. And since $1\text{AU} = 1.496 \times 10^8$km, $1\text{pc} = 3.086 \times 10^{13}$km. The speed of light is $c = 2.997\,925 \times 10^5\ \text{km s}^{-1}$. Because a year contains 365.2422 days and there are 86 400 seconds in a day, a beam of light will travel $365.2422 \times 86\,400 \times 2.997\,925 \times 10^5 = 9.461 \times 10^{12}$km in a year, which is the length of the light year. Given $3.086 \times 10^{13}\ \text{km pc}^{-1}$, there must be $30.86/9.461 = 3.262$ light years per parsec (and 63 240AU per light year). All the units are now related. Moreover, we can now relate the distances of the stars to that of the neighborhood grocery.

to an *absolute parallax*. The mean distance of the program stars can be found spectroscopically [4.4.1] or, if they have some common characteristic, through statistical parallaxes [4.3.3] found from their proper motions.

If the reference stars average closer than the program star (which may happen if the program star happens to be distant, too), the parallax will come out negative. Even if the reference stars are more distant than the program star – if the inevitable errors of observation are comparable to the parallax – the parallax can still come out negative. The important point is to determine the error of measurement (Δp) relative to the measurement itself (the "signal to noise"): the ratio of $p/\Delta p$. The larger it is, the more reliable the parallax (or any other measurement for that matter). If $p/\Delta p$ is greater than say 5, the measurement is called "reliable," though astronomers are so desperate for distance measurements that 3 (or even less) may be used.

Traditional errors of older photographically-derived parallaxes were rarely better than a few hundredths of a second of arc, limiting the ability to derive distances to only a few tens of parsecs. The use of electronic detectors allows for much better positioning of stars, resulting in errors as low as half a milliarcsecond (mas, where $1\text{mas} = 0.001$ second of arc). Even so, the derivation of ground-based parallaxes is labor-intensive, the diligent astronomer finding the distance of one star at a time. Space telescopes have the advantage of no disturbing atmosphere. The "fine guidance sensor" of the Hubble Space Telescope is able to measure parallaxes with errors as low as 0.2 milliarcseconds, though still just one at a time.

4.2.3 Hipparcos

All that changed with the Hipparcos satellite, which orbited Earth between 1989 and 1993. Two scanners simultaneously observed the sky in directions separated by about 1 radian (near 58°). As the satellite slowly rotated, it measured the exact separations of program stars that appeared in the separate fields of view, each observed star thus serving as both a

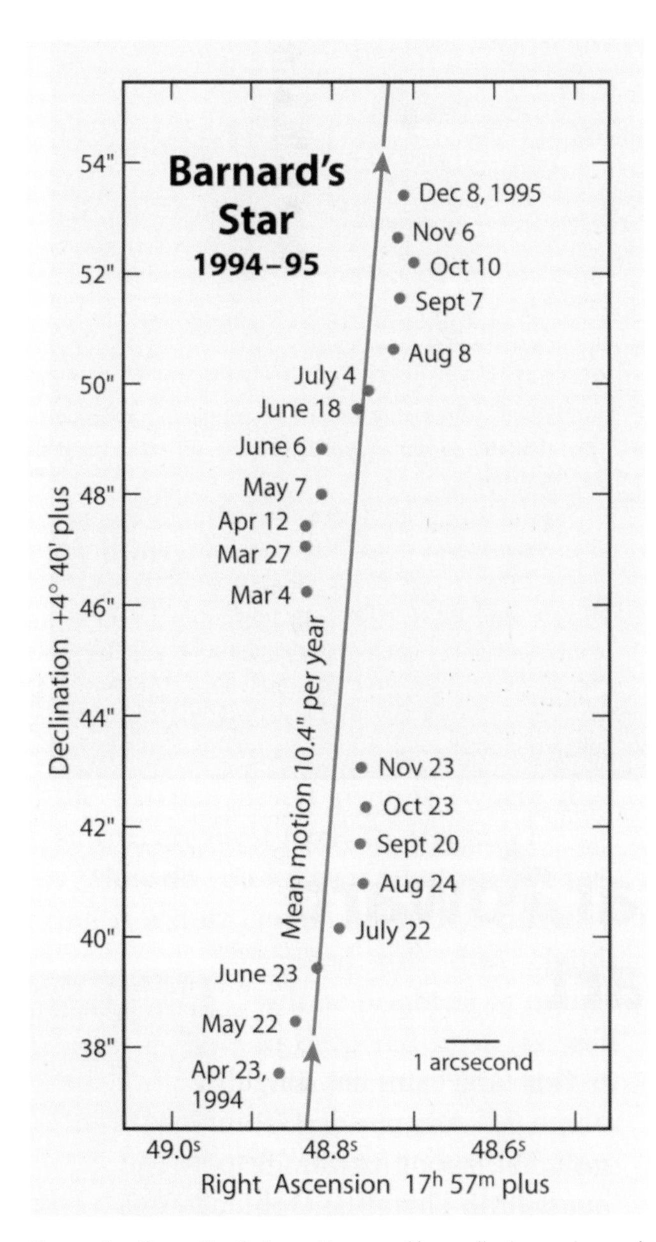

Fig. 4.5 **Parallax motion** Stellar motion caused by parallax is superimposed upon the straight-line track of proper motion, causing the star to appear to wobble as it plows ahead. The dots represent ground-based observations of nearby Barnard's Star, and the straight line the proper motion from Hipparcos. (DiCicco, D. (Sept. 2003). *Sky and Telescope*, p. 105.)

program and a reference star at the same time. By slowly also changing the axis of rotation, Hipparcos managed to observe the entire sky several times during the mission. The results of a massive calculation at the end were the precise right ascensions and declinations (tied to the International Celestial Reference Frame, and as closely as possible to the ground-based FK5 [2.6.2]), the proper motions in right ascension and declination [5.2.1], and the parallaxes of 118 000 stars, 15 times more than had been entered in the ground-based *Yale Parallax Catalogue*.

Not only were the errors reduced to just under a milliarc-second, but since the set of stars also served as self-consistent reference stars widely separated in the sky, the parallaxes are all absolute. The satellite allowed accurate measures of distance to around 100pc, and with concomitantly larger uncertainties even farther, for most stars brighter than $V = 9$, and for some others down to $V = 12.4$. The Tycho mission aboard the satellite used the separate "star-tracker" system, which monitored the positioning, to determine parallaxes of a million stars down to $V = 11.5$, with errors ranging from 7mas at the bright end to 25mas at the faint. Future satellites are expected to measure parallaxes accurate to a microarcsecond (0.001mas), thus allowing us to measure the distance to any star in the Galaxy.

Parallaxes are not limited to the optical domain. Very Long Baseline Arrays can determine the parallaxes of bright point radio sources (such as pulsars) with errors comparable to those from Hipparcos.

4.3 Distances from motions

Parallaxes involve the motion of the Earth. But the Sun moves too, as do all the other stars. The resulting proper motions [5.2.1] are a source of "noise" in parallax measurement, and must be removed from them. However, as in life, what may be bad in one context may be good in another: the relative motions are in themselves useful to obtain distances in instances where parallaxes fail.

4.3.1 The moving cluster method

Microarcsecond parallaxes are far in the future, while astronomers need distances now. Moreover, they need methods that can stretch to other galaxies. Any scheme is consequently fair game. Among those of great historical importance, and equally useful today in a new guise, is a method that can be applied to a nearby moving open cluster of stars [9.2]. Clusters are bound together under their mutual force of gravity. Though their stars are loosely circulating about a common center of mass, the orbital motion is slow, and to a good approximation can be ignored. The cluster is rather like a parade's marching band coming at you on a street. As it approaches, the band members collectively appear to come out of a point in space, a "divergent point," and as a group grow angularly larger, then pass by. When they march off toward a "convergent point" in the opposite direction, the group gets angularly smaller and seems to gather together once again.

In Fig. 4.6a, the stars of the cluster are moving parallel to each other, but because of the perspective, the observed proper motion vectors are directed toward the convergent point (Fig. 4.6b). By plotting the vectors and noting where they meet, we can find the right ascension and declination (A and D) [2.3.3] of where the cluster will be after an infinite time in the future. The angle θ between where the cluster is now (α and δ) and where it will ultimately appear (A and D) is also the angle formed by the average space velocity vector $\langle v \rangle$ [5.2.3] and the measured average radial velocity vector $\langle v_r \rangle$ [5.2.2] (where $\langle\rangle$ implies the average properties of the cluster's member stars, and all velocities are in km s^{-1}). The angle and the average radial velocity allow the calculation of both the cluster's gross space velocity ($\langle v \rangle=\langle v_r \rangle/\cos\theta$) and its tangential velocity ($\langle v_t \rangle=\langle v_r \rangle\tan\theta$). The tangential velocity of any celestial body is found from its proper motion (μ, in $''$year^{-1}) and distance (d, in pc) from the standard relation [5.2.1]

$$v_t = 4.74\mu d\,,$$

or

$$d = v_t/4.74\mu.$$

The average proper motion of the cluster, found from the individual proper motions of its members combined with the calculated mean tangential velocity, gives the cluster's collective distance.

Since in large open clusters internal orbital motion is slow, the spread in individual velocities – the velocity dispersion – about the average is low. We can therefore safely assume that the individual stellar tangential velocities are similar to the average. Enter the proper motions of individual stars into the above relation and we find their individual distances, enabling us to examine the cluster's three-dimensional depth and structure.

Though applied to several local clusters (initially to the Ursa Major moving group), the method works well only for the Hyades, which was (with some argument) measured to be 45 to 48pc away (see Fig. 4.6b and Fig. 4.7). In pre-Hipparcos days, when parallaxes were not all that accurate, the Hyades distance from the moving cluster method provided the first step up the ladder of the cosmic distance scale. The biggest problems involve the assurance that only real cluster members are included and the accuracy of the individual radial velocities.

4.3.2 The reversed moving cluster method

Hipparcos parallaxes of the member stars refine the Hyades' mean distance to a much more accurate 46.3pc, and thereby seem to vitiate the method, relegating it to history's closet. However, the Hipparcos parallaxes and proper motions are so good that the moving cluster method gains new life if it is turned around. As long as the errors associated with individual Hipparcos measurements are independent of each other (random and not in any way systematically related), the error in the mean parallax will be much smaller (by the square root of the number of individual measures) than the mean of the individual errors. In a compact cluster like the Pleiades (see Fig. 4.7), the stars are so close together that the satellite observed many at one time, and the parallaxes indeed exhibit systematically-related errors. But the Hyades stars are more spread out and the errors are random. We can therefore calculate a highly accurate mean transverse velocity from the cluster's average distance and proper motion. Once the average v_t is known, the proper motion of each star again yields its individual distance. Since an individual Hipparcos proper motion is so much more accurate than an individual

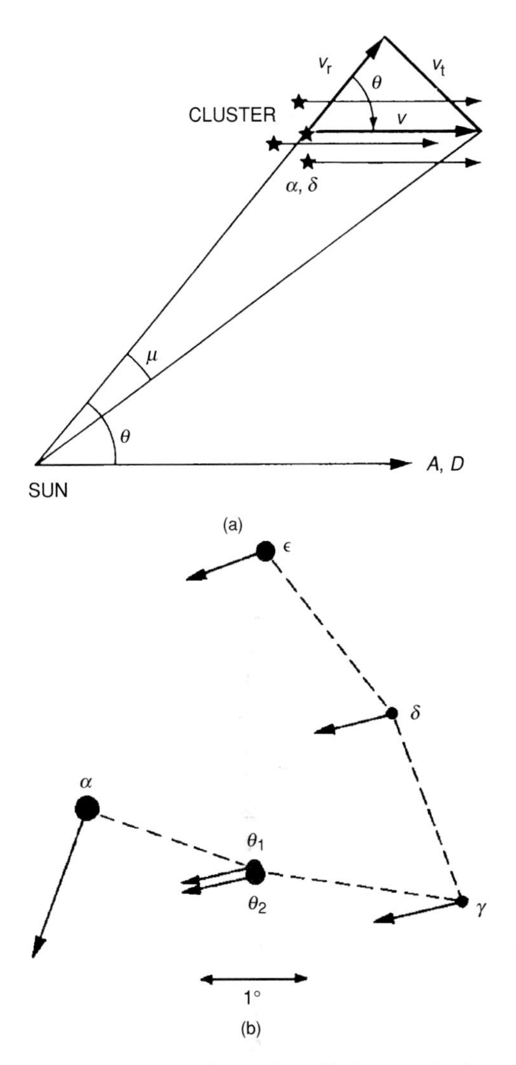

Fig. 4.6 **The moving cluster** Above (a), the stars of a cluster are actually moving parallel to each other, where the arrows represent the common individual space velocities. As seen below (b), perspective makes the proper motion vectors of the Hyades cluster[Figs. 1.12 and 1.13] appear to converge to coordinates (A, D), found by extending the individual proper motion vectors until they cross. The angle θ between (A, D) and the current position (α, δ) is the same as that between the radial (v_r) and space velocity (v) vectors, allowing the tangential velocity (v_T) and distance to be found. (Aldebaran is not part of the Hyades.) (Wyatt, S.P. and Kaler, J.B. (1974). *Principles of Astronomy: A Short Version*, Boston: Allyn and Bacon.)

Fig. 4.7 **Taurus and its open clusters** Taurus proudly displays the best-known open clusters of the sky, the Pleiades (Seven Sisters) toward upper right, and the Hyades, the vee-shaped grouping to the right of center left that makes the mythological Bull's head. Well spread out and nearby, the Hyades cluster is the best candidate for the application of the moving cluster method of distance. More than a dozen other open clusters dot the picture (as seen in a labelled version in Appendix 5). Southern Auriga is at upper right. Saturn shines brightly at lower left. (J.B. Kaler.)

parallax, the distance so calculated from proper motion is much improved. Not only does this procedure provide a vastly better knowledge of the cluster's three-dimensional structure, but it allows for a superior HR or color–magnitude diagram [6.3.3]. The method is so powerful that it can be applied to other large groups, such as moving OB associations.

We can go one better. Since for each star $v_r = v_t/\tan\theta$, the "reversed moving cluster method" also determines each star's radial velocity without the use of spectra and the Doppler shift. Comparison with traditional radial velocities from stellar spectra consequently give the opportunity to study effects in stellar atmospheres that can produce absorption line shifts not caused by the bulk motions of the stars.

4.3.3 Statistical parallaxes

In a broad sense, the moving cluster method is a special case of a more general method that uses proper motions to obtain distances. Parallaxes are limited by the size of the baseline, that is, by the diameter of the Earth's orbit. Were we to take our telescopes to Jupiter, which is 5.2 times farther from the Sun than Earth, the parallaxes would be 5.2 times greater. However, Jupiter's orbital period is 12 years, so to achieve a concomitant level of accuracy, we would have to observe for many Jovian orbits.

There is a simpler way to achieve a greater baseline: let the Sun do it for us. The Sun has its own motion through the surrounding swarm of stars with a velocity (depending on the set of stars adopted) of $V_0 = 20\,\mathrm{km\,s^{-1}}$ toward a well-defined "Apex" [5.3.1] that lies roughly in the direction of Vega. As the Sun moves toward the Apex, the stars will reflexively appear to move in the other direction, toward the "Antapex." Stars will therefore have proper motion shifts due to the solar motion alone that depend on their location relative to the direction of solar motion. Solar-induced proper motions of stars will be greatest in the direction perpendicular to solar motion (that is, perpendicular to the direction to the Apex). In one year, the Sun will move 6.3×10^8km or 4.2AU, which is 2.1 times the parallax baseline. Over a 10-year period, the baseline is 21 times as great. The longer we observe, the higher the precision of the measure, and the farther into the Galaxy we can probe.

For a star set perpendicular to the Apex direction, the transverse velocity is just the reverse of the solar velocity, so $d = v_t/4.74\mu = V_0/4.74\mu$, and since the parallax is $1/d$, $p = 4.74\mu/V_0$, where p is now called the *secular parallax* ("secular" used in the sense of "ongoing"). If the star

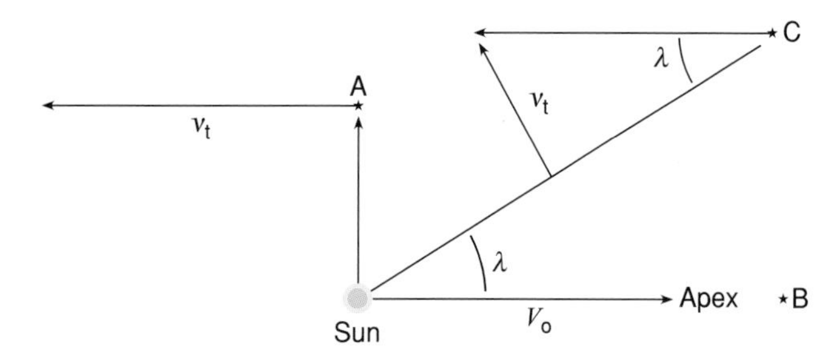

Fig. 4.8 **Reflexive solar motion** The Sun moves toward its Apex at speed V_0 and the three stars are stationary. Star A, perpendicular to the direction of solar motion, appears to have a transverse velocity $v_t = -V_0$, while Star B, in the direction of solar motion, will have no transverse velocity and no proper motion. If the angle between star C and the Apex is λ, then $v_t(C) = V_0 \sin\lambda$.

lies along the direction of solar motion, however, toward the Apex or Antapex, solar-induced proper motions disappear. In general, if the angle between the Apex and the direction to the star is $\lambda°$, the solar-induced transverse velocity is $v_t = V_0 \sin\lambda$ (Fig. 4.8). The secular parallax is therefore

$$p = 4.74\mu / V_0 \sin\lambda.$$

It would now seem that the stellar distance problem is solved. The difficulty is that the star has its own "peculiar motion," one that is peculiar to it alone, amidst the local stars. That is, the stars in Fig. 4.8 are moving too. Peculiar motions are quite independent of solar motion. We therefore have no prior idea of what a star's peculiar motion might be. We cannot disentangle solar motion from the star's peculiar motion unless the distance is already known, rendering individual secular parallaxes useless.

Sets of stars are another matter. Ejnar Hertzsprung noted that Antonia Maury's narrow-line stars had systematically smaller proper motions than her broad-line stars [6.3.1]. As a set, the narrow-line stars must be farther away and much more luminous, and he thereby discovered the existence of supergiants. The proper motion of a star will be the sum of its own peculiar proper motion and that induced by solar motion. The observed proper motion has a particular direction in the sky. Break the observed proper motion of a star into two components, one that lies along the direction to the solar Apex (called the "upsilon component," υ), the other perpendicular to the Apex (the "tau component," τ). Make the assumption that stellar motions are random relative to the local standard of rest [5.3.1] (the mean motion of all local stars), against which the Sun is moving, too (and which is in fact defined by our calculation of solar motion). There are just as many stars going one way as there are another. Therefore the peculiar motions of the stars average out to zero. With the appropriate weightings of the stars (determined by their angles relative to the direction to the Apex), we can now use the upsilon components to calculate the average parallax and thus the average distance to the set. As an example, we might pick as a set a particular kind of variable star, and thereby derive their average distance and therefore their average luminosity. The tau components can be used as well by assuming that the mean velocity in the tau direction is the same as the mean measured radial velocity.

As elegant as the method may look, it is rather one of last resort. Stars of different kinds are *not* randomly distributed, but have their own average peculiar motions relative to the Sun. That is, they may be moving systematically, or the Sun may have one Apex relative to one set and a different Apex relative to another set and we cannot find the appropriate solar vector. As real parallaxes (and other methods) improve, statistical parallaxes become more and more an historical artifact, but one that needs to be understood to make sense of past results.

4.3.4 Galactic rotation

When more accurate methods fail, under the right circumstances galactic rotation provides a decent means of assessing minimally accurate distances. The rotation curve of the Galaxy's disk – velocity as a function of distance from the Galactic center – is initially found from radial velocity observations of the neutral hydrogen 21-centimeter line relative to the solar orbital velocity [5.3.3; Box 11.2]. The Sun has a fairly circular orbit about the Galactic center. Another star with a circular orbit will have a different orbital velocity. Project the velocities of both the Sun and the distant star along the line of sight to the star. The radial velocity of the star is the difference between these projected velocities, and will depend on the nature of the rotation curve, on the star's distance, and on its Galactic coordinates [2.3.5]. For any Galactic longitude, we can construct a graph of distance against radial velocity, and from the star's measured radial velocity can determine the distance: the same method that is used to construct a map of the Galactic spiral arms.

This particular procedure works only for true young Population I stars [5.1] those that should actually fit the predetermined rotation curve, and is therefore limited to the youthful O and B stars (and related objects). Even these offer difficulties, as most belong to OB associations [9.3] and therefore have significant peculiar velocities that can lead to erroneous distances. We therefore need to examine a whole association so that we can average out the peculiar velocities. Moreover, a large fraction of O and B stars are high-speed "runaways" for which the method will not work at all [8.6; 14.3.2].

A variation on the theme uses the interstellar medium. Neutral hydrogen defines the thin Galactic disk almost better than anything else. Many kinds of stars radiate, or are associated with objects that radiate, in the radio region of the spectrum. If a distant object in the Galactic disk produces a sufficiently bright radio background, the neutral hydrogen clouds along the line of sight will superimpose the 21-centimeter absorption line [Box 11.2]. From the Doppler shifts in the absorptions we find the distances to the intervening clouds, and therefore a lower limit to the object's true distance.

4.4 Photometric methods

If we can infer the absolute magnitude of a star and then measure its apparent magnitude and extinction [3.5.1], we can calculate the distance from the magnitude equation [3.2]. There is a considerable number of methods for doing so, some that work for normal stars like the Sun, others that work for different kinds of variables. All we need is some way with which to calibrate the absolute magnitudes, and therein lies the rub.

4.4.1 Spectroscopic parallax

The word "parallax" has worked its way so deeply into the astronomers' language that it is difficult to expunge. We emphatically do not measure parallaxes from spectra, but measure distances. Yet since one is the inverse of the other, the term "spectroscopic parallax" is endemic. "Spectroscopic distance" is better. The method works the magnitude equation [3.2], $M = m + 5 - 5 \log d$, backwards. Rewrite it as

$$\log d = (m - M + 5)/5.$$

All we require to infer distance is the absolute magnitude. Note that distance d (rather its logarithm) scales with the difference between absolute and apparent magnitudes, $y = m - M$, called the *distance modulus*, such that $\log d = (y + 5)/5$. The distance modulus, as commonly used as distance itself, can be defined for any kind of magnitude, *U*, *B*, *V*, *R*, etc. If there is any interstellar absorption [3.5.1] in the line of sight, the absorption in magnitudes must first be subtracted from the apparent magnitude. If the magnitudes are visual,

$$\log d = (V - M_V - A_V + 5)/5 = (y_0 + 5)/5,$$

where A_V is the visual absorption. Distance moduli are usually expressed as corrected for interstellar extinction, $y_0 = m - M + A$, but not always, so the reader must be cautious.

To measure a spectroscopic distance, first determine the star's MKK class – dwarf, giant, etc. [6.3.2] – then locate it on the lower axis of a calibrated HR diagram (one with, say, M_V plotted against spectral class with MKK loci indicated on it) and read off M_V (Fig. 4.9). From the $B - V$ color determine the color excess and then M_V, and out drops the distance. For example, observe a star to have a V magnitude of 10.53 and a color of $B - V = 1.2$. The spectrum reveals it to be a K0 giant, K0 III. The calibration of true color against spectral class shows the true $B - V$ to be 1.0 (see Fig. 3.11), leading to a color excess of 0.2 magnitudes. The standard ratio of total-to-selective absorption [3.5.1], $R = 3.2$, gives an extinction $A_V = 0.2 \times 3.2 = 0.64$ magnitudes. From Fig. 4.9, $M_V = 0.74$, so $\log d = (10.53 - 0.74 - 0.64 + 5)/5 = 2.83$. Since the logarithm of a number is the power to which you have to raise 10 to get the number, the distance is then $d = 10^{2.83} = 676$pc. An F5 supergiant with $V = 5.05$ suffering two magnitudes of extinction would have $M_V = -8.14$, whence $\log d = [5.05 - (-8.14) - 2 + 5]/5 = 3.23$. The distance is therefore 1730pc.

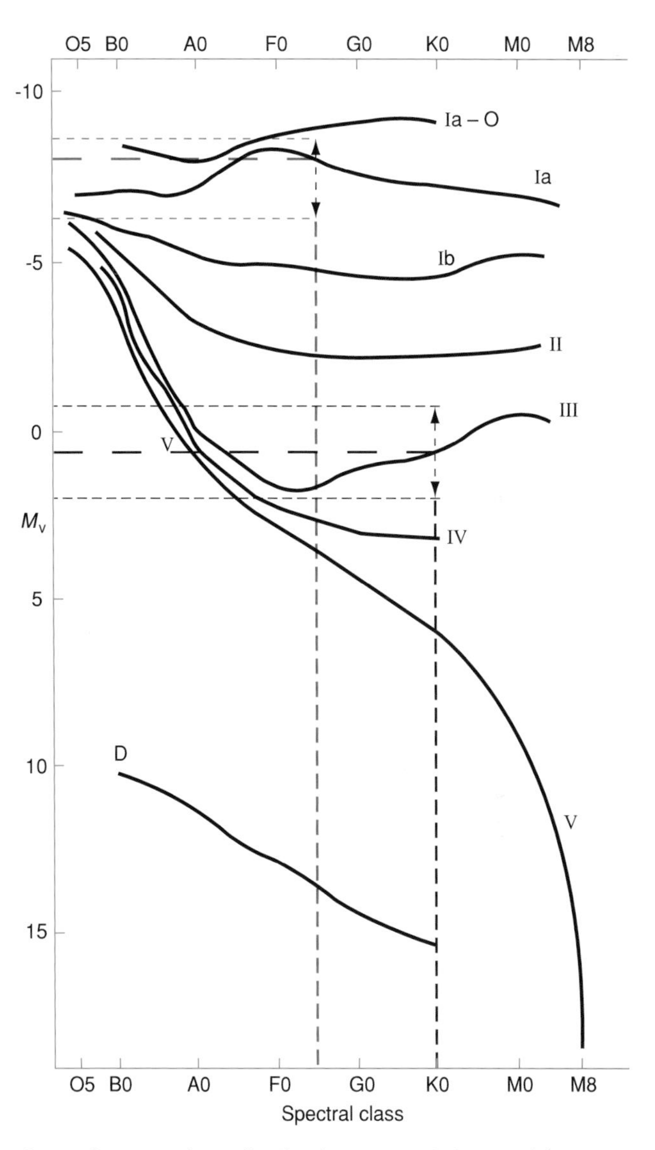

Fig. 4.9 **Spectroscopic parallax** If we know a star's MKK spectral class, we can, within limits, find its absolute magnitude. The limits of the K0 giant and F5 supergiant fall within the arrows on the vertical lines. (Adapted from Kaler, J.B. (1989). *Stars and their Spectra*, Cambridge: Cambridge University Press.)

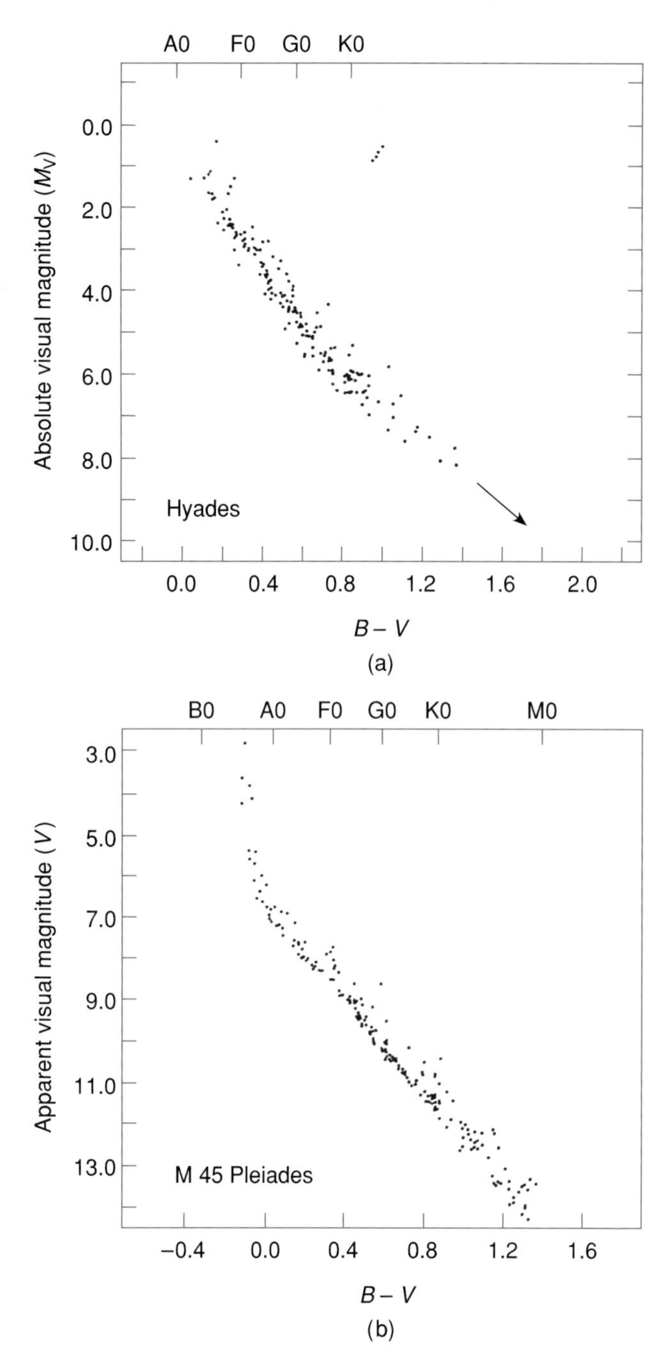

Fig. 4.10 **Main sequence fitting** The HR diagram of the Hyades, where M_V is plotted against spectral class and color index, is presented at the top (a). At the bottom (b) is the HR diagram of the Pleiades, in which apparent magnitude is plotted against color and class. In (c) (upper right), the two main sequences are superimposed, with blue dots for the Hyades, red for the Pleiades, allowing the determination of the distance modulus of the Pleiades. (Hagen, G. (1970) *An Atlas of Open Cluster Colour–Magnitude Diagrams*, David Dunlap Obs. and Kaler, J.B. (1994). *Astronomy*! New York: HarperCollins, reprinted by permission of Pearson Education, Inc.)

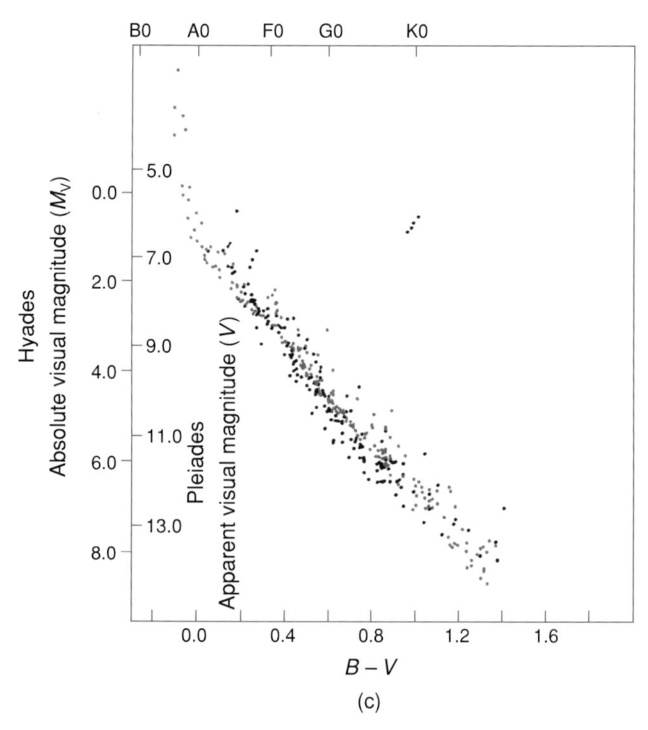

Fig. 4.10 *(cont.)*

The principal problem is that any MKK class has a considerable range in absolute visual magnitude. From the graph, the K0 giant could be as bright as $M_V = -0.8$ before being classified as a K0 II bright giant, or as faint as $V = 2.13$ before being called a subgiant, K0 IV. If on the luminous side of average, the star is 1.54 magnitudes brighter than before, so its distance comes out to be as great as 1370pc. It could similarly be as small as 356pc, a range from top to bottom of a factor of nearly 4. The precision can be improved by going to finer MKK divisions, for example K0 IIIb, or by using various spectral line ratios that are continuously correlated with luminosity. It is important also to note that the classification of stars is not dependent on luminosity, but on spectral features. A star classified as a giant might actually appear within the HR diagram's supergiant realm and vice versa. The range in error could therefore be even larger than expressed above.

The results improve for dwarfs, as the main sequence is rather tightly restricted in absolute magnitude. Take a G2 dwarf like the Sun that has an apparent visual magnitude of 13.05. Assuming it has the solar $M_V = 4.83$, and ignoring interstellar extinction, the spectroscopic distance is 440pc. The error in M_V for such a star is roughly plus or minus one magnitude, so it could be as far as 798pc or as near as 278pc, still not all that good.

Worse, many kinds of star have no decent calibrations of spectral class with luminosity, cases in point being low-metal halo stars [5.1], to which the standard spectral classes do not apply, the central stars of planetary nebulae [13.4.4], Wolf–Rayet stars [6.4.4], R Coronae Borealis stars [10.6.2], and many others of odd chemical composition. At the end, for all its usefulness, this method is limited, and is generally employed only when others fail.

4.4.2 Main sequence (and other) fitting

Rather like the case for statistical parallaxes vs. secular parallaxes, when a group of stars is used, the results dramatically improve. Rather than determining the absolute magnitude of one star from its spectral class, which can result in rather large errors in distance, we determine the distance of a cluster of stars from the sum of its many main sequence stars. In pre-Hipparcos days, the method, applied to open clusters [9.2], formed the backbone of the calibration of the HR diagram, and for the determination of the properties of luminous and rare stars still plays a powerful role.

Figure 4.10a shows the pre-Hipparcos HR diagram of the Hyades, in which absolute visual magnitudes (for a distance of 45pc) are plotted against spectral class and main sequence color. To use it to measure the distance to the Pleiades, first construct the Pleiades' HR diagram with apparent visual magnitude instead of absolute (Fig. 4.10b). Make the reasonable assumption that the main sequences of the clusters are intrinsically the same (which they would be for similar stellar metallicities). For the sake of argument, assume no interstellar extinction. In your mind, place the Pleiades diagram on a transparency. Lay it on top of the Hyades diagram and line up the spectral classes (or colors). Slide the Pleiades diagram up or down until the main sequences fit over each other (Fig. 4.10c). On the left-hand side, read off the difference between the absolute and apparent magnitudes, which is the distance modulus $y = V - M_V = 5.6$, from which we find $d = 132$pc.

Hipparcos parallaxes quickly gave a mean Pleiades distance of 116pc, well below expected, suggesting that the stars were dimmer than calculated by theory (thus throwing real fear into stellar theoreticians). The problem was resolved upon realization that in tight clusters, Hipparcos parallax errors correlate with parallaxes themselves, and are therefore not as reliable as thought. Use of outlying and low-error Pleiades yields near-perfect agreement with the main-sequence fitting method. Rather than being improved upon, the photometric distance provided the tool with which to check the supposedly excellent satellite results. The Pleiades therefore remain at close to 130pc.

At an age of around half a billion years, the Hyades has lost its upper main sequence O and B stars and begins its main sequence in class A. The Pleiades cluster, however, is much younger, and still has class B stars that range from dwarfs to giants. Main sequence fitting therefore not only tells the Pleiades' distance, but extends the calibration of the HR diagram to hotter stars that, because of their greater distances, have larger parallax errors (or no parallaxes at all). Continue the process to younger clusters, notably to the Double Cluster in Perseus, h and χ Persei [Fig. 9.3], which after a necessary correction for interstellar extinction derivable from a color–color plot [3.4; 3.5.1] yields a distance of about 2200pc and absolute magnitudes of class O stars and M supergiants. At these distances, however, cluster membership (excluding foreground and background stars), errors in interstellar extinctions (including the illusive ratio of total to selective extinction, R [3.5.1]), even unresolved binaries all contribute significantly to the uncertainty in distance. Nevertheless, the method allows decent distances to any properly-observed open cluster to be calculated and the absolute magnitudes of the kinds of stars that they contain. Fitting the white dwarf sequences of clusters works as well, providing the clusters are close enough so that these dim stars are visible and their apparent magnitudes accurately measurable.

The method is also applied to globular clusters [9.4]. These stellar systems are so old that they have no bright main sequence stars, and are generally so distant that their intact lower main sequences can be hard to reach. Moreover, they are so populous that they can be difficult to resolve into individual stars. However, the main sequences of the nearer globulars, M13, Omega Centauri, and the like, are indeed accessible.

The real problem is that globular cluster main sequence stars are essentially low-metal subdwarfs [5.1, 6.4.2], and the standard main sequence derived from open clusters is not suitable. Instead, we must use Hipparcos parallaxes to construct a special subdwarf main sequence, and there are not all that many of these stars, which renders globular cluster distances (hence ages) more problematic than we would like. Globulars are also so distant that binaries can rather easily masquerade as single, leading to overly bright stars that confuse distance determinations. Moreover, subdwarfs and globulars are not homogeneous, as we find the chemical compositions from one to the next have a large range. To achieve the best distances, the compositions of the calibrators must be matched to those of the clusters, and there are not enough calibrators to do a complete job.

Once the true distances of a few of the nearer globular clusters are known, we can work our way to more distant clusters by fitting the horizontal branches of their color–magnitude diagrams [9.4.9] against those whose absolute magnitudes have been secured. We can also find the distant clusters' distance moduli directly, by comparing the average apparent magnitudes of their RR Lyrae stars [10.3.3] (which inhabit horizontal branches) with their known absolute magnitudes.

4.4.3 Cepheid variables

In 1912, Henrietta Leavitt of Harvard College Observatory discovered that the longer-period Cepheid variable stars [10.3] in the Magellanic Clouds were also the brighter, thus starting us on the road to determining the distances of galaxies and the nature of the Universe, even before astronomers fully realized that there *were* galaxies other than ours (Fig. 4.11).

A century of research has confirmed and developed the *Cepheid period–luminosity ("P–L") relation* (Fig. 4.12), in which the

Fig. 4.11 **The Large Magellanic Cloud** Though 50.1kpc away, the Large Magellanic Cloud, a satellite of our own Galaxy, is easily resolved into stars. Since all its stars are at about the same distance, it is a natural laboratory for the study of relative stellar properties. The large reddish spot to the left is the Tarantula Nebula, 30 Doradus. (Courtesy of Lok Kun Tan, Starryscapes.)

average absolute magnitude of a star (M) is closely related to the logarithm of the period (log P). Such a relation allows an instant determination of distance. One need only measure the period and the average apparent magnitude of the star, derive the average absolute magnitude from the relation, and calculate the distance by the magnitude equation, of course including the degree of interstellar extinction. (The average magnitude of a Cepheid is not the average of the minimum and maximum magnitudes, as these are logarithmic quantities. Instead, we convert the magnitudes to intensity units, average the maximum and minimum intensities, and convert the results back into a magnitude.)

As brilliant Population I (disk population) supergiants [5.1], Cepheids can be seen very far away, allowing us to

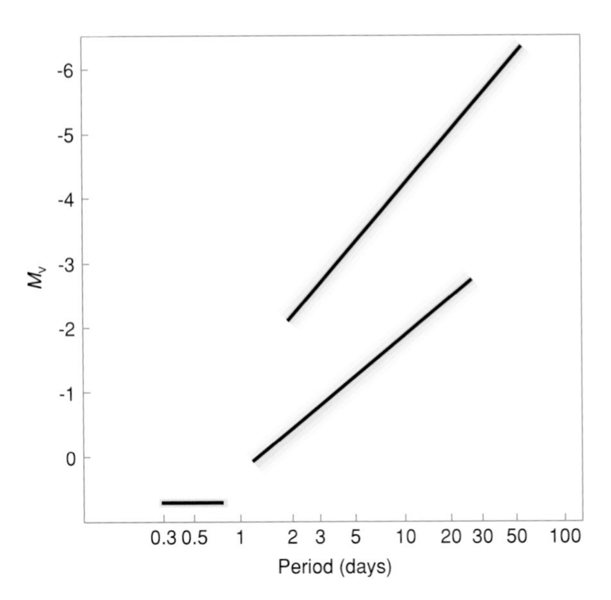

Fig. 4.12 **Cepheid period–luminosity relation** The period–luminosity relation for classical Population I Cepheids rises quickly with increasing period. A similar curve for Population II W Virginis stars runs parallel to it, but at fainter magnitudes. At lower left lie the RR Lyrae stars. (Data from Cox, A.N. ed. (1999). Allen's Astrophysical Quantities, AIP/Springer.)

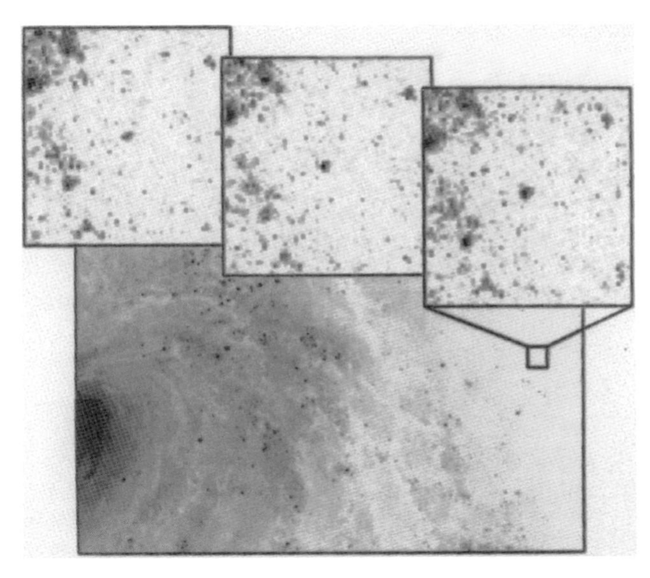

Fig. 4.13 **A distant Cepheid** A faint Cepheid varies at the edge of the spiral galaxy M100, 15.3Mpc (50 million light years) away. (Courtesy of Hubble Space Telescope, W.L. Freedman, Observatories of the Carnegie Institution of Washington, and NASA.)

determine the distances of spiral galaxies. The first such measure was by Edwin Hubble in 1924, when he used Cepheids in the nearby Andromeda Galaxy, M31 [Fig. 5.1], to find its distance (now known to be 750kpc) and to show beyond doubt that it was an external system, one much like our own. We can now locate Cepheids in spirals (which have strong Population I components) to well over 20Mpc away (Fig. 4.13).

The slope of the P–L relation is readily found from the numerous Cepheids in the Large Magellanic Cloud (LMC). Since all the LMC's stars are at nearly the same distance, we derive a tight relative correlation between apparent visual magnitude and log P. (The Small Magellanic Cloud, the SMC, could be used as well, but the concentration is on the much larger LMC.) To make use of the P–L relation, however, we need to map *absolute* visual magnitude against log P, that is, we must calibrate the relative relation in terms of absolute magnitude. If we can find the distance to at least one Cepheid in our Galaxy, we immediately find the difference between V in the LMC and M_V for that particular pulsation period, and thereby calibrate the entire relation in terms of absolute magnitude. In effect, we determine the distance modulus ($V - M_V$) for the LMC, which results in the LMC's actual distance in parsecs. Since all measurements have errors inherent to them, the more calibrators available, the better the results. Cepheid calibration is an industry in itself. The result of all the effort is an LMC distance of 50.1kpc (Box 4.3).

Cepheids change their characteristics as they evolve across the instability strip [10.3.1]. As a result, the surface temperature, and therefore color (C), plays a small role as well. The P–L relation is therefore really a P–L–C relation. A deeper look shows that metallicity is also a factor that must be taken into account. For a "Key Project" in which the Hubble Space Telescope was used to determine extragalactic distances, the P–L–C relation is collapsed back to a P–L relation, to wit:

$$M_V = -2.760\ (\log P - 1) - 4.218.$$

The expected errors are only ±0.16 magnitude, which corresponds to a distance error of under 8%. Like any photometric method, Cepheid distances must be corrected for interstellar extinction, which can be hard to assess. The P–L relation extends all the way across the spectrum. A magnitude of any color will work. Though Johnson V is the most commonly used, infrared magnitudes reduce the error considerably because of lower reddening and less dependence on temperature, hence color.

Mira variables [10.5.1] follow a period–luminosity law, too, though its application is less accurate.

4.4.4 Population II variables

In the early days of Cepheid research, astronomers failed to distinguish between two different kinds of Cepheid, the massive Population I [5.1] stars (the *classical Cepheids*) that are evolving more or less horizontally through the Cepheid instability strip, and lesser-mass Population II Cepheids that are brightening on the Asymptotic Giant Branch (the AGB) and that are commonly known as W Virginis Stars (the type incorporating RV Tauri stars as well) [10.3.2]. W Vir stars exhibit a period–luminosity relation that is useful for distance determination, but one that is 1.5 magnitudes below the P–L relation for classical Cepheids (see Fig. 4.12). Early Cepheid studies mixed the two, resulting in a severe underestimate of luminosity for a given period. The difference, discovered in 1951, led to a near-doubling of the known scale of the Universe.

At the short end of the stick are the short-period RR Lyrae stars [10.3.3], which run off the end of the P–L relation for the W Virginis stars (see Fig. 4.12). At an earlier stage of evolution than W Vir stars, RR Lyrae stars, with periods of under a day, occupy a special position on the low-metal Population II horizontal branch, and are easily identifiable in globular clusters [9.4] and in the general field (where they are epitomized by RR Lyrae itself). With rather consistent absolute magnitudes of $M_V = 0.7$ or so, RR Lyrae stars make good distance indicators. But like classical Cepheids, they are rare and tend to be distant, RR Lyrae itself over 300pc away. As a result, the absolute magnitude calibration is not as good as astronomers would like.

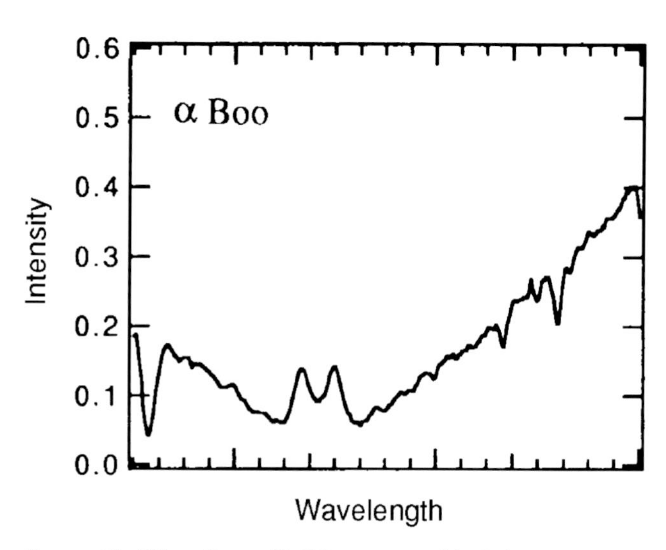

Fig. 4.14 **The Wilson–Bappu effect** The spectrum of the K giant Arcturus contains a weak chromospheric emission line at the bottom of the broad K line absorption of ionized calcium (which overfills the picture). The emission has a narrow absorption at its center. (Wallerstein, G., Machado-Pelaez, L., and Gonzalez, G. (1999). *Pub. of the Astr. Soc. of the Pacific*, **111**, p. 335.)

4.4.5 Novae and supernovae

Exploding stars range from fair to excellent candidates for distance determination. Among the first clues that the Andromeda Galaxy was a distant external system came from observations of novae [8.5.4] within it. At maximum, novae reach absolute visual magnitudes of −6 to −10, depending on the speed of eruption. The light curve gives us the apparent magnitude of an exploder, and therefore its distance, though the error is significant (in large part because of the uncertainty in interstellar extinction).

Box 4.3 Cepheid calibration: the hunt for the LMC's distance

Nearly two dozen methods have been used to calibrate the Cepheid P–L relationship and to find the distance to the Large Magellanic Cloud, the two activities going hand in hand. The LMC is vastly too far for parallaxes. One approach is to measure the parallaxes of the nearer Galactic Cepheids to find their absolute visual magnitudes. Though many Cepheids are within parallax range (with both Hipparcos and advanced ground-based techniques), the errors are high enough that the method is not entirely unsuitable. A major problem is the "Lutz–Kelker effect," in which we undervalue the data for which random errors make the distances appear too large, and thereby underestimate the average luminosity of the set. The classic calibration technique uses the handful of Cepheids found in open clusters, whose distances can be determined from main-sequence fitting, but again the small number leaves us with relatively large errors.

Innovation from old ideas adds spice. The "Baade–Wesselink method" finds the difference between maximum and minimum radii ($R_{max} - R_{min}$) of a Galactic Cepheid from the surface radial velocity, and the ratio (R_{max}/R_{min}) from luminosity and temperature variations. The difference and the ratio are then solved for the radii themselves and the mean radius, $R_{mean} = (R_{max} - R_{min})/2$ (converted to units of AU). Interferometric measures of mean angular diameter $\theta'' = 2R_{mean}/d$ (pc) then lead to distance. If the interferometry is good enough, we could also derive the distance simply from the difference in angular radius $(\theta_{max} - \theta_{min})/2$ compared with the calculated difference in true maximum and minimum radii. (In practice, the problem is more complex, and is solved by tracking angular size variations with real size variations, where R_{max}/R_{min} are computed from detailed model atmospheres.)

Similarly, the "Barnes–Evans technique" uses angular diameters of a variety of non-variable stars to find the relations among effective temperature (hence color index), V magnitude, bolometric correction, and stellar surface brightness. Applied to a Cepheid, the waltz of $B - V$ over time gives the change in angular diameter, the changing radial velocities give the change in true diameter, and again out comes distance. The method can be applied to any kind of pulsator, including RR Lyrae stars.

Several methods determine the LMC's distance modulus directly from stars whose absolute magnitudes are determined from Galactic observations. Examples are the stars at the tip of the red giant branch [13.2.2, 13.3.1], "red clump" stars (helium-burning stars that clump together on the HR diagram), RR Lyrae variables [10.3.3] whose distances are found from direct parallaxes, statistical parallaxes, and their memberships in globular clusters (whose distances are found from subdwarf main-sequence fitting or similar means), and long-period variable stars [10.5.1]. To this set we add a direct distance to Supernova 1987A [Box 14.1] as found from its light echo.

The immensity of detail is not so important as the idea of the great effort being expended, which currently finds the LMC to be 50.1kpc away. Once the Cloud's distance is known from the average of all of the techniques, we can turn the tables and use it to re-calibrate the methods that went into this much-sought number.

Of much greater interest are supernovae [14.2], which tend toward common maximum visual luminosities that depend on their type. Moreover, they are brilliant, allowing the astronomer to assess distances into the range of billions of light years. While Type II core-collapse supernovae [14.2.2] have a rather large scatter in maximum brightness, white-dwarf Type Ia supernovae [14.2.5] do not. Rare within a particular galaxy, a large number of such event has been observed in other galaxies whose distances are known from studies of their Cepheid variables. Nearly all of them reach a common absolute visual magnitude M_V near -19.45, rendering them excellent "standard candles" with which to measure distances of galaxies far beyond Cepheid range. Most of the small divergence from commonality can be assessed from the shape of the light curve. The powerful combination of Cepheid and Type Ia supernovae were responsible for the discovery of the acceleration of the expansion of Universe. Closer to home, this calibration allows the distances of historic supernovae to be determined, notably for Tycho's and Kepler's stars, for which the light curves have been reconstructed from the writings of the original observers.

4.4.6 The Wilson–Bappu effect

By far the most powerful spectrum lines in the Sun are H and K of ionized calcium at 3968 and 3934Å [6.1.1]. The H line is contaminated with real H (hydrogen), but the K line is fairly clean. High-dispersion solar spectra reveal a much narrower emission line (called "K2") smack in the middle of the broad absorption (rendering the broad line "K1") that is produced by the thin gases of the solar chromosphere. In the middle of the emission is an even-narrower self-reversed "K3" absorption caused by the immense gaseous opacity at the center of the line. All cooler stars with chromospheres show such features (Fig. 4.14).

In 1957, O.C. Wilson and M.K.V. Bappu discovered that the widths of the central emissions for G, K, and M stars were tightly correlated with their absolute visual magnitudes, M_V. No one yet knows quite why. The correlation is good over an astounding range of absolute magnitudes, from eighth to brighter than zero, and therefore covers both dwarfs and giants – even supergiants – in spite of spectral class. Measure the width of the emission (W_0) at the half-intensity point in angstroms and convert to velocity units (km s^{-1}) using the Doppler formula. Dimmer than M_V of about zero, the absolute visual magnitude is

$$M_V = 27.03 - 14.89 \log W_0$$

in spite of spectral class. The relation for brighter stars is similar, but seems to depend on spectral class as well. To determine the distance of a star, the astronomer must observe the apparent magnitude and take a spectrogram with sufficient resolution to detect the inner emission. W_0 then gives M_V and, with correction for interstellar extinction, the distance.

4.4.7 Interstellar reddening

Interstellar extinction [3.5] usually causes problems in distance determination. However, if one is desperate enough, the liability can become an asset, allowing distance actually to be found. You wish to find the distance to a particular object but have no other means available: novae [8.5.4], planetary nebulae [13.4], and other objects that have analyzable spectra but whose absolute magnitudes (or other relevant parameters) you do not know.

Use the continuous or emission-line spectrum to determine the extinction constant or color excess. Then use an extinction map to find how the reddening increases with distance, enter the reddening, and read off the distance. Extinction is so spotty, however, that generalized maps do not work too well. A "pencil-beam" approach is better, in which the astronomer measures extinction vs. distance for ordinary stars along the line of sight in as tight a cone around the program object as possible. Even then there will be error produced by the chaotic nature of the interstellar medium. As usual, R, the ratio of total to selective absorption, enters into the matter as well. And of course the method works only as long as there is dust all the way along the sight-line, and therefore really only for objects within the Galactic disk.

4.5 Expansion parallaxes

An astronomer observes an expanding spherical shell of interstellar gas that radiates emission lines [Fig. 8.16]. The emission lines from the shell's forward side, which is coming at the Earth relative to the center of the object, are Doppler shifted [5.2.2] to shorter wavelengths (to the violet) relative to the shell's average velocity. The emissions from the receding back side are similarly Doppler shifted to the red. The average of the maximum velocity of approach and the maximum velocity of recession gives the bulk radial velocity of shell relative to Earth, while the difference between the two velocities gives twice the expansion velocity, v_{exp}.

Over a period of time, observe the transverse angular expansion rate caused by the physical expansion, and reduce it to seconds of arc per year. The annual angular expansion rate across the diameter of the shell is twice the proper motion [5.2.1] (μ_{exp} $''$yr^{-1}) of the visible edge of the shell relative to the bulk proper motion of the center. Applying the rule that governs proper motion, the distance $d = v_t/4.74\mu$, where v_t is the transverse velocity of the shell relative to the center, or just the expansion velocity, which we know from spectroscopy. The distance of the shell is therefore

$$d = v_{exp}/4.74\mu_{exp}.$$

As long as we can obtain both the expansion velocity and the angular expansion rate, we can derive the distance independently of interstellar reddening. All we need do is wait a few years until the expansion becomes visible and accurately measurable.

This powerful method was first used to determine the distances of novae from the parameters of their rapidly expanding optical shells [8.5.4; Fig. 8.16]. Application to planetary nebulae [13.4] has been more difficult because the shells are expanding slower and we must wait longer intervals to observe the expansion, although the accuracy derived from Hubble Space Telescope observations renders much of that problem solved, at least for closer objects. Modelling of an expanding ionization front is often needed for such work. Supernova remnants [14.3.4] are excellent candidates as well. The Crab Nebula's distance of 2kpc is nicely found from optical expansion rates. Expansion rates are also available for supernovae from both X-ray and optical observations. It is also common to use the movements of individual gaseous filaments, particularly for planetary nebulae and supernova remnants.

A highly accurate variation on the theme uses light travel time and measures of the delay of emission from the rear side of a shell relative to that from the forward side [10.5.2]. Another variation is applied to "light echoes" from supernovae. Shortly after Supernova 1987A [Box 14.1] exploded in the Large Magellanic Cloud, the optical blast illuminated a ring – really a shell – of dusty gas that had been ejected while the ex-star was a supergiant. The time delay and the speed of light gives the physical diameter of the surrounding shell which, with angular diameter, leads to distance.

The main problem with such methods is that they initially assume spherical symmetry, that the diameters of the shells and the velocities are the same in all directions, and that in the case of SN 1987A that the ring is not somehow tilted to the line of sight. Nova shells are clearly not projected against the sky as round, but are commonly prolate. Nothing really seems to have the ideal symmetry. Moreover, measured velocities are as seen "now," and accelerations may well have taken place. The only way out of the woods is to model the different kinds of shells and rings in three dimensions, based upon velocities and what we see projected against the sky, a procedure that still allows for plenty of error to creep in.

4.6 Binary stars

Double stars [8.2] present their own useful distance methodologies. The classic example is *dynamical parallax*. Express Kepler's third law for a double star in solar units,

$$P^2 = a^3/(M_A + M_B),$$

where P is the binary period in years, a is the semimajor axis in AU, and M_A and M_B are the masses of the two components [8.2.3]. The observed parameter is a in seconds of arc, a''. If you know the distance, d (pc), a (AU) $= a''/d$, and you can solve for the sum of the masses. If you do not, then you are stuck with the equation written as

$$P^2 = (a''/d)^3/(M_A + M_B)$$

and the mass remains unknown. Reverse the equation and solve it for distance, and find

$$d = a''/[(M_A + M_B)^{1/3} P^{2/3}].$$

If spectra reveal the stars to be dwarfs, take a guess at the masses, and calculate the distance. From d and the apparent magnitudes, find the absolute magnitudes, and from these and the main sequence mass–luminosity relation [7.2.4; 12.4.1], find new masses. Plug these back into the above equation, re-derive the distance, and do it all over again. Keep going in the cycle until the distances and masses converge on stable numbers. This iterative scheme works because the distance depends only on the masses to the one-third power. The method will not work if the stars are not dwarfs, as the mass–luminosity law does not apply to them. Since you have the spectra, you can get distances from spectroscopic parallax, but these can be subject to significant error, especially if you cannot nail down the exact class but know enough only to be sure that the stars are dwarfs.

More powerful is the combination of visual and spectroscopic orbits [8.3]. In older times, the use of the two approaches for one system was rare, because to have high speeds and a good spectroscopic orbit, the stars had to be close to each other, making visual observations difficult or impossible. With the advent of interferometric techniques, more and more systems are available. For a double-lined binary, the spectroscopic orbit yields the semimajor axis in AU times the sine of the orbital inclination to the plane of the sky, i, or ($a \sin i$). The visual orbit provides the semimajor axis in seconds of arc, a'', but also gives up the inclination, and thereby tells us a (AU); $a'' = a$ (AU)$/d$ (pc), which is solved for d.

A more common scenario is to have an eclipsing binary [8.4] for which you have a complete spectroscopic orbit. As before, the spectroscopic orbit gives ($a \sin i$). The light curve solution gives the ratio of the radii of the two stars to the semimajor axis, but also gives the inclination, i, which in turn yields a (AU) and therefore the physical stellar radii. Ideally, spectroscopy also gives the classes and temperatures of each of the pair, which with radius gives luminosity. Apply the bolometric correction [3.3] in reverse and you get absolute magnitudes, and from the magnitude equation [3.2], the distance. For such a binary in the Galaxy, distance does not much matter, as we already have enough information to solve for masses (a, P, the mass sum and the mass ratio). Put the binary in an external galaxy, however, notably in the Large Magellanic Cloud, use sophisticated modelling of the combined spectrum to get good solutions for the temperatures, and calibrate the Cepheid P–L relation.

There is more – measures that apply to specific cases – and they will be taken up as needed. For now, we can be comfortable in knowing to good accuracy how far away most things are, from the Moon to the farthest galaxy we can observe.

Chapter Five

The Galaxy in motion

Fig. 5.1 **Andromeda and its companions** Our great neighbor, a galaxy much like ours, the Andromeda spiral (M31), is only 725 000pc away and is easily visible to the naked eye. Though the flat disk of the galaxy is tilted at a 70° angle to the line of sight, the dusty spiral arms are readily visible, their blue color contrasting smartly with the reddish central bulge. Just below M31's center, and up and to the right, are two elliptical companions (M32 and NGC 205) that have no spiral arms. (Courtesy of Bill Schoening, Vanessa Harvey/REU Program/NOAO/AURA/NSF.)

We live in a Galaxy of stars (Fig. 5.1). Within it, everything moves. If stars were stationary relative to one another, gravity would draw them together into a lump and none of us could be. Our Galaxy exists as it is because all of its stars are orbiting the Galactic center on paths of various shapes that depend on their places of residence. Stars also constantly affect one another through their gravitational interactions, so no orbit is fixed, but undergoes continuous change. As a result, the Sun and stars slowly alter their relative positions, some moving past us, others left behind.

Nearly all the naked eye stars, those that make our constellations [1.1, 1.2], are local. Though the constellations appear invariable and unchanging throughout our lifetimes, they are all slowly disintegrating, while new ones are taking shape and forming patterns for the people of the far-distant future to admire and name. If we could speed up time, we would see today's first magnitude stars disappear into the distance, while others, now faint, would brighten to glory. Within a few tens of millions of years, hardly any of the naked eye stars you see now will be visible at all, while a host of those not even named in our catalogues will take their places.

5.1 The Galaxy

Fig. 5.2 **A spiral galaxy** The spiral galaxy M109 (15 megaparsecs away) presents itself more face-on than M31, allowing a fine view of its spiral arms. The center is pierced by a short "bar," rendering it a "barred spiral." (Courtesy of NOAO/AURA/NSF.)

Vast numbers of galaxies of different kinds dot the sky (Box 5.1). A special one, ours, bears no real name. Most commonly called simply "the Galaxy" (capital "G"), it is sometimes referred to as the "Milky Way Galaxy" from its most obvious manifestation. Before setting our Galaxy in motion, we need a plan, a map of sorts, to place measurements and deductions into context. So look first at a summary of a century of research and at the big picture, to which we shall return at the end.

Our Galaxy (Fig. 5.3) is a collection of 200 to 400 billion stars. No one is quite sure of the number, as we do not know the lower limit to the masses of star production or how many of the smallest, faintest stars there actually are. The vast majority of our Galaxy's stars are concentrated into a thin disk that slowly rotates and into which are embedded

Box 5.1 Galaxies

The stars of the Universe are gravitationally grouped together into *galaxies*. Though these systems come in a great variety of shapes and sizes, two broad forms stand out. *Spirals* (Figs. 5.1 and 5.2; see also Figs. 3.13 and 4.13) consist of thin but very populous disks surrounded by sparsely populated spheroidal halos. Within the disks are set beautiful, winding spiral arms that are the seats of star formation [Chapter 11] and that contain bright, young, massive blue stars, open clusters [9.2], and both dark and illuminated clouds of interstellar matter [11.1]. The halos are older than the disks and (in the standard view derived from our own Galaxy) contain the more ancient but massive globular clusters [9.4]. As stars age, they eject newly formed heavy elements into the star-forming interstellar matter, so older halos have a lower metal content than do the disks.

Ellipticals, on the other hand, have no disks (see Fig. 5.1), are reddish, contain no significant interstellar matter, no star formation, and no young massive stars. They are instead populated by low-mass stars, red giants, and globular clusters, and are rather like galaxy halos, though much more densely populated and higher in metal content. For reasons still not clear, ellipticals used their interstellar stuff quickly, whereas spirals have retained a great deal that settled into the rotating disks.

Spirals, which are seen in a variety of arm windings, are all relatively massive, whereas ellipticals range from giants far larger than our own to tiny dwarfs not much bigger than massive globular clusters. At the lower end of the galaxy scale are also numerous irregular systems that contain significant interstellar matter and active star formation (see Fig. 4.11). A significant fraction of galaxies are bound together into clusters, which combine into "walls" of systems, and these finally into the Universe of trillions of observable systems (see Fig. 5.17).

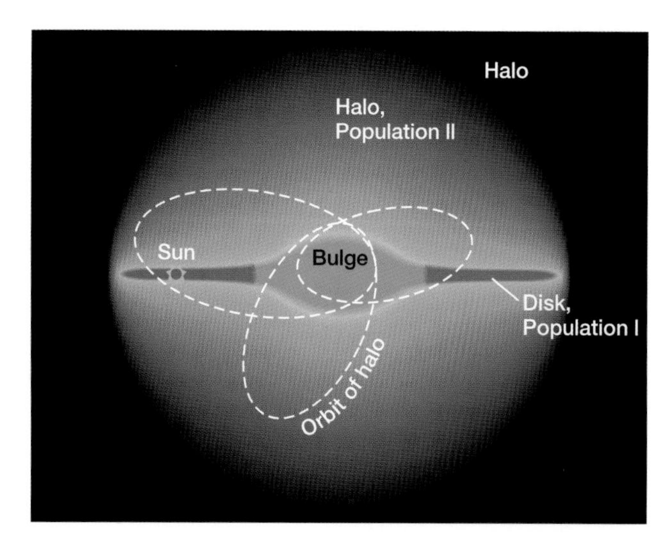

Fig. 5.3 **Our Galaxy edge-on** An edge-on plan of our Galaxy shows its "Population I" disk, the bulge, and the surrounding Population II halo. The disk's stars are edge-on near-circles, while the stars of the halo have highly elliptical orbits. (Kaler, J.B. (1994). *Astronomy!* New York: HarperCollins, reprinted by permission of Pearson Education, Inc.)

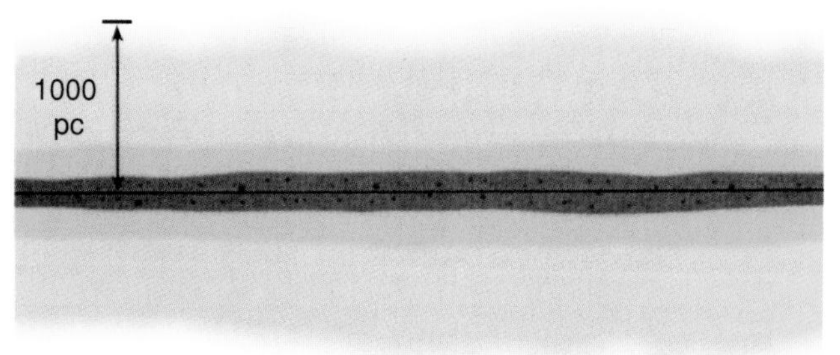

Fig. 5.4 **The Galactic disk** The disk is complex and multilayered. Like the Galaxy's "edge," there are no sharp lines of demarcation. A thin black layer of dust that contains the youngest of stars runs down the center. A thicker disk (yellow) of somewhat older and lower metal stars (like the Sun) spreads out from that. Thicker yet is a broader layer some two kiloparsecs deep that is more a transition from the disk into the halo. (Kaler, J.B. (1994). *Astronomy!* New York: HarperCollins, reprinted by permission of Pearson Education, Inc.)

spiral arms, which we can map. The disk's canonical diameter, that which incorporates 90 percent of the stars, is around 25 000 pc, rather 25 kiloparsecs (kpc), rendering the radius 12.5kpc. Unlike simplistic diagrams, the Galaxy has no real "edge." Rather like the Earth's atmosphere, its stars and the dusty gases of interstellar space just gradually fade away, and can be seen as far out as 25kpc from the center, or even more. The Sun is set close to the midline of the disk about 8kpc from the center (Fig. 5.3). Seen from a distance, the Galaxy might look something like M109 in Fig. 5.2 and (seen edge-on) like NGC 4565 in Fig. 3.13.

Since the stars are so highly concentrated to a disk, their combined light makes a bright band across the sky, a wide great circle called from ancient times the *Milky Way* (see Fig. 1.1), the centerline of which defines the Galactic equator [2.3.4]. Though we cannot ever get a true image of our Galaxy as seen from the outside, we are far enough out that in gazing off toward the Galactic center in Sagittarius it almost looks that way. Opposite the center, toward the *anticenter* (Appendix 5), where we look through the disk to the outside, the Milky Way dims through Taurus and Auriga.

Surrounding the disk is a huge, ellipsoidal, near-empty (but profoundly important) halo that contains but one or two percent of the Galaxy's stars (see Fig. 5.3). Pervading the halo and extending to unknown, but far greater, distances is the Galaxy's "dark matter halo," made of gravitating matter of unknown form. Both the halo and disk thicken toward the center and merge into a *Galactic bulge* a couple of kiloparsecs in radius (half its mass is contained within 1kpc) that has some of the characteristics of both, yet has its own unique properties. At the center, the stars densely encroach on one another and madly orbit an immense central body that is most likely a black hole [14.3.3] of three million solar masses.

The halo, disk, and bulge are differentiated in a variety of ways, the first of them discovered by the German-American astronomer Walter Baade (1893–1960) in our local analogue, the Andromeda Galaxy, M31 (see Fig. 5.1). He found the outlying regions of this galaxy to be more blue in color, and the central regions to be more red, which he respectively called *Population I* and *Population II*. While the terms are not clearly defined and their use is often challenged, they hang on and have been extended to represent the general disks of spiral galaxies and their surrounding halos. Intermediate (and contradictory) terms abound.

The disk is complex, and consists of various overlapping and ill-defined layers (Fig. 5.4). The very thinnest portion of the disk, "extreme Population I" (or the "young thin disk"), extends a mere 100 or so parsecs above and below the centerline, contains most of the Galaxy's gas and dust [11.1], young and massive O and B stars (born in the gas and dust) [11.2.4], a relatively high metal content, and evolved massive stars and their ultimate evolutionary state, core-collapse supernovae [14.2.2]. A thicker Population I disk (sometimes the "old thin disk") that extends roughly 300pc up and down contains the older and more stable parts of the stellar population, stars like the Sun and those of the general main sequence [12.4], and most of the open clusters [9.2]. Older and lower-metal stars spread out 1000pc or more into a thicker plane as they merge with the stars of the inner halo. Though often said to be "young" Population I and the Galactic disk may stretch back to as much as 10 billion years. And though the Sun actually happens to lie within the thin disk, it is conceptually part of the next-thicker disk. Face on, the disk displays glorious spiral arms, created by gravitational disturbances that set off "density waves" that temporarily pile stars into localized "traffic jams," which the Galaxy's rotation in turn sweeps into the graceful arms (see Fig. 5.2).

The halo, notably older than any segment of the disk, embraces the massive globular clusters [9.4] epitomized by

Omega Centauri [Fig. 9.13], 47 Tucanae [Fig. 9.12], and Messier 13 [Fig. 3.3], and which extend back to 12 billion years and more, to the creation of the Galaxy itself. Here we find low metals that range from a tenth of the solar content to below 1/10 000 solar. Since we are spatially also part of the halo, low-metal halo stars show up locally as "subdwarfs" [6.4.2; 7.3.3]. The halo is layered as well, with an inner flattened portion, an "intermediate Population II" whose metal content is higher than that found in the outer halo. The metal content of the inner disk also varies radially, and climbs from perhaps a tenth solar at its outer edges to nearly double solar as it runs into the bulge, which seems to contain a mixture of both the high metals of the disk and of the lower metals of the halo.

The different stellar populations are reflected in different dynamics. Disk stars tend strongly to move in roughly circular orbits in the clockwise direction as viewed looking "downward" from the north galactic pole. As a result, the whole disk can be said to be rotating. At the Sun's distance of 8000 parsecs and orbital speed of 220 km s^{-1}, the journey around takes about 250 million years. Population II orbits contrast smartly (see Fig. 5.3), and at the extreme are highly elliptical [Box 8.2]. The orbits are randomized such that the halo is not in the disk's high state of gross rotation. Stars from the halo crash through the disk, and appear as "high velocity stars" (that chemically appear as our local low-metal subdwarfs) as they and the Sun meet while going in different directions. This is our home.

5.2 Relative motion

If we know how stars of different kinds and locations circulate within the Galaxy, we can calculate its mass and be on the road to learning about its origin, history, and content. The measurements start right here, with local stars that are currently visiting the solar neighborhood. The fundamental observations are broken into two parts: angular movements across the line of sight and velocities along it. Their combination yields a three-dimensional picture of both stellar motions and the motion of the Sun through the surrounding swarm.

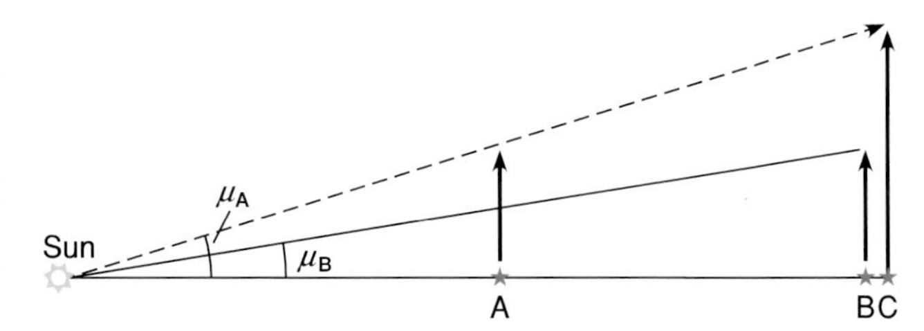

Fig. 5.6 **Stellar motion** Star A moves across the line of sight with velocity v_t km s^{-1}. Double the distance to star B and the proper motion (μ) is cut in half; double B's speed to that of star C and the proper motion doubles. (Kaler, J.B. (1994). *Astronomy!* New York: HarperCollins, reprinted by permission of Pearson Education, Inc.)

5.2.1 Proper motion

Over the years, the stars move across the line of sight relative to the infinite background, some quickly, others with great leisure; but they all do (Figs. 5.5 and 4.5). The sought-for datum is the *proper motion* (μ), the annual angular rate of motion in seconds of arc per year (μ ″yr^{-1}) and direction of motion against the grid of equatorial coordinates [2.3.3] after correction for general precession [2.4], parallax [4.2.1], and binary orbital motion [8.2.3]. Proper motion derives directly from multiple measurements of coordinate position, and is intimately related to coordinate determination. It can also be found through multiple measures of position made against background reference stars in much the same way as parallaxes are determined [4.2.2], for which proper motions are necessary by-products. Such relative proper motions need to be corrected to an absolute system based on fundamental stars for which absolute coordinates are available.

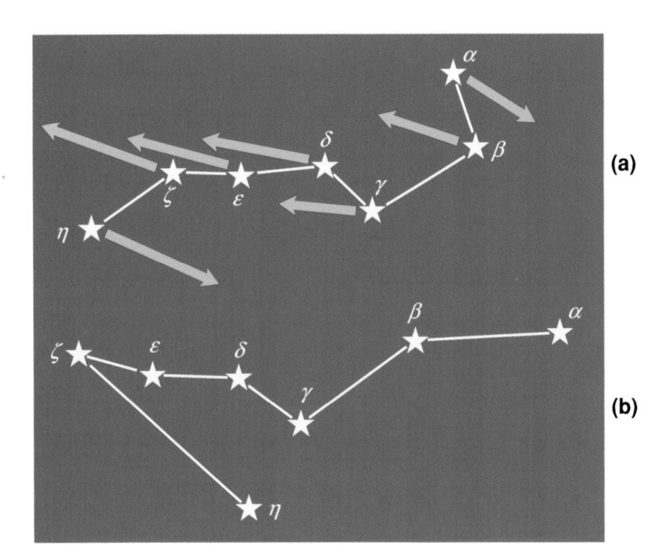

Fig. 5.5 **The changing Dipper** At the top, the core of Ursa Major, the "Big Dipper" or "Plough," is breaking apart. The arrows show the directions in which the stars are moving. In a quarter-million years the Dipper will look like the figure below, and we will have to give it a new name. (Kaler, J.B. (1994). *Astronomy!* New York: HarperCollins, reprinted by permission of Pearson Education, Inc.)

Ignore any motion toward or away from us. In Fig. 5.6, a star moves only across the line of sight at a *tangential velocity* v_t km s^{-1}. The proper motion is related to v_t through distance. When you drive a country road, the fence posts whip by quickly with high angular drift, but the cows in the distance seem to move only leisurely. Proper motion and distance are inversely related; for a given tangential velocity, double the distance and the proper motion is cut in half. If, however, you double the speed across the line of sight, the angular motion doubles too: that is, μ is directly proportional to v_t. Taking the various units of measure into account (Box 5.2) gives the formula that connects the three quantities,

$$v_t = 4.74\mu d\,,$$

where velocity (v_t) is expressed in km s^{-1}, proper motion (μ) in seconds of arc per year, and distance (d) in parsecs.

The ability to measure proper motions depends on the sensitivity of the equipment and time. The longer the period of observation, the farther the star will move, and the greater the resulting precision; indeed, the greater our ability to detect any motion at all. By far the best data set is from the Hipparcos mission [4.2.3], where the errors are in fractions of a milliarcsecond.

5.2.2 Radial velocity and the Doppler effect

Tangential velocity is one of two components of stellar motion. The star will also be moving toward or away from us at the same time, that is, it will have a velocity along the line of sight as well, the *radial velocity*, v_r. Unlike tangential velocity, radial velocity does not depend on distance, as we make

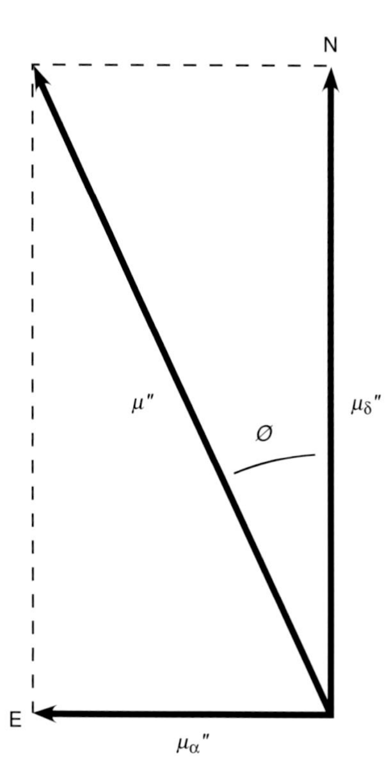

Fig. 5.7 **Components of proper motion** A star moves through angle μ''_δ in proper motion and in μ''_α in right ascension. The total proper motion is $\mu'' = \sqrt{(\mu''^2_\delta + \mu^2_\alpha)}$, and the direction or position angle ϕ is given through $\tan\phi = \mu''_\alpha/\mu''_\delta$.

direct measure by the Doppler shift of the spectrum lines. The Doppler concept is central to all of astronomy. A wave with wavelength λ and frequency ν moves past you. If you run into the wave, or if the source of the wave is moving toward you, you get hit with the wave crests faster than you would were you both at rest with respect to each other. The frequency therefore appears higher and the wavelength shorter. On the other hand, if the body is moving away from you, or you from the body, the frequency appears to be lower and the wavelength greater. The most common Doppler experience is in audio. As a car approaches you, the pitch of its sound is higher, and when it passes and recedes, it drops.

It does not matter what kind the wave; water, sound, and light, all behave the same. As long as the speeds are ordinary – that is, not approaching the speed of light so that relativity must be accounted for – the apparent shift in wavelength is directly proportional to the speed. As expressed by the Doppler formula,

$$(\lambda_{\text{observed}} - \lambda_{\text{true}})/\lambda_{\text{true}} = v_r/c,$$

where $\lambda_{\text{observed}}$ is the measured wavelength, λ_{true} is the wavelength that would be observed at rest, and c is the speed of light, whence

$$v_r = c(\lambda_{\text{observed}} - \lambda_{\text{true}})/\lambda_{\text{true}}.$$

The speed of light is so large that stars are not moving anywhere nearly fast enough to change their actual colors. Instead, we use tiny shifts in stellar absorption (or nebular emission) lines [7.1.4], wherein the reference, λ_{true}, is that observed in the lab. Given the great speed of light, the shifts are generally tiny. At a typical radial speed of 20km s^{-1}, a yellow–green spectrum line at 5000 Å will be shifted by

Box 5.2 Proper motion, distance, and tangential velocity

Proper motions are commonly measured independently in right ascension in seconds of time per year (μ_α s yr^{-1}) and in declination in seconds of arc per year (μ_δ ″yr^{-1}). Since the units of the two are different, they cannot be directly combined. At the celestial equator, each second of time corresponds to 15 seconds of arc. Hour circles converge to the north and south celestial pole. As a result, movement in right ascension near the pole as expressed in seconds of time corresponds to a smaller real angular shift in seconds of arc than will the same change near the equator. At the equator, there are 15 seconds of arc per second of time [2.3.2], so to find μ_α in seconds of arc per year (μ_α ″yr^{-1}), multiply μ_α s yr^{-1} both by 15 and by the cosine of the declination, or μ_α ″yr^{-1} = 15 (μ_α s yr^{-1}) cos δ. Hipparcos results are expressed directly in seconds of arc per year.

Since proper motions in right ascension and declination are perpendicular to each other, apply the Pythagorean theorem, such that (μ ″yr^{-1})^2 = (μ_α ″yr^{-1})^2 + (μ_δ ″yr^{-1})^2, or μ'' yr = $\sqrt{[(\mu_\alpha\,''\text{yr}^{-1})^2 + (\mu_\delta\,''\text{yr}^{-1})^2]}$. The direction of motion (the *position angle*, ϕ, as measured from north through east) is found from the inverse tangent of the ratio μ_α/μ_δ, or $\tan^{-1}\mu_\alpha/\mu_\delta$, where both are expressed in seconds of arc per year (Fig. 5.7).

The proper motion of a star is directly proportional to tangential velocity, and inversely proportional to the distance, or, turning the relation around, tangential velocity is directly proportional to both distance and proper motion. For use, the three must be connected by a constant of proportionality. In Fig. 5.6, star A moves through space at speed v_t km s^{-1}. Given 3.156 × 10^7 seconds per year, in one year the star will cover a distance of 3.156 × 10^7 v_t km. Since the angles are small and the triangle long and thin (so that the arrow that represents v_t resembles an arc of a circle), the angle μ ″yr^{-1} = 206 265 (3.156 × 10^7 v_t)/d, where distances are both expressed in kilometers [Box 4.1]. Since d (km) = 3.086 × 10^{13} d (pc), μ ″yr^{-1} = [(2.06265 × 10^5) (3.156 × 10^7)/(3.086 × 10^{13})] v_t (km s^{-1}) / d (pc) = 0.21094 v_t (km s^{-1}) / d (pc) = v_t (km s^{-1})/4.74 d (pc). We thus arrive at the standard proper motion formula for tangential velocity,

$$v_t = 4.74\mu d,$$

where the v_t, μ, and d are respectively in km s^{-1}, arcseconds per year, and parsecs.

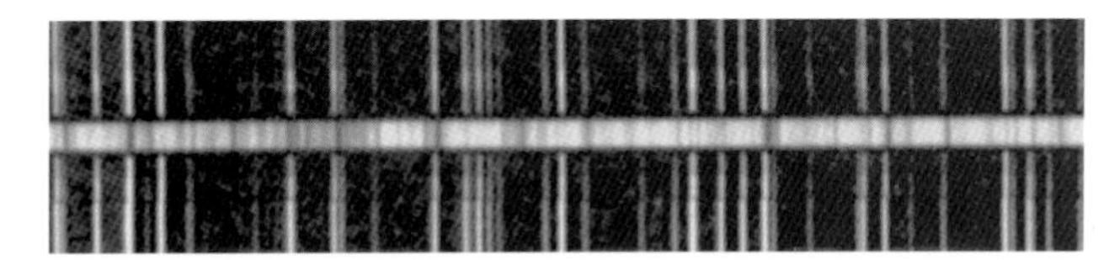

Fig. 5.8 **Radial velocity measure** The spectrum of this cool star is flanked by emissions from iron that calibrate the wavelength scale and allow the exact absorption line wavelengths to be measured. The result is the degree of Doppler shift and the star's radial velocity. Electronic calibration is similar, just less visible. (Kitt Peak National Observatory, courtesy of E.C. Olson.)

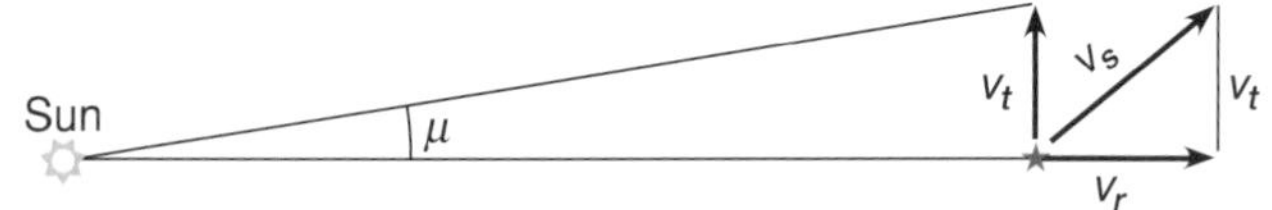

Fig. 5.9 **Space Velocity** A star speeds relative to the Sun along its space velocity vector v_s, which is broken into perpendicular components of radial velocity, v_r, and tangential velocity, v_t. From the Pythagorean theorem, v_s is the square root of the sum of the squares of v_t and v_r. (Kaler, J.B. (1994). *Astronomy!* New York: HarperCollins, reprinted by permission of Pearson Education, Inc.)

only a third of an angstrom unit. If the star is moving away from us, $\lambda_{observed}$ is larger than λ_{true}, and v_r is positive; if the star is moving at us, $\lambda_{observed}$ is the smaller of the two, and v_r is negative.

To measure radial velocity, the astronomer takes a spectrogram of a star (or nebula) along with that of a laboratory wavelength calibrator built into the spectrograph. Measurements of the positions of the star's absorption lines (or a nebula's emission) against the calibrator give the wavelength shift and the radial velocity. Older photographic spectrograms are flanked by spectra of iron from an electric arc, or by the spectra of noble gases – helium, neon, argon – from a discharge tube (Fig. 5.8). Electronic spectra (CCD spectrograms for example [3.1.3]) are calibrated by taking separate spectra of noble gases that then calibrate the pixel array.

Application of the Doppler formula from the raw data gives the radial velocity relative to the observer. But the observer is moving too, riding an Earth that orbits the Sun with a speed of 30 km s^{-1}, which is comparable to the stellar radial velocities. Could we speed up time, we would watch the spectrum lines of a star on the ecliptic [2.3.3] swing back and forth through a good fraction of an angstrom unit. Turned around, the annual variations of stellar radial velocities are (along with parallaxes) fundamental proofs that the Earth really does go around the Sun and that Copernicus was right. Even the variation in speed caused by the eccentricity of the Earth's orbit and speed of the observatory around the Earth's rotational axis (which depends on latitude) must be accounted for.

Radial velocities are of two kinds, relative and absolute. The first involves the changing velocity of a particular star (a spectroscopic binary for example). They are vastly the more accurate, as it is only necessary to watch the absorption lines shift their relative positions (after correction for terrestrial motions). That is, we look only at change. The precision of modern spectroscopy has improved the accuracy of such relative velocities to below 3 meters per second. The techniques are now so good that variations superimposed on the gross absolute radial velocities of the stars can detect sub-Jupiter-sized planets [11.4.3]. Absolute velocities require knowledge of the actual central wavelengths of the absorptions, and involve numerous other corrections that make for considerably lower accuracy.

5.2.3 Space motions

The point of the observations is not so much in the individual movements of tangential and radial velocity, but to find the combination of the two, the space velocity vector, v_s (Fig. 5.9). A star moves through space relative to the Sun. Its velocity vector is aligned both across the line of sight (the tangential velocity, v_t) and along the line of sight (the radial velocity, v_r). The three quantities are related to one another through the Pythagorean theorem:

$$v_s^2 = v_r^2 + v_t^2, \text{ or}$$
$$v_s = \sqrt{(v_r^2 + v_t^2)},$$

where $v_t = 4.74\mu d$. The direction of the space velocity vector v_s is determined from the position angle of the v_t vector and from the relative sizes of v_t and v_r. With these simple formulae, we open the window and look out on the company of stars, on the Galaxy.

5.3 Solar motion and Galactic rotation

We live in a crowd of stars, none of which are on the same Galactic orbits as we, or have the same motions. How do we relate to our surroundings? Where are we heading? The topic is traditionally treated in two parts, our motion relative to the local stars (the *solar motion*) and our large-scale motion relative to the Galaxy as a whole, that is, our *orbital* motion. Combination of the two gives a fundamental reference point in the grand rotation scheme of the Galaxy.

5.3.1 Solar motion

Consider only Population I, the stars of the disk that share the properties of the Sun. The few local Population II stars that are just crossing the disk from the halo will be ignored. The solar motion, the Sun's own space velocity vector, can be found from the observed proper motions and radial velocities of the stars that immediately surround us. Assume the Sun to be moving through a crowd of stationary stars. All we will see is reflexive motion. Stars in the direction of the solar motion (both in the forward and rearward directions) can have no proper motions. In the direction of solar motion, stars come right at us and the radial velocities will be maximum negative, while in the opposite direction, v_r will be maximum positive. At right angles to the solar motion, we see the reverse. Since we are traveling perpendicular to the directions to these stars, there will be no radial velocities, but instead the proper motions will be maximized. The situation is much the same as for the discussion of secular parallax [4.3.3; Fig. 4.8].

By locating coordinates in the sky at which the radial velocities are maximized and the proper motions minimized to zero (or by finding the circle of maximum proper motions), we can quickly find the direction of solar motion, or the *Apex of the Sun's Way*. The direction opposite is the *Antapex*. From the position and value of the maximum negative and positive radial velocities, we can find not only the Apex but the speed of the Sun. Speeds can also be found from the maximum tangential velocities, though we must first have the distances to the stars so that the reflexive tangential motions can be calculated. The solar speed will then just be the reverse of the stellar tangential velocity.

The stars, however, are not stationary, but have their own motions that are "peculiar" [4.3.3] to them. This problem is overcome by making the reasonable assumption that relative stellar motions are common to all types and that they are random, that there are just as many stars moving in one direction as there are in another. We therefore get the same result by calculating statistical averages of their motions. Toward

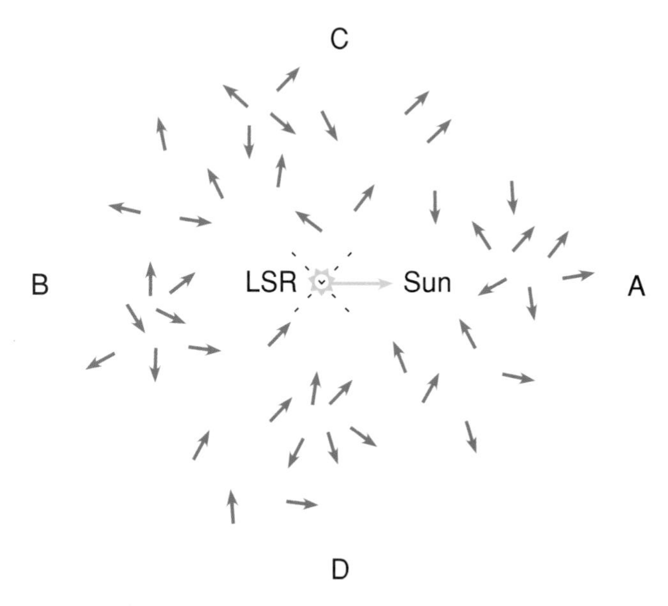

Fig. 5.10 **Stellar and solar motion** The local stars move randomly relative to the Sun. The average radial velocity will be maximum negative toward point A, the Apex of the Sun's Way. In the opposite direction, toward B, the average radial velocity will be maximum positive. The average proper motion toward A and B will be zero. Toward C and D, however, the proper motions will be maximized in the direction opposite the solar motion, while the average radial velocity will be zero. (Kaler, J.B. (1994). *Astronomy!* New York: HarperCollins, reprinted by permission of Pearson Education, Inc.)

point A in Fig. 5.10, some stars are indeed moving across the line of sight. But since there are just as many going "up" in the figure as going "down," their peculiar proper motions (and tangential velocities) average to zero. Conversely, toward point A the average of the radial velocities will be maximum negative, toward B maximum positive, and toward C and D, it will be zero.

To find the solar motion vector (both speed and direction), we use large numbers of stars and search for the direction and values of the maximum positive and negative mean radial velocities, the maximum and minimum mean proper motions (which by themselves give only the solar direction), the maximum and minimum mean tangential velocities (which give the complete solar motion, speed and direction), or best, we use the collection of space velocities. The last two of these have been the most difficult because of the lack of good parallaxes; the problem was finally resolved by Hipparcos [4.2.3]. In practice, the calculations are complex, since all stars are used at all directional angles from the Apex, for which we take into account the projections of contribution of solar motion onto their proper motion and radial

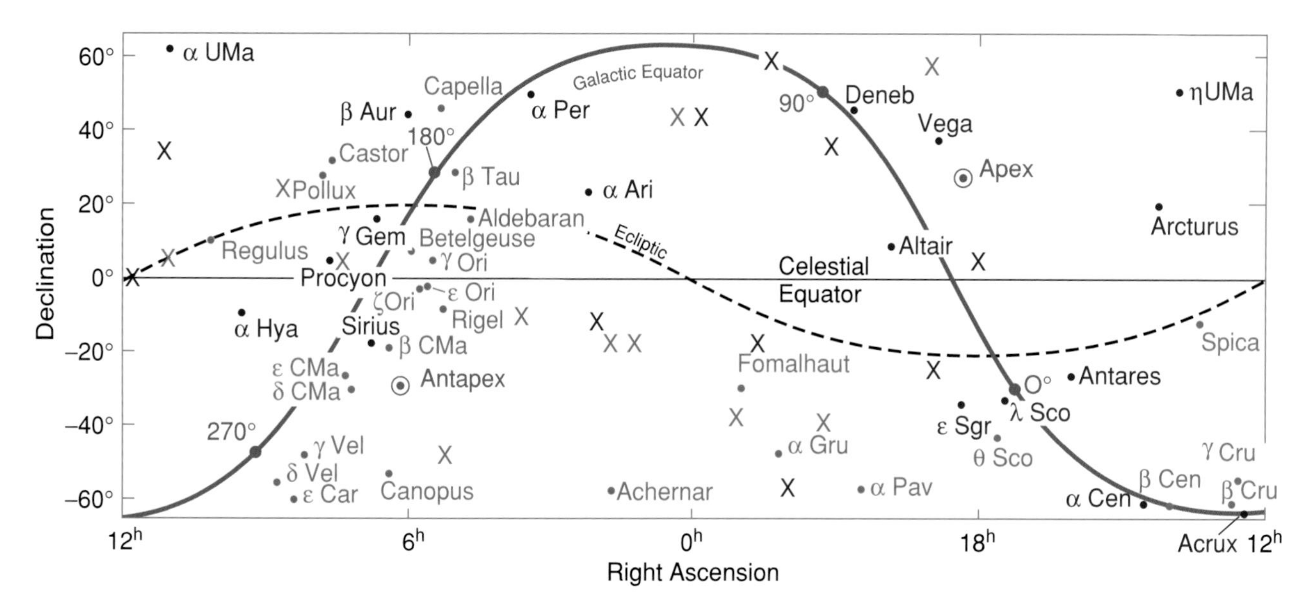

Fig. 5.11 **Radial velocities and the Apex** The brightest stars from Appendix 2 are plotted on a map of the sky, along with their names. The closest stars from Appendix 3 that do not overlap those from Appendix 2 are plotted without names. Receding stars with positive radial velocities are colored red, approaching stars with negative velocities, blue. While there is significant scatter, the trend is obvious, with approaching stars clumping about the Apex (defined by the standard solar motion), the receding stars gathering more around the Antapex.

velocity vectors (much as in Fig. 4.8). Ideally, the result is the same as if the stars were indeed stationary.

From the class A, F, and G stars listed in catalogues of radial velocities, the Sun is moving toward right ascension $\alpha = 18^h\,04^m$ and declination $\delta = +30°$ (equinox 2000 [2.4]), in eastern Hercules about 10° to the southwest of Vega (Galactic coordinates $l = 56°$, $b = 23°$) at a speed of $V_0 = 19.5\,\mathrm{km\,s^{-1}}$, which defines the *standard solar motion*. The determination of the vector of solar motion in turn defines a *local standard of rest* (the LSR), which should ideally be in circular orbit about the Galaxy at a constant speed, and against which all the stars including the Sun should have random local motions.

The general effect is seen in the simplest of data, and is illustrated in Fig. 5.11, in which the sets of the brightest and nearest stars in Appendix 2 and 3 are plotted by right ascension and declination, and are color-coded as to whether their radial velocities are negative (approach, blue) or positive (recession, red). Although there are numerous individual exceptions, the stars of the northern ecliptic hemisphere [2.3.3] are on the average approaching us, while those in the southern ecliptic hemisphere are moving away. The blue points are clustered around the Apex, the red ones about the Antapex. The Apex is only about 35° south of the north ecliptic pole, explaining the rough and coincidental alignment with the ecliptic. Both it and the Antapex are close to the Galactic equator. The Sun's motion is more or less in the Galactic plane, but is also directed a bit to the north of it.

Reality intrudes into this simple picture. The Galaxy is complex, with many different kinds of stars with different ages and distributions. The assumption of common random motion cannot be entirely correct, as different stellar types will have systematic motions relative to each other. For example, O and B stars [6.2.1] commonly belong to OB associations [9.3] that have bulk velocities of their own, so the motions of their members are related and consequently different from those of G dwarfs. Class A dwarfs are younger on the whole than M dwarfs, and their average Galactic locations and Galactic orbits will be somewhat different as well.

As a result, the determination of solar motion depends on the particular selection of stars used in the calculation. Use a different set, and out comes a different solar velocity vector and Apex. The one used depends on the particular research problem. A calculation with regard to local stars, those within only about 100pc, gives the *basic solar motion* of $V_0 = 15.4\,\mathrm{km\,s^{-1}}$, with an Apex at right ascension $17^h\,56^m$ and declination $+26°$ ($l = 51°$, $b = +23°$), somewhat to the southwest of that for the standard solar motion [2.3.3, 2.3.4]. The difference between standard and basic solar motions roughly parallels the use of the tables of the brightest and nearest stars in Fig. 5.11.

To see where the Sun is heading, look at the three components of its local motion as defined by Galactic directions, U, V, and W (Fig. 5.12). U is directed toward the Galactic center at Galactic longitude $l = 0°$, latitude $b = 0°$ [2.3.4], V is the component in the Galactic plane toward the direction in

Definition	$U_\odot$	$V_\odot$	$W_\odot$	V_0	α (2000)	δ (2000)	l	b
Standard	10.4	14.8	7.3	19.5	$18^h\,04^m$	+30°	56°	23°
Basic	9	11	6	15.4	$17^h\,56^m$	+26°	51°	23°
Dynamical[b]	9	12	7	16.6	$17^h\,48^m$	+28°	53°	25°
Hipparcos Dynamical	10.0	5.25	7.17	13.4	$16^h\,46^m$	+10°	28°	32°

[a] See Sections 2.3.3 and 2.3.5 for discussion of coordinates.
[b] Pre-Hipparcos.

Table 5.1 ***Solar motion***[a]

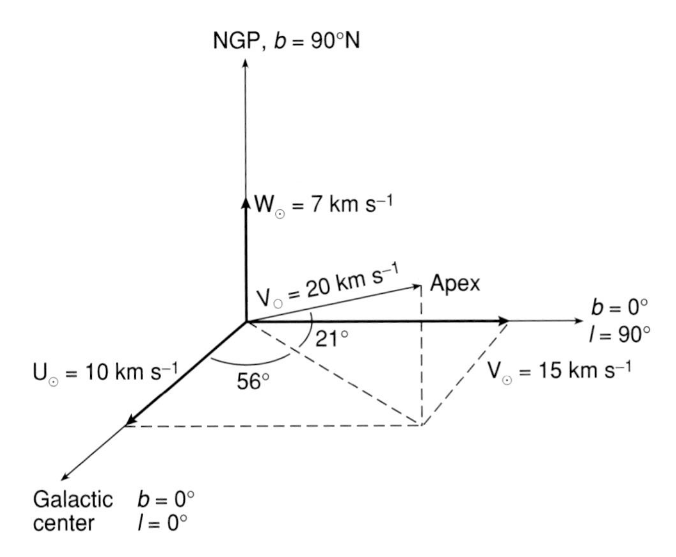

Fig. 5.12 **Solar direction** Relative to Galactic coordinates, the standard solar motion shows the Sun to be moving in the direction of Galactic rotation at $l = 90°$, but also a bit northward through the Galactic plane and significantly inward. Hipparcos observations reduce the $V_\odot$ vector to 7km s^{-1} and V_0 to 13 km s^{-1}.

which the Galaxy is rotating ($l = 90°$, $b = 0°$), and W is the perpendicular component toward latitude 90° N. The values of U, V, and W are taken to be positive in these directions. Once we know the position of the Apex and the solar velocity V_0, we can find the values of U, V, and W for the Sun, $U_\odot$, $V_\odot$ and $W_\odot$. Or if we calculate $U_\odot$, $V_\odot$ and $W_\odot$ directly from stellar motions, we can find the Apex and V_0 from $V_0^2 = U_\odot^2$, $V_\odot^2$ and $W_\odot^2$ (Table 5.1). (The symbols are confusing. Do not confuse $V_\odot$ with V_0, the former a component, the latter the full velocity.)

The components of motion set out in Fig. 5.12 clearly show that the Sun's "standard motion" path is not directed along a circle around the Galactic center, but is pointed rather noticeably inward at an angle of 34° off the rotation direction toward $l = 56°$, and toward the north out of the plane of the Galaxy at an angle of 21°. Basic solar motion is rather similar. Corrections for non-randomness give yet different directions, as do modern Hipparcos measures (Box 5.3).

5.3.2 The solar orbit

The solar motion defines the Local Standard of Rest. "Local" is the key word, as the LSR is hardly at "rest." It and all the stars that define it are madly orbiting the Galactic center. Though it does not always work out quite that way because of the non-random stellar motions that serve to define it, we ideally take the LSR as having uniform circular motion at the solar distance from the Galaxy's core. What then is the speed? The problem is the same as that for determining the solar motion, except on a much larger scale (see Fig. 5.10). But we cannot use local stars, or any stars of the disk, as they are partaking in the general rotation of the Galaxy as well. Instead, we must look outside the Galaxy, or at least outside that part of it that is included in the general rotation of the disk.

Three principal candidates raise their hand for election: globular clusters in the halo [9.4], general halo field stars, and the galaxies of the Local Group (our small local cluster of galaxies, which includes M31 and three dozen or so others). Assuming random motion and a net motion of zero relative to the LSR, we look for the maximum average approach and recession velocities (after correction for the solar motion). The result is, as usual, somewhat problematic, as different kinds of objects give different results that range from under 200km s^{-1} to 300km s^{-1}. The International Astronomical Union (IAU) standard is 220km s^{-1}, which is taken as the circular rotation velocity at the distance of the Sun.

The Sun thus provides the absolute reference point for the Galaxy's *rotation curve*, the manner in which the ideal circular orbital revolution of the stars in the disk varies with distance outward from the center. The solar orbital speed (rather that of the LSR) is of little use unless we know the distance to the Galactic center, R_0. The technique is again to use objects that are outside the disk. The classic method adopts the system of globular clusters [9.4.2], whose distances are determined by main sequence fitting or from the apparent magnitudes of their horizontal branches. A plot of the cluster distances locates the Sun relative to the center of the distribution of the clusters, which is taken as identical to the Galactic nucleus. We therefore know how far away we are from a point we cannot optically observe. Any kind of halo object will do as long as we can determine the distances of its members. RR Lyrae stars [10.3.3], which are bright enough to see far away and also have relatively fixed absolute magnitudes, also work well. The major problem is in accounting for objects within the group that are too far, or too obscured by dust, to see. While there is considerable argument about the exact number, the Sun seems to lie somewhere between $R_0 = 7.5$ and 9kpc outward from the center, and probably closer to 8kpc. The radius gives a circumference of the "solar circle" of 50kpc, or (given 3.08×10^{13}km pc^{-1}) 1.6×10^{15}km. Dividing by the circular LSR velocity of 220km s^{-1} gives an orbital period of 7.5×10^{12}s, or (with 3.16×10^{7}s yr^{-1}) 240 million years. The orbital period of the Sun will be similar. The Sun is 4.5 billion years old. With such a period, we have traveled around nearly 20 times. The neighborliness of the local stars is clearly very temporary.

Since the solar velocity vector is directed inward relative to the local stars, the Sun is moving closer to the Galactic center and toward its *perigalacticum* point roughly half a kiloparsec inward (*apogalacticum* about half a kpc out). But though we speak readily of the elliptical orbit of the Sun, the true path significantly deviates from an ellipse [Box 8.2]. While an ellipse is a decent approximation of the instance, the gravitational forces of the Galaxy rotate the ellipse backwards, in the direction opposite the Galactic rotation, resulting in an orbit shaped rather like a rosette. Though we take 240 million years to orbit, we catch up with our perigalacticum (or apogalacticum) points every 170 million years, which is the *epicyclic period*. Our Galactic path is also slightly inclined, causing us to now be moving northward through the disk. As found from the asymmetrical distribution of Galactic objects, we are now about 15pc north of the centerline. However, the gravitational pull of the disk makes us oscillate back and forth through it with a period of only 62 million years, far less than the orbital period, hardly what we would expect for a proper ellipse.

Still, the stars of the disk all share basic more-or-less circular orbital paths, which keep the relative motions small. Halo stars, even the stars of the thicker disk, are another matter. Halo stars – low-metal subdwarfs [5.1, 6.4.2, 7.3.3] – and the halo globular clusters do not share in the general Galactic rotation, but are on long elliptical orbits. The differing paths give these low-metal stars their high relative velocities, which can reach over 200 kilometers per second as they speed past (see Fig. 5.3).

5.3.3 The rotation curve

To find the Galaxy's rotation curve we must use only young stars and other objects that we know partake of circular motion: O and B stars, Cepheids, interstellar matter. A distant star lies at point C in Fig. 5.13a. Measure its radial velocity and distance (using the Cepheid period–luminosity relation, spectroscopic parallax, etc.). Project the orbital velocity of the Sun onto the line that connects the two. From the Galactic longitude of the star we know all the angles of the small triangle at the Sun and can find the component of solar orbital motion that lies toward the star. The radial velocity of the star relative to the Sun gives the projection of the *star's* orbital velocity onto the connecting line. Since the angles of the small triangle at star C are known too, so is the rotation

Box 5.3 Apex saga continues

The complexity of the Galaxy reveals itself through the search for the Apex and the solar velocity vector. When the Local Standard of Rest, the LSR, is specifically defined as having uniform circular motion, the *V* component of solar motion presents a distinct problem. Even if some stellar orbits are fairly circular, all are really ellipses (and because of the constant gravitational interactions with other stars, not very good ones). The local stars are therefore on different parts of their elliptical orbits. Some are near their perigalacticum points (where they are closest to the Galactic center), others near their apogalacticum (farthest) points, and yet others are in between. The perigalacticum stars are coming from outside the solar orbit, while the apogalacticum stars are coming from inside it. Following Kepler's second law [8.2.3], those near their apogalactica will be going more slowly than the LSR and will fall behind, while those near perigalacticum will be going more quickly. If the Galaxy's stars were uniformly distributed, there would be as many perigalacticum as apogalacticum stars. But they are not uniformly distributed, and as a result, motions are quite distinctly not random.

Star density falls dramatically with distance from the Galactic center. The star density inside the solar orbit is much greater than outside it, so there are many more stars near their apogalactica than near their perigalactica. As a result, local stars on the average lag behind the circular velocity of the LSR in the *V* direction, a phenomenon called the *asymmetric drift*. The Sun therefore appears to be going too fast relative to these stars. Correction for asymmetric drift lowers $V_\odot$ and defines the *dynamical solar motion* (see Table 5.1).

Ignoring O and B stars, the cooler the spectral class, the greater value of the drift. But the later the spectral class, the more older stars it contains, and the greater the scatter in the velocities (as older stars have more chance to encounter others and move away from the central plane of the Galaxy). A graph of velocity against a measure of the scatter extended back to zero scatter gives the dynamical value for $V_\odot$ – or so we hope.

Here there is argument. The generally accepted total dynamical solar velocity is 16.6 km s^{-1} near right ascension $17^h 48^m$, declination +28°, for which $V_\odot = 12$ km s^{-1} (as expected, less than the standard value). Newer results from Hipparcos observations lower $V_\odot$ to 5.2 km s^{-1} (while slightly changing $U_\odot$ and $W_\odot$), giving a much lower solar velocity of 13.4km s^{-1} toward a very different apex (as witness Table 5.1). There seems to be no single "truth" to solar motion; as in so many things, it is all in the definition.

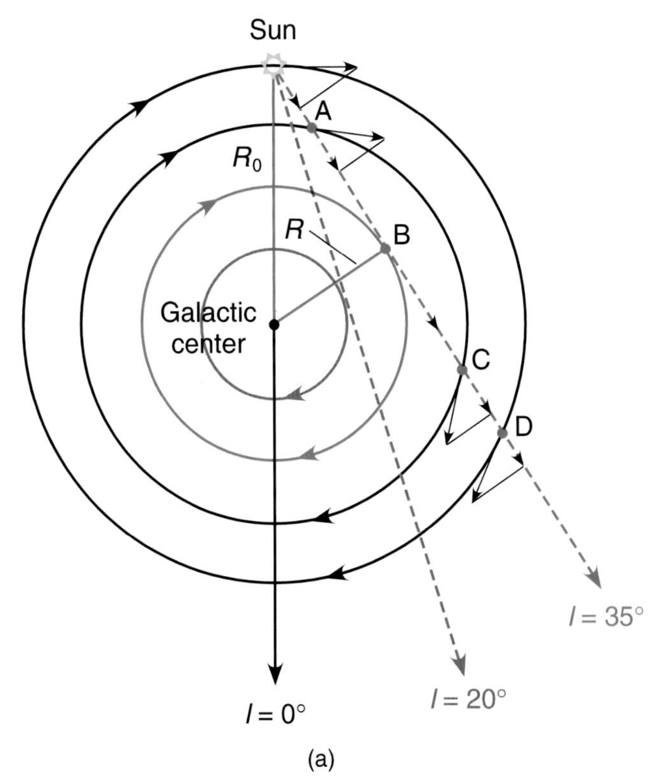

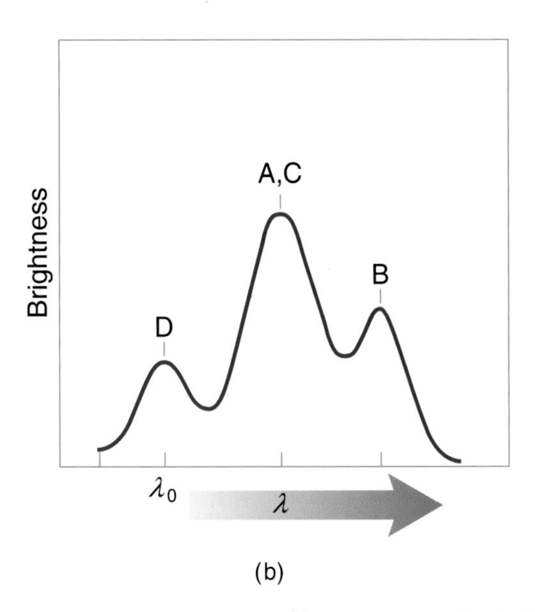

Fig. 5.13 **The rotating Galaxy** In (a), stars and interstellar clouds lie along any Galactic longitude. Their radial velocities are the circular velocities as projected onto the line of sight. Here, the $l = 35°$ line passes through spiral arms at points A, B, C, and D, each of which provides a Doppler-shifted 21-cm hydrogen emission line (b). Since B is a tangent point to a circular orbit, the radial velocity equals the circular velocity, and the velocity and Doppler shift are maximized. The distance of B is known from the right triangle, giving one point on the rotation curve. Different longitudes give different points. (Kaler, J.B. (1994). *Astronomy!* New York: HarperCollins, reprinted by permission of Pearson Education, Inc.)

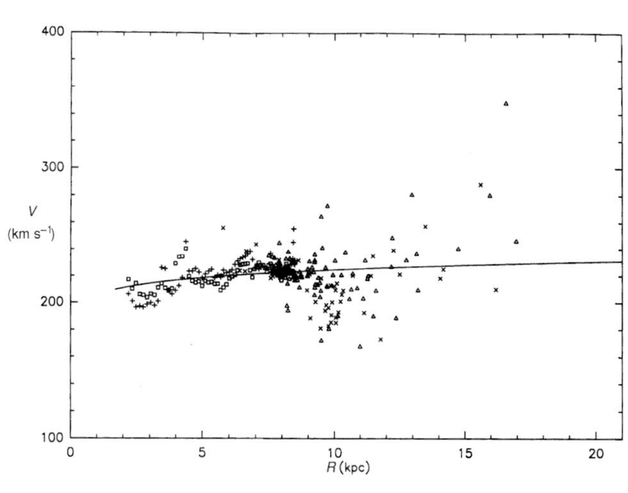

Fig. 5.14 **The Galaxy's rotation curve** The Galactic rotation curve, the circular orbital velocity (V) graphed against distance from the center (plotted with individual data points), on average is relatively "flat" with distance, exhibiting only a slow rise from 2kpc outward to nearly 20kpc. (Brand, J. and Blitz, L. (1993). *Astronomy and Astrophysics*, **275**, 67.)

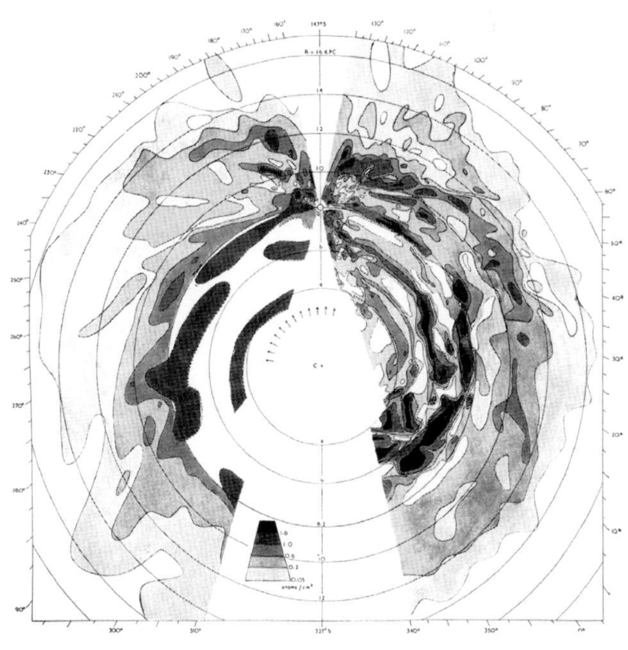

Fig. 5.15 **Spiral arms** The first detailed map of the Galaxy's spiral structure was made using the 21-cm line of neutral hydrogen. (Oort, J.H., Kerr, F.T., and Westerhout G. (1958). *Monthly Notices of the Royal Astronomical Society*, **118**, 379.)

velocity at star C's distance from the center (after the subtraction of the solar component). Application of the technique to stars A and D give the rotation velocity at other distances from the Galactic nucleus, as will observations at other Galactic longitudes. The result is a plot of rotation speed as a function of distance from the center, or the Galaxy's *rotation curve*.

What, however, about interstellar dust? There is no way that stars C and D can be seen, as there is too much of it in the way. The technique works only with nearby stars (one at point A) or for stars or nebulae outside the solar circle where the dust drops off and we can see for good distances. Deep inside the solar circle, however, we seem to be stuck. The problem can be avoided by going to longer wavelengths that can punch through the dust. But in the far infrared and radio there is a dearth of objects whose distances we can know.

The interstellar medium comes to the rescue. The disk is filled with a complex, lumpy mixture of dusty clouds of hot and cold interstellar gas that concentrate strongly in the spiral arms [11.1]. Part of this mix is in the form of a thin sheet filled with cool clouds of neutral hydrogen that radiate a powerful emission line at a radio wavelength of 21 centimeters [11.1.4]. As a radio line, it ignores the dust and can be seen all across the Galaxy.

In Fig. 5.13a observe at 21 cm along the red line at Galactic longitude 35°. The clouds at each point along the line have different radial velocities relative to the Sun, and will radiate at different Doppler shifts. Since neutral hydrogen is concentrated to the spiral arms, the radiation will be strongest coming from the points at which the line of sight crosses an arm, that is, at points A, B, C and D. The observed 21-cm line will be a combination of the individual Doppler-shifted emissions coming from the four points (Fig. 5.13b). We thus have the differential cloud–Sun velocities, but not the distances – save one.

The highest velocity cloud (see Fig. 5.13b) will be the one that lies closest to the Galactic center, cloud B, where the line of sight is tangent to the rotational circle and all the rotational velocity therefore lies along the line of sight. Moreover, the line from us to the cloud and from the cloud to the Galactic center makes a right triangle. The angle between the cloud and the Galactic nucleus is the Galactic longitude. If cloud B is distance R from the Galactic center, the tangent of the longitude is just R/R_0, from which we quickly find R ($R = R_0 \tan l$). That gives one point on the rotation curve. Direct the radio telescope to another longitude

Box 5.4 The mass of the Galaxy and dark matter

An estimate of our Galaxy's mass can be found by a simple application of Kepler's generalized third law [8.2.3]. Newton showed that the Earth's mass behaves gravitationally as if it were all concentrated to a point in the center. If the Galaxy's mass is symmetrically distributed (which it really is not), we can assume the same thing. The portion of the Galaxy that is interior to the solar circle behaves as if it were all concentrated to the center, with the Sun (actually the LSR) orbiting at a distance of 8kpc. The matter outside the solar circle is small and is more or less symmetrically distributed, so it has little effect. The Sun and the inner Galaxy therefore behave something like a simple two-body system to which Kepler's generalized third law scaled to the Earth and Sun works nicely, so that

$$M_{\text{Galaxy}} + M_{\text{Sun}} = a^3/P^2,$$

where P is the orbital period of the Sun about the Galaxy in years, a is the orbital semimajor axis in AU, and M_{Galaxy} and M_{Sun} are the inner Galactic and solar masses in $M_\odot$, the sum the same as M_{Galaxy}.

Plugging in the numbers, $M_{\text{Galaxy}} = (8000 \times 206\,265)^3/(240 \times 10^6)^2 = 7.8 \times 10^{10} M_\odot$. Studies of the distribution of stars show that about 70% of the light of the Galaxy is inside the solar circle, 30% outside, so multiply the interior mass by 1.4 to derive a mass of $1.1 \times 10^{11} M_\odot$. Ignoring brown dwarfs [6.2.2, 7.2.4, 12.4.1], the average stellar mass is around $0.5 M_\odot$, so the Galaxy (also ignoring interstellar matter) contains around 200 billion stars. More sophisticated modelling of the Galaxy's gravitational field through the details of the rotation curve lead to more accurate results, but the basic idea stays the same, that the Galaxy contains somewhat over 10^{11} solar masses worth of stars and interstellar matter.

The speed of an orbiting body is the orbital circumference divided by the period, or $v = 2\pi a/P$. Express everything in years, AU, and solar masses. Since $P^2 = a^3/M$ and therefore $P = a^{3/2}/\sqrt{M}$, $v = 2\pi a\sqrt{M}/a^{3/2} = 2\pi\sqrt{M}/\sqrt{a}$. That is, orbital speeds decrease in proportion to $\sqrt{(M/a)}$. If the circle at 12.5kpc from the center contains 90% of the stars, while the solar circle contains 70%, a star orbiting at 12.5kpc feels roughly 1.3 times more mass than we do, and should have an orbital speed of $\sqrt{1.3}/\sqrt{(12.5/8)} = 0.9$ times that of the Sun, or 200 km s^{-1}, which is not inconsistent with the rotation observations in Fig. 5.14.

Farther out we would expect the rotation curve to continue to decline, but instead it increases gently to more than 20kpc. If we get far enough out, so that nearly all the mass is interior to any orbit of increasing size, the orbital velocity should decline in pure "Keplerian" fashion according to $1/\sqrt{a}$ (as do the speeds of the planets of the Solar System). But at great distances from the center, that does not happen. If for simplicity the rotation curve stays flat with distance, M must be proportional to a. At a distance twice ours, an orbiting star feels twice as much matter as we do and there must be twice as much matter inside that circle as within ours. But we do not see it! There is no glow coming from stars or anything else consistent with that amount of mass.

The Galaxy therefore seems to contain matter that does not radiate or make itself known – *dark matter*. Dark interstellar clouds do not count; they are dark only to the eye. In the infrared and radio they radiate brightly. Dark matter is totally dark. Dark, too, in the sense of hidden, dark in the sense of our ignorance of what it is. For all the suggestions – white dwarfs, odd atomic particles, black holes, and so on – no one knows what it is.

Additional evidence comes from the motions of our small satellite galaxies, which are moving too fast for the mass accounted for by the Galaxy's bright matter only. The Galaxy may contain 10 times more mass than we see, raising it to a trillion solar masses. Dynamical studies suggest that most of it is in a vast *dark matter halo* that completely surrounds the visible portion of our Galaxy, dwarfing the Population II halo that we have come to know so well. Other galaxies, as well as their clusters, show the same phenomenon. The stuff is everywhere, its nature and origin a mystery.

(say $l = 20°$, the green line in Fig. 5.13a) and derive the velocity for another point. By going all the way around through the Galactic center from $l = 270°$ to $90°$, this *tangent point method* builds up the entire rotation curve of the inner Galaxy (Fig. 5.14). The tangent method cannot work in the outer Galaxy. But here we see more clearly, allowing the construction of the rotation curve to as far as 25kpc from the Galactic nucleus, three times the solar distance.

Now reverse the process. The radial velocities of clouds A, B, C, and D in Fig. 5.13a depend on their distance from us, so we can tell how far away they are, and can build up a picture of the inner Galaxy's spiral structure (Fig. 5.15). (Indeed, radial velocities coupled with the rotation curve can give the distance to any object as long as it is young and in circular motion; see Section 4.3.4). These data, combined with distance data from a great variety of other youthful objects, allows the construction of a detailed model of the spiral structure both in and outside the solar circle. Non-circular velocities inside the solar circle, as well as a variety of other observations, provide evidence of a central bar (see Fig. 5.2). Our Galaxy is not just a spiral, but a *barred* spiral.

Once we have the rotation curve, we can also derive the mass and mass distribution of the Galaxy through Kepler's laws [8.2.3] to find that our grand system contains over 100 billion solar masses of stars and interstellar matter and some 200 billion individual stars (Box 5.4). And much more of a mysterious nature.

5.4 Evolution

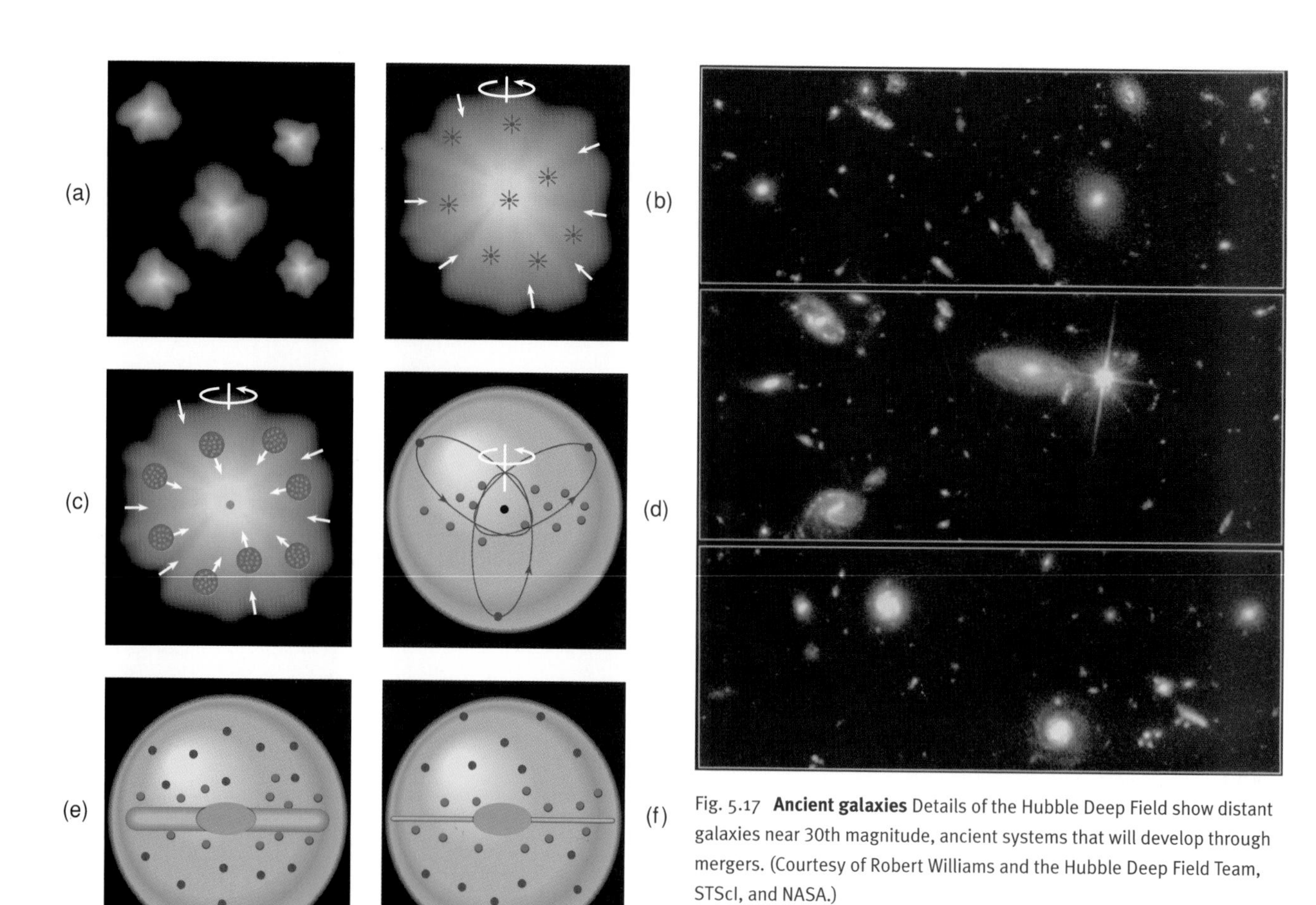

Fig. 5.16 **Simple evolution of the Galaxy** The panel shows the old, simple picture of the formation of the Galaxy, starting with (a) parent protogalactic clouds, then the creation of: (b) Population III; (c) the outer globular clusters with their elliptical orbits; (d) a flatter disk of globular clusters (seen as green dots, the older ones in red); (e) the thick disk; and (f) the current thin Population I disk. (Nothing is to scale.) (Kaler, J.B. (1994). *Astronomy!* New York: HarperCollins, reprinted by permission of Pearson Education, Inc.)

Fig. 5.17 **Ancient galaxies** Details of the Hubble Deep Field show distant galaxies near 30th magnitude, ancient systems that will develop through mergers. (Courtesy of Robert Williams and the Hubble Deep Field Team, STScI, and NASA.)

We began with the overall picture of the Galaxy, delved into the details, and now end with a wider view of the grand scheme. How did this complex structure come into being? How did the Galaxy evolve? The old picture – which is being highly modified, if not destroyed, by new data and thinking – is one of simple gravitational contraction from a large seminal cloud of hydrogen and helium that was a product of the shredded gases created in the Big Bang (Fig. 5.16). (The dark matter discussed in Box 5.4 seems to track the bright matter, at least to some extent.) As the cloud contracted and fragmented internally, the first stars formed within it. The by-products of their evolution and death seeded the giant collapsing blob with the first "metals" (used in the astrophysical sense of anything heavier then helium). The next generation of stars was the first to have metals in place [7.3.3]. Some of the fragments of the proto-galaxy were even big enough to condense into proto-globular clusters, which in turn condensed into stars. Since this generation was born during the proto-galaxy's collapse, these and free stars had large velocities of injection toward the Galactic nucleus. The contracting cloud left them behind, and so we see them moving on their highly elliptical orbits in the ancient Population II halo. The globular clusters' HR diagrams date this time of early collapse at around 12 to 13 billion years ago.

As the parent cloud contracted it also spun faster as a result of the conservation of angular momentum [Box 8.3], slowly flattening itself into a thick and then into an ever-thinner disk, while the orbits of the stars forming from it became more circular. Each new generation of stars also contained more metals that were given out by the earlier generations. The last generation fell into a truly thin disk where the parent – the interstellar medium – resides today. The oldest open clusters

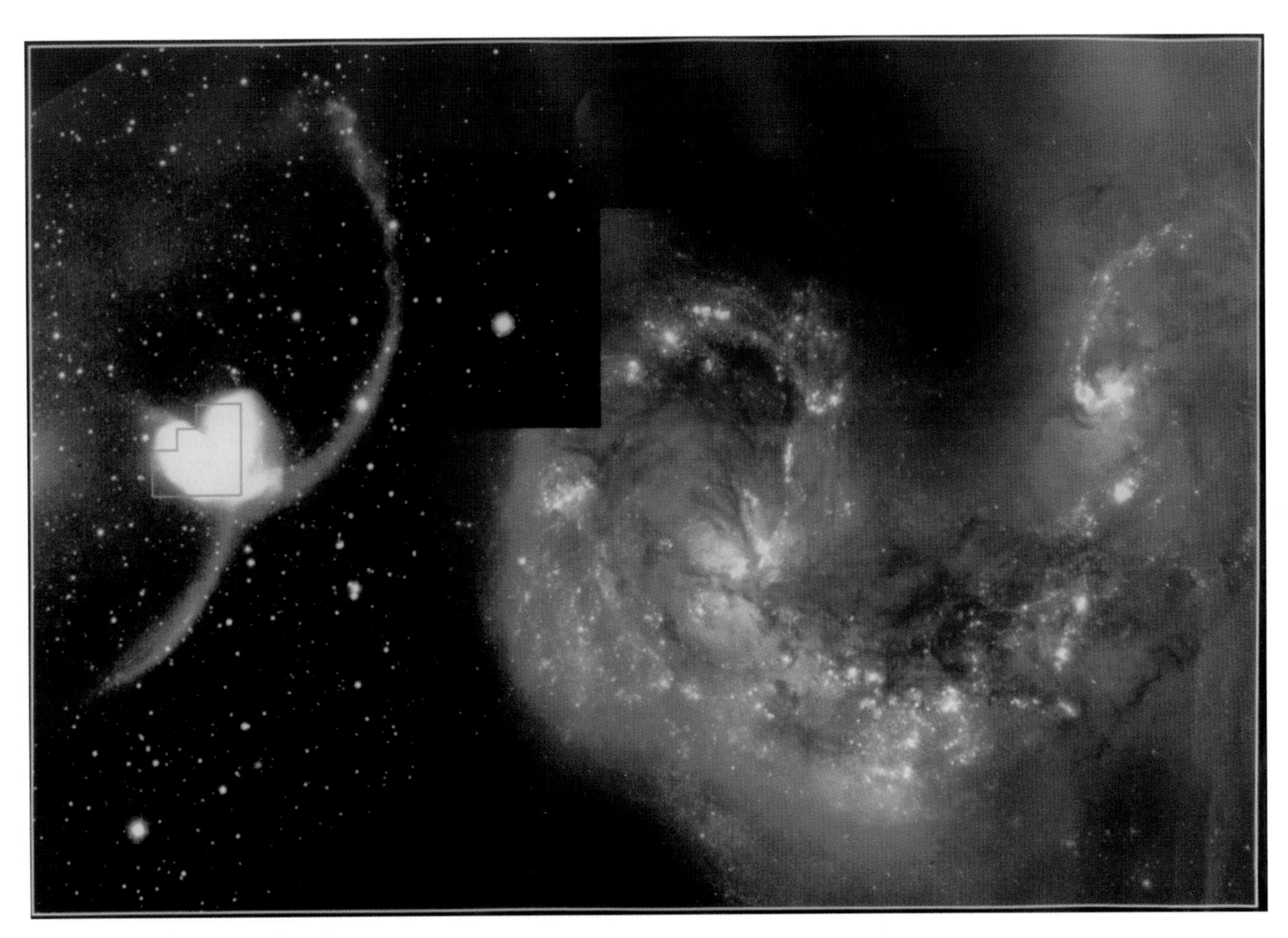

Fig. 5.18 **Galactic collision** A pair of galaxies (NGC 4038 and 4039) only 20 megaparsecs away collide and begin to merge. The resulting tides and interstellar shock waves enormously enhance star formation, as witness the blue light from hot young OB associations. (Courtesy of Brad Whitmore (STScI) and NASA.)

and the faintest white dwarfs tell us that the disk is between 7 and 10 billion years old. The concept explains why today we see a relatively young metal-rich Milky Way surrounded by an ancient metal-poor halo from which the old stars shoot past us on their elliptical orbits [14.4].

This pretty picture does not really work all that well. Orbital eccentricity, metallicity, and age do not correlate nearly as nicely as one might expect. Indeed, except for the gross differences between halo and disk, they hardly correlate at all. Much more telling, Hubble Space Telescope and other observations show that galaxies so far from us that we see them when they were truly young just do not look like what the simple theory suggests. Instead they seem to be unorganized, without the grand structures we see today (Fig. 5.17).

The new picture superimposed on the old is one of the early creation of galaxies that over time merged to become the ones that surround us today. Powerful support comes from observation of the collisions of galaxies both in the past and present, the celestial crashes enhancing star formation and creating myriad small fragments that sail away to merge back at some distant future (Fig. 5.18). Our Galaxy therefore is a mixture of many smaller ones, each of which had their own paths of evolution before the collisions took place, which explains the messy correlations. And so it still goes on, the "Sagittarius Dwarf galaxy" currently passing through our Galactic disk. Someday the Magellanic Clouds will collide with us – even our great neighbor, the Andromeda Galaxy, M31, may have a shot at us.

The details of the puzzle pale beside a bigger mystery. Under any scenario, we cannot escape the strong conclusion that the Big Bang gave us only hydrogen, helium, and a salting of lithium. The first stars must somehow have been devoid of metals, their number counted under *Population III* [11.3.4]. Where are they? We find none. The most metal-deficient, and presumably the oldest, stars of the halo, have metal contents under 1/100 000 that of the Sun. But we can go no farther. We find none with no metals at all.

This negative observation at first puts the entire picture at risk. The most distant (and therefore the youngest) galaxies we can observe, and the most ancient and youthful intergalactic medium between them, have significant metals, showing that the first heavy elements were created very quickly. Since the Big Bang could have made none, stars must still somehow be the culprit. The fact that we see no Population III stars implies that the members of this original generation were all massive and exploded quickly as supernovae [14.2.2], leaving none of its members behind. It would have taken very few of them to make enough metals to create the amounts seen in the current stars of extreme Population II.

The early lack of star-forming dust implies a different means of star formation from what we now find, a means that made only massive stars. Once there were enough metals to create dust, lower mass stars could form that would linger on until we could see them in lonely orbit. It is even possible that the stars of Population III may have been born from the gas of intergalactic space, without the need of galaxies to make them. The earliest assemblies we call galaxies would therefore have already had metals in place.

Such ad hoc assumptions are bothersome, as they are based not on data, but on lack of data. Strong support, however, comes from the abundance patterns in the oldest stars. The patterns are just what you would expect from the rapid neutron capture (the r-process) [13.3.3, 14.2.3] that is somehow associated with massive core-collapse supernovae, showing that there *had* to be an ancient generation of massive stars. As we go to younger Population II stars, we begin to see a rapid increase in iron content and in the iron-to-oxygen ratio, just what you would expect once the Type Ia white-dwarf supernovae [14.2.5] show up after the final evolution of the lower mass stars of the second generation is complete.

If it takes a galaxy to raise a star, the stars are the galaxies' children. But our Galaxy is the creation of its stars, of its interstellar matter, of its dark matter. So the Galaxy is also the child of the stars, much as the child is father to the man.

Chapter Six

Spectra and the HR diagram

"Never, by any means, will we be able to study [the stars'] chemical compositions, their mineralogic structure . . . I persist in the opinion that every notion of the true mean temperature of the stars will necessarily always be concealed from us." So wrote the French philosopher August Comte (1798–1857) in his *Cours de Philosophie Positive*. He could hardly be blamed. In 1835 no one, not the best of astronomers, was able to reckon the power of the spectrum.

All the studies of magnitudes, distances, motions, mean little without analysis of a star's spectrum, which allows the determination of its physical properties: temperature, density, chemical composition, age, magnetism, everything that ultimately leads us to the knowledge of how stars are born, live, and die. It is not just the Sun that supports us, but the stars that gave of themselves, that provided the raw material out of which the Earth was created. Without the power of spectral analysis, we would never have known the stars' true personal significance.

But analysis is the second step, not the first. Before we can make any sense out of stellar spectra, we must classify the data and then correlate them with other empirical properties, such as color, motion, and luminosity. From the patterns that emerge we can begin to apply theories that make sense of the observations. The closer we look, the finer our classification divisions become, until we see a continuum in which each star is distinguished from all others. It is not the individuals we seek here, however, but their families, which are as diverse as any found on Earth.

6.1 The solar spectrum

Whether fair or not, the Sun is the standard for all stars, the base against which everything else is compared, whether it be temperature, mass, luminosity, or spectral characteristics. In nature, the solar spectrum appears most commonly as the everlastingly beautiful rainbow (Fig. 6.1). Credit for deliberately producing a solar spectrum with a refracting prism goes to Isaac Newton. Galloping quickly on history's horse, we next encounter William Wollaston (1766–1828), who first found the dark gaps in the solar spectrum – the *absorption lines* – that would open the way to discovery. Shortly thereafter, Josef von Fraunhofer catalogued the most prominent of them and extended the discovery of absorption lines to the stars (Fig. 6.2, Fig. 6.3).

Less than 50 years after Wollaston's discovery, Robert Wilhelm von Bunsen (1811–1899, of burner fame) and Joseph Kirchhoff (1824–1887, of Kirchhoff's laws [7.1.5]) were occupied in identifying the absorptions with emissions

Fig. 6.1 **The rainbow** A solar spectrum arches across the fields, the rainbow, produced by refraction of sunlight through water droplets. (The inner bow is caused by two refractions in the raindrops separated by an internal reflection. An additional reflection produces the outer bow and reverses the order of colors.) (J.B. Kaler.)

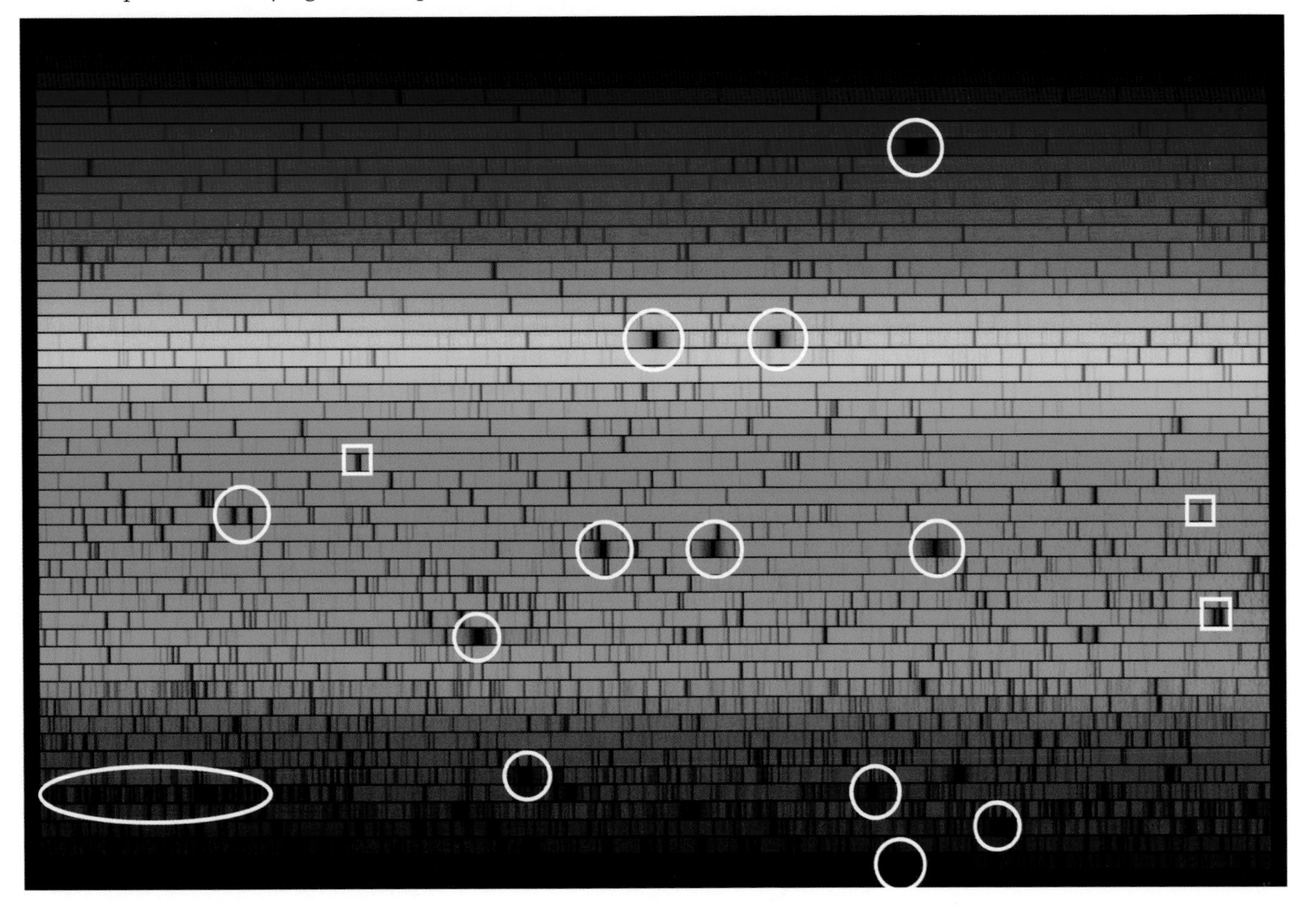

Fig. 6.2 **Fraunhofer lines** When the solar spectrum is sufficiently spread, the absorption spectrum emerges from the array of colors. The circles indicate the lettered Fraunhofer lines that fall between 7000Å at upper right and 4000Å at lower left. From top to bottom: C (Hα); the D doublet (Na I); E (Fe I); the b triplet (Mg I); F (Hβ); d (Fe I); the G-band (CH, the ellipse) and G (Hγ) both on the same line; g (Ca I); and h (Hδ). The squares indicate other iron lines. (Courtesy of Nigel Sharp, NOAO/NSO/Kitt Peak FTS/AURA/NSF.)

Fig. 6.3 **Fraunhofer H and K.** In the deep violet lie the strongest absorptions of the solar spectrum, Fraunhofer H and K of ionized calcium (Ca II). Compare their strength with that of Ca I, which lies to the right of the label and appears in Fig. 6.2. (Courtesy of NOAO/AURA/NSF.)

radiated at the same wavelengths by pure samples of chemical elements. Along with the beginnings of theory that explained how the solar absorptions are produced, these discoveries completed the introduction of spectroscopy into astronomy. The next steps involved the hard, grinding work of mapping and chemically identifying all the detectable solar absorption lines and explaining them in detail in terms of quantum theory [7.1]; acts that would occupy the next century or more.

6.1.1 Names

Much of the lure and charm of astronomy is its persistence in keeping its history alive through the definitions and nomenclature of the past. We use magnitudes invented by Hipparchus over 2000 years ago [Chapter 3], and still use – to the confusion of beginners – the designations that Fraunhofer applied to the solar (hence stellar) absorptions. Though he counted a total of nearly 600 lines, he catalogued only 324 of them. The most prominent he lettered in order of descending wavelength, red through violet, with upper-case Roman letters that have nothing to do with the symbols of the chemicals that were later found to create the lines (Table 6.1). A few out-of-order lower-case letters and a short extension of the upper case followed later.

The first two lines, Fraunhofer A and B (as well as "a") are of terrestrial origin, caused by atmospheric oxygen and water vapor superimposing their spectra onto solar and stellar spectra. "A" is very strong. *Telluric* (pertaining to Earth) lines such as these invade the whole spectrum, can be in either absorption or emission, and are a serious hindrance to ground-based spectroscopy. The letters C, F, G and h are the first four members of the hydrogen Balmer series (Hα, Hβ, Hγ, and Hδ [7.1.4]), and, along with E and d for neutral iron, are no longer in use. The other letters are part of astronomical tradition; references to "sodium D," "calcium H and K," and the "G band" (a complex mixture of absorptions from the CH molecule, iron, and titanium) are common in both parlance and the literature. Even "D_3" for helium and "g" for neutral calcium are occasionally seen.

Beyond the Fraunhofer designations, spectrum lines are most typically named by chemical element (or compound), ionization state, and wavelength. While the ionization state of an atom is denoted by plus signs [Box 7.4], the spectrum of the ion is called out by Roman numerals: "I" denotes the spectrum of the neutral species, "II" that of the first ionization state, and so on. The spectrum of triply-ionized iron, Fe^{3+}, is Fe IV. Thus calcium's K line (at 3934Å) is also expressed as Ca II 3934, the D_3 line as He I 5876, or in nanometers, He I 587.6. (While the international wavelength standard for the optical domain is nanometers, astronomers persist in the use of more-convenient angstrom units [see Box 7.1]). Higher order identification of a spectrum line gives the wavelength and the technical names of the upper and lower energy states that give rise to it [7.1.4].

Letter Designation	Wavelength (angstroms)	Identification
A	7594	terrestrial oxygen (O_2)
a	7165	terrestrial water vapor
B	6867	terrestrial oxygen (O_2)
C	6563	atomic hydrogen (Hα)
D_1	5890	neutral sodium (Na I)[1]
D_2	5896	neutral sodium (Na I)[1]
D_3	5876	neutral helium (He I)[2]
E	5270	neutral iron (Fe I)
b	5167	neutral magnesium (Mg I)
b	5173	neutral magnesium (Mg I)
b	5184	neutral magnesium (Mg I)
F	4861	atomic hydrogen (Hß)
d	4384	neutral iron (Fe I)
G	4340	neutral hydrogen (Hγ)
G-band	4300	"banded" CH, neutral titanium and iron
g	4227	neutral calcium (Ca I)
h	4102	neutral hydrogen (Hγ)
H	3968	ionized calcium (Ca II)
K	3934	ionized calcium (Ca II)
h	2802	ionized magnesium (Mg II)[3]
k	2795	ionized magnesium (Mg II)[3]

[1] D_1 and D_2 are together called "sodium D."
[2] D_3 is not seen in absorption in the Sun, but in emission from the chromosphere and in absorption in hot stars.
[3] Named late by analogy with H and K of Ca II and not seen from the ground; do not confuse with older Fraunhofer h.

Table 6.1 **The Fraunhofer lines**

6.1.2 Beyond Fraunhofer

A sketch of the Fraunhofer spectrum is but the beginning. Hundreds of thousands of absorptions break the continuum of the solar spectrum. Some, like Ca II H and K, remove great chunks, while other lines are barely detectable as mere dips

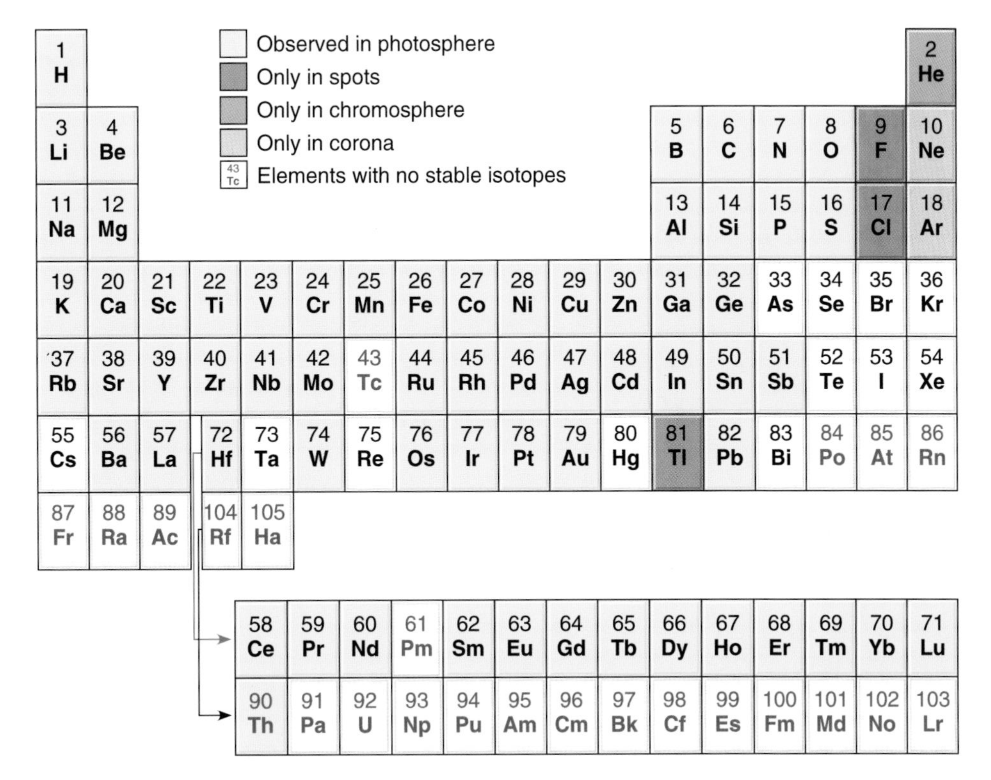

Fig. 6.4 **Elements in the Sun** Elements discovered in the Sun are in filled boxes, colored according to the origins of their spectra. Aside from rare radioactive elements, only a few elements remain to be found. Uranium has been seen in other stars, though not yet in the Sun. (Kaler, J.B. (1994). *Astronomy!* New York: HarperCollins, reprinted by permission of Pearson Education, Inc.)

in the background (see Figs. 6.2 and 6.3). There are so many lines that they can severely overlap or blend with each other, enough to make the identification of rare elements quite difficult, especially at shorter wavelengths. The vast majority of solar lines are caused by neutral and once-ionized atoms, particularly metals, but here and there are also molecular bands, as the solar temperature is not quite high enough to make all the molecules go away. The G-band shortward of 4313Å is loaded with lines of the CH molecule, and CN is prominent shortward of 3883Å and 4216Å. Both molecules are important in classification.

While iron lines outnumber all the other kinds, the strongest lines are by far those of the light metals and hydrogen. The D lines of neutral sodium and the Mg I triplet are strong, though nothing comes close to the power of deep violet H and K of Ca I (Fig. 6.3) except the analogous h and k lines of Mg II, which (because of atmospheric absorption) are not visible from the surface of the Earth. The Balmer lines begin powerfully with Hα, and though weakening, remain strong through Hδ. Hε is lost in calcium H (Fraunhofer "H," not "H" as in "hydrogen"), and the rest disappear quietly into the ultraviolet. Mixed in with these are great numbers of lines of other metals (vanadium, chromium, nickel, titanium), rare earths (cerium, neodymium), and other elements.

The solar spectrum is hardly limited to the Sun's photosphere (its apparent surface [12.1.1]). Spectra of the low-density chromosphere [12.1.3] show the Fraunhofer lines in emission rather than absorption, including the D_3 line of He I, as well as lines of more highly ionized atoms. The transition layer to the corona and the corona itself [12.1.4] produce emissions that are tied to the temperature of the layer from which they arise, from strong Si II to Ni XVII. These pervade the spectrum, but are especially strong in the far ultraviolet. When the entire solar spectrum is accounted for, of the 81 non-radioactive elements up to bismuth, we find 69, leaving but 12 to be found (Fig. 6.4). They are surely there, just too rare to make much of a dent in a crowded spectrum.

The absorption lines are the foundation for the study of other stars and for their comparison with the Sun. They are our basis for classification and for the structuring of the most important of all stellar tools, the Hertzsprung–Russell (HR) diagram in any and all of its many forms, to which we shall turn below.

6.2 The spectral sequence

To the surprise of early spectroscopists, few stellar spectra looked like that of the Sun. Many are dominated completely by hydrogen lines, whereas the H and K lines are weak. In others, these Ca II features are gone altogether, while in still others the helium absorptions are prominent. A noted number of stars sport complex molecular spectra. Though no one at first understood what was going on, some astronomers knew enough to classify.

6.2.1 Harvard classification

The first attempts at classification, made in the mid nineteenth century, led to a five-type scheme pioneered by Father Angelo Secchi (1818–1878): I (strong hydrogen like Vega); II (numerous metallic lines, as in the Sun); III (spectra with bands shading darker toward the blue: Betelgeuse); IV (deep red stars with bands shading in the other direction); V (stars with emission lines). The bands of class IV were quickly

Class	Characteristic	Criterion	Stars
O	H, He II, He I C III, Si IV	He II/He I	λ Ori, σ Ori
B	H, He I, Si IV–II, C III, Mg II	Mg II/He I	τ Sco, γ Ori, Regulus
A	H, Ca II, Fe II, Fe I, Sr II	Ca II "H"/Hϵ+Ca II "K"	Vega, Fomalhaut, Altair
F	H, Ca II, G-band, Fe I, Ca I	Ca I/Hδ	γ Vir, α Per, β Vir
G	Ca II, G-band, H, Ca I, Mn I, Fe I	Ca I/Hδ	η Cas, β Her, ϵ Vir
K	Ca II, Ca I, G-band, Mn I, H	Cr I/Fe I, Ca I/Hδ	α Cas, Aldebaran, 61 Cyg
M	Ca I, TiO, Ca II, Cr I	Cr I/Fe I	δ Vir, 30 Her, Mira
L	TiO, VO, FeH, CrH, H_2O, K I, Cs I	CrH/TiO, H_2O depth	...
T	CH_4, H_2O	line depths	...
	(Carbon stars)		
R	CN, C_2, G-band		
N	CN, C_2, CH		
S	ZrO, Ca I		

Class	Color	Color Index	Temperature (K)[a]
O	blue	−0.32 to −0.27	47 500–31 000
B	blue–white	−0.27 to −0.02	30 000–10 000
A	white	−0.02 to 0.24	9800–7300
F	yellow–white	0.24 to 0.64	7200–5800
G	yellow	0.64 to 1.02	5700–4900
K	orange	1.02 to 1.5	4800–3900
M	orange–red	>1.5	3800–2200
L	red–infrared	...	2100–1200
T	infrared	...	<1200

[a] Degrees kelvin, equal to degrees celsius above absolute zero, −273°C.

Table 6.2 **Characteristics of the spectral sequence**

The upper table gives typical spectral characteristics, one good indicator that best fits the whole class and includes both giants and dwarfs, and some typical stellar examples. The lower table gives color and temperature averaged over all luminosities. L and T stars are dwarfs only. The temperatures of class R stars roughly parallel those of class G5 through K, while those of S and N roughly follow those of class M. See Fig. 6.12.

Fig. 6.5 **Objective prism spectrogram** Commonly used for early classification, an objective prism spectrograph produces low-dispersion spectra for each star in the field of view. Absorption lines are faintly visible. Early images could not record the red part of the spectrum. (Courtesy of Department of Astronomy, University of Michigan.)

identified with carbon compounds, while those of class III were later seen to be from titanium oxide, TiO. (Do not confuse these Roman numerals with the MKK classes used in Section 6.3.2 to denote luminosity.)

This early scheme led directly to a more detailed one developed from 1890 onward by E.C. Pickering and his three prominent assistants (Williamina Fleming, Antonia Maury, and Annie Cannon) at the Harvard College Observatory. They developed a system that used letters A through Q based mostly on the strengths of the hydrogen lines, wherein Secchi's class I became A and B, II went to E through L, III and IV became M and N, and V (with its emissions) broke into a variety of others.

The sequence did not take long to evolve. Several of the original letters were dropped or merged with others, while Maury and Cannon showed that better sequencing of other absorption lines was had if B came before A and O before B. The result was the famed seven-letter Harvard Spectral Sequence, OBAFGKM (Table 6.2). The Sun falls in the middle: class G.

Cannon did all the classification herself from objective prism plates (in which a prism is placed over the telescope objective to create spectral images, as in Fig. 6.5), and found that the system was not fine enough, so she began to decimalize it. The modern version begins at O2, goes to O4 . . . O9, B0 . . . B9, and so on down to M9 (and occasionally M10, which is no longer in much use). Cannon's classification of 225 300 stars was published between 1918 and 1924 in the *Henry Draper Catalogue* (named for a prominent physician–astronomer), and was later expanded in the *Henry Draper Extension* to 359 082 stars. There are so many stars included that HD and HDE numbers are still among the most-used stellar names. In the refined system, the Sun is a G2 star.

The early classifiers quickly saw that the sequence correlated with stellar color [3.4]: O and B stars blue, A stars white, K orange, M orange–red, and N red. Since blackbody radiation laws [7.1.2] show that the hotter the body, the bluer it is, the spectral sequence is clearly a temperature sequence, and in fact stands in well for both color index [3.1.2; 3.4] and temperature. O stars are hot, M and N stars very cool. An early (and wrong) theory of stellar evolution held that stars began their lives as hot, and later cooled. Hot stars were therefore called *early* and cool stars *late*. Class M is "later" than K, class G0 earlier than G2. The terms are still in common use. History lives.

The most powerful way of presenting the sequence is to lay images of the spectra of the different classes adjacent to each other in order of color (blue to red), which allows the peruser both to compare the classes and to look at the overall spectral variations caused by the continuous changes in temperature (Fig. 6.6). In the panel, note the way that the hydrogen lines begin as weak in the O stars, strengthen through class B to A, then weaken again, disappearing in class M (and even becoming emission lines [10.5.1; 10.7]). He I

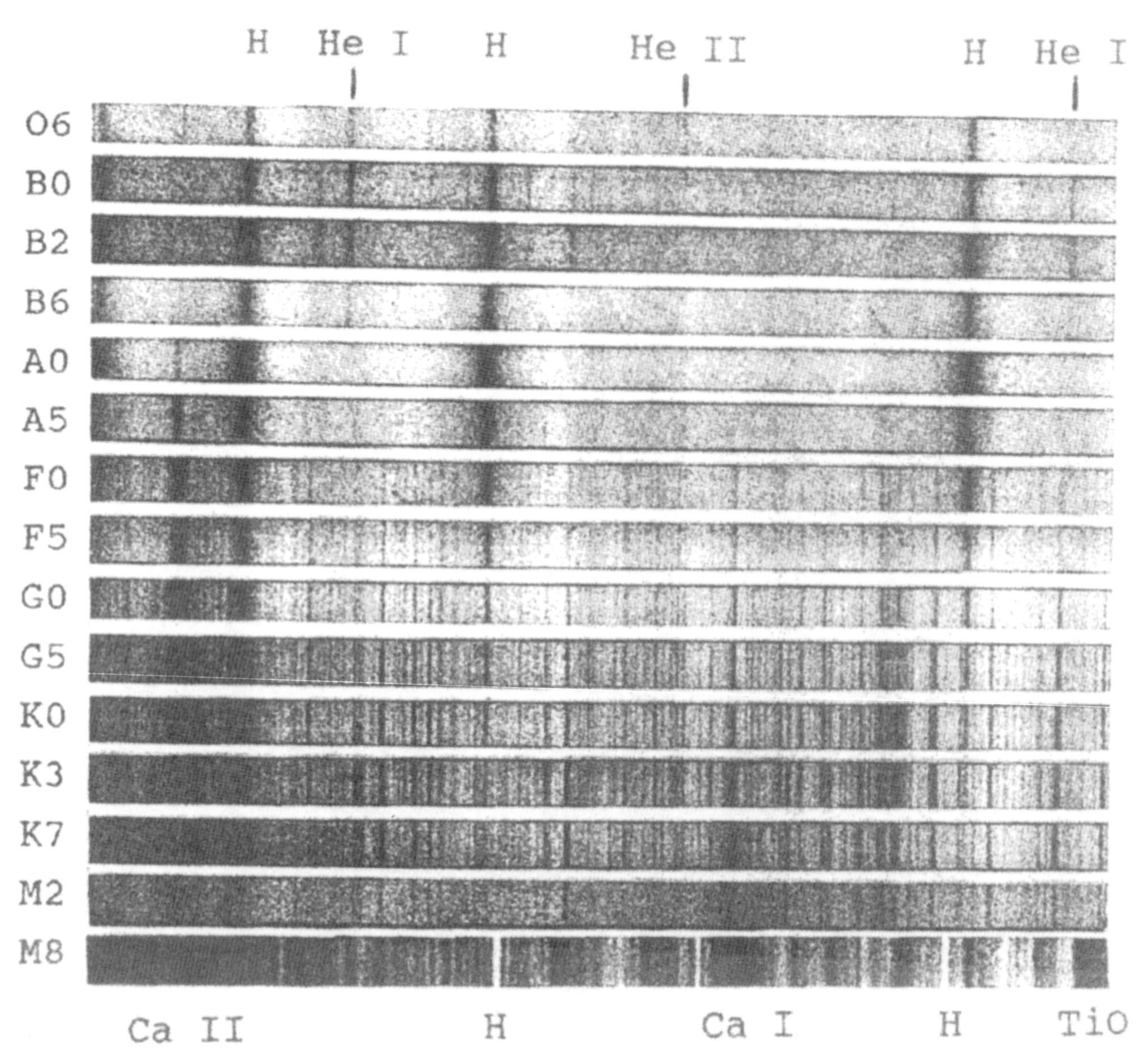

Fig. 6.6 **The spectral sequence** The spectral compilation displays a pure dwarf sequence from O6 to M2, to which is attached the spectrum of an M8 giant (which is not quite on the same wavelength scale), and that extends from Hβ through Hδ and into the near ultraviolet to both Ca II H and K. Hydrogen lines reach their peaks in class A. The He II lines come out nicely in the O star, He I in class B, and a piece of TiO in the M8 star. (Ginestet, N., Carquillat, J.M., Jaschek, M., Jaschek, C., Pédoussaut, A., and Rochette, J. *Atlas de Spectres Stellaires*; Observatoire Midi-Pyrénées and Observatoire de Strasbourg (1992); bottom spectrum from Yamashita, Y., Nairai, K., and Norimoto, Y. (1978). *An Atlas of Representative Stellar Spectra*, University of Tokyo Press, 1978.)

is prominent in class B, He II in class O. The Ca II lines are very weak in class B, strengthen powerfully into class K, and then weaken. Ca I, on the other hand, begins weakly in the A stars, and strengthens all the way down, complementing Ca II rather nicely. A wealth of metal lines appear in class F and then dominate in G and K. Metals are strong in class M, but come in second to the grand TiO bands, which at the bottom simply take over.

Since the photographic plate is far more sensitive in the blue and violet (and ultraviolet) than it is in other wavelength domains, the blue–violet became the primary band of choice. Better plates, the electronic revolution in observing capability, and the ability to carry spectrographs into space allowed the expansion of classification into the ultraviolet, the red, and the infrared (Fig. 6.7).

A century after the system was first developed, astronomers had sufficiently good infrared detectors to be able to find stars redder, and therefore cooler, than even the coolest class M stars, resulting in two more classes, L and T (see Table 6.2). L and T spectra cannot be displayed like those of the other classes because the stars radiate little or nothing in the optical, and their spectra are difficult, if not impossible, to photograph. Instead, we rely on digital infrared spectra that show us powerful lines of alkali metals, water, iron hydride, carbon monoxide, and in class T even methane, a molecule that dominates the spectra of our giant planets (Fig. 6.8).

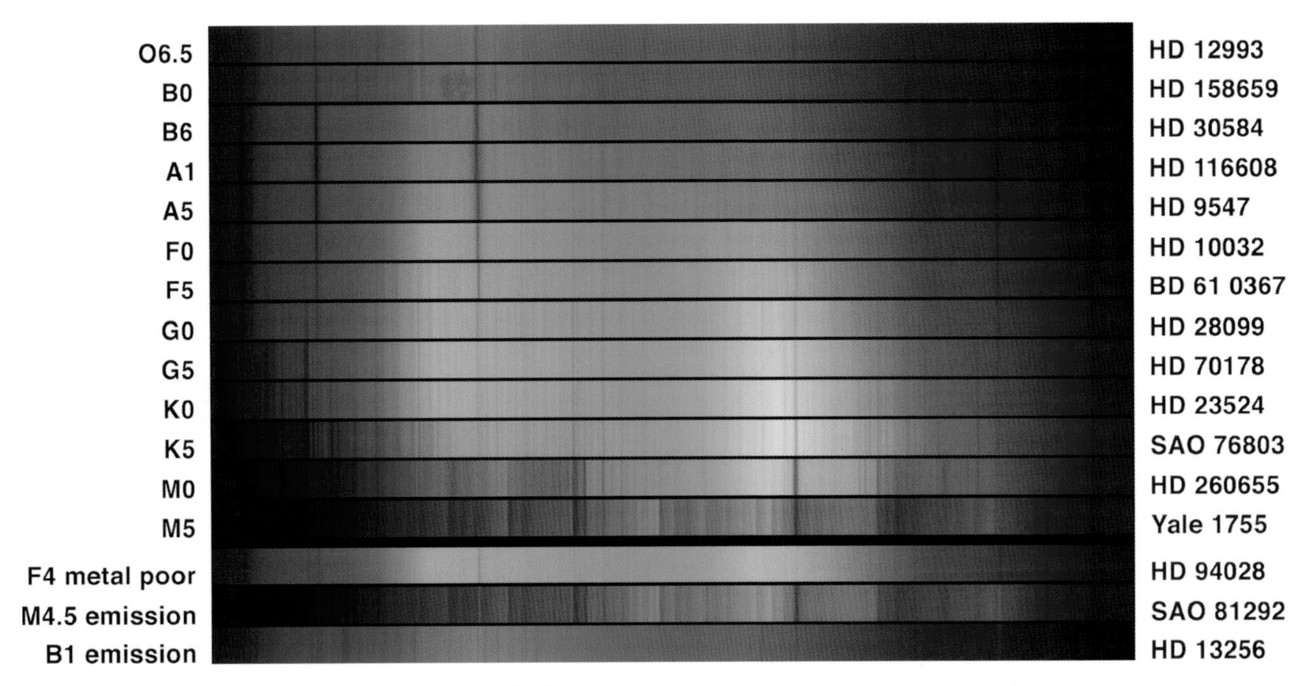

Fig. 6.7 **Full spectrum sequence** An all-dwarf sequence in full color extends across the visual band from deep violet at 4000Å to redward of Hα. The telescopic power is sufficient to bring out the TiO bands at M5. At far right is Hα. The prominent absorption in the yellow–orange is a blend of the sodium D lines. Three special cases are added at the bottom. While the color is realistic, it was added later to digital observations constructed to look like photographs. (Courtesy of NOAO/AURA/NSF.)

T stars are so cool that they are invisible to the eye and have no *B–V* color indices. Class L contains a large number of brown dwarfs [12.4.2, 13.1], while class T consists of nothing but; the full sequence is now OBAFGKMLT. Add three classes for carbon-rich stars (R, N, and S) and we are done. (Carbon stars will be addressed in Section 6.4.3 below). Much more information about a star is expressed through a system of prefixes and suffixes appended to the Harvard classes (Box 6.1, Table 6.3).

6.2.2 Explanation

Immense effort has been expended in correlating spectral class with both color and temperature (Fig. 6.9), which plummets from around 50 000K at class O2 to under 1000K in class T. The laws of atomic physics show that the standard sequence (sans carbon-rich R, N, and S) can be explained entirely by temperature-dependent changes in excitation and ionization of a common chemical mix: 90% hydrogen, 10% helium, 0.1 percent everything else, all in (or close to) solar proportion [7.3.1].

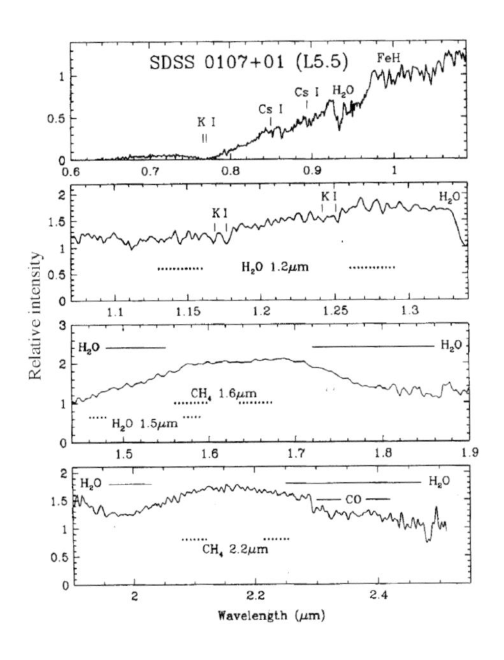

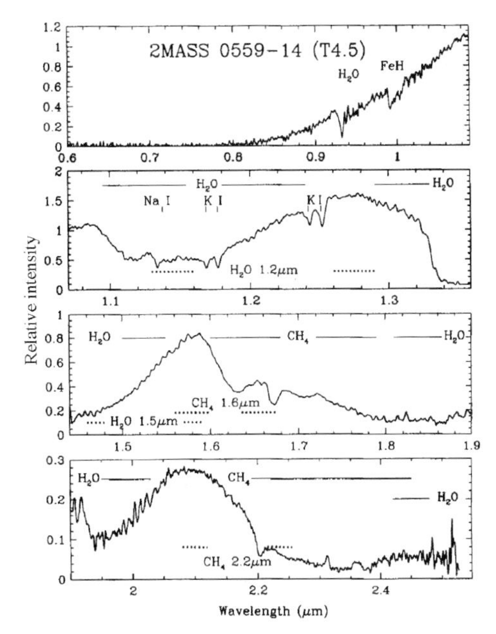

Fig. 6.8 **L and T stars** The infrared spectrum of the L5.5 star on the left (plotted in microns, μm; multiply by 10 000 to get angstroms) displays strong lines of alkali metals, here potassium (K) and cesium (Cs), and strong molecular bands of hydrides (here FeH) and water. In the spectrum of the cooler T4.5 dwarf on the right, while these lines all strengthen, they are overwhelmed by huge bands of methane (CH_4) and water. (Geballe, T. R. *et al.* (2002). *Astrophysical Journal*, **564**, 4660.)

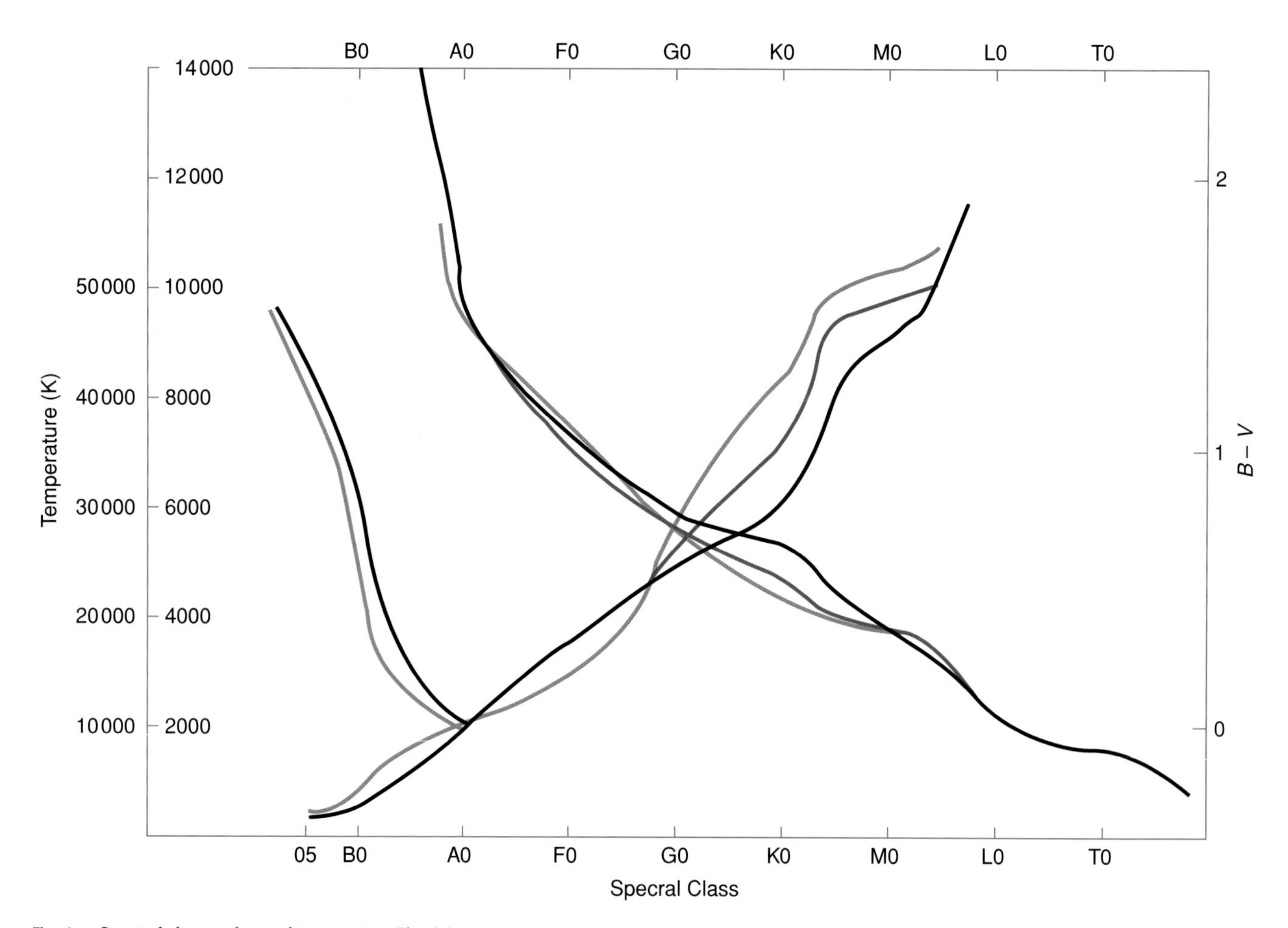

Fig. 6.9 **Spectral class, color, and temperature** The rising curves represent *B–V* colors that are read from the right-hand side of the graph. The falling curves represent temperatures read on the left. The left- and right-hand temperature scales go with the left- and right-hand temperature curves. (Color does not correlate here with temperature; for that, see Fig. 3.11). As the spectral sequence progresses from O to M, *B – V* increases (reddens) and the temperature drops. The relation for dwarf stars like the Sun are in black, while the relations for giants and supergiants are respectively in red and green. At high temperatures and low colors, the relations for giants and dwarfs overlap. Classes L and T are not shown. (Colors from Schmidt-Kaler, T., *Landolt-Börnstein Tables*, Springer. Temperatures compiled from a wide variety of research papers.)

Spectral classification was originally something of an empirical art, the class based on judgement by experts trained to match the spectra of program stars with standards. While quantitative measures go back a long way, they have only recently taken the ascendancy. Each different kind of ion has its own peculiar behavior relative to temperature (Fig. 6.10). For that matter, so does each line. The spectral class is therefore told from the ratio of the strengths of two lines that have different temperature dependencies, or from the absolute strength of a particularly strong feature (see Table 6.2).

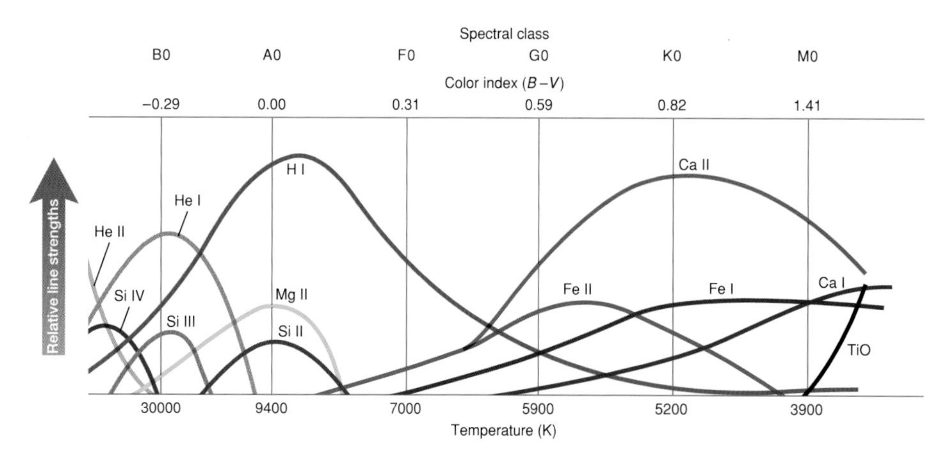

Fig. 6.10 **Ions and temperature** The curves schematically show the dependencies of different ionic spectra on temperature. (Kaler, J.B. (1994). *Astronomy!* New York: HarperCollins, reprinted by permission of Pearson Education, Inc.)

Hydrogen lines, for example, are effectively gone in class M and cooler. In many M stars they in fact reverse themselves, and are seen in emission [10.5.1, 10.7]. At these low temperatures, the collisional energies are not enough to raise electrons into the second orbit, from which the optically visible absorption lines arise [7.1.4]. As temperature climbs, hydrogen's second orbit becomes more populated, and the hydrogen increases in strength [7.3.1]. Above about 10 000K, however, the collisions are so violent that hydrogen becomes ionized, so its absorptions begin to disappear, at which point the neutral helium and then ionized helium (both of which require very high temperatures to excite the levels from which the lines arise) come onto the scene. At the highest temperatures, hydrogen and ionized helium can both reverse into emission. As another example, in M and class K, the neutral calcium 4227Å line is strong. As temperature climbs, neutral calcium becomes ionized, and Ca II H and K build at the expense of Ca I. We see similar sequential progressions in all the elements, including prominent silicon and iron.

Molecular spectra of cool stars are related to temperature as well. Molecules are frail constructions that break up easily as a result of temperature-dependent collisions. They become increasingly more prevalent from early class M on down through cooler L and T, more fragile ones taking over at lower temperatures. In class L, TiO disappears to be replaced by metallic hydrides, while in chilly class T even methane and water can appear.

The changes from one class to the next are compounded by temperature- and density-dependent differences in the opacities of the gaseous photospheres where the lines are formed, causing us to see to different depths along the sequence. The details depend specifically on the range of spectral classes assumed, and thus each class, even just a set of subclasses (B0 to B5 for example), has its own set of criteria. See Chapter 7 for a more detailed explanation.

6.3 The second dimension

Spectral classification by itself reveals little of the physical natures of the stars and how the stars are linked together in the cycle of stellar life and death. Temperature through spectral class provides but one dimension in the lives of stars, but there is far more to consider. We need to add higher dimensions, the first given by absolute magnitude, or luminosity.

6.3.1 The HR diagram

The first foray into the second dimension was made by Henry Norris Russell (1877–1957) and Ejnar Hertzsprung (1873–1967). In 1913, Russell took the few distances of stars derived from rather poor parallaxes and plotted them against spectral class (Fig. 6.11). Most of the stars fell into a band in which stellar luminosity increases with temperature. The luminosity of a spherical star is given by

$$L = R^2T^4,$$

where L is the luminosity, R the radius, and T the temperature, all in solar units (that is, $T = T_{star}/T_{Sun}$, and so on) [7.1.2]. Since L depends on such a high power of the temperature, such a relation should come as no surprise.

However, there is another band that increases up and to the right, luminosity increasing even as the temperature goes down. The only way to produce high luminosity at low temperature is to have an offsettingly large radius. Nearly a decade before, Hertzsprung had studied an earlier spectral classification designed in 1897 by Antonia Maury, in which she appended letters to her now-defunct classes that indicated line width, wherein a, b, and c denoted "normal," "broad," and "narrow" absorptions (Box 6.1 and Table 6.3). Hertzsprung noted that the cooler c stars had much lower proper motions

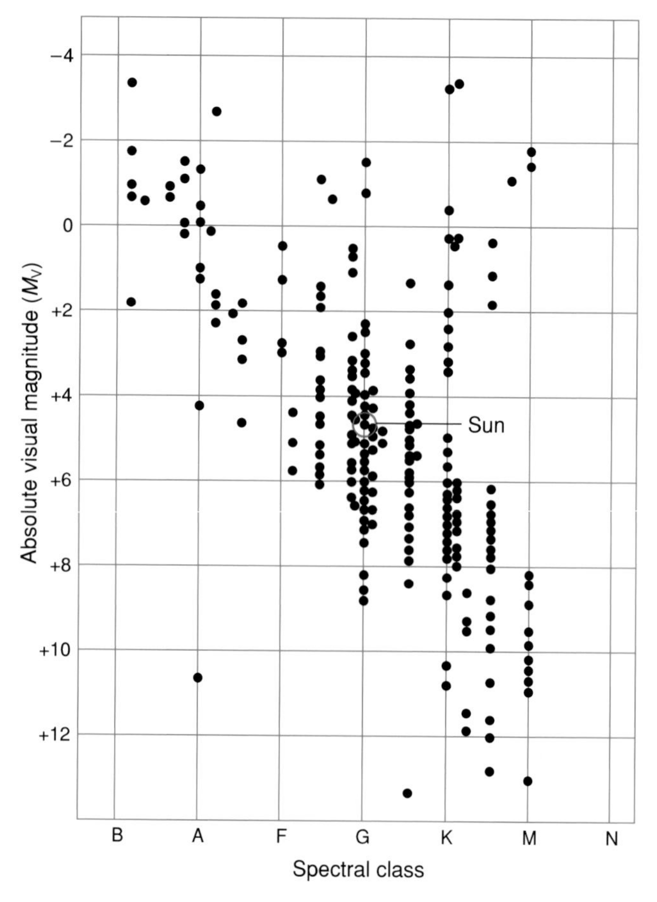

Fig. 6.11 **The first HR diagram.** The first HR diagram shows a band of dwarfs climbing up and to the left, while the giants stomp up and to the right. (Russell, H.N., as rendered in Kaler, J.B. (1994). *Astronomy!* New York: HarperCollins, reprinted by permission of Pearson Education, Inc.)

Box 6.1 Detailing the spectrum

A great deal of fine structure is seen in stellar spectra, and is described with the use of both prefixes and suffixes to the spectral class (Table 6.3). They are important, as they can lead to significant discoveries about how stars work. Maury's a, b, and c prefixes, which described line widths, helped birth the HR diagram. Because the c stars turned out to be supergiants, the letter was commonly used to describe these luminous stars. While the MKK system has taken over luminosity classification (Section 6.4.1), we still use these lower case letters in such creations as the Lc (irregular supergiant) and SRc (semi-regular) variables [10.5.3]. While Maury's class b stars are now known to be rapid rotators, and since c was used for supergiant, a and b were redefined to indicate giant SRa and SRb (semi-regular) stars; a and b are now also used as suffixes to help indicate fine structure in the MKK class. Before MKK revolutionized the classification system, dwarfs and giants were called out by d and g prefixes, while sd still stands for "subdwarf" and D for "degenerate white dwarf."

Eight suffixes are used to indicate various emission lines, three are used for line widths, three for chemical peculiarities, and another for spectral variations. For example, an O star with hydrogen emission is listed as Oe, while one with helium and nitrogen becomes Of, and if "peculiar" in any way, becomes Ofp. More detailed chemical differences can be added through numerical indexing on the basis of the strengths of the lines.

Prefixes (Maury)	
b	wide lines (no longer in use)
a	normal lines (no longer in use)
c	narrow lines (now meaning supergiant; Betelgeuse is a cM1 star; no longer common)
Intermediate types denoted by ab, bc; these are no longer in use.	
Prefixes (Other; older usage to indicate luminosity class; mostly replaced by MKK system)	
d	dwarf (main sequence; the Sun is a dG2 star; occasional use)
g	giant (Arcturus is a gK1 star; occasional use)
sd	subdwarf (wide use, both for true subdwarfs to the right of the main sequence, such as Kapteyn's Star, a high-velocity sdM1 star, and for evolved subdwarf O and B stars, sdO, sdB)
wd	white dwarf (uncommon)
D	white dwarf ("D" for "degenerate"; common)
Luminosity class suffixes	
Maury letters a and b are now commonly used to indicate MKK luminosity on the high or low side of average. A K giant that is a little brighter than normal but not bright enough to be a class II bright giant would be K3IIIa, one a little fainter than normal would be K3IIIb. Care must be exercised, as not all authors use them consistently.	
Suffixes denoting emission lines	
e	emission lines (now hydrogen emission, usually for class M and O; Mira is M7e)
em	emission from metallic lines (uncommon)
ep	peculiar emission
eq	P Cygni emission (emission flanked by blueward absorption)
er	reversed emission (central absorption within emission as seen for Be stars; uncommon)
f	helium and nitrogen emission for O stars (Zeta Pup, O5f, or with luminosity class, O5Iaf)
(f)	N III emission, but no He II 4686 (modern)
((f))	N III emission, but He II 4686 in absorption (modern)
Other suffixes (alphabetic)	
k	interstellar lines present (uncommon)
m	strong metallic absorption (in particular use for Fm and Am stars with chemical anomalies)
n	diffuse lines (moderately fast rotators)
nn	very diffuse lines (fast rotators)
p = pec	peculiar ("p" in particular use for Fp, Ap and Bp stars with strong magnetic fields and chemical anomalies)
s	sharp lines (other than Maury's "c"; rare)
v	variation in spectrum other than caused by velocity effects
wk	weak lines
!	marked characteristics (rare)
Suffixes are commonly combined. An O star with emission and interstellar lines with a variable spectrum might be O5fkv; a supergiant according to MKK class might be O5Iafkv.	
Chemical indices	
If a particular element or molecule is enhanced, the spectral class is followed by the chemical symbol, and sometimes by a numerical index that indicates line strength. A K giant mildly enriched in barium might be K4IIIBa0.5, a star rich in mercury might be A5VpHg, while one enhanced in cyanogen would be K0IIICN+2, or if deficient, K0IIICN−1. Carbon stars are often classed with a double index (C7,4), the first denoting temperature class, the second the carbon richness.	

Table 6.3 **Fine structure prefixes and suffixes**

The letters above give more detailed information on the spectrum of any particular star. They are divided into five groups: prefixes; suffixes that help indicate luminosity; suffixes that indicate emission lines; other suffixes: and suffixes that provide some kind of indexing of chemical abundance. Defunct terms are given as they are encountered in older literature.

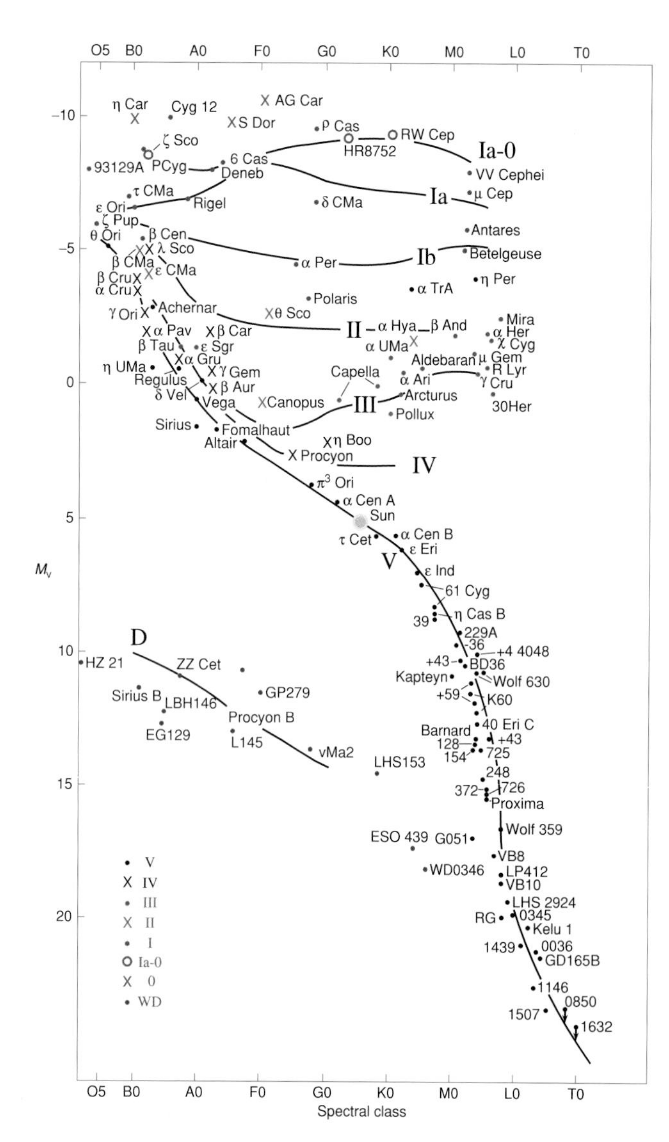

Fig. 6.12 **A modern HR diagram** The brightest and closest stars in Appendix 1 and 2 are plotted according to their classified luminosities on this updated HR diagram in which M_V is plotted against spectral class. The curves show the mean loci of the luminosity classes. They are supplemented from the tables of red dwarfs, cool red giants, brightest stars, and largest stars from the author's *Extreme Stars*, with a few others shown to fill in the blanks. White dwarfs (D) are plotted by temperature according to the scale of ordinary dwarfs. The density of points at any given position does not indicate true numbers of stars. T stars cannot be plotted as they have no *V* magnitudes. (Adapted from Kaler, J.B. (1989). *Stars and their Spectra*, Cambridge: Cambridge University Press; additional data for L dwarfs from Dahn, C.C. *et al.* in the *Astronomical Journal.*)

[5.2.1] than the others, and must statistically be much farther, more luminous, and therefore – again – larger than the others.

Russell and Hertzsprung needed contrasting names for the different kinds of stars. Russell's group of larger stars became *giants*, the others *dwarfs*, starting an anthropomorphism whose charm has had great staying power and is in part responsible for the public's fascination with stars. "Giant" is apt. Say a class M giant is 5 magnitudes [3.1.1], or 100 times, brighter than the Sun, but has only half the temperature. Since (from above) $R = \sqrt{L}/T^2$, the radius must by $10/(1/4) = 40$ times that of the Sun or 0.19AU, half the size of the orbit of Mercury. That is but a start, as some giants can extend to well past the orbit of Mars. Then factor in Hertzsprung's discovery. A really luminous M star might be another 1000 times brighter than the giant, 100 000 times brighter than the Sun, whence $R = 316/(1/4) = 1265$ or nearly 6AU in radius, larger than Jupiter's orbit. The extreme, represented by μ Cephei and VV Cephei, approach the orbit of Saturn. These are not mere giants, but *supergiants*. Smack in the diagram's middle, the Sun is a G2 dwarf.

In their honor, the graph of absolute visual luminosity against spectral class is known as the *Hertzsprung–Russell diagram*, or the *HR diagram*, a term that is also applied to various adaptations of the original. A greatly expanded view that employs modern parallaxes is displayed in Fig. 6.12, where the dwarfs form a long *main sequence*. Yet the term "dwarf" lingers on, it and "main sequence" synonymous. A class B main sequence star may be 10 times larger than the Sun, yet in astronomy's peculiar parlance it remains a dwarf.

At the lower left of Russell's diagram in Fig. 6.11 is a star far fainter than the dwarfs of the same temperature. To be faint and hot, it must be small, application of the above equation showing it to be roughly the size of Earth. In Fig. 6.12 we see a string of such stars. Since the first ones found were white in color, they became known as *white dwarfs*, even though some are bluish and others red. They are quite distinct from the "ordinary dwarfs" of the main sequence.

6.3.2 Luminosity classes: introducing MKK

The natures of the stars – whether dwarf, giant, etc. – were originally given by lower case letters appended to the spectral classes (see Box 6.1 and Table 6.3). A vastly improved system was presented in 1943 by W.W. Morgan, P.C. Keenan, and E. Kellman (MKK), who appended Roman numerals to the spectral classes of the different kinds of stars. Supergiants are class I, giants III, dwarfs V. The MKK class of the Sun is therefore G2 V. The stretch in supergiant luminosity is so great that these vast stars later had to be divided into brighter Ia and dimmer Ib. In between the supergiants and the giants there is still enough room to accommodate *bright giants*, class II; between the giants and the dwarfs range the *subgiants*, class IV (Table 6.4). The various classes converge at the upper left of the HR diagram, where there is little difference in luminosity among them (Fig. 6.12).

The MKK system is far more than a set of curves. It is a strict empirical classification scheme in which a broad selection of combined Harvard luminosity classes are defined by a large set of selected standard stars. Classification is done through comparison of a program star with the standards, interpolating where necessary. For example, the spectrum of class A2 V is set by ζ UMa, A5 V by β Ari, and so on (Fig. 6.13). In what has become known as the "MK Process," no physical parameter is allowed to influence the class. Only

Class	Description	M_V			
		O	G	M	
0	Hypergiants	−9	−10	−9	S Dor, η Car
Ia–0	Extreme supergiants	−8	−9	−8	ρ Cas, P Cyg
Ia	Luminous supergiants	−7	−8	−7	Betelgeuse, Deneb, ζ Pup
Ib	Less luminous supergiants	−6	−5	−5	Antares, ζ Ori A, δ Cep
II	Bright giants	−6	−2	−2	Canopus, α Hya, ϵ CMa
III	Giants	−5	+1	−1	Aldebaran, Arcturus, Capella
IV	Subgiants	−5	+3	. . .	β Cep, α Tri, δ Gem
V	Main sequence, dwarfs	−5	+5	+12	Sun, α/Proxima Cen, Sirius, 61 Cyg
VI = sd	subdwarfs	. . .	+6	+13	Kapteyn's Star
VII D	white dwarfs	+10	+14	. . .	Sirius B, Procyon B, 40 Eri B

Table 6.4 **The luminosity classes.**
Classes I through V are original MKK classes. Class "0" (Arabic number zero) was added later. Classes "VI" and "VII" for subdwarfs and white dwarfs are non-standard.

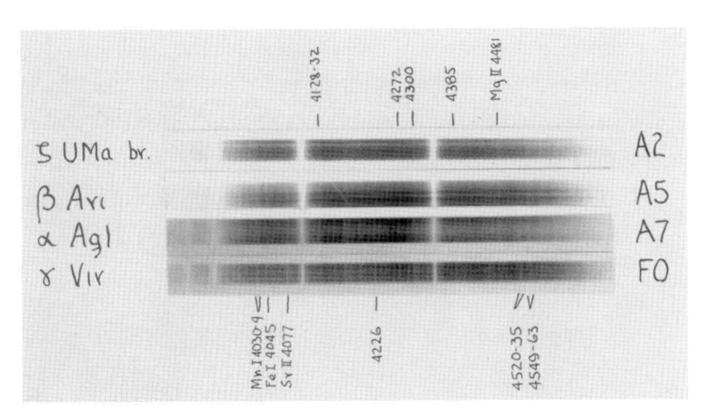

Fig. 6.13 **MKK standards** The panel shows the standards for class A2 through F0 taken from the original MKK atlas. As in the original, the image is presented as a photographic negative. Note again how dramatically the hydrogen lines weaken from top to bottom, and how the Ca II K line (far left) strengthens. (Ca II H is blended with Hϵ.) (Morgan, W.W., Keenan, P.C., and Kellman, E., *An Atlas of Stellar Spectra*, Yerkes Observatory.)

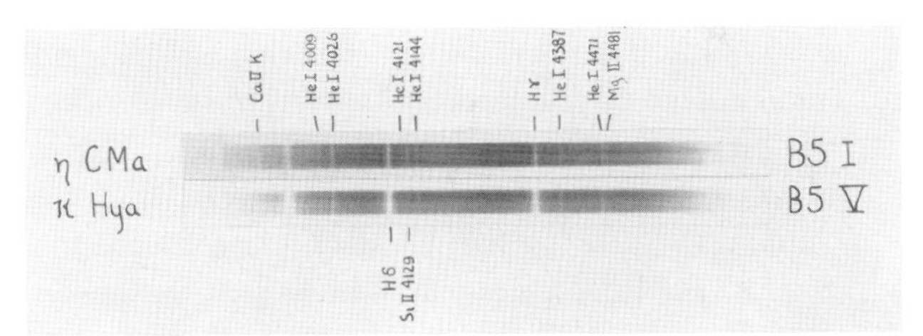

Fig. 6.14 **Luminosity effect** The best-known spectral effect of differing stellar luminosities is the dramatic narrowing of the hydrogen lines from dwarf (V) to supergiant (I). (Morgan, W.W., Keenan, P.C., and Kellman, E., *An Atlas of Stellar Spectra*, Yerkes Observatory.)

the observed characteristics of stellar spectra are to be used, with no judgements being made as to why they appear as they do. Classification is thereby put on a permanent basis, no matter what physical discoveries and correlations are later made.

Nevertheless, to understand luminosity classification, one needs to look at physical principles. In any given range of spectral class there are criteria that depend on luminosity, not through temperature changes, but through density changes. While the mass range across the stellar landscape is large, it is nothing compared to the range in radius, hence volume. Though a class K giant is more massive than a class K dwarf, it is much larger, so that the spectrum-forming photosphere is much more tenuous. Some spectrum lines are especially sensitive to density, hence to luminosity. As a result, giants, supergiants, and dwarfs all have somewhat different relations to temperature (Box 6.2).

The clearest example of luminosity effects is the striking change that appears in the widths of the hydrogen lines among the A and B stars. The lines are broad among the dwarfs and narrow among the supergiants, and is the effect that led Hertzsprung to discover the supergiants in the first place (Fig. 6.14). Atoms do not exist in isolation. After all, it is the collisions among them that raise hydrogen's electrons to the second orbit and that allow the Balmer lines to be seen in the first place [7.3.1]. If the electric field of a close-passing atom or ion disturbs the energy level of an orbiting electron, making it a bit higher or lower, the electron can absorb at a wavelength that is offset from the center of the line. The statistical result from an ensemble of atoms will be a broadened absorption. At high density and pressure, as in dwarfs, the atoms are closer together, and the broadening is greater than it is in supergiants, where the density is low [7.1.6]. We then say that the hydrogen lines show a "negative luminosity effect," since they get weaker as luminosity increases.

Box 6.2 Giants, dwarfs, and temperature

Giants, dwarfs, supergiants, and the intermediate types are separated by relatively subtle spectral effects caused in large part by differing densities. As a related matter, different luminosity classes have different relations to temperature and color, especially in classes G and K (see Fig. 6.9). The ionization balance for a given atom depends upon the rate of ionization by collision, and by absorption of radiation relative to the rate of recombination between ions and electrons. For a given temperature, the lower the density, the less the recombination rate, and the higher the ionization level. To achieve the ionization balance, and thus the spectral class, of a high-density dwarf, a lower density F or G giant must therefore be somewhat cooler and redder. The effect can be reduced, even reversed, by opacity differences among the different kinds of stars to create the differences seen in Fig. 6.9.

The hydrogen effect is much less noticeable among the early B stars. Here, MKK show a strong positive luminosity effect in the Si IV 4089 line. At class K, the hydrogen lines are weak and insignificant, but the CN bands show a positive effect, as does Ca I 4226 in early M. It is most important to note that none of these variations is theoretical or even needs to be understood. The system is empirical, with line changes matched to known luminosities. Explanations can come later. With this knowledge we can roam through "luminosity space" at will. These criteria form the foundation for "spectroscopic parallax" [4.4.1]. As for the Harvard class, the luminosity class can be quantified in terms of measured line strengths or their ratios. In principle, these give finer scales for both parameters, and improved distances.

The simple MKK class leaves out a great many spectral features that we might like to record, such as the presence of emission lines, the appearance of the absorption lines, and various other features. These can be dealt with by again adding suffixes to the spectral class, a process that predates MKK, but that MKK easily accommodates (see Box 6.1). Suffixes are also useful in indicating finer luminosity structure, the Ia–Ib division in supergiants extended to giants and even dwarfs (a IIIa giant brighter than a IIIb). Fine divisions can also be indicated by combining classes: an F star that falls between bright giant and supergiant is classed F5Ib-II.

Like any good classification system, MKK's empiricism supports good theory. The different kinds of stars are seen to be linked together by means of stellar evolution calculations [Chapters 13 and 14]. Main sequence class V dwarfs are core hydrogen burners [12.2.2]. Class III giants are either core helium burners [13.2.2] or are pre- or post-helium burners, while class IV subgiants are in transition to becoming giants. Giants and supergiants are clearly separated by spectrum and luminosity, but the terms have respectively also come to mean stars of intermediate mass (born below about 10 solar masses) that will become white dwarfs, and high-mass stars that can be in a variety of nuclear-burning states, that evolve from class O and early B dwarfs, and that will explode as supernovae [14.2.2]. The density of stars on the

Box 6.3 Stellar numbers

The density of stars across the HR diagram reveals a combination of where the stars have stable (what ranges of temperature and luminosity), long-lived internal configurations, and in what state they are most likely to be born. The meaningful numbers are the percentages of stars within a unit volume that fall into different classes, and here we run into a serious problem. O and B stars are very rare, and in order to incorporate a statistically significant number, we must observe over a huge volume. The distances are then so great that we cannot begin to see the faint red dwarfs, and must use a smaller volume to evaluate how populous they are, within which there are no luminous stars. Even then, huge numbers of dim stars, both white and red dwarfs, escape detection. Given the newness of the explosion in the exploration of L and T dwarfs, we still have only a limited idea of just how many there are.

Nevertheless, we can yet draw some potent conclusions. Within broad limits, 95% of all stars lie on the main sequence, 5% are white dwarfs, and 1% are giants and supergiants. (The numbers are rounded, so they do not add to 100%.) The results clearly show that the main sequence stars endure far longer than giants. Between the dwarfs and giants is the *Hertzsprung gap*, in which rapid evolution results in very few subgiants. (As an end product of evolution, of course, white dwarfs last the longest: eternity. The Galaxy is just not old enough to have built up their numbers.)

Along the main sequence, the number of dwarfs within each spectral class starts out very high at low luminosity among the M stars, and drops precipitously toward class O. Nature not only does not like to make luminous (and as we will see, massive) O stars, but their lives are short to boot. Ignoring L and T stars, whose count is not yet known, 75% of the stars on the main sequence fall into class M. Just 25% fall into classes F, G, and K, and but 1% in class A and 0.1% in class B. Then the bottom really drops out, class O containing only 0.0001%. There may well be as many L and T stars as M stars, yielding some 90% from class M on down, and making the upper main sequence even more exotic.

Yet the constellations are made largely of A and B dwarfs and a variety of giants. The obvious reason is the stars' brilliance, allowing them to be seen from great distances. No dwarf later than class K7 is visible to the naked eye. This powerful "observational selection effect" is seen in many HR and color–magnitude diagrams as well. In spite of the fact that late-type dwarfs dominate the celestial vault, in the diagrams they drop away toward zero, late M dwarfs perhaps appearing as rare as B dwarfs. The Hipparcos satellite, from which Figure 6.15 was made, had a limit of magnitude 11 or so, and most red dwarfs are simply fainter than that and were not picked up. Most HR or CM diagrams will show a similar phenomenon when the detectors reach their limits. We have yet to count the critters at the bottom of the well. We do not even know how deep it is.

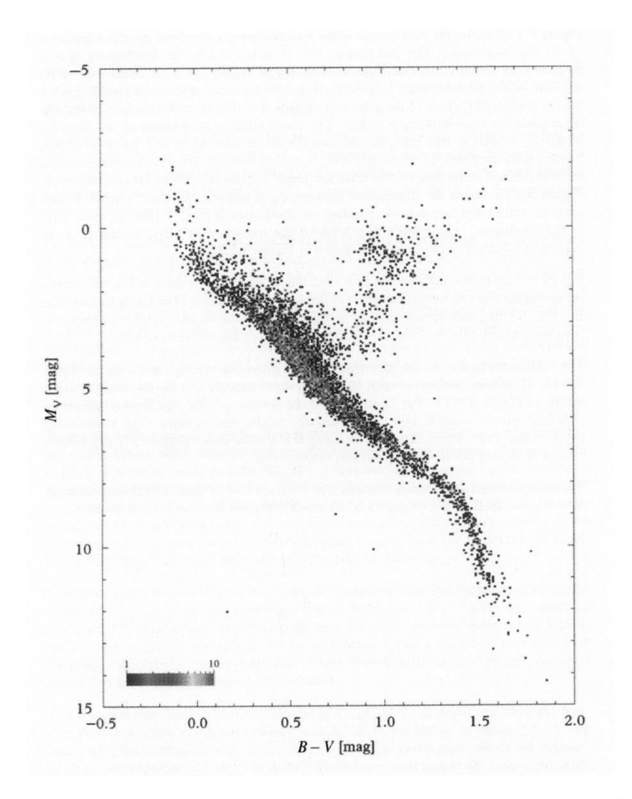

Fig. 6.15 **Color–magnitude diagram** The absolute visual magnitudes (M_V) of Hipparcos stars with accurate parallaxes are plotted against $B-V$. The falloff in star numbers toward the lower main sequence is a selection effect, as Hipparcos was limited to brighter stars. In reality, the numbers dramatically increase toward the later dwarfs. (Perryman, M.A.C., ESA Space Science Dept. and the Hipparcos Science Team, *The Hipparcos and Tycho Catalogues*, ESA Publications Division.)

HR diagram gives information on the lifetimes and formation probabilities of the different kinds (Box 6.3). The data taken together then provide a check on basic nuclear physics. Observation of nature comes first, then comes theory in explanation and support, illustrating the classic and effective scientific method.

6.3.3 Other expressions of the HR diagram

The classic HR diagram plots absolute visual magnitude against spectral class. There are, however, two useful ways of expressing the same kind of information: as a *color–magnitude* (CM) diagram, and as the *log L–log T plane*. Color–magnitude diagrams, in which absolute or apparent magnitudes are plotted against stellar colors [3.4], belong to observers. Broad-band colors are easier and quicker to acquire than spectra and do not require interpretation. The most common and traditional CM diagram is a plot of M_V against $B-V$ (Fig. 6.15), but other magnitude systems and colors (particularly infrared) can and are easily available for use. Indeed, infrared color–magnitude diagrams (M_J vs. $J-K$ for example) must be used for diagrams that display cool L and T stars.

CM diagrams are subtly different from classic HR diagrams because of the non-linear conversion between spectral class and color, and because of the differences in the conversions between giants, supergiants, and dwarfs, as seen in Fig. 6.9. Different magnitude systems will also give quite different appearances. Color–magnitude diagrams are in very common use in assessing the evolutionary status of clusters and of newly formed stars, far more than are traditional HR diagrams. A drawback is that $B-V$ is compressed at high temperatures. While the O stars separate out very nicely spectroscopically, they all have about the same $B-V$, -0.30 to -0.34, not much of a range. Another is that $B-V$ is subject to interstellar reddening [3.5.1], while spectral classes are not. On the other hand, so is M_V. Other magnitude–color systems are similarly affected.

The log L–log T plane belongs to the theoreticians, who calculate luminosity, temperature, and evolutionary tracks from stellar models [12.2, 13.2]. These are used to test theory through comparison with observations and to find the ages of clusters and the Galaxy. Since magnitudes are logarithmic, and spectral class is very approximately linear in log T, the HR diagram and log L–log T distributions will again appear somewhat the same. But because of different temperature dependencies, the relative positioning of the different kinds of stars on the HR diagram and the log L–log T plane will still be different. The biggest problem in transference from one to the other lies in the bolometric corrections [3.3], which are well known for stars in the center of the HR diagram, but become increasingly uncertain at the edges. The log L–log T plane is particularly useful for stars that are not classifiable on the classic Harvard system, or whose colors are compressed because of high temperature, such as the central stars of planetary nebulae [13.4] and white dwarfs [13.5].

6.4 The third dimension

The major problem with standard spectral classification is that it is set up for stars of solar (or Population I) chemical composition. Accommodation to different compositions adds a third and difficult dimension to the system, one that is far from complete.

6.4.1 Metallicity variations

"Metallicity" is an overall term for the abundance of all the elements heavier than helium. Many are the variations, which fall into a number of categories, and are expressed by suffixes attached to the spectral class (see Box 6.1). In some cases, such stars can still be placed on the HR diagram, whereas for others it becomes problematic. There are no fixed rules. One case in point is the set of "metallic line stars," those in which certain metals are enhanced and others depleted (as a result of atmospheric diffusion), leading the astronomer to classify a star as too early or too late depending on the criteria used. Metal-weak stars obviously pose a similar problem (note the third spectrum up from the bottom in Fig. 6.7). The metallic line stars are lumped mostly in classes A and F, where they are known as Am and Fm stars [7.3.3]. Other warm stars that display peculiar spectra and odd chemical compositions, as well as high magnetic field strengths, are dubbed Ap, Bp, and Fp [7.3.3].

6.4.2 Subdwarfs

The composition problem is especially acute for the metal-weak Population II subdwarfs of the Galactic halo [5.1; 5.3.2; 7.3.3]. A subdwarf's low metal content causes low atmospheric opacity and, as a result, the star is smaller than it would be with normal (solar) metal content. As a result, for the same luminosity as a Population I star of the same mass and evolutionary state [12.4.1; 13.2.1], the subdwarf will be hotter, bluer, and of earlier spectral class. The main sequence of subdwarfs (of a given metallicity) runs alongside, but somewhat to the left of the Population I main sequence in Fig. 6.12. By extension they are sometimes called MKK class VI, but more usually go by the term "sd," a classification that is also used for highly developed subdwarf O and B stars [13.4.4].

6.4.3 Carbon stars

Delving into the classification of carbon stars takes a strong constitution. While there are carbon-rich dwarfs (probably the result of mass transfer from a binary companion when

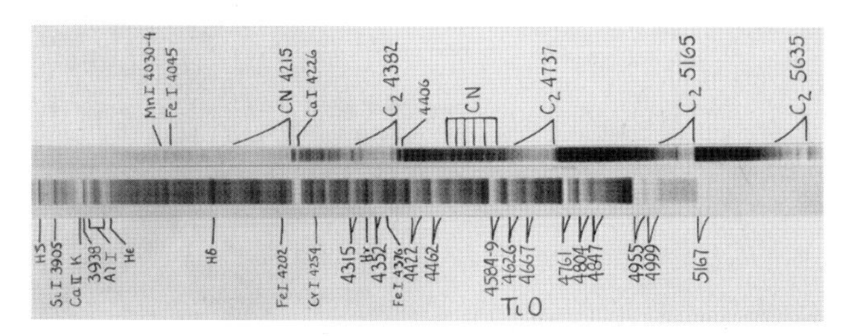

Fig. 6.16 **Carbon star spectrum** The spectrum of a carbon star, HD 52432, at the top, rich in C_2 and CN, is compared with TiO-rich Mira at the bottom. The images are photographic negatives. (Morgan, W. W., Keenan, P. C., and Kellman, E., *An Atlas of Stellar Spectra*, Yerkes Observatory.)

it was a giant [8.5.10]), classical carbon stars are generally giants and display powerful bands of the C_2 molecule (Fig. 6.16), as well as those of CH, CN, and silicon carbide (the so-called Merrill–Sanford, or MS bands; do not confuse this "MS" with actual MS stars, which are hybrids between class M and class S; see below). The warmer R and cooler N stars, which are respectively distinguished by the presence and absence of blue continuous radiation around and below 4000Å, were originally classed on the basis of their carbon bands, much as M stars are classed by TiO. Since R and N seemed to join in some kind of continuous sequence, they were combined into class C. C0 begins at R0 with a temperature roughly the equivalent of G5, and then progresses to C9, where the equivalents to M subclasses and to the older N subclasses are not straightforward.

The titanium abundance across the subclasses of class M is constant, which allows the TiO bands to be excellent signatures of temperature. The compositions of the carbon stars, however, have been changed by the processes of stellar evolution (indeed, some are still being changed [13.3.3]), and therefore the relative carbon abundance is *not* invariable from one star to the next. In the Sun and most other stars, oxygen is about three times as abundant as carbon. In the carbon stars, the ratio is turned around and highly variable from one star to the next. C_2 band strength is therefore a product of both temperature and abundance, and by itself can be a misleading temperature indicator. As a result, carbon stars are commonly classed by two indices as "Cx,y," where "x" gives temperature subclass as found primarily from atomic lines, and "y" gives a qualitative measure of carbon content as found from C_2. TX (19) Piscium, for example, is classed as C7,2, Chi Cancri as a (possibly) warmer but more carbon-rich C5,4. A subset, the *J stars*, have an anomalously high abundance of the heavier isotope of carbon, ^{13}C, the "J" added as a subscript to the class.

Fig. 6.17 **A carbon star** With an amazing $B-V$ color of just over 3.9, the red class C4 carbon star (temperature 2800K) and semi-regular variable AW Cygni practically jumps off the page. (J.B. Kaler.)

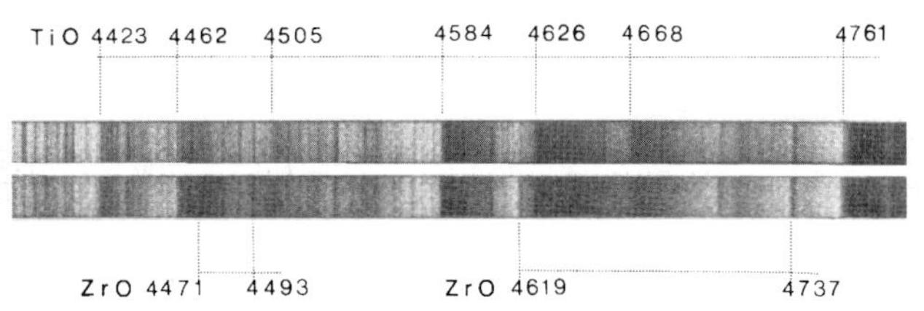

Fig. 6.18 **An S star** The S7 star Chi Cygni (bottom) displays zirconium oxide bands not present in the M6 giant 30 Herculis (top). (Ginestet, N., Carquillat, J.M., Jaschek, M., Jaschek, C., Pédoussaut, A., and Rochette, J. (1992), *Atlas de Spectres Stellaires*, Observatoire Midi-Pyrénées and Observatoire de Strasbourg.)

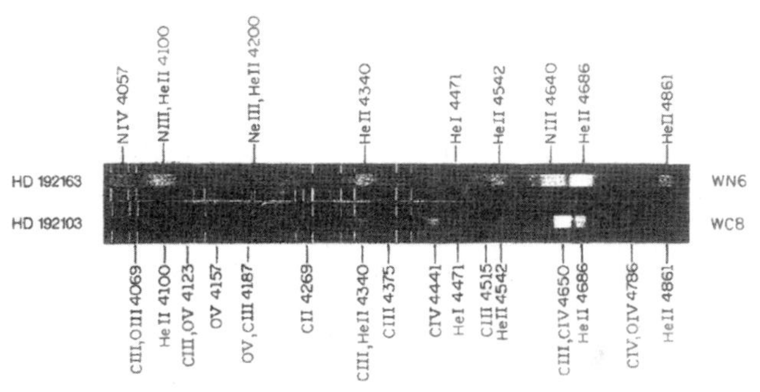

Fig. 6.19 **Wolf–Rayet stars.** At the top is a nitrogen-rich Wolf–Rayet star, at the bottom one that is carbon rich. (University of Michigan photos from Aller, L. H. (1954). *Astrophysics: Nuclear Transformations, Stellar Interiors, and Nebulae*, Ronald Press Co.)

A small fraction of the warmer carbon R giants displays strong CH bands, and rather logically became known as "CH stars." An upgraded classification scheme divides the carbon stars into three groups: C–R (old R or early C), C–H (the CH stars), and C–N (related to the old N stars). The C–R and C–H groups are subdivided 0 to 6 and parallel G4 to M2, while the C–N group is divided 1 to 9. Overlapping C–R and C–H, it parallels G7 to M8. The C–H stars have a strong propensity to be binary, leading us to believe that the atmospheres were contaminated by their companions while the latter – now white dwarfs – were giants themselves [8.5.10]. They are apparently preceded by the dwarf carbon (dC) stars and the subgiant CH stars.

Because the blue portions of cool carbon star spectra are so heavily blanketed with metallic and molecular lines and most likely by dust grains, carbon stars are amazingly red. They stand out on color photographs like droplets of blood (Fig. 6.17). $B-V$ colors can be as high as 5! Though effective temperatures are difficult to extract, they cover about the same range as do the M giants, from about 3500K to 2000K. Temperatures do not correlate very well with class, however, as the spectra are so heavily influenced by differing carbon abundances.

To all of these, add back class S, which are mild carbon-rich giant stars for which the carbon-to-oxygen ratio is close to unity and that parallel the M giants in temperature (Fig. 6.18). Since the carbon abundances in these stars is enhanced through evolutionary processes, there must be a continuum that stretches from oxygen-rich class M to carbon-rich class N. In between the M and S stars we find a few hybrids with both characteristics, the MS stars (not to be confused with the Merrill–Sanford SiC_2 bands), and between S and C are the SC stars, the spectral (and evolutionary) sequence going M–MS–S–SC–C in increasing carbon.

6.4.4 Wolf–Rayet stars

Carbon-rich stars come in a dazzling variety that far transcends the classical R and N "carbon stars." All that is needed to create an extreme example is somehow to strip off the overlying hydrogen-rich layers and reveal the inner layers that have been enriched by thermonuclear processing [14.1.2]. Among the most intriguing of them are the Wolf–Rayet (WR) stars, named after the French discoverers Charles Wolf (1827–1918) and Georges Rayet (1839–1906). These stars actually come in two distinct flavors (Fig. 6.19). Both have immensely broad emission lines that imply strong outflowing winds. The carbon-rich (WC) variety displays powerful emissions of carbon in the form of C II, C III, and C IV (particularly the 4659Å line) and of He II 4686. Oxygen is present, but there is neither hydrogen nor nitrogen. A nitrogen-rich WN variety, defined by broad emissions of N III, N IV, and He II, reverses the roles. Here we see no carbon, though a little hydrogen sneaks through. The WC stars have an offshoot that is commonly considered a third kind, the WO stars, which are similar to the WC stars, but have more oxygen.

The WR stars are subdivided WN2 through WN11 and WC4 through WC9, depending on the ratios of the strengths of the lines of high ionization relative to those of low ionization. Temperature correlates with class, though the values, which range from perhaps 100 000K down to 50 000K, are hard to determine because of the surrounding windy ejecta. In spite of the great strengths of the nitrogen and carbon lines, all three kinds are really hydrogen-deficient helium stars in which carbon or nitrogen is the secondary element. WR stars are all massive, belong to Population I of the Galactic disk [5.1], and (because of their lack of hydrogen) are prime candidates to make Type Ib supernovae [14.2.1]. Related to WR stars are the highly evolved B[e] stars that have emission lines and surrounding disks (and are not to be confused with more ordinary Be stars [10.6.3], which also have emissions from surrounding disks).

6.4.5 Planetary nebula central stars

As so often happens in astronomy, entirely different scenarios can produce highly similar phenomena: the central stars of some planetary nebulae closely mimic the Wolf–Rayet phenomenon [13.4.4]. A major difference is that while Population I WR stars divide into WN and WC, the planetary nuclei are (with one exception) all WC, the result of their particular evolutionary origins. These WC stars are subdivided in much the same way as their Population I cousins, so it is imperative upon seeing a WC class that the particular kind of WC star be known, whether massive Population I or low-mass planetary nebula nucleus. Confusing the issue is that some WR stars have surrounding nebulae created by their own ejecta that can mimic the planetary nebulae.

Such borrowing continues with the use of "O" for the cooler central stars with pure absorption spectra, and Of (or Ofp) for stars with helium and carbon emission (see Box 6.1). Those that are in the process of cooling are commonly described in terms used for white dwarfs.

6.4.6 White dwarfs

More borrowing is used for the dense, degenerate white dwarfs [13.5]. The spectra of the first white dwarfs found – 40 Eridani B, Sirius B, Procyon B – exhibited broad, dark hydrogen lines (Fig. 6.20). They looked like stars of class A, and were so categorized. A different variety showed helium lines and were classed as B stars, while yet others had the ionized helium of the much more luminous O dwarfs and giants. To distinguish these from main sequence and other kinds of stars, the prefix "D" is used (for "degenerate"), Sirius B and 40 Eri B, for example, becoming DA stars; others became DB and DO. To these are added the "DC" stars, whose spectra are continuous.

The similarities between the spectra of white dwarfs and those of the main sequence are superficial. DA white dwarfs can be much hotter or cooler than ordinary A stars. There are

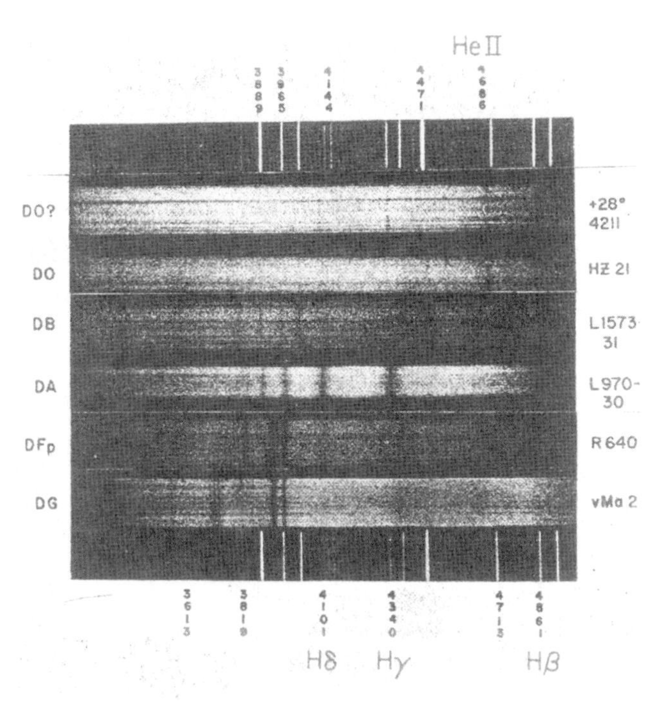

Fig. 6.20 **White dwarfs** White dwarf classification depends on both temperature and composition. DA stars have no helium, whereas DO and DB stars have only helium. Other classes badly mimic other stars of the main sequence. (Courtesy of Palomar Observatory/California Institute of Technology, J. L. Greenstein.)

few, if any, other absorption lines that can discriminate other classes. Sirius B has a temperature of 27 000K, which would place it well into the realm of early B. Yet no characteristic helium lines are found. And while ordinary O and B stars all display strong hydrogen lines (hydrogen being by far the most popular celestial element), the DO and DB stars have no hydrogen lines at all.

Most of the stars on the HR diagram have chemical compositions at least reasonably similar to that of the Sun, which is really why the HR diagram works at all. Classes O through T are caused by differences in temperature, not composition. In spite of this general rule, even reasonably large differences in composition, such as those exhibited by carbon stars, can find a place. The white dwarfs, however, are quite the extreme. The DA stars have no photospheric helium, and have nearly pure hydrogen atmospheres. Above 7000 or 8000K they will all look like A stars. Toward lower temperatures, as the hydrogen absorptions weaken, we do see some metals, particularly the ubiquitous Ca II H and K lines, so these can be classed as DF, DK, and so on. Yet the spectra bear little actual similarity to those of main sequence stars. Because the atmospheres of the hydrogen-less white dwarfs are nearly pure helium, their low-temperature versions take on continuous spectra, or have weak carbon lines, and are at least recognizable by the absence of hydrogen, unless the temperature is so low that not even that element would be visible.

The result is that white dwarfs divide into two broad classes that imply not temperature, but chemical composition. DA stars have no helium, while DB stars have no hydrogen. Since

DO stars are really hot DB stars, the terms "DA" and "non-DA" are in common use. A more realistic system is now in place that keeps the original DA, DB, DO, and DC, but adds DZ (metal lines only), and DQ (carbon present). Four suffixes indicate magnetic fields (P, with polarized light, and H, without), peculiar (X), and variable (V). By extension from subdwarfs, white dwarfs are sometimes seen referred to as distinctly nonstandard MKK luminosity class VII.

Overall, DA white dwarfs are three to four times more common than non-DA. The ratio, however, is oddly dependent on temperature. Above about 15 000K, there are six times as many DA as non-DA, while below that mark they count about equal. Odder still, between 30 000 and 45 000K there is a "non-DA desert," with none of them at all. While the reason must somehow lie in the realm of stellar evolution [13.5], no one yet knows what it is.

Without the pursuit of physical parameters – temperature, luminosity, mass, chemistry – there is no point to any classification scheme. Classification, on the other hand, simply and naturally leads to the physical properties and to the correlations among them, much as this chapter leads to the next.

CHAPTER SEVEN

Stellar properties

Once the different kinds of stars are defined, we can examine their physical natures, all the properties that are common to stars and that at the same time make each star unique. To learn the nature of a rock, bring it into the laboratory and analyze it. However, with the exceptions of pre-solar interstellar dust grains found in Solar System meteorites [11.4.1], particles from the solar wind [12.1.4], and cosmic rays [14.3.5], we can learn of stars only by remote sensing, from the radiation they emit and absorb. Interpretation of the radiation first requires that we examine the laws that govern it and the way in which atoms and ions interact with it.

7.1 Radiation laws

Astronomy and physics are an intimate couple. In the hands of Galileo and Newton, much of physics grew from astronomy. Yet without the physics required to interpret observation, astronomers would still be debating constellation boundaries. In reality there are no compartments to human knowledge, as one branch flows smoothly into another. Radiation laws, as well as all the other laws that govern our world, are just as much a part of what we call astronomy as they are of what we call physics.

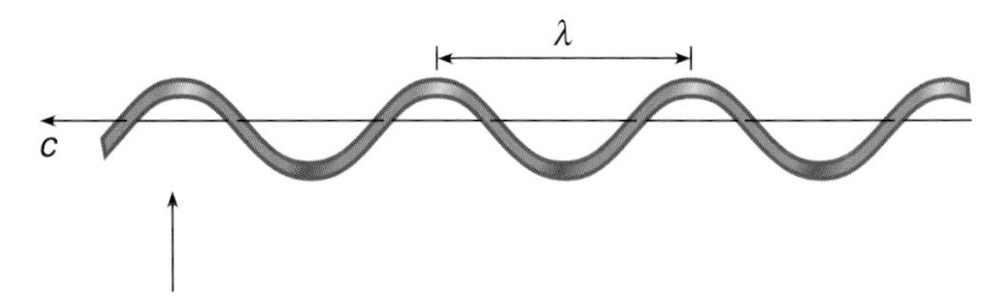

Fig. 7.1 **Light as a wave** A light wave with wavelength λ travels through space at the speed of light, *c*, with ν cycles per second – the frequency – passing the arrow. (Kaler, J.B. (1994). *Astronomy!* New York: HarperCollins, reprinted by permission of Pearson Education, Inc.)

7.1.1 Inverse square

The electromagnetic force, carried by protons and electrons, declines in strength by the inverse of the square of the distance from its source. So do light and all the other forms of electromagnetic radiation (Box 7.1). The *luminosity*, L, of a source of light is the amount of energy that it radiates per second into space. In standard meter–kilogram–second (mks) units, the unit of energy is the *joule*, while the unit of power – the rate at which energy is used (joules per second) is the familiar *watt*. (Astronomers commonly use centimeter–gram–second, or cgs, units, in which the unit of energy is the *erg*, where 1 erg $= 10^{-7}$ joule). The luminosity of the Sun, for example, is 4×10^{33} ergs s^{-1} $= 4 \times 10^{26}$ watts.

A star lies at the center of a sphere of radius R (Fig. 7.3). Given a steady state, the amount of radiation that leaves the star per second must also pass through the surface of

Box 7.1 Light

The known forces of nature are four in number. The most familiar is the weakest, *gravity* [8.2.3, Box 5.4]. What we call "light" is a manifestation of the far stronger *electromagnetic force*. (Between the two is the "weak force" that governs radioactivity, and at the top is the strong force that ties atoms together; see Box 7.4). Consisting of a flow of alternating electric and magnetic fields, light is only a small portion of the more general concept of *electromagnetic radiation*, which moves in vacuo at the "speed of light," c, 299 792.458 kilometers (186 282 miles) per second. The "kind" of electromagnetic radiation is determined by its *wavelength* (λ), the separation between wave crests, or by its *frequency* (ν or f), the number of wave crests that pass a given point per second (Fig. 7.1). Multiply the two together and you get how far the wave has moved in a second, and recover the speed: $\lambda\nu = c$.

The full range of wavelengths defines the *electromagnetic spectrum*. For historical reasons, different vaguely defined ranges of wavelengths carry different names. *Visual* (or *optical*) radiation falls between about 4×10^{-5}cm and somewhat longer than 7×10^{-5}cm, or (given that 1 angstrom, Å, equals 10^{-8}cm) between 4000 and 7000Å. Within that band, different wavelengths define different visual colors from red to violet (Fig. 7.2). Radiation with waves longer than that of red light up to around a tenth of a millimeter is called *infrared*, and beyond that *radio*. Shortward of violet lies the *ultraviolet*, which makes a transition to the *X-ray* at around 100Å. Shorter than around 1Å such radiation is called *gamma ray*. There are no known limits. Radio waves can have wavelengths of kilometers, gamma rays of billionths of an angstrom.

In some contexts, electromagnetic radiation also behaves as if it were a flow of particles, and in fact it is both at the same time. The concept is embodied in the *photon*, which is a particle that in a crude sense can be thought of as a distinct packet of energy carried by an internal wave. The energy carried by a photon is directly proportional to its frequency, or inversely proportional to its wavelength:

$$E = h\nu = hc/\lambda,$$

where h is Planck's constant, equal to 6.6×10^{-34} (in mks units), or 6.6×10^{-27} in cgs. Radio photons carry relatively little energy and are benign to humans; infrared is felt as heat; visual radiation ("light") is energetic enough to activate the retina; ultraviolet burns the skin; X-rays are dangerous and gamma rays are deadly. Individual optical photons do not carry much energy: a hundred-watt light bulb emits nearly 10^{20} of them per second.

The Earth's atmosphere is more or less transparent to visual radiation, allowing us to admire and examine the stars from the ground. Much of the infrared and some of the radio, however, is absorbed or refracted back into space. The air fortunately blocks everything shortward of 3000Å, else we could not exist in the intense high-energy radiation bath from the Sun. To examine both low- and high-energy processes, astronomers must therefore go into space. The greatest story of twentieth-century astronomy was the opening up of the entire electromagnetic spectrum for scrutiny, which has allowed the study of how stars are born, live their lives, and die.

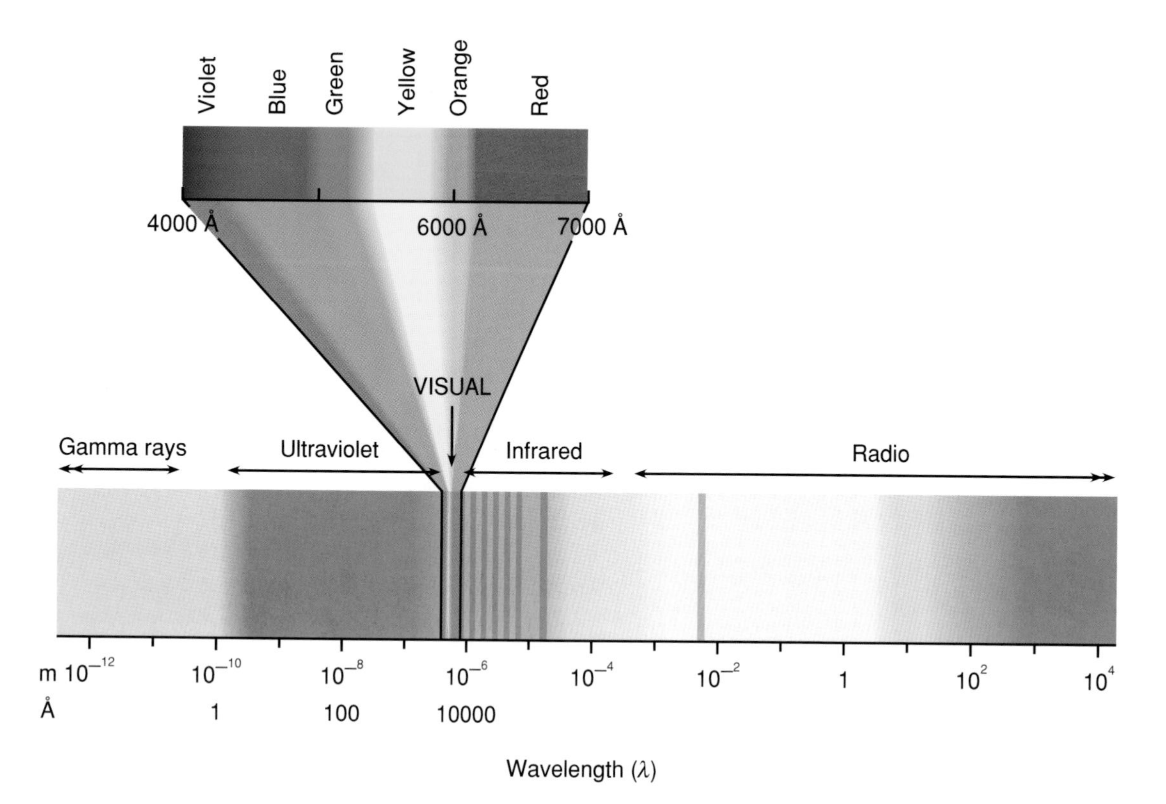

Fig. 7.2 **The electromagnetic spectrum** The spectrum runs from gamma rays at left to radio radiation at right. The visual spectrum lies in the middle, and is expanded above to show the colors seen with the human eye. (The dark bands show regions in which the Earth's atmosphere is opaque.) (Kaler, J.B. (1994). *Astronomy!* New York: HarperCollins, reprinted by permission of Pearson Education, Inc.)

the sphere per second. The amount of radiation that passes through a unit area of the sphere is called the *flux*, F (with units of watts per square meter). The flux through any sphere is simply L divided by the surface area of the sphere (which is $4\pi R^2$) or $F = L/4\pi R^2$. In the simplest terms, the apparent brightness of a star depends on how much light enters the aperture of your eye, or upon the received flux. The apparent brightness therefore drops off as $1/R^2$. If you increase your distance from a star by a factor of 10, the star looks 100 times (5 magnitudes [Box 3.3]) fainter.

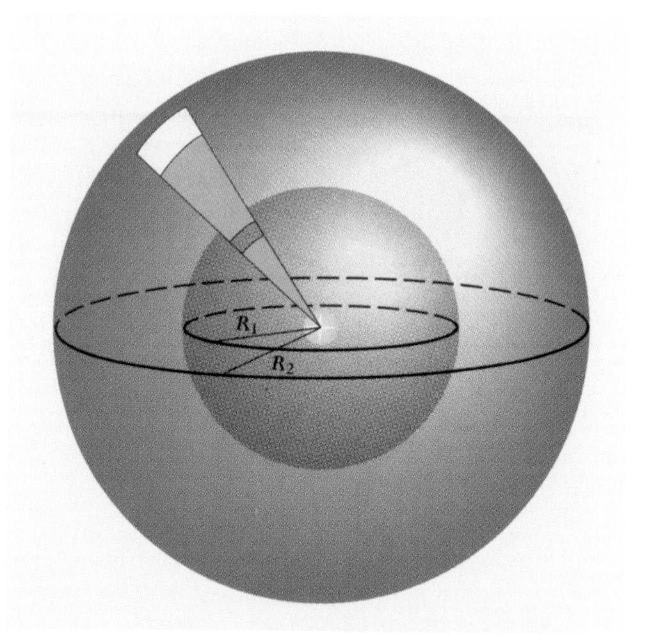

Fig. 7.3 **Inverse square law** The light radiated by a star in the center of a sphere must spread itself over the sphere's surface area, which depends on the square of the radius. The outer sphere (R_2) has twice the radius of the inner sphere (R_1), so the energy passing through a unit area of the outer sphere per second (the flux) is only a quarter that passing through a unit area of the inner sphere. (Kaler, J.B. (1992). *Stars*, Scientific American Library, New York: Freeman.)

7.1.2 Blackbodies and radiation laws

In normal parlance, "black" is the absence of color, signifying no reflection. A body that is black would therefore be invisible except in relief against an illuminated body. A *blackbody* is similarly defined as one that reflects no radiation, that is, it absorbs all the light that falls upon it. But the definition says nothing about the blackbody's ability to radiate on its own. Blackbodies can therefore be very bright. While there is no perfect blackbody in nature, some physical systems come quite close. Even the Sun can be reasonably approximated by one; if you could shine a light on it, the radiation would be absorbed.

If a blackbody absorbs radiation, it gains energy, and its temperature (a measure of internal energy) must increase. To maintain a constant temperature, the blackbody must emit just as much radiation as it absorbs. The nature of the output, however, need not be the same as the nature of the

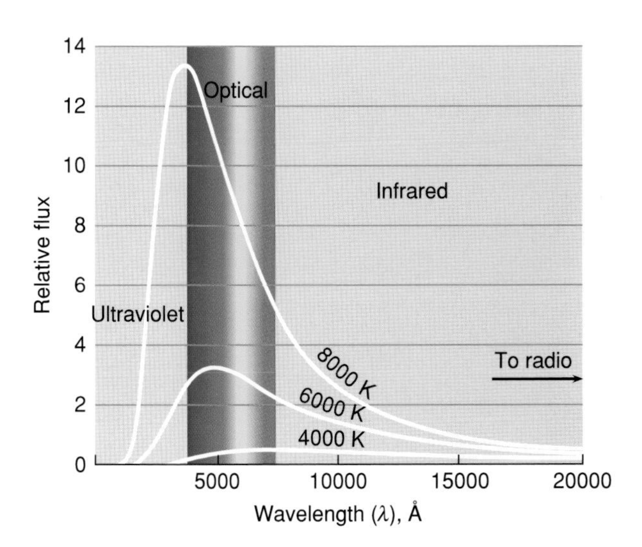

Fig. 7.4 **Blackbody radiation** The three curves show the intensity of radiation plotted against wavelength as emitted by a blackbody of a given size at three temperatures. The mathematically similar curves go to infinite wavelength at right, and have sudden cutoffs at left. The color seen by eye is an amalgam of the radiated colors. (Kaler, J.B. (1994). *Astronomy!* New York: HarperCollins, reprinted by permission of Pearson Education, Inc.)

Box 7.2 The blackbody law

At some point all astronomers encounter the "Planck function" or "blackbody law." Derived by Max Planck from the tenets of quantum mechanics, it matches the observations of how blackbodies radiate. Write the flux F (watts per square meter) per unit wavelength interval (that is, per angstrom) as F_λ:

$$F_\lambda = 2hc^2/\lambda^5(1 - e^{hc/\lambda kT}),$$

where h is Planck's constant and k is Boltzmann's constant (in mks units equal to 1.38×10^{-23}). Though not an overt presence in these pages, scientists would be lost without calculus. If you "integrate" the Planck function, that is, determine the areas under the curves in Fig. 7.4, you obtain the Stefan–Boltzmann law, $F = \sigma T^4$. "Differentiating" the function gives the slope of the curve at all wavelengths, which allows you to find that wavelength where the curve is horizontal to the lower axis. You therefore have λ_{max} and the Wien law. To a reasonable approximation, stars radiate according to the Planck function, which then provides a good starting point for theoretical analysis.

input. A cold blackbody only a few degrees above absolute zero (−272°C, −459°F) has little internal energy, and cannot possibly emit high-energy radiation (see Box 7.1). It will radiate only radio waves and will be quite invisible to the eye. At a few tens or hundreds of kelvin (kelvins, K, defined as Celsius degrees above absolute zero), there is enough energy to radiate at shorter wavelengths. Now the body glows both in the radio and the more energetic infrared. At a few thousand K, we see it glow in the visual or optical, and at several thousand K it becomes a good high-energy ultraviolet emitter; at the same time it also easily radiates long-wave radiation. At hundreds of thousands of K, X-rays, then gamma rays pour out.

The concept leads to two powerful radiation laws. Perfect blackbodies emit according to a specific equation discovered early in the twentieth century by the German physicist Max Planck (1858–1947) under the rules being developed for the science of quantum mechanics, the idea that energy comes in discrete packets, hence the concept of "photons" [see Box 7.1]. Though the law is mathematically complex, the predictions are graphically clear, and describe what is observed (Fig. 7.4; Box 7.2). Going from short to long wavelengths, at any one temperature the amount of radiation suddenly erupts from near zero, climbs to a peak, and then declines asymptotically back to zero at infinite wavelengths. Radiation of all kinds is present at wavelengths longward of the well-defined peak, whereas shortward of it, the radiation quickly disappears. Cold bodies do not radiate high-energy photons, while the hottest bodies radiate photons of all wavelengths.

As temperature climbs, the peak of the blackbody curve (or *Planck curve*), called λ_{max}, shifts to shorter wavelengths according to

$$\lambda_{max}(\text{meters}) = 2.898 \times 10^{-3} T^{-1} \text{ (K)},$$

called the *Wien law*, after another German physicist Wilhelm Wien (1864–1928). The law explains star colors. The color of a blackbody as seen by eye depends on the wavelengths emitted combined with the eye's sensitivity to different wavelengths. The result will not be pure spectral color, but subtle shades that depend on what spectral color dominates. Starting at around 2000K, stars appear reddish, then turn to yellow–orange in the low thousands of K, to white near 8000–10 000K, to bluish-white at high temperatures. The fact that stars are *not* perfect blackbodies can enhance the colors, carbon stars [6.4.3] in particular appearing very red (see Fig. 6.17).

As its temperature increases, a blackbody emits more radiation at all wavelengths up to the high energy cutoff, and consequently also becomes brighter. The flux (F) of radiation per unit area from the surface of a blackbody is given by the *Stefan–Boltzmann law* (after the Austrian physicists Josef Stefan, 1835–1893, and Ludwig Boltzmann, 1844–1906) which states:

$$F \text{ (watts per square meter)} = \sigma T^4,$$

where the Stefan–Boltzmann constant $\sigma = 5.67 \times 10^{-8}$ (in mks units) or 5.67×10^{-5} (cgs). The quantity of radiation therefore has a powerful dependence on temperature. Double it, and the surface of a blackbody shines 16 times brighter; triple it and the flux increases by a factor of 81. The total luminosity of a spherical blackbody of radius R is just equal

Box 7.3 Radiation and kinetic temperature

In a dense, high-pressure gas, temperature defines not only the radiative output, but also the velocities of the atoms, ions, electrons, molecules: whatever the gas is composed of. The particles, which are constantly colliding, do not have a single velocity, but a distribution of velocities. Some collisions stop the particles, such that for a brief moment they are motionless, while other collisions will cause the particles to go screaming away at very high speed. Even in a thin gas, the collisions are sufficiently numerous to establish a "Maxwellian" distribution (after James Clerk Maxwell, 1831–1879).

N is the number of particles per unit volume. If N f(v) is the number of particles moving at a specific speed within a given speed range (v),

$$N\,f(v) = N\,4\pi(M/2\pi kT)^{3/2}v^2\exp(-Mv^2/2kT),$$

where M is a particle's mass, and "exp (x)" = "e^x" (where e is the base of the natural logarithms = 2.718 . . .). Nf(v) starts off at zero, climbs quickly to a maximum, and then slowly declines toward high velocities with a long "tail" that goes toward infinity, and that in practice allows for very fast moving particles (Fig. 7.5). The equation gives both the "most probable" and average velocities, which climb as the temperature increases. If we can find a measure of the speeds, which is frequently possible, we can reverse the procedure to find what is called the *gas kinetic temperature*, T_{kin}. In a dense gas, this temperature will be the same as the temperature derived from the two blackbody laws, T_{bb}. In a low density gas, that is, in a non-blackbody, T_{kin} does not predict the radiation flux, and we have to go to more detailed theories that involve exactly how the radiation is produced.

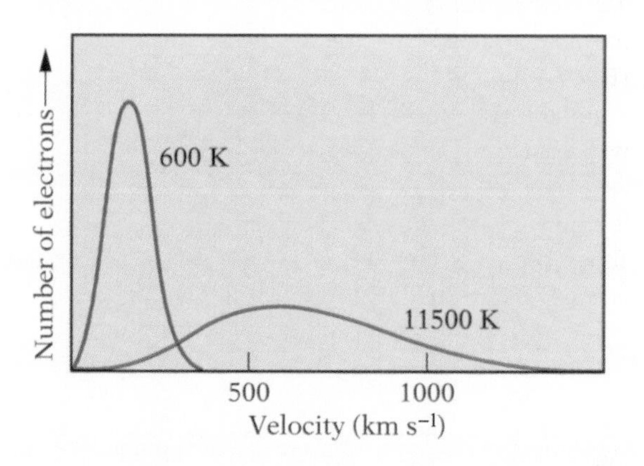

Fig. 7.5 **Maxwellian velocity distribution** The curves show the number of electrons moving at a given velocity for two temperatures. At each temperature, some particles are moving at zero velocity, while others move at very high speeds. As temperature increases, average velocity increases, along with the number on the extended high-speed tail. (Kaler, J.B. (1997). *Cosmic Clouds*, Scientific American Library, New York: Freeman.)

to the flux (radiation per unit area) multiplied by the surface area, or

$$L = 4\pi R^2\sigma T^4,$$

which, along with Kepler's third law [8.2.3] and the magnitude equation [Box 3.3], is among the most important relationships in astronomy.

Blackbodies are just the beginning, as stars only approximate them. We thus need to see how deviations can be factored in.

7.1.3 Non-blackbodies

Blackbodies radiate as a result of their heat, and thereby produce *thermal* radiation. There are many *non-thermal* ways in which radiation can be produced, to which the blackbody laws do not apply. A dentist does not use a 100 000 kelvin heat source to take an X-ray image. A prime example is the solar corona [12.1.4] which, though at a temperature of two million kelvin, is invisible to the eye in a blue sky – a clear violation of the Stefan–Boltzmann law. Blackbodies are opaque to the radiation they produce. The corona, on the other hand, is a thin gas that is heated by the release of energy derived from solar magnetism. A blackbody's temperature describes both

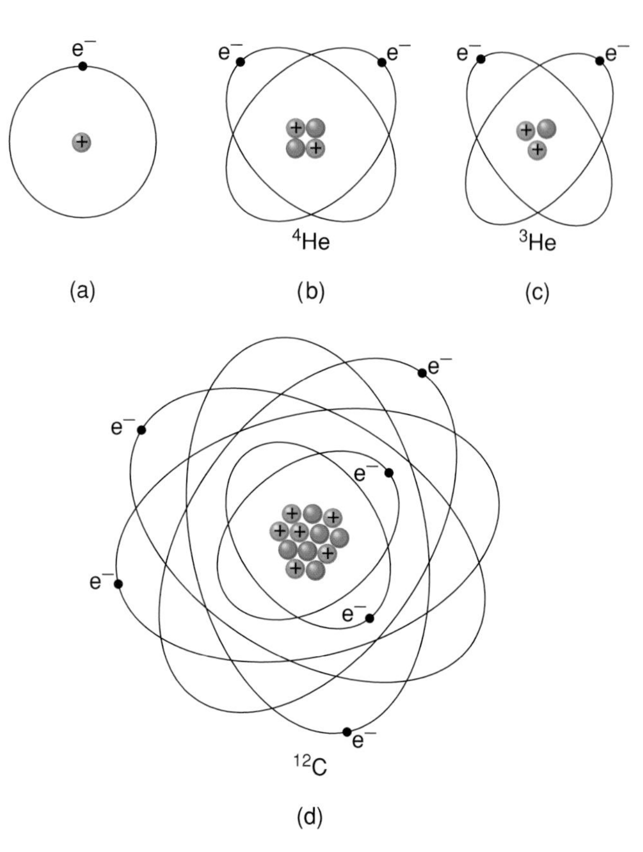

Fig. 7.6 **Atoms** The kind of atom is determined by the number of nuclear protons (red), the simplest hydrogen (a); the neutrons (blue) of heavier atoms help provide nuclear binding energy that keep the nuclei together. Two helium isotopes, ^{4}He and ^{3}He (b) and (c), are shown, along with common carbon, ^{12}C (d). (Kaler, J.B. (1994). *Astronomy!* New York: HarperCollins, reprinted by permission of Pearson Education, Inc.)

the radiation emitted and the velocities of the atoms in the gas. In the corona, however, the temperature describes only the velocities and has little to do with the emitted radiation (Box 7.3).

Another example is *synchrotron radiation*, which is produced when electrons moving near the speed of light are accelerated around embedded magnetic field lines. Such radiation is emblematic of the outer planets (Jupiter in particular) and supernova remnants [14.3.4]. The kind of radiating source is easily determined from the spectrum of the radiation.

7.1.4 Absorption and emission lines

Blackbody radiation laws (as well as those that describe non-blackbody radiation) involve *continuous radiation*, radiation that changes smoothly and continuously with wavelength. They exclude *absorption* and *emission* lines which, when introduced, by their very natures cause deviations from blackbody behavior. Absorptions and emissions are produced as a result of the interaction of radiation with atoms, their ions, and molecules (Box 7.4).

Box 7.4 Atoms

Everything, said Democritus in the fourth century BC, is made of uncuttable parts, *atoms*. He was close. The atom can indeed be cut, but only a bit farther down the chain, until we reach what apparently really is uncuttable. All matter is made of different kinds of atoms that are in turn made of protons, neutrons, and electrons. (Protons and neutrons are made of still smaller particles called quarks, but these need not be addressed here.) The proton carries the positive electric charge, the electron the negative charge. The neutron has about the same radius (10^{-13}cm) and mass (6×10^{-27} grams) as the proton, but is neutral. The negative electron has 1/1800 the mass of the proton and no measured radius.

Protons and neutrons combine to form the core, or nucleus, of an atom. Like charges repel, unlike charges attract. Negative electrons are therefore attracted to the positive nucleus, and are loosely said to "orbit" (much as gravity causes the planets to orbit the Sun) with a negative *binding energy* that depends on the kind of atom involved. (The analogy is not that good, because like photons, these particles also have wave natures, which drastically changes all the rules.)

The kind of chemical element is determined by the number of protons in the nucleus, the "atomic number," *A* (Fig. 7.6). Hydrogen, the simplest atom, has but one proton. Two protons make helium, three lithium, 14 nitrogen, and so on all the way to number 92, uranium, and beyond, all of them sequentially listed in chemistry's periodic table (see Figure 6.4). Since protons are like charges, they need another force, the *strong force*, to keep them together. This force, unlike gravity and electromagnetism, does not behave according to an inverse square law. It is powerfully attractive over a distance of the radius of the nucleus, and then drops quickly to zero. If protons are close enough together, they stick (with a negative binding energy far greater than that which holds the electron) in spite of their positive charges. Neutrons also carry the strong force, and are added in to make the binding even tighter. Two protons alone cannot exist side by side, but add one or two neutrons, and you float a helium balloon.

In a neutral atom, there are just as many electrons as protons. Take away an electron (which requires removing it from its "energy well"), and you have a positively charged ion, take away two, and the atom is doubly ionized, and so on. Of necessity, we live in a largely neutral world, as the electromagnetic force is so powerful that unbalanced ions and electrons can be exceedingly dangerous (as anyone who has ever been electrically shocked will tell you). Ions are indicated by superscripts: hydrogen with one electron gone (a bare proton) is positively charged one unit, and is therefore H^+, while iron with 3 electrons removed is Fe^{+3}.

Ions are but one variation on the atomic theme. Most chemical elements exist in a range of stable *isotopes*, in which the atoms carry different numbers of neutrons. The kind of isotope is indicated by a superscript that gives the total number of nuclear particles. Ordinary hydrogen – a proton – is ^1H. Add a neutron, and you get a heavier isotope, deuterium, ^2H, add a second and there is tritium, ^3H. Only the first two are stable, however. The extra neutron makes tritium unstable and causes it to fall apart with the release of energy, that is, it is *radioactive* (a by-product of the *weak force*). Most helium is ^4He (two neutrons and two protons), but a small fraction also exists as ^3He (see Fig. 7.6). Tin (Sn, 50 protons) has 10 stable isotopes, while technetium (Tc, 43) has none at all.

Above bismuth (Bi, 83) there are no stable isotopes: all are radioactive. Particle and electromagnetic radiation from decaying radioactive isotopes is dangerous. Just how much depends on how quickly a radioactive isotope disintegrates. Disintegration time is told by the isotope's *half-life*, the time it takes a given quantity to cut itself in half. That of ^{238}U (uranium) is 4.5 billion years, so the isotope is relatively benign. The longest-lived isotope of radium, ^{226}Ra, has a half-life of only 1600 years, making the stuff very dangerous.

Atoms can share electrons and can combine into an almost infinite variety of *molecules* that are expressed by linking the elemental symbols and using subscripts to count their number. The very simplest molecules join atoms of the same kind: molecular hydrogen, H_2, consists of a pair of hydrogen atoms, O_2 of two oxygens. O_2 behaves entirely differently from atomic oxygen or from ozone (O_3). Water (H_2O) has two hydrogen atoms and one oxygen, methane (CH_4) a carbon and four hydrogens. The isotopic makeup of a molecule can be distinguished with the usual superscripts. Ordinary carbon monoxide (CO) is $^{12}C^{16}O$, while isotopic versions might be ^{13}CO or even $^{13}C^{18}O$.

The world, stars, the Universe, are made from a mixture of it all. It is the astronomer's job to figure how much of each kind of matter there is and where it all came from.

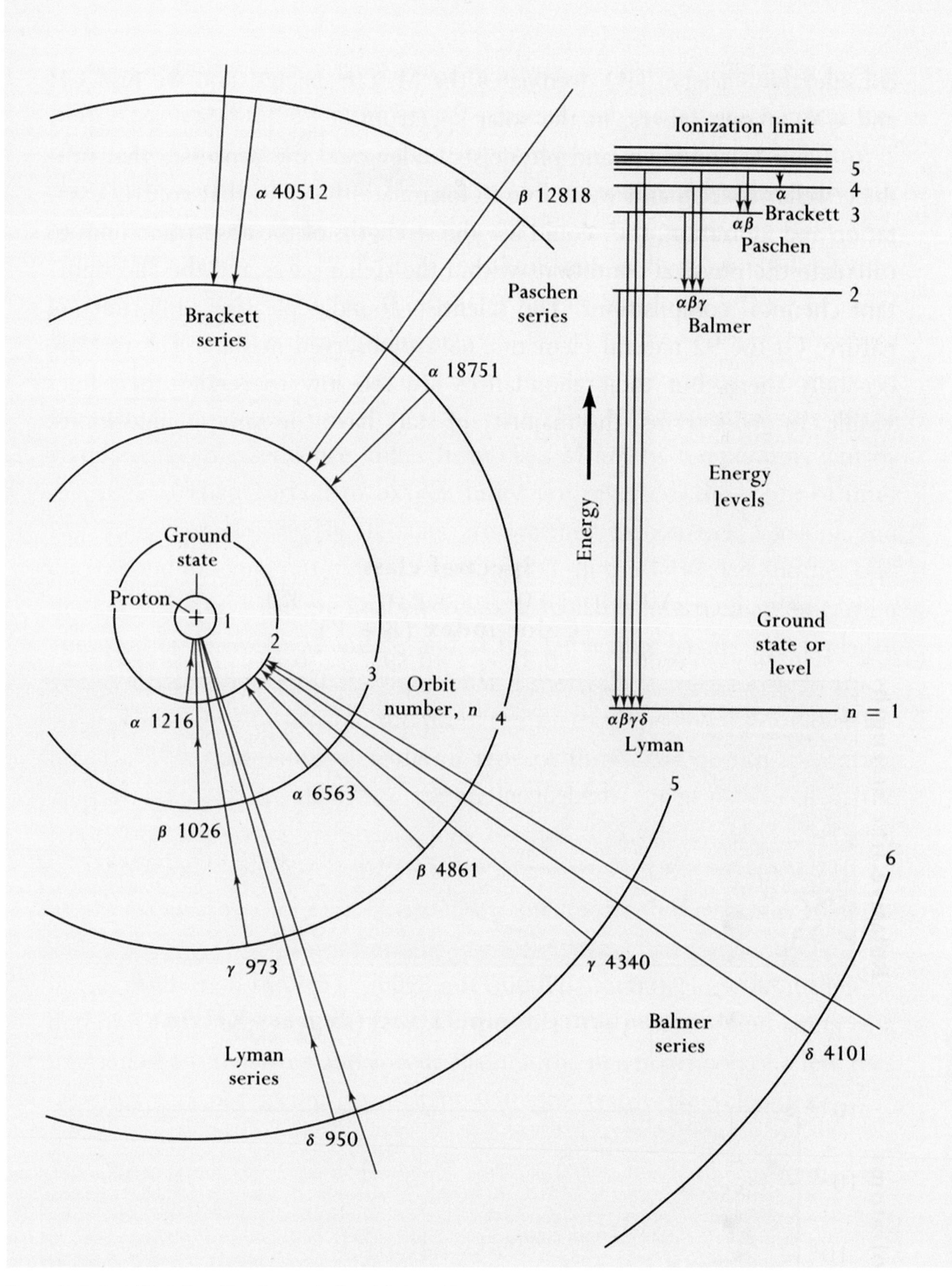

Fig. 7.7 **Hydrogen orbits and transitions** Hydrogen's allowed electron orbits scale in radius according to $n^2 r$, where r is the radius of orbit 1. The electron can move back and forth through collisions with other atoms or by absorbing or emitting photons. The arrows show sets of downward transitions that form series of emission lines (and in reverse, series of absorption lines). The energies of the orbits are shown in the inset. (Kaler, J.B. (1992). *Stars*, Scientific American Library, New York: Freeman.)

The electrons are the culprits behind the lines. Unlike planets, bound electrons can take on only specific orbital energies, and in the classical view, specific orbital radii. Use hydrogen as a simple example. The minimum orbital radius of hydrogen's single electron is about 10^{-8}cm. Since it is bound to the proton, it has a negative energy of -13.6 *electron volts* (eV), that is, it takes 13.6eV to rip it away, or to ionize the atom (where 1eV is 1.6×10^{-12}erg). As discovered by the Danish physicist Niels Bohr (1885–1962), hydrogen's electron can exist in orbits with radii that are n^2 times the minimum radius, where n is a number from 1 to (in principle) infinity (Fig. 7.7). Since the orbital radii are fixed, so are the orbital energies, which are now said to be *quantized*. An input of energy is required to move an electron from a tightly-bound inner orbit to a more loosely-bound outer one, so electrons in outer orbits have more energy than those in inner orbits. However, the increase in energy from one orbit to the next one outward steadily diminishes (see the inset to Fig. 7.7). The energies, or *energy levels* of each orbit therefore converge on a limit, the *ionization limit*, at which point the electron becomes unbound and flies free.

Physical systems seek their lowest energies. So will the electron, as it will try to reside in orbit 1, the *ground state*. It can gain energy and move to an outer orbit through a collision with another atom or by absorbing a photon from a flow of passing radiation. If the latter, it must absorb all of the photon's energy or none of it. Consequently, it can absorb only photons with energies equal to energy differences between orbit 1 and the upper orbits (Fig. 7.8). The transition from hydrogen's orbit 1 to orbit 2 requires a photon of 1216Å, orbit 1 to 3 of 1026Å, and so on. As a result, in an ensemble of atoms some of these photons are removed from the flow of radiation, producing absorption lines. Elevation to higher orbits requires successively less energy, and consequently the allowed photon wavelengths form an absorption series (here the Lyman series, after its discoverer) that converge on a limit, 912Å, which carries 13.6eV of energy. Photons with wavelengths shorter than 912Å have enough energy to ionize the atom off the ground state. The excess energy after ionization goes into the free electron's speed. A free electron can then be recaptured by a proton (or another kind of atom altogether) in any orbit.

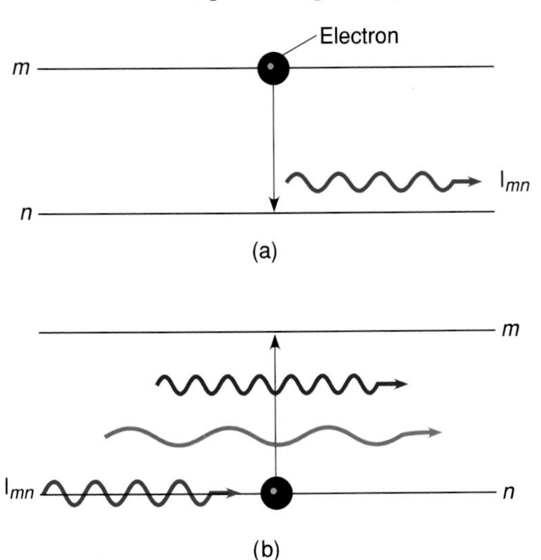

Fig. 7.8 **Photon absorption** A generic energy level m (with energy E_m) lies above energy level n (energy E_n). At the top, the electron drops from level m to n with the production of a photon with energy $E_m - E_n$. At the bottom, to be absorbed, a passing photon must have exactly (or near so) the same energy. (Kaler, J.B. (1994). *Astronomy!* New York: HarperCollins, reprinted by permission of Pearson Education, Inc.)

Once the electron is in an upper orbit (through photon absorption, collision, or electron–proton recombination), it can again move around by collision, by absorption of another photon of the right energy that will carry it outward, or by jumping downward and releasing its energy as a photon with an energy equal to the energy difference between the two energy states, producing (if conditions are right) an emission line. All combinations of orbits are possible. The whole set of emissions that ends on orbit 1 is also called the Lyman series, all electron transitions that end or begin on level 2 the Balmer series, 3 Paschen, 4 Brackett, and so on. Within a series, the emission or absorption lines are given Greek letters starting at the longest wavelength. The optically visible lines are those at the beginning of the Balmer series that arise or end on orbit 2, Balmer α (6563Å), β (4861Å), γ (4340Å), δ (4101Å), which because of their visual prominence are commonly called Hα, Hβ, Hγ, and Hδ (and are also Fraunhofer C, F, G, and h [6.1.1]).

Other atoms and their ions behave similarly. Two or more electrons, however, interact with each other, and cause vastly more complicated orbital structures. For every level of hydrogen, such atoms may have whole sets of multiple levels, each of which may be split into multiple sublevels, all with different energies (Fig. 7.9). Though the resulting spectra can look hopelessly complicated, with thousands of absorption or emission lines, systematic schemes have been developed to name every one.

And atomic spectra are as nothing compared with molecular spectra. Unlike a single atom, molecules can also rotate and vibrate, and their electron bonds can stretch and bend. Each mode results in further splitting of the already complicated electronic states. Molecular spectra can be mind-bendingly complicated, each atomic line now split into sets of bands of lines. Just witness TiO in M stars and CN in carbon stars [Figs. 6.7, 6.16].

7.1.5 Kirchhoff's laws

Three basic rules for the creation of spectra were laid out in the nineteenth century by Gustav Kirchhoff (Fig. 7.10). First, an incandescent solid or gas under high pressure will produce a continuous spectrum (so can synchrotron or other sources).

Second, allow the light from the incandescent source to pass through a box containing a heated gas under low

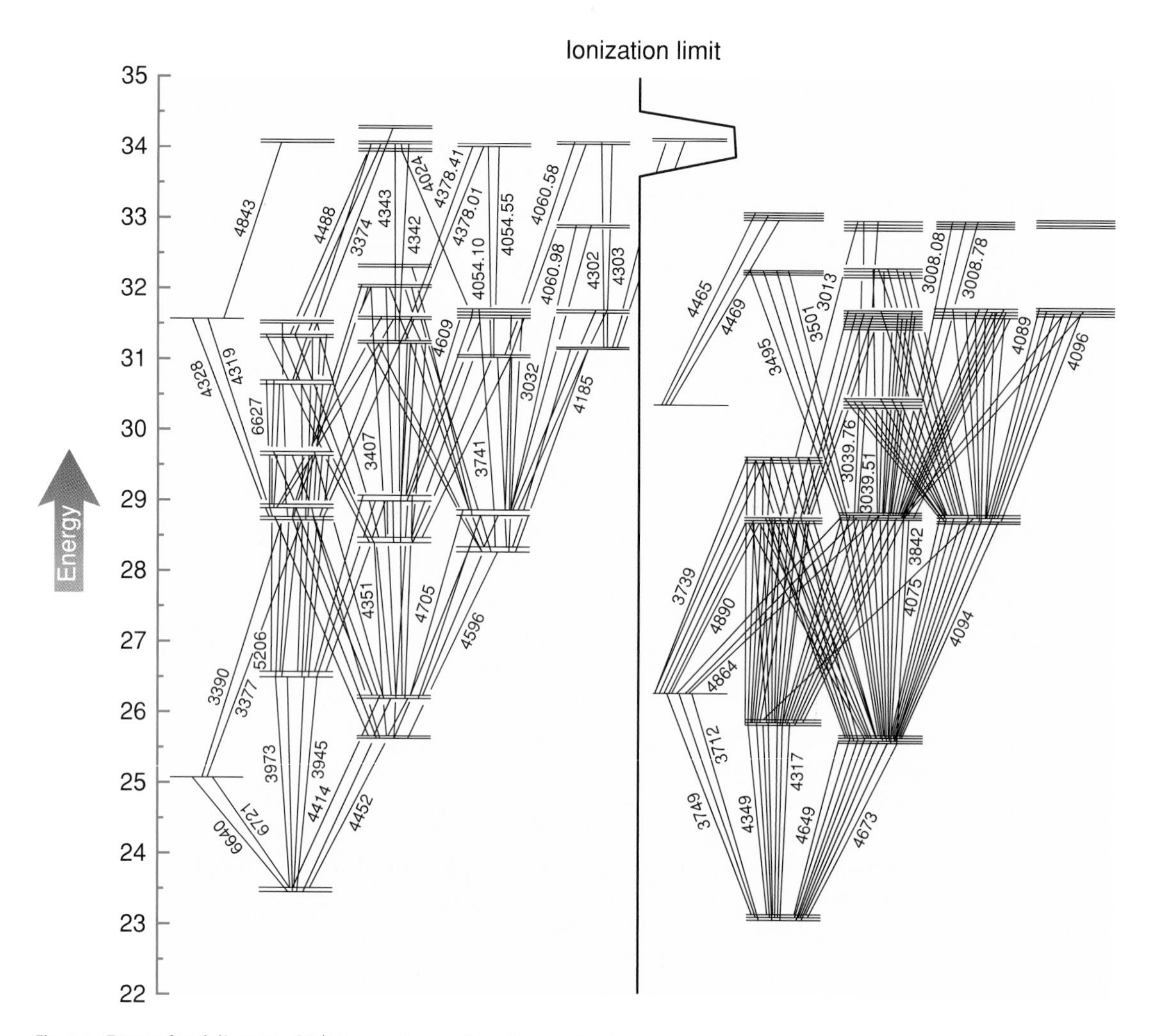

Fig. 7.9 **Energy-level diagram of O^+** An amazing number of energy levels contribute to the optical spectrum of singly ionized oxygen. (Kaler, J.B. (1994). *Astronomy!* New York: HarperCollins, reprinted by permission of Pearson Education, Inc.)

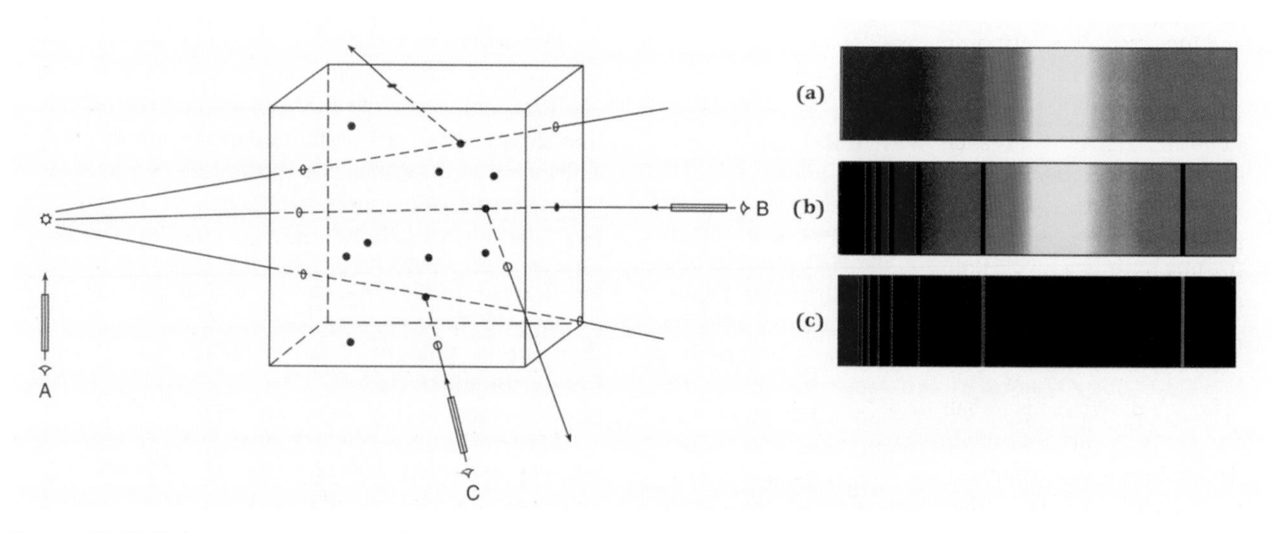

Fig. 7.10 **Kirchhoff's laws** Spectroscope A at far left looks at an incandescent light source and sees a continuous spectrum (a). Spectroscope B looks at the light through a box filled with hydrogen and sees the superimposed absorption spectrum of hydrogen (b). Spectroscope C looks only at the box and sees the reverse, an emission spectrum of hydrogen (c). (Kaler, J.B. (1994). *Astronomy!* New York: HarperCollins, reprinted by permission of Pearson Education, Inc.)

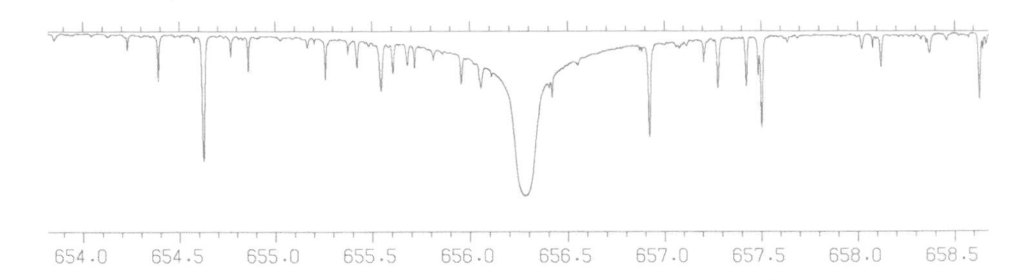

Fig. 7.11 **Line profiles** The solar Hα line has a deep central core flanked by broad wings that stretch many angstroms to each side. Iron and other weaker lines are superimposed upon it. The shape, or profile, is determined by many line-broadening processes. The scale is in nanometers; multiply by 10 to get angstroms. (Kurucz, R.L. *et al.* *Solar Flux Atlas from 296 to 1300nm*, National Solar Observatory, NOAO, NASA.)

pressure. Continuous collisions among the atoms kick electrons into orbits above the ground state; at least up to a certain level that depends on the gas's temperature. Considering all the atoms in the gas, at any given time each of these higher orbits will contain a number of electrons that are capable of absorbing photons from the continuous radiation bath. Therefore, when we look at the light source through the box, we see absorption lines.

Third, to maintain equilibrium, excited electrons must snap back to lower levels, which they can do through collision or by releasing photons at the same wavelengths as those of the absorption lines. But the photons can be radiated in any direction. So when we look at the heated box by itself, we see emission lines.

7.1.6 Real spectra

Continuous blackbody radiation (or something close to it) pours out of the depths of the Sun where the gases are highly opaque. In the cooler outer layers where the gases start to become transparent and the radiation begins to escape into space, in the solar photosphere [12.1.1], individual atoms and ions create the solar absorption spectrum. If you look at the extreme solar edge (or limb) by itself, you see an emission spectrum, the "flash spectrum" of the chromosphere [12.1.3]. All stars have such photospheres, which are also commonly referred to as "stellar atmospheres."

Absorption lines are not simple, narrow black cutouts in the spectrum. They have depth and width that depend on the line itself and on the physical conditions under which it is formed (Fig. 7.11). The depth is given by the fraction of residual light left at the center of the line, and the width is the length of the line in angstroms drawn between the two points at which the line falls to half its depth. Some lines are wide and black in the center, while others are barely detectable depressions in the continuum. Photons with wavelengths of the absorption lines *may* be absorbed, but they do not *have* to be. The *strength* of an absorption line refers to the total amount of energy drawn from the flow of continuous radiation, which is given most simply by the *equivalent width*, W. Assume the line to have a rectangular shape, an outline that goes directly from the continuum, immediately to zero, and then back up again. W is the width in angstroms (or milliangstroms) of such a line that would extract the same amount of energy as does the real line. It depends on the number of atoms or ions that have electrons in the right orbits, on the odds that a passing photon will actually be absorbed, and on the path length over which the relevant ion exists (see Section 7.3.1 below).

The width and detailed *profile* of a line (its graphical shape, as seen in Fig. 7.11) depends on a number of phenomena that allow photon absorption off the line's center: inherent quantum uncertainty and the multiplicity of unresolved sublevels; the degree to which neighboring atoms jiggle the energy levels ("pressure broadening"; the phenomenon that allows the determination of luminosity in A and B stars [6.3.2]); Doppler effects that include simple atomic motions ("thermal broadening"), turbulence, and stellar rotation [7.2.5]. Added to these is the *Zeeman effect*, which broadens or even splits absorption lines created by atoms and ions that are affected by magnetic fields [12.1.2]. Since each of these has a different effect, we can in principle untangle them to measure rotation speeds, magnetic field strengths, turbulent velocities, and so on.

Kirchhoff's laws are an oversimplification. The solar continuum and the absorption lines are actually formed in more or less the same layers, the absorptions on the average just a bit farther out. Within the absorption line, photons in the right span of wavelengths are collectively impeded in their outward flow, a certain percentage stopped, the rest getting through. The solar gases are more opaque within the line rather than outside of it, in the continuum. The greater the odds of a photon being blocked, the higher the opacity. When we look at filtered sunlight right at absorption line wavelengths, we do not see as deeply as when we look at light from the surrounding continuum. Since the temperature of the photosphere steadily declines in the outward direction, the amount of light emerging in the center of the line must be less, in accordance with the Stefan–Boltzmann law. The line is therefore darker than its spectral surroundings. The greater the opacity, the darker the line. If the opacity is exceedingly high, as in the centers of the H and K lines, we cannot see

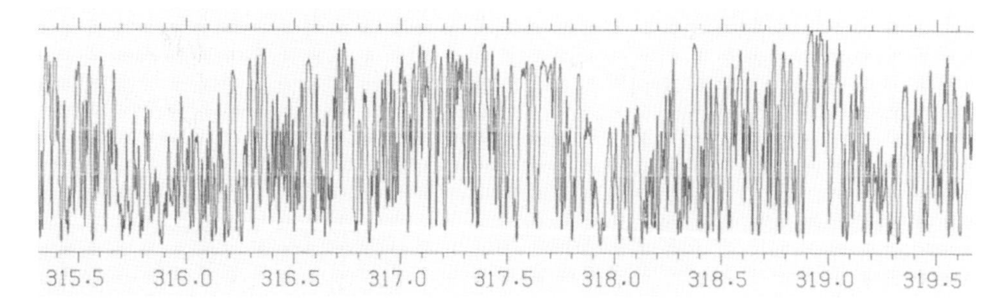

Fig. 7.12 **Line density** The solar ultraviolet spectrum is so thickly packed with absorption lines (mostly of common metals) that the continuum begins to disappear in overlapping line profiles. The scale is in nanometers; multiply by 10 to get angstroms. (Kurucz, R.L. *et al. Solar Flux Atlas from 296 to 1300nm*, National Solar Observatory, NOAO, NASA.)

into the photosphere at all, but see the chromosphere instead, which provides the H and K lines with central emission peaks [4.4.6; 12.1.3].

The solar gases are a mixture of the elements of nearly the entire periodic table, most of which exist in more than one ionization state. Add in molecules and the result is a fantastic mix of strong and weak lines that can be so dense that it can depress the whole continuum and cause significant, if not severe, deviations from a blackbody (Fig. 7.12). All stars behave in much the same way; the differences are mostly caused by different states of ionization. The absorptions in cooler stars are so densely packed that the continuum can disappear altogether.

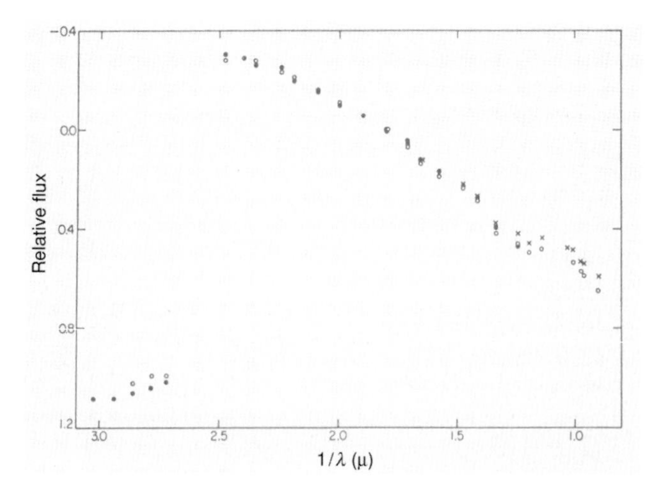

Fig. 7.13 **Absorption limits** The curve shows the continuous spectrum of Vega plotted against the inverse of the wavelength in microns ($1/\lambda$), where $1\mu = 10\,000$Å. The huge Balmer absorption edge at 3636Å ($1/\lambda = 2.74$) is obvious toward the left-hand side. The Paschen edge at 8200Å ($1/\lambda = 1.22$) is smaller, testimony to the greater number of electrons that reside in the second level. (Oke, J.B. and Schild, R.E. (1970). *Astrophysical Journal*, **161**, 1015.)

7.1.7 Continuous absorption and emission

As the upper level n of the Balmer series increases, the lines get closer and closer in wavelength until they pack together at the *Balmer limit* at 3646Å. This wavelength corresponds to a photon with just enough energy to ionize a hydrogen atom that has an electron in the second level. Beyond the limit, absorption is unquantized, so any photon shortward of 3646Å is capable of being absorbed. The result is *continuous absorption* of radiation that diminishes as wavelength decreases. While the limit is a "wall" of sudden absorption, observationally it sets in at longer wavelengths because of overlapping of the Balmer lines. The *Paschen limit*, which corresponds to absorptions from level 3, lies at 8200Å, while the *Lyman limit* at 912Å corresponds to the ground state and the full ionization energy of hydrogen. The Balmer limit (also called the "Balmer discontinuity" or the "Balmer jump") is hidden in the solar spectrum by hosts of absorption lines, but is very prominent in hotter stars of classes A, B, and O (Fig. 7.13). It is the cause of the hook in the color–color plot in Fig. 3.12. All atoms and ions have ionization limits relative to their various energy levels, and all contribute significantly to the total continuous absorption, or gaseous opacity. The process also works in reverse, as recaptures of free electrons onto various energy levels can produce continuous emission.

These absorption continua are not enough to explain the general opacity of photospheres of cooler (solar-type) stars, however, which is due mostly to the negative hydrogen (H^-) ion. Though one electron neutralizes hydrogen, the proton can actually support a second electron. In terms of classical electron orbits, when the first electron is on one side of the proton, the other side has a net positive charge because the proton is a bit closer, enough that the second electron can be grabbed by it. Since the bond is weak, the second electron can be torn away by any photon with a wavelength shortward of 10 200Å. The result is a powerful continuous absorption that goes through the whole optical spectrum, makes the photospheres of the Sun and similar stars very opaque, and gives the Sun its razor-sharp edge. Other opacity mechanisms act in different kinds of stars. In cool stars, molecular absorption almost behaves as if it were continuous, whereas in hot stars with high ionization states, the scattering of light by free electrons becomes important.

7.2 Gross properties

Stars are described by a rather large number of physical properties (temperature, luminosity, mass, and so forth) that are correlated to define the variety of stars. Some are easy to obtain, while others are quite difficult. Several are linked together, in that you cannot find one without the other, though different observational quantities allow for different linkages.

7.2.1 Luminosity

Three important quantities – temperature, luminosity, and radius – are intimately linked through the luminosity equation, $L = 4\pi R^2 \sigma T^4$. If any two are known, the third can be found. Yet each can also be independently determined outside the confines of the equation, in some instances in conjunction with one of the other quantities, sometimes by itself. Because there is a complex mixture of methods, the three have to be taken together, or at least in sequence.

The definition of stellar luminosity is straightforward. Luminosity (L) is the star's total power output in watts, or as more commonly expressed in astronomy, ergs per second (where $1\ \mathrm{erg\ s^{-1}} = 10^{-7}$ watt). It is best found by observing the flux from the star at the Earth per unit wavelength interval, f_λ, in watts per square meter per angstrom (or ergs per square centimeter per second per Å). Integrating f_λ, that is, summing the flux in each wavelength interval, gives the *bolometric flux* at the Earth, f_{bol}, in watts per square meter. Since the same flux must pass through each unit area of a sphere centered on the star, $L\ (\text{watts}) = f_{bol} 4\pi D^2$, where D is the stellar distance in meters. Dividing by the solar luminosity of $L = 4 \times 10^{26}$ watts gives the stellar luminosity in solar units, $L_\odot$, where $\odot$ stands for "Sun."

The luminosity can (as in Chapter 3) also be found from the absolute magnitude [3.2], the bolometric correction (BC) appropriate to the star's temperature [3.3], and the calibration of magnitude against radiative luminosity (that is, from the solar absolute bolometric magnitude and the solar luminosity). This approach works fine for M_V in the mid temperature range where the bolometric correction is small, but not so well for cool or hot stars, where it is large. For cool stars, infrared magnitudes can be used, and the energy distribution found by measuring magnitudes in overlapping (or near-overlapping) wavelength bands. For hot stars, some of the ultraviolet flux can be measured from spacecraft, and the rest extrapolated from computer models of the photosphere that predict the observed emergent radiation.

The ability to calculate emergent flux depends on the opacity of the stellar gas, as produced by continuous absorption, the resulting continuous emission, and by hosts of absorption lines, all of which depend strongly on wavelength. As a result, at different wavelengths we see to different depths within stellar photospheres, which in turn prevents real stars from being real blackbodies (as noted in Section 7.1.7). Stellar opacities are important not just for understanding stellar photospheres, but for determining the internal structures of stars as well [12.2.1]. Unglamorous as they may be, huge efforts have gone into their calculation.

7.2.2 Radius

Like luminosity, radius (R) is commonly measured in solar units, $R_\odot$. The definition of radius seems obvious for a spherical star, though not quite so much for stars distorted by rotation or tides, for which some kind of average must be used. But even for spherical stars the definition is not really so clear, as stars are gaseous and do not possess truly sharp boundaries. Smaller stars like the Sun present no problem as their photospheres [12.1.1] are opaque and very thin, but those of giants and supergiants are not, as the low-density surface gases are so transparent. Because opacity can change with wavelength, such stars can have different radii as seen in different colors. And if examined at the wavelength of an absorption line or band, where the opacity is very high, a star can appear much bigger than if observed in the light radiated by the continuum. In these cases we have to specify not only the wavelength, but also some measure of the position within the photosphere that we specify as an "edge."

Distances below the top of the photosphere can be measured in kilometers, but are more usually specified by *optical depth*, τ, where τ represents the degree of absorption of the light as it passes outward through higher absorbing layers. Light from a layer in the photosphere at optical depth τ is dimmed by $e^{-\tau}$ (where e is the base of the natural logarithms, 2.14159 . . .). A stellar radius might be specified at an optical depth of unity (the light absorbed by 1/e). As a related matter, a giant or supergiant can be limb-darkened [12.1.1] to the point at which it is hard to tell where the star actually ends and "space" begins, which makes radius specification problematic. When necessary, the meaning of the radius must be specified along with the value.

The radius is most easily found by simply inverting the luminosity relation to get

$$R = \sqrt{L}/T^2,$$

Box 7.5 Deriving effective temperatures

The luminosity L of a star in watts is $4\pi d^2 f_{bol}$, where d is distance and f is the flux at the Earth. But L also equals $4\pi R^2 F = 4\pi R^2 \sigma T_{eff}^4$, where R is the stellar radius and F is the flux at the stellar surface. Consequently, $T_{eff}^4 = d^2 f_{bol}/R^2\sigma$. Since the star's angular diameter $\theta = 2R/d$, $T_{eff}^4 = 4 f_{bol}/\theta^2\sigma$. Expressing angular diameter in milliseconds of arc rather than in radians (and using mks units),

$$T_{eff} = 1.316 \times 10^6 (f_{bol}/\theta^2)^{1/4}.$$

As long as we can get angular diameter and f_{bol}, we can find T_{eff} without reference to distance. (Watch the units – astronomers frequently use cgs or even mix them.)

However, angular diameters are available only for large and/or nearby stars. One way around the problem is to calculate the angular diameter from infrared observations. Instead of bolometric fluxes, the "infrared flux method" (the *IRFM*) uses monochromatic fluxes, f_λ and F_λ, where both are now in watts per square meter per angstrom. From the above relationships, $F = 4f/\theta^2$, $F_\lambda = 4f_\lambda/\theta^2$, and $\theta = 2\sqrt{(f_\lambda/F_\lambda)}$. The infrared spectrum contains relatively few absorption lines. Moreover, since it is on the long-wave tail of the blackbody curve for all but ultracool stars, the continuum flux is not sensitive to temperature. As a result, a rough estimate of T_{eff} leads to a good value of F_λ, which with a measurement of f_λ gives θ. That and f_{bol} yields a better value of T_{eff}, which gives better θ, and so on.

The remaining problem is f_{bol} for hot stars, for which we cannot get ultraviolet observations from space. For these, we are stuck with modelling the ultraviolet spectrum from what we know about the accessible visual spectrum, leading to some significant errors in the determination of effective temperature. Effective temperatures are given for different spectral classes in Table 6.2 and Fig. 6.9. White dwarfs and neutron stars (the latter having temperatures that hover in the million kelvin range) are not included.

where all quantities are now expressed in solar units (or $R = \sqrt{L}/(T/5780)^2$, where T is in kelvin). By using the blackbody equivalent effective temperature (Section 7.2.3, Box 7.5), R becomes a blackbody effective radius for which limb darkening and optical depth play no roles.

There are several ways of actually measuring radius. The classic method uses the eclipsing binary [8.4], wherein the absolute radius in kilometers or solar units is found from the duration of the eclipse, or from the duration of ingress or egress combined with orbital velocity. All manner of different kinds of stars are within such binary systems, and the results are not dependent on distance, though there are complications from tidal effects, mass exchange, and so on.

Equally useful is measurement of angular diameter, θ. A tiny handful of stars (Mira, Betelgeuse) are angularly large enough for them to be imaged directly. Much better, the orbiting Moon is constantly passing in front of, or *occulting*, stars. Since we know the Moon's angular rate of motion (0.55 arcsec per second of time), angular diameters can be found from the time it takes the star to disappear (or reappear from) behind the lunar disk. Since the lunar orbit is tilted by five degrees to the ecliptic, and also precesses over an 18.6-year period, all the stars within that angle of the ecliptic will eventually be occulted.

Most far-reaching is *interferometry*. The technique goes back to 1920, when A.A. Michelson placed a 10-foot beam across the 100-inch Mt. Wilson telescope. Light from a star was reflected from two moveable mirrors set on opposite ends of the beam into two fixed mirrors over the telescope, and then into the telescope's optics, where the pair of light rays created an interference pattern (one set of light waves interfering with the other, which produces visible "fringes"). Since a star is not a point source, the pattern is smeared. By moving the outer mirrors and finding the position at which the pattern was filled in, Michelson and his colleagues could determine angular diameters of larger stars; Betelgeuse for example was found to be 0.047″ across. Numerous vastly improved variations on the theme have been built, the best now able to find angular diameters (θ) to an accuracy of 0.1 milliarcsec. Such diameters are often given "as measured" and with an uncertain correction for limb darkening. Even dwarfs down to Proxima Centauri have been so examined.

Combination with distance gives physical radius. In radian measure, $\theta = 2R/d$, where both are in AU [Box 4.1]. Converting to seconds of arc, $\theta''/206\,265 = 2R/d$, or $\theta'' = 206\,265(2R/d) = 2R/(d/206\,265)$. But d (AU)/206 265 is distance in parsecs. Therefore, $\theta'' = 2R\,(\mathrm{AU})/d\,(\mathrm{pc})$, so $R\,(\mathrm{AU}) = 0.5\theta'' d\,(\mathrm{pc})$. Since the solar radius is 0.00465AU, $R/R_\odot = R\,(\mathrm{AU})/0.00465$. Betelgeuse has a measured parallax [4.2.1] of 7.63milliarcsec, which yields a distance of 130pc, and from Michelson's old measurement, a radius of 3.1AU or $660R_\odot$.

Stellar radii, as calculated from the inverted luminosity equation, the visual magnitudes in Fig. 6.12, the temperatures in Fig. 6.9, and the bolometric corrections in Fig. 3.10 (combined with recent data on cool dwarfs), are plotted against spectral class in Fig. 7.14. The end products of stellar evolution, white dwarfs [13.5], are left out. White dwarfs cool at a constant radius, typically 0.01 that of the Sun (roughly the diameter of Earth), the radii dependent on mass (larger masses having smaller radii as a result of increased gravitational compression), while neutron stars [14.3.1] have radii of about 10km.

7.2.3 Temperature

Temperature is necessary to find the chemical compositions of the stars, which in turn lead us toward understanding their evolution, the Galaxy, and the Universe. When used by itself, the "temperature" of a star refers to that of its photosphere. (Interior temperatures, used in stellar structure models, are given as a function of distance from the center; core temperatures are so specified.)

The simplest definition is the *color temperature*, which is found by locating the peak of the continuous energy distribution (the wavelength at which the star emits most of its light) and

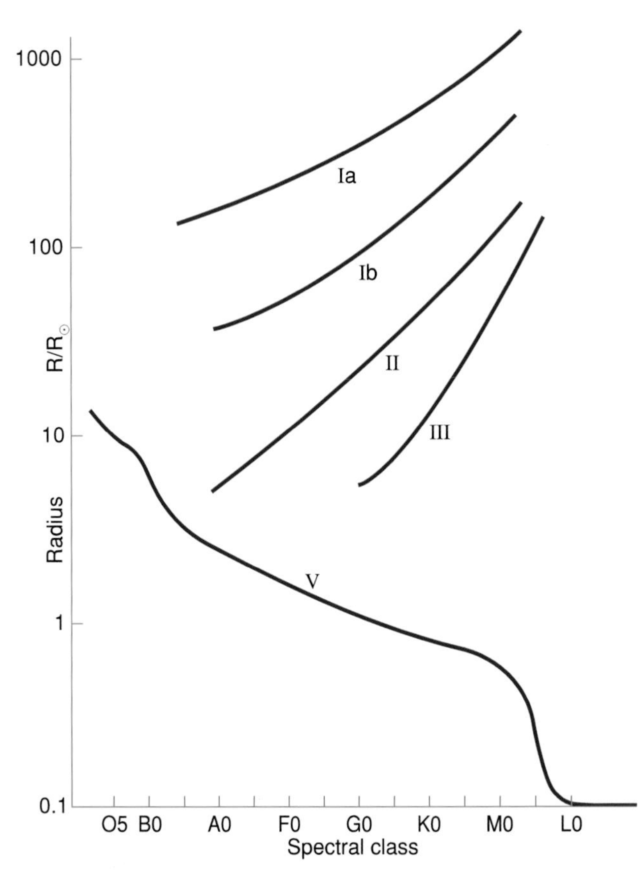

Fig. 7.14 **Stellar radii** Dwarf radii drop quickly with increasingly later spectral class (decreasing temperature), especially later than class M, while those of evolved giants and supergiants climb quickly with increasingly later class. The difference between evolved stars and unevolved dwarfs is even greater than it is on the HR diagram.

applying the Wien law. Somewhat more complex is the *ionization temperature*, which from the equations of atomic physics describes the relative abundances of different ionic states of an atom, or the *excitation temperature*, which describes the balance of electrons among energy states. If stars were blackbodies, all would be the same.

Stars, however, are not the ideal of Kirchhoff's first law. Instead, a star has a photosphere of finite thickness made of layers of semi-transparent gases in which temperature climbs with depth. We see down into it to a layer where the opacity is so high that no radiation escapes directly, but is instead transferred to a higher layer for absorption and re-radiation. If each layer emits something like a blackbody, the light that emerges from the star is a combination of blackbody radiation at different temperatures, each of which is modified by the wavelength-dependent opacity of the upper layers (see Section 7.2.1 above). Consequently, the color temperature from the Planck equation does not necessarily predict the flux of radiation at any given wavelength. T might be specified for optical depth $\tau = 1$, or even given as a table listed against τ, allowing the specification of ionization at different optical depths (which themselves are wavelength dependent).

Even specifying optical depths is a problem, as the stellar photosphere may not be uniform. For example, the solar surface is granulated with millions of hot rising convection bubbles surrounded by dark, cooler interstices. Stellar surfaces may be spotted as well. When in addition we factor in the energy-redistributing absorption-line spectrum, the end result is that since stars are not blackbodies, and that because different absorption lines are formed at different depths, no single temperature can fit the equations needed to describe the wavelength distribution of energy and of atomic excitation and ionization.

That said, we can still derive a measure of gross temperature by pretending that stars *are* blackbodies. The *effective temperature* of a star is the temperature of an equivalent blackbody, one with the same radius and luminosity of the star. That is, we take the luminosity equation and solve it for T to find

$$T_{\text{eff}} = \sqrt[4]{(L/R^2)} = L^{1/4}/R^{1/2}.$$

To apply this equation directly, we must know both the luminosity and the radius of the star, both of which are determinable without reference to (or at least with minimal reference to) temperature. A deeper look into the problem shows that distances are not actually necessary: angular diameter and flux observed at Earth will suffice (Box 7.5).

7.2.4 Masses, densities, and gravities

The life cycle of a star is determined by its mass, chemical composition, and to some degree rotation [Chapter 13]. Of the three, mass is by far the most important, composition second. While theory can help, there is only one way to determine mass independently: from binary stars [Chapter 8]. From the mutual orbit, each star is used to test the gravitational field of the other star at a known distance, from which each stellar mass is quickly derived [8.2.4]. Observed masses range from somewhat below one percent solar (excluding very low mass objects that may be either planets or brown dwarfs) to about 80 times that of the Sun. Main sequence masses follow a strict mass–luminosity relation [12.4.1]. Extending the relation theoretically to the highest luminosities suggests maximum masses of around $120M_\odot$ (and quite possibly more). Extension in the other direction into the realm of the brown dwarfs [12.4.1, 13.1] shows that the minimum mass may indeed approach that of the giant planets.

Masses of giant stars [6.3.1] are in general above one solar mass, while those of supergiants are comparable to the high-mass main sequence stars from which they evolved [14.1.2] minus mass loss (which can be very large). White dwarfs, which are the end products of intermediate-mass (solar type) evolution [13.5], have masses that range between 0.5 and $1.4M_\odot$ (the latter the Chandrasekhar limit, beyond which white dwarfs are not allowed to go). Given the masses and radii of white dwarfs, their typical densities are the order of 10^6gcm^{-3}: a metric ton per cubic centimeter. Neutron stars

[14.3.1] have masses from around 1.5$M_\odot$ to a similar upper mass limit of between 2 and 3$M_\odot$, which with radius yields densities of 10^{14}gcm^{-3}, which is effectively the density of nuclear matter.

Masses can also be found indirectly from the photospheric absorption spectrum. The profile of an absorption line is set in part by the pressure in the stellar atmosphere or photosphere [7.1.6], hence by the gravitational acceleration, $g = GM/R^2$ [8.2.3]. The gravity g can be found by fitting theory to the exact profiles of hosts of absorption lines in which there is enough information to constrain all the causes of line broadening. If the radius is known through one of the many methods outlined above, we derive mass. Unfortunately, the masses so derived are significantly smaller than those found from the much more reliable double stars. No one knows why. The parameters are clearly not being constrained as well as one would like.

Gravities of stars are usually expressed as log g, where g is in cgs units. The solar gravity of $g = 2.74 \times 10^4$ (log g = 4.44) is 33 times that of Earth. The range of stellar gravities, simply scaled from solar according to mass and radius, is huge, ranging from a hundred-thousandth that of the Sun (log $g = -0.5$) for the greatest supergiants, through 5000 times solar for white dwarfs (log g = 8), to 10^{19} solar for neutron stars [14.3.1], for which log g = 23.

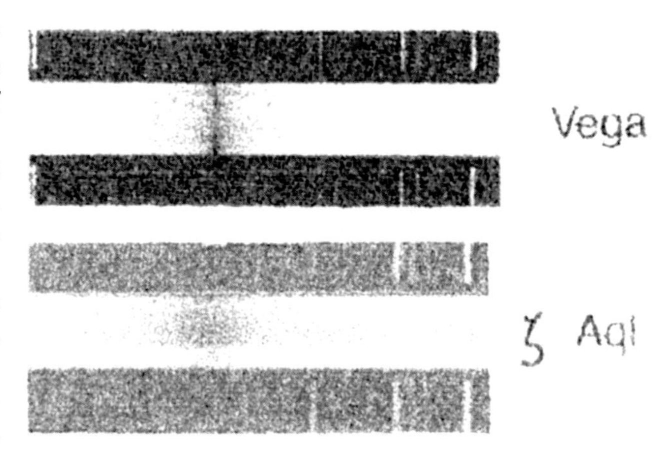

Fig. 7.15 **Stellar rotation** Class A0 Vega rotates slowly and has a narrow Hδ line, while Zeta Aquilae, of the same spectral type, spins at 345km s^{-1}, the resulting Doppler broadening washing out the line. (Vega is probably rotating nearly pole-on). (NOAO, courtesy of E.C. Olson.)

7.2.5 Rotation

Watch the Sun and see the sunspots [12.1.2] move across the surface as the solar rotation carries them along. Other stars should rotate too. A century ago, Antonia Maury noted line-width differences that led Hertzsprung to help discover supergiants. While her class "a" stars had average-width lines, those of class "c" were narrow and "b" were broad. These last are the fast rotators. Ignoring the gross radial velocity of the star, one hemisphere comes at you while the other recedes. Each slice of the star cut parallel to the rotation axis gives absorption lines that are Doppler shifted by a different specific amount. A slice at the rotation axis is at zero velocity, but provides the most light, while a slice at the limb gives the maximum velocity (and shift) but sends the least light. Adding up all the slices yields a broadened line with a specific kind of profile that is readily identifiable. Rapidly rotating stars have broad, washed-out lines, while slowly rotating ones have lines that are sharp and narrow (Fig. 7.15). Once the other sources of line broadening are extracted, all that is left is the rotation speed.

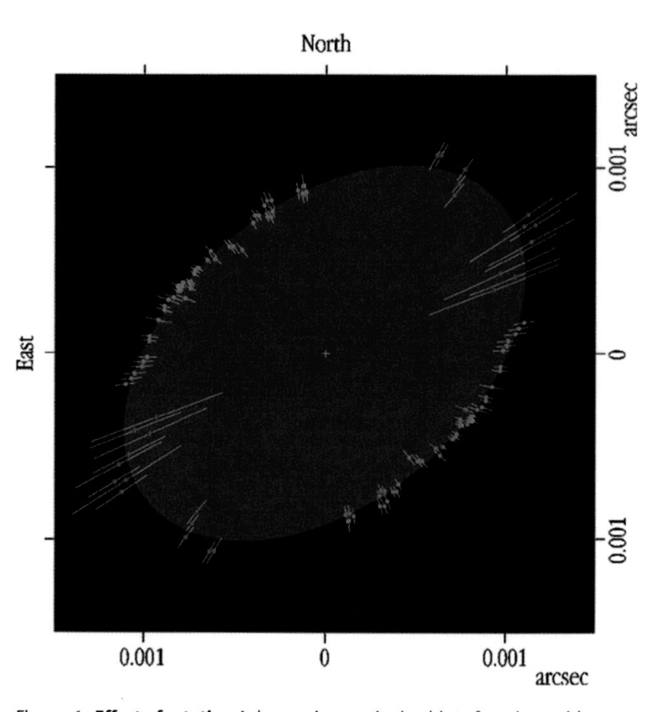

Fig. 7.16 **Effect of rotation** Achernar is amazingly oblate from its rapid rotation. The longest diameter is 12 times bigger than the Sun. (Courtesy of A. Domiciano de Souza (Lab. Univ. de Astrop. Nice-LUAN) *et al.* and ESA.)

What we observe, however, is the projected radial velocity. If the rotation axis points directly at us, there can be no rotational Doppler shift, and we derive 0 km s^{-1}. Only if the pole is perpendicular to the line of sight will we be privileged to see the full rotational action. In between we measure a lower limit to rotation speed. If v_{rot} is the true equatorial velocity, and if the axis is inclined by i° to the line of sight (and the equator inclined by the same angle to the "plane of the sky"), the measured (projected) rotation speed will be v_{rot}sin i. If the radius of the star is known, the upper limit to the sidereal rotation period P is $2\pi R/v_{rot}$ sin i (where units are in kilometers and seconds) or $P_{days} < 50/(V_{rot} \sin i)$, where R is the stellar radius in solar units. The inclination is not readily available from the spectrum. We can, however, assume that stellar inclinations are randomly oriented, which allows for statistical assessment of the rotation speeds of different kinds of stars.

If the star has features associated with magnetic fields, rotation periods can be observed directly. For stars that have starspots, the larger ones will survive for more than one rotational period. The star then becomes variable as the spots move in and out of view, so the variation period equals the rotation period. Magnetic patches on rotating Ap stars [6.4.1; 7.3.3 below] work equally well, the period now derived from variations in gross magnetic field strength. Periodic variations in the emission cores of the Ca II K lines can also be used [4.4.6, 12.1.3]. Comparison with v_{rot}sin i from Doppler broadening (if available) then yields the axial tilt. Groups of spots at different stellar latitudes can even yield information on differential rotation.

Late-type dwarfs rotate slowly, by only a few km s^{-1}. Around F5, however, the spins speed up to hundreds of kilometers per second in class B [12.4.5]. Rapid spins can have serious impacts on stars. Interferometry shows that Altair, with v_{rot} sin i $= 210$km s^{-1} is spun out into an oblate figure that has an equatorial radius at least 1.14 times bigger than the polar radius. Achernar (225km s^{-1}) is even more oblate, the equatorial axis nearly 60 percent larger than the polar (Fig. 7.16). As a result of gravity darkening (the von Zeipel effect) [8.4.3], the bulged equatorial region becomes cooler than the polar caps, which complicates spectral analysis [7.3.1]. Among the larger B stars, rapid rotation is related (in ways still not understood) to the Be star phenomenon [10.6.3].

As might be expected from the conservation of angular momentum, expanded evolved stars [Chapter 13] rotate more slowly. As the progeny of high-mass, high-speed rotators, early giants can still spin relatively quickly. At G0 to G5, however, as deep convection sets in, spins drop dramatically to near zero. White dwarfs, the cores of middling-mass main sequence stars, typically spin 20km s^{-1}, their small radii giving them very short periods of half an hour or so. Variations in observed magnetic field strengths, however, show a fraction of magnetic white dwarfs rotate with periods up to 20 days [7.4]. Newly formed remnants of high-mass evolution, neutron stars [14.3.1], spin at vastly greater rates. With a period of 30 milliseconds, the Crab pulsar's [14.3.1] equatorial velocity is about 2000km s^{-1}.

Binary stars add additional flavoring, as tidal and magnetic locking will take place at small separations, speeding up or slowing down individual rotations.

7.3 Chemical composition

Ever since the British-American astronomer Cecelia Payne-Gaposchkin (1900–1979) figured out that stars were mostly hydrogen, hordes of other astronomers have worked on the problem of stellar compositions, their variations, and their various correlations with other properties.

7.3.1 Calculation from line strengths

The base problem of chemical studies is the conversion of absorption-line strengths [7.1.6] into relative abundance ratios. The solar Ca II H and K lines are far more powerful than hydrogen's Balmer lines, yet calcium has only a millionth the abundance of hydrogen. The laws of physics tell us why.

The equivalent width of an absorption line [7.1.6] depends on five different parameters:

1. The abundance (usually relative to hydrogen) of the chemical element that produces the line;
2. The state of ionization, which determines the fraction of the element that can hope to absorb the line;
3. The excitation state of the ion (that is, the distribution of electrons among energy levels), which determines the fraction of the ion that can absorb the line;
4. The path length through which the photons travel (that is, the number of impeding ions of the right kind and excitation);
5. The odds that an ion capable of absorbing the line will actually do so.

Work from last to first. An electron sits in energy level n below energy level m, both of which belong to a specific ion of a specific chemical element (see Fig. 7.8). A photon of the appropriate wavelength, of energy $E_m - E_n$, passes by (actually, given that the line has breadth, within the right energy range). The probability of the photon actually being absorbed depends on how the mathematical description of the wave nature of the electron in the lower level (were it there) interacts with that were it in the upper level. Probabilities can be very high, approaching unity, or very low, indeed zero. Given by "oscillator strengths," f, probabilities can in simple instances be calculated from theory, or they can be determined under controlled conditions in the laboratory. Or they may not be known at all. Huge lists of oscillator strengths exist in the literature.

The line's equivalent width now gives the relative number of atoms with electrons in level $N(n)$ that lie along the path through which the photons travel. To find the number of ions, N_i, we have to calculate the ratio N_n/N_i. For any atom or ion, the vast majority will have their electrons in the ground state (level 1), so we really need to find only N_n/N_1 (for that ion). The distribution of electrons among energy levels depends on temperature. At low temperatures and low collision energies, electrons can barely make it out of the ground state, while at high temperatures, very energetic states can be populated, though if the energies are large, the populations (number of atoms per unit volume with electrons in a particular level) will always be very small. In an ensemble of atoms or ions, the ratio N_n/N_1 is given by the Boltzmann equation,

$$N_n/N_1 = (g_n/g_1)\exp -[(E_n - E_1)/kT],$$

where the g represents a "statistical weight" (which among other things depends on the number of substates of n), and $(E_n - E_1) = h\nu = hc/\lambda$. The important quantity is the exponential $\exp(x) = e^x$. Plug in h (Planck's constant) = 6.63×10^{-27}, k (Boltzmann's constant) = 1.38×10^{-16}, and the speed of light (c), and find that the exponent is $-1.44 \times 10^8/\lambda T$, where λ is in angstroms.

Now we can compare ionic abundances with one another. Hydrogen's Balmer lines arise from electrons in the second level, for which ratio of statistical weights is 4. The wavelength that corresponds to the level 1 to level 2 transition is that of Lyman α. (Remember, though, the Balmer lines are those we observe; Lyman α serves only to give the level 2 to level 1 energy difference.) Assume the Sun, so $T = 5780$K and therefore $N_2/N_1 = 1.3 \times 10^{-9}$. The astonishing result is that for every billion hydrogen atoms, at any one time only *one* is capable of absorbing the Balmer lines. That is, to get the abundance of the hydrogen atom, we need multiply the abundance on level n, found by observing the Balmer lines, by 1.3 billion. No wonder the solar hydrogen lines are relatively weak. The Ca II H and K lines, on the other hand, arise from the ground state, $n = 1$. While only one in a billion H atoms can absorb Hα, *all* the Ca^+ ions can absorb the K line. When the N_2/N_1 factor is applied to hydrogen, we see there are a million times more H atoms than Ca^+ ions. Applications of these rules to stellar spectra allow the calculation of abundances relative to hydrogen for all observed ions.

The final step is to find the elemental abundance by summing the abundances of its various ions, which is simple if we can observe them all. If we do not, we can calculate the abundances of the others from the ionization equation that was developed by the Indian scientist, M. Saha (1893–1956), which gives the ratio of the number of each of a certain kind of ion q to the number of the next lower ionization stage $q - 1$ as a function of ionization energy (the energy required to go from stage $q - 1$ to q), electron density, and temperature.

The final result is the ratio of the numbers of all observed atoms to those of hydrogen.

These detailed concepts and equations provide the real explanation of the spectral sequence [6.2.2]. At the low temperatures of M stars, the exponential term in the Boltzmann equation crashes, leaving the population of hydrogen atoms with electrons in the second level vastly smaller than it is in the Sun. We therefore see no hydrogen lines. Helium provides another fine example. The optical absorption lines of neutral helium arise from a complex second level that is about twice the energy above the ground state as hydrogen's. As a result, the solar N_2/N_1 ratio drops from hydrogen's 10^{-9} to 10^{-41}! There are effectively zero solar helium atoms capable of absorbing photons, and we therefore see no helium lines, in spite of the fact that helium constitutes 10 percent of solar matter. (The solar helium abundance is derived from the helium emissions radiated by the chromosphere [12.1.3], the solar wind [12.1.4], solar oscillations [12.3], and if nothing else, by analogy with other stars.) To see the helium lines, the temperature must be about double that of the Sun, close to 12 000K, where, in the realm of the B stars, we indeed find them. At higher temperatures, helium becomes singly ionized. The relevant level of ionized helium is double the energy of neutral helium, so again we must double the temperature to see the He II lines, where at 25 000K they appear in the O stars. Other elements and ions follow the same patterns but at different temperatures, and all (including Fig. 6.10) is now explained. Analysis of molecular spectra in cool stars is similar, though more difficult.

As usual, the real world is complex. Absorption lines are formed in a semi-transparent medium that has a temperature gradient (change of temperature with depth). As a result, for highest accuracy, we model the photosphere mathematically to predict not just line strengths, but line profiles as well, while including depth of formation, the total photon path length, temperature gradients, rotational gravity darkening (the von Zeipel effect) [7.2.5, 8.4.3], opacities, and all the line-broadening effects. The parameters are adjusted until the predicted spectrum matches the observed. Mathematically, it is a solution of many unknowns (all the parameters) in many equations (the absorption line profiles).

And that is not yet the end of it. Simple modelling assumes a concept of "local thermodynamic equilibrium" ("LTE"), that each layer acts like a blackbody and that all processes are balanced by their opposites, and that the absorbing gases at each temperature level are in flat layers parallel to each other. But given the semi-transparent nature of a real low-density, spherical stellar photosphere, LTE can be a poor approximation. As a result, the "ups" may not balance the "downs," forcing each atomic process to be examined in detail.

7.3.2 Solar (standard) composition

The results of decades of work provide detailed chemical compositions of great numbers of stars. Since dying stars

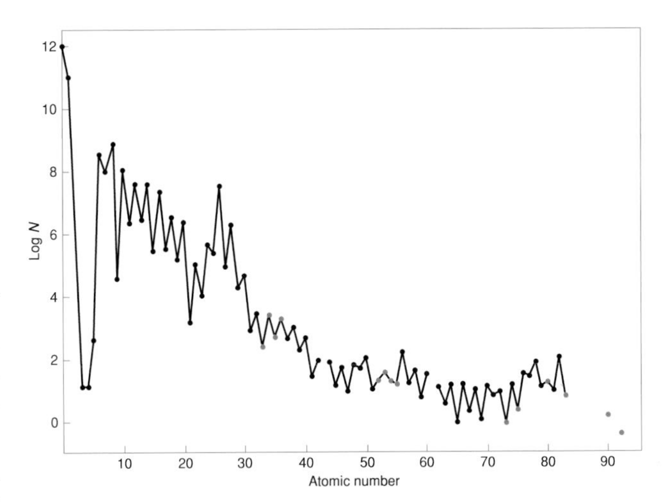

Fig. 7.17 **Solar chemical composition** The graph shows the logarithm of the abundance of each chemical element found in the Sun (from Appendix 4), filled in with meteorite data (in red). Note the low abundances of elements 3, 4, and 5, the iron peak, the drop-off after element 30, the plateaus centered around elements 50 and 70, and the alternation in the abundances of odd–even elements.

feed their chemical production into the interstellar medium, from which new stars are born, each set of compositions is a "data point" in the evolution of the Galaxy. The chemical compositions of all stars are referred to the composition of the Sun, which is taken as the Population I [5.1] standard. The solar chemical composition (filled in with the chemical compositions of meteorites, primitive asteroids that hit the Earth) is given in Appendix 4.

The abundances from Appendix 4 are plotted in Figure 7.17, where they show a variety of revealing relationships. Hydrogen and helium totally dominate. In general, as atomic number increases, the abundance decreases. The light elements lithium, beryllium, and boron (numbers 3, 4, and 5), however, are very rare compared to those around them. There is also a saw-tooth pattern in which even-numbered elements are more abundant than the odd-numbered ones that lie next to them. We then see a peak around iron (number 26) (not surprisingly called just that, the *iron peak*), a precipitous drop beyond it, a plateau between numbers 40 and 60, and then another drop and plateau between 60 and 75. These patterns are explainable by theories of stellar evolution [Chapters 13 and 14]. In a broad sense, the solar composition is both a fundamental reference and a goal for theoreticians who try to replicate it, beginning with the Big Bang.

The range in the abundances of different elements is immense, of the order of a trillion or more. As a result, observational astronomers traditionally list abundances by the relative number of atoms, ϵ (element), with hydrogen taken as 10^{12}. Theoreticians, on the other hand, commonly use mass fractions of the different elements, that is, the fraction of the total mass held by a particular element. For ease of listing, Appendix 4 Tables A and B adopt 10^6 for hydrogen.

7.3.3 Other stars

The chemical mixes of other stars are presented in one of two ways, as ϵ (element) or, since the Sun is the standard, as relative to the solar composition. If E1 and E2 represent two elements,

$$[\mathrm{E1/E2}] = \log\left[\epsilon(\mathrm{E1})/\epsilon(\mathrm{E2})\right]_{\mathrm{star}}/\left[\epsilon(\mathrm{E1})/\epsilon(\mathrm{E2})_{\mathrm{Sun}}\right].$$

Thus if $[\mathrm{Fe/H}] = -1$, the star's iron content relative to hydrogen is 1/10 solar. An abundance listed as [E1] alone assumes hydrogen as E2 by default. It is often convenient to average all the elements heavier than helium, which together are lumped into generic "metals," reducing the real world to three elements, hydrogen (H), helium (He), and "metals" (M). $[\mathrm{M/H}] = -2$ just means that the total heavy element content of the star, M/H, is about 1/100 solar.

Differences from the solar chemical mix fall chiefly into one (or more) of four main categories.

1. *Very old stars.* Ancient Population II stars, the stars of the globular clusters and the subdwarfs of the general field of the Galactic halo [5.1, 5.3.2, 6.4.2], were formed before stellar evolutionary processes could have had much effect on enriching the star-forming cosmos. They are not a single set, but range from stars with M/H of 0.1 to below 10^{-5} solar ($[\mathrm{M}] = -1$ to -5). The standard view is that the Big Bang gave us only hydrogen, helium, and a small bit of lithium. The rest was all created in stars. The first stars – those of Population III [5.4] – therefore, had no metals. They seem no longer to exist, and were most likely massive stars that exploded as Type II (core-collapse) supernovae [14.2.2] to create the first metals that salted the next generation, those of extreme Population II that we do see [5.4]. The scenario is supported by the fact that the lowest-metal stars have abundance patterns typical of those produced by rapid neutron capture [13.3.3, 14.2.3], which is somehow related to supernovae.

As the Galaxy aged and lower-mass stars could turn into white dwarfs, Type Ia explosions began to take place [14.2.5]. Since these create more iron than Type II, the Fe/O ratio increased and stars began to approach the mildly metal-poor state that we see in some globular clusters and subdwarfs. As the Galaxy collapsed into its disk [5.4], the new Population I stars had more or less solar compositions. Stars toward the inner Galaxy, where chemical evolution worked at a faster pace, have by now become metal-rich, with M/H greater than solar, $[\mathrm{M/H}] > 0$. This simple scene was greatly complicated by mergers with other systems in different states of evolution and by mixing of stars within the Galaxy, even by low-metal gaseous flows entering from the outside.

2. *Highly evolved stars.* As stars age they inevitably change their internal constitutions and therefore their surface compositions. Some giants turn into S stars, and then into genuine carbon stars in which the C/O ratio is reversed [6.4.3; 13.3.3]. These will also possess elements created by slow neutron capture. Even if the fresh chemistry does not make it to the giant's surface, eventually the whole envelope of the star will be stripped away to reveal the highly enriched inner regions and then the nearly-exposed cores themselves to produce the different categories of white dwarfs, the DA variety with nearly pure hydrogen atmospheres, and the non-DA with nearly pure helium atmospheres [6.4.6]. Higher mass stars strip themselves down even more quickly to create Wolf–Rayet stars [6.4.4, 14.1.2], R Coronae Borealis stars [10.6.2], and a variety of related breeds.

3. *Chemically peculiar stars.* From late B to early F, main sequence and subgiant stars can exhibit bizarre chemical compositions. The subject is complex, with different groups and subgroups that are not always agreed on. The "metallic line stars" (or "CP1" for "chemically peculiar class 1") are enriched in metals such as copper and zinc by up to a factor of 10 or more, as well as in the "rare earths" (the elements between atomic numbers 57 and 71), but are low in calcium and scandium. They run throughout class A to mid F, where they are called Am and Fm stars [6.4.1]. The best known example is the brightest star of the sky, Sirius. At higher temperature, among the B stars, they change their form and become hugely enriched in mercury and manganese (as well as other elements), the former by factors up to 100 000. Logically called HgMn (or CP3) stars, they range from around B6 to A0. Good examples are Alpha Andromedae (B8 IV) and Chi Lupi (B9 IV).

Long debated, and once thought to have been created by enrichment from nuclear processes, these odd characters are almost certainly the products of element diffusion in quiet atmospheres. As hotter stars, Am, Fm, and HgMn stars do not have convective envelopes [12.2.1], and are all slow rotators, so the atmospheres are not much stirred by any circulation of gas produced by stellar spin. As a result, some elements sink under their own weight, while others that have strong spectrum lines at wavelengths where the stars are most luminous rise upward under the action of radiation pressure.

From mid class B to mid class F we find a magnetic version, the CP2 or Bp, Ap, Fp stars (commonly called just Bp–Ap stars [6.4.1]). These have intense magnetic patches on their surfaces (detected via the Zeeman effect [7.1.6]) near their poles, in which for the cooler set – strontium, chromium, and rare earths (particularly europium) – can be thousands of times enhanced. The "p" stands for "peculiar." Do not confuse them with the Am–Fm stars, where the "m" stands not for "magnetic," but for "metallic." In the hotter set we see surface enrichment of silicon. Again, diffusion, now affected by the powerful magnetism, is the culprit. The magnetic axes of these stars are tilted, or oblique to the rotation axes. As the stars rotate, the magnetic patches swing in and out of view, and the spectra vary with time, allowing the measurement of rotation speed, hence period [7.2.5]. Since different parts of the star are coming at us at different velocities and Doppler shifts, the line profiles can be dissected to construct a picture of the stellar surface, the technique called "Doppler imaging" (Figure 7.18). The best-known example is Alpha-2 Canum Venaticorum (A0p), which has fields comparable to those in

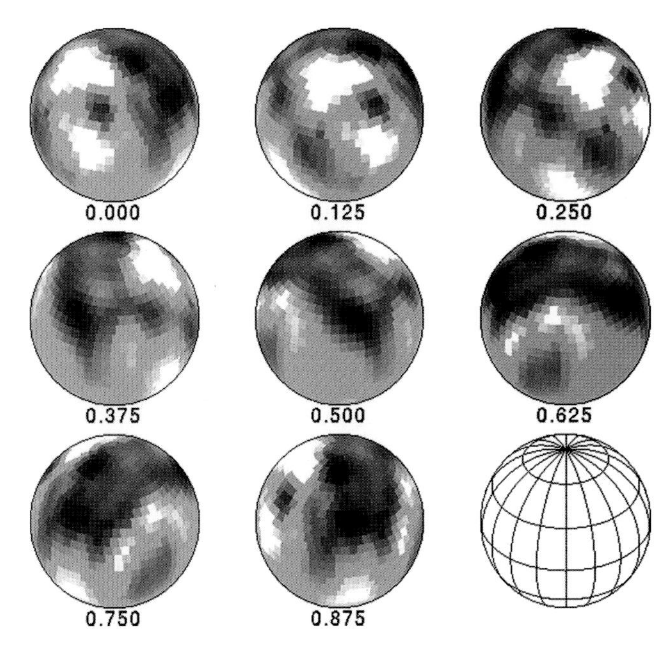

Fig. 7.18 **Doppler imaging** As Epsilon Ursae Majoris (A0p) rotates, regions of enhanced (white) and depleted (dark) chromium swing in and out of view. Since the regions are Doppler shifted by different amounts, the absorption line profile changes with rotation, which in turn allows the construction of the image. (Courtesy of A.P. Hatzes.)

sunspots. In other Ap stars, the fields can be tens of thousands of times stronger than Earth. A subset of the Ap stars, the "rapidly-oscillating Ap" (roAp) stars, oscillate in brightness by a few thousandths of a magnitude with periods of a few minutes [10.4.1]. The CP2 stars are also slower rotators, but whether the magnetic fields are the cause of the slowness through magnetic braking [13.2.1], or are fossilized fields left over from the time of stellar formation is not known.

Thrown in with these are the helium-weak (He-weak, or CP4) stars of class B (which may be high-temperature versions of HgMn stars), the early-B He-rich stars, and the very peculiar and rare Lambda Boötis (class A0) stars. The metal lines of the Lambda Boo stars are oddly faint. The current theory for the Lambda Boo phenomenon is diffusion combined with the accretion of metal-poor gas from interstellar dust clouds, the metals having condensed onto dust grains that are kept from accreting onto the star by radiation pressure.

4. *Mass transfer in binaries*, in which an evolved star throws matter enriched by nuclear processes onto its companion; leading examples, the barium and CH stars [6.4.3, 8.5.10].

7.4 Magnetism and winds

The combination of observations of Ca II (H and K) emission line cores [4.4.6; 12.1.3], ultraviolet emission lines, X-ray fluxes, and the Zeeman effect [7.1.6] show that magnetic fields cover the HR diagram. Late-type dwarfs can have magnetic fields and activity similar to that of the Sun [12.4.4]. Hot coronae [12.1.4] must then generate winds that are similar to the solar wind. Ignoring Ap and Fp stars, magnetism generally disappears in class A [12.4.4]. However, X-ray radiation picks up again through B and O, where it is caused by violent instabilities within powerful winds. Colliding winds from hot-star binaries can create X-rays, too. While winds among the cooler dwarfs are magnetically driven, the winds of these early-type stars are driven by the outward pressure of radiation as it is picked up by absorption lines.

Evolved stars are a different matter. As a star expands to become a giant, the rotation must slow. X-ray radiation more or less disappears at the "X-ray dividing line," which on the HR diagram [6.3.1] stands more or less vertically at class K3. Earlier giants show the emission, while later types do not. Among more luminous stars, we observe emission from low-ionization chromospheres that suggest much cooler 10 000K winds rather than 100 000 kelvin transition regions and hot coronae. In between are the *hybrid stars* that exhibit both sets of phenomena. Winds from cooler stars, in particular the AGB (second-ascent) Mira variables [10.5; 13.3.1], which can exceed 10^{-4} solar masses, are driven neither magnetically nor by line absorption, but by the action of radiation pressure on condensing dust grains.

Hot luminous stars, highly evolved giants, supergiants, even developing protostars [11.3] can possess immensely powerful winds, which at the extreme can reach mass loss rates billions of times that of the Sun. Such potent winds make themselves known through *P Cygni lines* (after the prototype, P Cygni, a "luminous blue variable" [10.6.1; 14.1.2]), in which emission lines from gaseous outflows are flanked to the short-wave side by Doppler-shifted absorptions caused by matter flowing from the star directly to the viewer (Fig. 7.19).

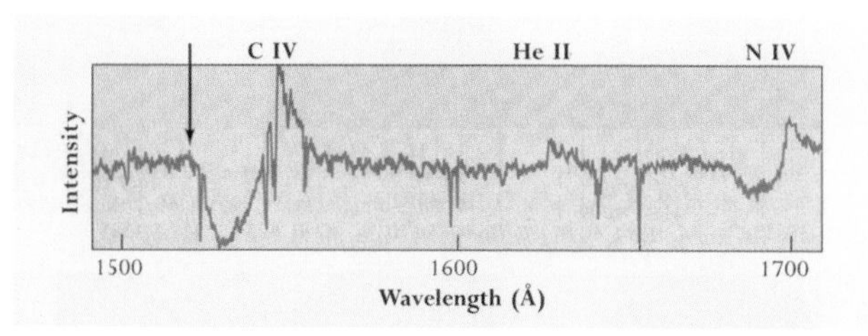

Fig. 7.19 **P Cygni lines** A graphical representation of the ultraviolet spectrum (taken from space) of the luminous O4 supergiant, Zeta Puppis, reveals combined emission–absorption P Cygni lines. N IV, He II, and C IV emissions come from a dense outflowing wind. That part of the wind coming directly at the observer is seen in relief against the star, and for N IV and C IV produces absorptions that are Doppler-shifted toward shorter wavelengths. (NASA, IUE, Goddard Space Flight Center from Kaler, J.B. *Stars*, Scientific American Library (NASA/Goddard).)

Just under five percent of all white dwarfs (either DA or non-DA [6.4.6]) exhibit magnetic fields with strengths that range from 10^5 to 10^9 that of Earth. White dwarfs are topped only by neutron stars, which hold the record with fields up to 10^{14} Earth's. Neutron stars also produce immense winds that flow at near the speed of light, all these to be discussed in context.

Chapter Eight

Double and multiple stars

To understand our Sun, astronomers concentrate their major efforts on single stars. Moreover, single stars, which are not under the control of outside influences, are easier to study. Most stars, however, are not single. Whether in the Galactic halo or disk, two-thirds or more are in double (**binary**) or multiple systems. Duplicity actually becomes an advantage when the component stars are far enough apart that they act like single stars and do not influence each other's behaviors and evolutionary paths. Then double stars can be used to determine normal stellar masses and to establish the main sequence's mass–luminosity relation [12.4.1]. But when the binary members are sufficiently close, as is common, each star affects the other, and the concepts of stellar evolution must be massively modified to deal with the system as a pair.

8.1 Organization

Stellar couplings are classified in a variety of ways. An *optical double* is an uninteresting line-of-sight coincidence. At the simplest level is the two-member *binary*. Like the planets to the Sun, the two are locked in gravitational embrace, and if there is no outside disturbance, they orbit on elliptical paths around a common center of mass that lies between them. Double stars are traditionally grouped by the techniques used to observe them. *Visual binaries* are separated into two components directly through the telescope. The observer can see the two individual stars and, if they are close enough together, watch them orbit. The traditional separation limit for visual binaries is about half a second of arc, below which the pair seem as one.

If the two stars are physically connected but very far apart, the orbital periods can be thousands of years or more, which disallows observation of orbital motion. Gravitational linkage and true duplicity can be assumed, however, if they share a *common proper motion* (CPM) and radial velocity [5.2]. If the fainter of the two stars is too dim to be seen in the glare of the main component, it may still deflect the brighter through a proper motion wobble, making the system an *astrometric binary*.

Spectroscopic binaries are in general too close together to be resolved by traditional Earth-based methods. In the ideal case, each stellar spectrum is discernible from the combined spectrum, that is, we see the spectrum lines of the individual stars (yielding a *composite spectrum*). Since the members of these *double-lined spectroscopic binaries* are physically near each other, their orbital velocities are naturally high, as are the induced Doppler shifts. As a result, we easily see the individual spectra swinging back and forth in wavelength and can find orbital radial velocities. If one star is much brighter than the other, only one set of lines will be seen, such *single-lined spectroscopic binaries* allowing at least the inference of a companion. If the orbital plane lies close to the line of sight, then the two stars can eclipse each other, and the system periodically dims. *Eclipsing binaries* can produce partial or total eclipses. The three different kinds of binaries are not mutually exclusive, but strongly overlap: interferometry [7.2.2] pushes the separation limits into the realm of the spectroscopic binary, and eclipsers are so close together and speedy that they are invariably spectroscopic binaries as well.

Surrounding each member of a binary is a teardrop-shaped "tidal lobe" at whose surface the gravity is effectively canceled by the other star. The lobes touch each other at the points of the teardrops. The components of a *detached binary* are well within their respective lobes and are thereby separate from each other. In a *semi-detached binary*, one of the components is large enough to fill its lobe and thereby transfers mass to the other. In a *contact binary*, both stars fill their lobes and thus connect. Proximity leads to a variety of violent effects and a huge collection of *cataclysmic binaries* (or *cataclysmic variables*), even to supernovae [14.2.5].

Gravitationally-connected stars are also organized by multiplicity. Most are simple binaries. A smaller fraction, however, is made of triples, and a smaller fraction yet of quadruples, and so on. In a *hierarchical multiple*, a distant star or double orbits another double (or multiple) such that each gravitationally feels the other as a unit. The members of *trapezium systems*, on the other hand, are chaotically thrown together in close complex, unstable orbits. Each of these will be addressed and illustrated in turn.

8.2 Visual binaries

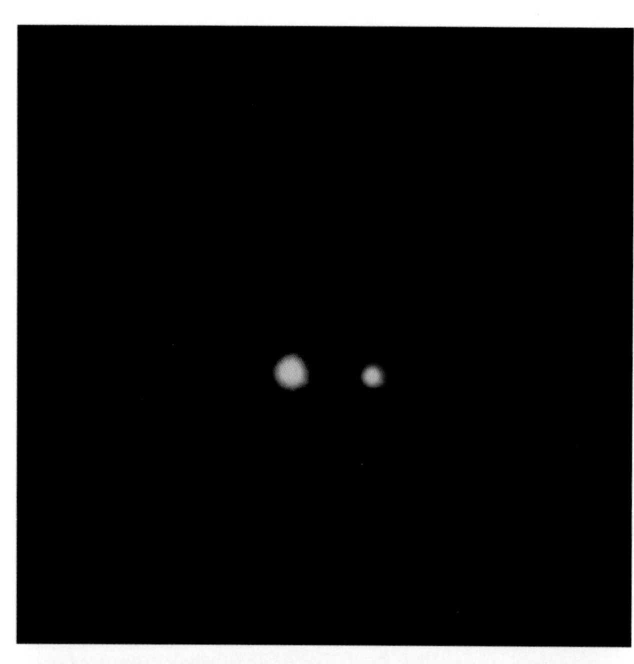

Fig. 8.1 **Visual doubles** On top is the classic wide double star Albireo, Beta Cygni. Its stars, 35 seconds of arc apart, shine with subtle shades of blue and orange. At bottom is Polaris, separated from its ninth-magnitude companion by half that of the Albireo pair. (Courtesy of Jack Schmidling.)

There are few things in the sky prettier than a tight double star (Fig. 8.1). While star colors are subtle, even small contrasts in color and brightness are magnified by the eye. Nineteenth-century astronomers made a small industry of trying to assess double-star colors, thinking that binary members were in fact differently colored than single stars: "topaz . . . sapphire blue" (Albireo, Beta Cygni), "pale orange . . . sea green" (Epsilon Boötis), "bright white . . . pale grey" (Gamma Arietis). Great numbers of such stars dot the sky, so many that nearly all their variety is on view on any given night (Table 8.1).

8.2.1 Names

The brightest star within a visual binary, the *primary star*, is called "A," the other *secondary star*, "B," and so on. Beta Cyg A is the third magnitude orange primary, while Beta Cyg B is the fifth magnitude blue secondary. If the "A" star of a double is later found to be binary, the members might be Aa and Ab. The scheme is used for spectroscopic binaries as well.

The first double, Mizar, was found in 1650, not long after Galileo turned his telescope to the sky. By the time of William Herschel (1738–1822) – who discovered Uranus and, with his sister Caroline, large numbers of clusters and nebulae – nearly 300 doubles were known. Though suspicion abounded that they were real pairs and not optical coincidences, Herschel was the first to see one star orbit another, and thus claims discovery of true binary stars. The work of the Russian astronomer F.G.W. Struve (1793–1864) and his son Otto (1819–1905) greatly extended the list with catalogue numbers indicated by Greek sigma (Σ). The remarkable American amateur S.W. Burnham discovered many more, and compiled the known systems within 121° of the north celestial pole in 1906 in his *General Catalogue of Double Stars* (the *BDS*). The work was expanded by Robert Aitken in the 1932 *New General Catalogue of Double Stars*, *ADS* numbers becoming standard. This work, melded with the *Southern Double Star Catalogue* (*SDS*), became the basis of a huge punch-card catalogue that migrated to the U.S. Naval Observatory, where it is now incorporated electronically into the *United States Naval Observatory Double Star Catalogue* (of visual binary stars) that includes positions, orbits, and photometric data.

8.2.2 Magnitudes

If the stars of a visual binary are well separated, the magnitudes of the components can be measured just like those of

	α (2000) δ						
Name	**h m**	**° ′**	**Magnitudes**	**Spectra**	**sep**	**P. A.**	**remarks**
γ And	02 04	+42 20	2.26–4.84	K3 II–B8 V+A0 V	10	63	gold, blue
γ Ari	01 54	+19 18	4.75–4.83	A1p –B9 V	7.8	360	both white
ϵ Boo	14 45	+27 05	2.70–5.12	K0 II–A2 V	2.9	338	orange, white
ζ Cnc	08 12	+17 39	5.44–6.20	F9 V–G9 V	0.9	337	1968, P=60 yr
ζ Cnc A–C[a]			6.01	G5 V	5.8	83	triple star
α Cen	14 40	−60 50	−0.01–1.33	G2 V–K1 V	8.7	8	1946, P=81 yr
α Cru	12 27	−63 06	1.58–2.09	B0 IV–B1 V	4.4	115	A is spect. bin.
β Cyg	19 31	+27 58	3.08–5.61	K3 II+B0 V–B8 V	34	54	gold, blue
ν Dra	17 32	+55 11	4.87–4.88	A4 Vm–A6 V	62	312	both white
α Gem	07 35	+31 54	1.58–1.59	A2 Vm–A1 V	1.8	140	both spect. bin.
α Gem A–C			9.5	M0.5	73	164	spect. bin.
γ Leo	10 20	+19 50	2.61–3.80	K1 III–G7 III	4.4	123	
α Lib	14 50	−16 01	2.75–5.80	A3 IV–F4 V	231	314	naked eye
ϵ Lyr	18 44	+39 39			208	173	naked eye
ϵ^1 Lyr	18 44	+39 40	5.06–6.02	A4 V–F1 V	2.8	359	
ϵ^2 Lyr	18 44	+39 37	5.37–5.71	A8 V–F0 V	2.2	98	
σ Ori A–B	05 39	−02 33	4.1–5.1	O9.5 V–B0.5 V	0.2	199	multiple
σ Ori A–C			8.79	A2 V	11	236	
σ Ori A–D			6.62	B2 V	13	84	
σ Ori A–E					42	61	
θ^1 Ori A–B	05 35	−05 23	6.73–7.96	O7– B0 V	8	32	The trapezium
θ^1 Ori A–C			5.13	O6	12	132	
θ^1 Ori A–D			6.70	B0.5 V	21	96	
ζ UMa[b]	13 24	+54 55	2.27–3.95	A1 V–A1 Vm	14	151	both spect. bin.
γ Vir	12 42	−01 27	3.65–3.68	F0 V– F0 V	4.7	306	

[a] Multiple star. This row gives information on the C component and its relation to A, as it does for C, D, or E components below.
[b] CPM with Alcor. Quintuple star.

Table 8.1 **A sampling of visual binaries**

any other star, and the absolute magnitudes calculated from distance [Chapter 3]. However, as the binary members draw together, the light from each one affects the measurement of the other, and magnitude measures become increasingly difficult. At some point, it is the combined magnitude that is observed, and the astronomer then determines a brightness ratio to separate the two. The magnitude of star A cannot be just added to that of star B, however. Instead, each magnitude must be converted to an intensity, the two intensities added, and the sum re-converted to a magnitude (Box 8.1).

8.2.3 Simple orbits and Kepler's laws

The stars of a well-separated binary behave according to Kepler's laws of planetary motion. In 1610, Johannes Kepler (1571–1630) discovered that the orbit of a planet is (to a good approximation) an ellipse, with the Sun at one focus (Box 8.2). He then noted that the line connecting a planet to the Sun (the "radius vector") sweeps out equal areas in equal times, requiring that a planet move fastest when closest to the Sun (at *perihelion*) and vice versa (at *aphelion*). Ten years later, he found that the squares of the orbital periods in years (P) equal the cubes of the semimajor axes (a) in Astronomical Units (AU), or as famously put:

$$P^2 = a^3.$$

Newton re-derived Kepler's laws from his laws of motion [force (F) = mass (M) times acceleration (A), etc.] and his law of gravity. Two spherical bodies attract each other with a force directly proportional to the product of their masses

Box 8.1 Summing magnitudes

The member stars of a binary have magnitudes m_A and m_B. What is the observed magnitude if the stars are so close together that they are seen as one? To find the combined magnitude, we add the intensities of radiation from the two stars, not the magnitudes. Magnitudes are logarithmic, and can be expressed without scaling or calibration as $m = -2.5 \log I$, where I is the actual intensity [Box 3.1]. The intensity of a star (on a relative scale) is therefore $I = 10^{-0.4m}$. From the magnitudes m_A and m_B find the summed relative intensities $I_{A+B} = I_A + I_B$. The combined magnitude is then $m_{A+B} = -2.5 \log I_{A+B}$. The only requirement is that the magnitudes be of the same kind: visual, blue, absolute, bolometric, etc.

Box 8.2 The ellipse

An ellipse is a closed curve defined by two points called *foci* (plural of *focus*), such that the sum of the distances of each point along the ellipse to the two foci is the same (Fig. 8.2). The line through the foci and the center, which is also the longest line that can be drawn in the ellipse, is the *major axis*; the line through the center perpendicular to the major axis is the *minor axis*. The size of the ellipse is defined by the length of half the major axis, the *semimajor axis*, *a*. The shape is given by the eccentricity, *e*, which is the ratio of the distance from the center of the ellipse to the focus divided by *a* ($e = CF/a$). If the foci come together at the center, the result is the circle ($e = 0$), an extreme example of the general ellipse. As the ellipse stretches out, *e* approaches, but does not reach, 1.

Box 8.3 Angular momentum

Tie a weight to a string and whirl it about in a circular "orbit." The weight's *angular momentum* (L) is the product of its mass (M), its velocity (v), and the string's length (r), that is, $L = Mvr$. The angular momentum of a rotating body is simply the sum of the angular momenta of all its parts as they go about the rotation axis. In the absence of an outside force, angular momentum is constant, that is, it is "conserved." Get the weight going, shorten the radius, and the velocity goes up. The conservation of angular momentum is crucial to numerous concepts in astronomy, including orbits, star formation, and stellar evolution.

(M_1 and M_2) and inversely proportional to the separations between their centers (R), or

$$F = G\, M_1 M_2 / R^2,$$

where G is the universal gravitational constant and the units are physical, either kilogram–meter–second or gram–centimeter–second. Your weight (*W*) is the force $W = G\, M_{\text{Earth}} M_{\text{you}} / R_{\text{Earth}}^2$. Given $F = MA$, the acceleration of gravity g at the Earth is $g = G\, M_{\text{Earth}} / R_{\text{Earth}}^2$, where R_{Earth} is the distance from your center to the center of the Earth, effectively the Earth's radius. Similarly, the gravity at the surface of

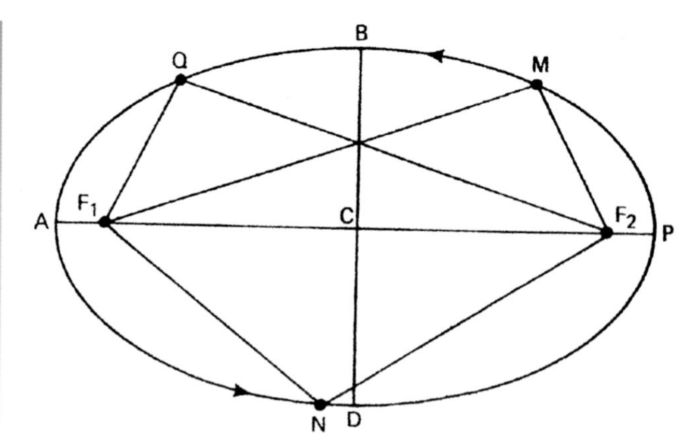

Fig. 8.2 **The ellipse** The sum of the distances from the ellipse to each of the foci (F_1 and F_2) is constant ($MF_1 + MF_2 = NF_1 + NF_2 = QF_1 + QF_2$ etc.). The size of the ellipse (*a*) is given by the semimajor axis AC = CP, the eccentricity (*e*) by $CF_1/a = CF_2/a$. The eccentricity of this ellipse is $e = 0.84$. If the Sun is at F_2, then P is perihelion, the point of closest approach of an orbiting body, while A is aphelion, the point of farthest separation. (F_1 then has no physical meaning.) (Kaler, J.B. (1996). *The Ever-Changing Sky*, Cambridge: Cambridge University Press.)

a star [7.2.4] is $g = G\, M_{\text{star}} / R_{\text{star}}^2$. Newton's generalizations of the laws apply to any two bodies:

(1) orbits are "conic sections" that can include circles, ellipses, parabolas, and hyperbolas (the latter two open-ended);
(2) the conservation of angular momentum (Box 8.3);
(3) $P^2 = (4\pi^2/G) a^3 / (M_1 + M_2)$,

where *a* is now in meters or centimeters and M is in either grams or kilograms.

The gravitational pulls of the other planets make perfect elliptical orbits impossible. Two isolated bodies, however, come much closer to that ideal. The best example is a double star with components sufficiently separated that tides [8.5.1] play no role. The stellar orbits are then almost perfect ellipses and prime examples of Kepler's laws.

One star of a binary does not actually orbit the other. Gravity is a mutual affair. Two bodies locked in gravitational embrace will have identically shaped elliptical orbits relative to a center of mass that lies between them and at the focus of each orbit (Fig. 8.3). The semimajor axes of the two elliptical orbits are in inverse proportion to the individual stellar masses, such that $a_A/a_B = M_B/M_A$. Planetary orbits seem to focus on the Sun because planetary masses are so small compared to the solar mass. In reality, the Sun moveth too, just not by very much.

It is simpler, however, if we take one star (usually the primary "A" component) as the reference point and mathematically pretend that star "B" has an elliptical orbit with A at the focus. In that case, the semimajor axis of star B about star A, *a*, is the sum of the semimajor axes of the individual orbits, $a = a_A + a_B$. The eccentricity of the pretend orbit, *e*, remains the same as before. As planets pass through perihelion and

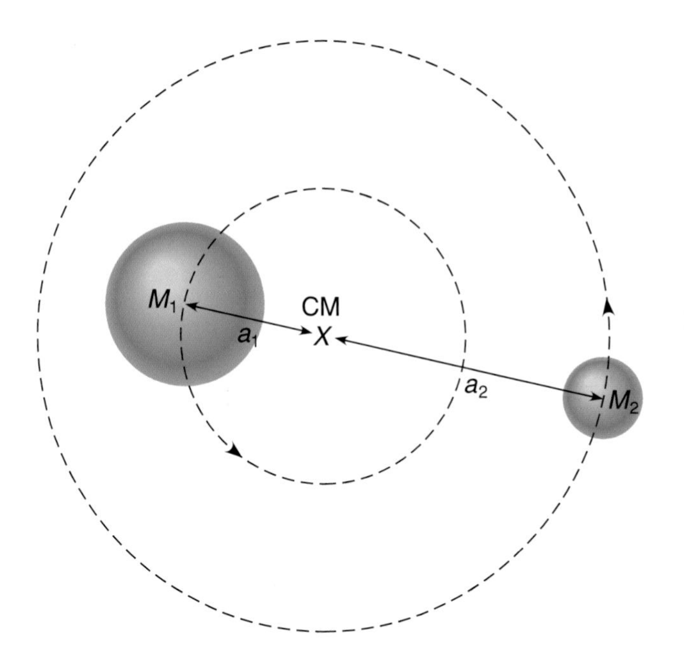

Fig. 8.3 **Center of mass** Two stars orbit a center of mass on a line between them. Here, the mass of the primary star (M_1) is double the mass of the secondary M_2, so the secondary orbit is twice as large as the primary's orbit. (Kaler, J.B. (1994). *Astronomy!* New York: HarperCollins, reprinted by permission of Pearson Education, Inc.)

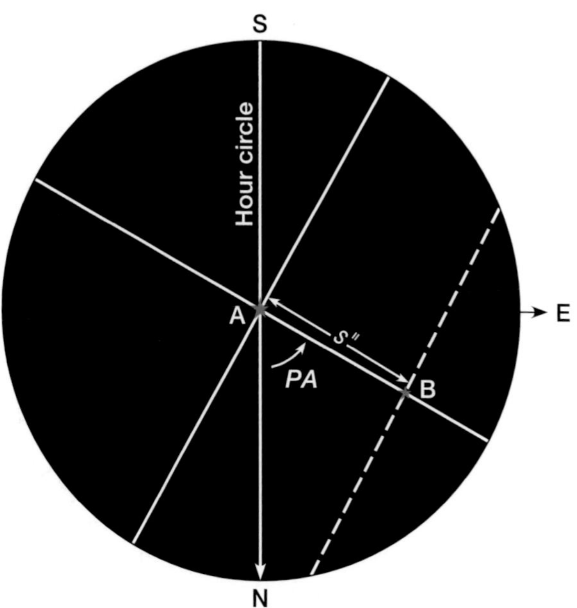

Fig. 8.4 **The visual double** The view through an astronomical telescope is inverted, so south is up, north down. The crosshairs of a micrometer are set to determine the position angle (PA) and the separation (s) of the secondary star (B) relative to the primary star (A). (Kaler, J.B. (1994). *Astronomy!* New York: HarperCollins, reprinted by permission of Pearson Education, Inc.)

aphelion, double stars pass through *periastron* and *apastron*. The periastron distance is given by $a(1 - e)$ and the apastron distance by $a(1 + e)$.

We can get rid of the constants in Kepler's generalized third law by putting it in terms of the Earth and its orbit (that is, just divide the law as written above by the same as written for the Earth). The law then just becomes

$$P^2 = a^3/(M_1 + M_2),$$

where the period is now in years, the semimajor axis in AU, and the masses are in solar units, $M_\odot$ (where $\odot$ is the universal symbol for "Sun"). Application to "one star going about the other" thereby gives the sum of the two stellar masses $(M_1 + M_2) = a^3/P^2$ (in $M_\odot$). The task of the double-star astronomer is to find the orbital period P in years and the length of the semimajor axis, a, in AU. If the individual orbits are observed and the center of mass located (which is considerably more difficult), then the mass ratio is known from the ratio of semimajor axes. Combination of the sum and ratio of masses then yields the gift of the individual masses and a step toward the mass–luminosity relation [12.4.1].

8.2.4 Observation and analysis

Classical observation of visual double stars begins with the view through the telescope (Fig. 8.4). Two quantities determine the location of the secondary star (B) relative to the primary (A): the angular separation in seconds of arc (s) and the *position angle* (PA), the angle that the line from star A to star B makes with the line between A and north, as measured through east. Both s and PA were originally measured by eye with the aid of a "filar micrometer" attached to the telescope. Because the observer could take advantage of momentary stillnesses in the twinkling of stars, double-star astronomy was among the last bastions of true visual observing. Photography provided a permanent record, and electronic recording easily trumped both.

If the stars are sufficiently close, both s and the PA will noticeably change, allowing the astronomer to construct an elliptical orbit of B around A and to find the period, the semimajor axis (in seconds of arc), and the eccentricity (Fig. 8.5). If the period is long, only a partial orbit might be observed, but that may still be enough to fit an ellipse to the data, allowing the full orbit to be extrapolated so as still to derive P, a, and e.

This discussion is fine if the orbit lies on the "plane of the sky" such that the orbital axis points toward Earth. In general, however, the orbit will be tilted to the plane of the sky through *inclination* i (the orbital axis similarly tilted through the same angle against the line of sight). What we actually observe is the projection of the true ellipse onto the plane of the sky (which is still an ellipse), which alters the observed a and e. As an extreme example, with $i = 90°$ (the orbital plane in the line of sight), the secondary star just goes back and forth in a straight line past A, and the observed eccentricity $= 1$, no matter what the true value of e might be. (P, e, a, and i constitute four of the six or seven "elements" of the orbit, which collectively define it. T gives the time of periastron passage, the angle Ω describes the orientation of the orbit

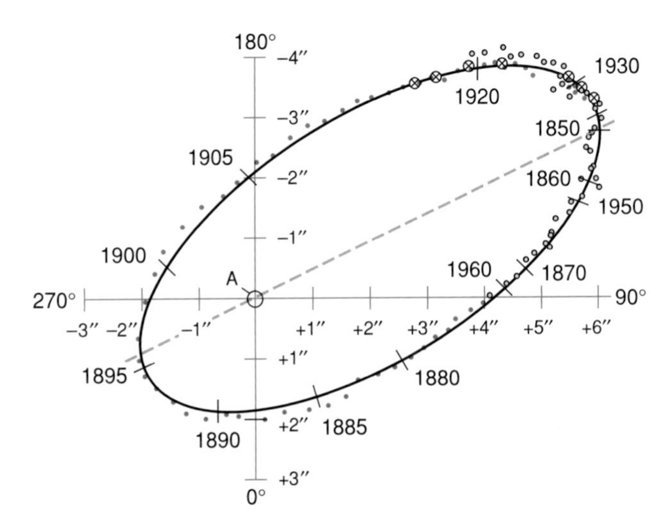

Fig. 8.5 **A double-star orbit** The elliptical orbit (as projected onto the plane of the sky) of 70 Ophiuchi is fitted through the observations. Note that the primary star is not on the major axis of the projected ellipse. The dashed red line is the true major axis. (Courtesy of Yerkes Observatory and Kaler, J.B. (1994). *Astronomy!* New York: HarperCollins, reprinted by permission of Pearson Education, Inc.)

against the plane of the sky, and ω gives the orientation of the orbit in its own plane.)

Though the observed orbit is still an ellipse, for any inclination but 0° the primary star will not be at the focus, nor even lie along the major axis. From the position of the primary, which we know *must* lie at the focus of the true ellipse, the actual orbital orbit can be recovered; that is, the observed ellipse can be "rectified" to a face-on appearance.

The most important result is the true semimajor axis of the secondary's orbit relative to the primary star in seconds of arc. From Box 4.1, $a'' = 206\,265\, a/d$, where d is distance and both quantities are in AU. But d (AU) is 206 265 d (pc) [Box 4.2], so $a'' = a$ (AU)/d (pc) and $a\ (AU) = a''d$ (pc). To find the semimajor axis in AU thus requires that we know the parallax [4.2]. Because binary orbital motion can be seen much farther than the reach of accurate parallaxes, the semimajor axes of double stars are usually expressed as a'' (for the rectified orbit), allowing the user to put in new distances as they become available. Once a (AU) is known, the sum of the masses $M_A + M_B$ immediately falls out.

Finding the orbit of star B around star A (with A as the reference) avoids having to deal with the actual motions of the stars. To find the center of mass requires that the stars be tracked against those of the distant background (Fig. 8.6). The motion of each binary component now becomes a combination of proper motion, parallax, and orbital movement [4.2.2], all of which must be separated from each other, a challenging job that is not always possible. When it is, we find the ratio of the true semimajor axes (which is not dependent on orbital inclination) and, with the sum of masses already

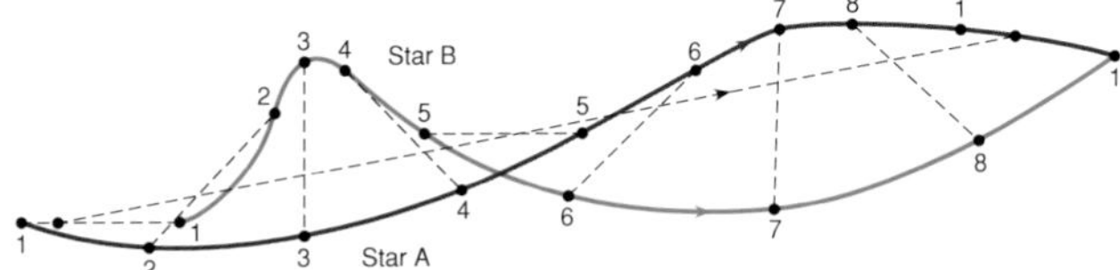

Fig. 8.6 **Double star in space** As the center of mass of a double star moves through space, the components move sinuously about it. The ratio of the amplitudes (maximum displacements from the center line) gives the ratio of stellar masses. Superimposed on these curves is another wobble produced by annual parallax. (Kaler, J.B. (1994). *Astronomy!* New York: HarperCollins, reprinted by permission of Pearson Education, Inc., after S.P. Wyatt.)

known from the orbit of B around A, again the individual stellar masses.

Though the number of observations of binaries is staggeringly great and the number of orbits large, the number of *accurate* orbits that are useable in the study of stellar masses is still relatively small, although constantly growing. Each year that passes provides an increasing database for binaries of longer period. Moreover, a variety of techniques are being used that enable the astronomer to observe to much smaller angular separations than are possible to achieve directly through the telescope. Indeed, the term "visual binary" is slowly being replaced by "resolved binary," which, ever-more, is encroaching on the realm of the "spectroscopic binary". Speckle interferometry (in which very short exposures are combined to overcome the smearing effects of the Earth's atmosphere) can resolve to a few hundredths of a second of arc. Repeated observations by Hipparcos established both binary natures and orbits, extending separations into the milliarcsecond realm. Standard interferometry that combines the output of two or more mirrors or telescopes can do even better. Even lunar occultations get into the act. A close pair will wink out twice when covered by the Moon, a triple thrice. Observations over periods of years can reveal the orbit.

Wider separation leads to weaker gravitational bonds and longer orbital periods, until we get to the point of not being able to detect orbital motion at all. Binarity is then known only through common distance and proper motion. Two solar mass stars orbiting with a period of 10 000 years would be 600AU apart. At a distance of 100pc they would be separated by 6 arcsec. Such systems abound. At very large separations, orbits become influenced by passing giant molecular clouds [11.1.1] and stars, and even by tides [8.5.1] raised by the whole Galaxy. The members of such *fragile binaries* will slowly leave each other's company. The limit seems to be a separation of about 20 000AU (0.1pc), which for solar mass stars yields an orbital period of two million years. Proxima Centauri, the closest star to the Earth [4.2.1], is a good example, taking roughly a million years to orbit Alpha Centauri (which is itself double; see Table 8.1) at a distance of some 15 000AU.

8.3 Spectroscopic binaries

The sky abounds in close doubles that can be separated only by spectrographic analysis. When enough spectrographic power is brought to bear, stars thought for ages to be single are suddenly found to have companions. The spectrograph has led us to find ever-lower masses down to the point of discovering planets, which abound as well. Look first, though, to the stars.

8.3.1 The ideal

Figure 8.7 presents an ideal spectroscopic binary. In Fig. 8.7a, identical stars A and B orbit in circles about a common center of mass halfway between them, the orbits lie perpendicular to the plane of the sky with an inclination of 90°, and the radial velocity of the system (that of the center of mass) is zero. Each star has velocity v about the center of mass. Since the stars are identical, the combined spectrum is the sum of two equal contributors (Fig. 8.7b). At time T_1, the stars are both moving across the line of sight, the spectra will exactly overlay each other, and the combined spectrum will have the appearance of that of a single star. As the stars orbit, A begins to move toward the observer, while B recedes. As a result, the spectrum of A is Doppler shifted to shorter wavelengths, that of B to longer wavelengths, and the combined spectrum separates into two individual spectra. After a quarter of an orbit, at time T_2, star A is coming directly at the Earth, while B directly recedes, and both display their maximum absolute radial velocities. After half an orbit, the spectra merge, and after three quarters, they are again fully separate, but in the opposite direction. At any given point, we can calculate the velocities of the stars and construct a *velocity curve* (velocity vs. time) as in Fig. 8.7c.

The relevant parameter of a velocity curve is its *semiamplitude*, K, which is the maximum velocity of approach or the maximum velocity of recession. In this ideal example, $K_A = v_A$ and $K_B = v_B$, where v is the orbital velocity. The orbital circumference of star A is the period multiplied by the orbital velocity, $v_A P$, but is also $2\pi a_A$, where a_A is the orbital radius in kilometers. Since $2\pi a_A = v_A P$, $a_A = v_A P/2\pi$. Similarly, $a_B = v_B P/2\pi = a_A$, whence both are converted to AU. If we consider star A as the focus of the orbit of star B, then the semimajor axis of B's orbit is $a = a_A + a_B$. As before, the sum of masses $M_A + M_B = a^3/P^2$, where the masses are in solar units. Since the stars are identical (as given by identical velocities), the individual masses are half the sum.

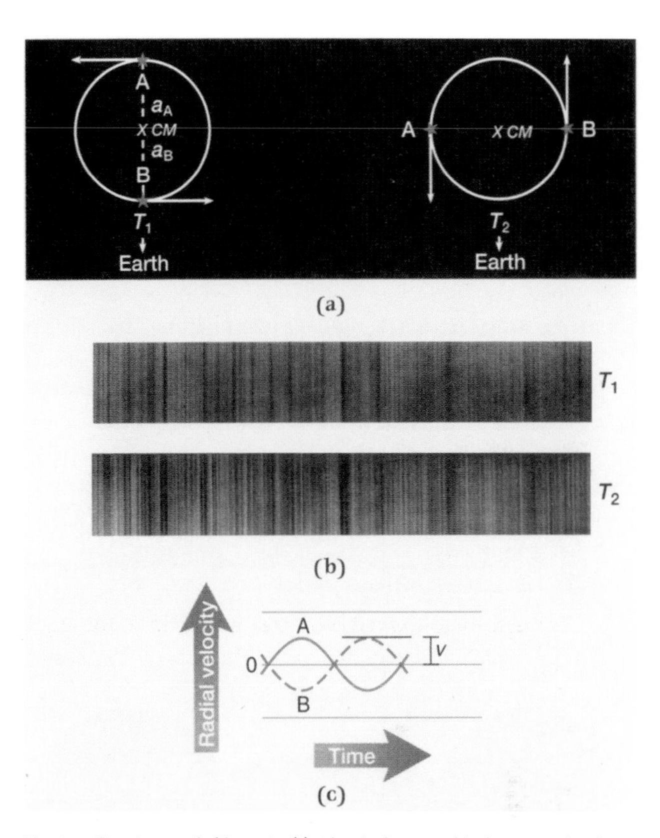

Fig. 8.7 **Spectroscopic binary** In (a), identical stars orbit the same circular path about a common center of mass. The spectra in (b) are presented for times T_1 (when there is no Doppler shift) and T_2 (when star A approaches and B recedes). The resulting velocity curve is in (c). (Kaler, J.B. (1994). *Astronomy!* New York: HarperCollins, reprinted by permission of Pearson Education, Inc., after S.P. Wyatt.)

8.3.2 Real systems

Several variations and complications alter this simple view. First, the system will have its own radial velocity that must be removed. The maximum orbital velocity – the semiamplitude of the velocity curve, K – is then the observed maximum velocity minus the systemic velocity (Fig. 8.8). The stars will also rarely have identical masses. Since the duo must share the same period, their orbital velocities are proportional to their semimajor axes. The ratio of masses is therefore the inverse of the ratio of velocities, that is, $M_A/M_B = K_B/K_A = v_B/v_A$. With the sum and ratio known, we derive the individual masses.

Second, the orbital planes are unlikely to be in the line of sight (Fig. 8.9); if they are, the stars will eclipse each other and we will then know that the inclination is close to 90°.

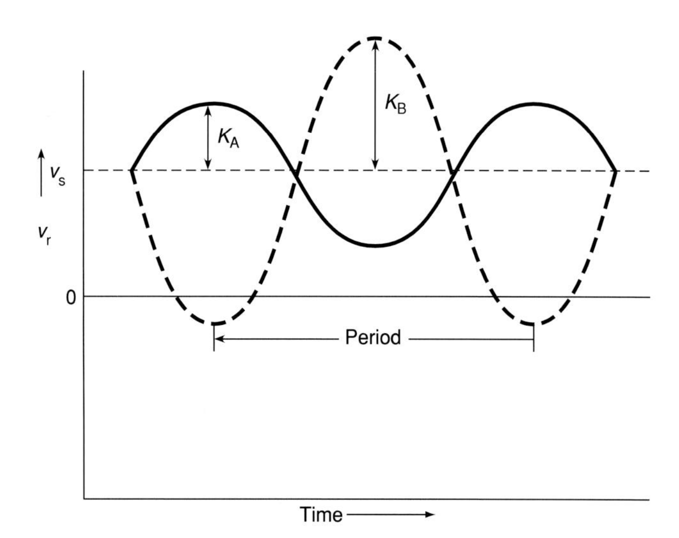

Fig. 8.8 **Double-line velocity curve** The system has a radial velocity v_S. Star A has twice the mass of star B, so its velocity curve has half the amplitude (K_A) of star B (K_B). If the orbital plane is in the line of sight, $K_A = v_A$ and $K_B = v_B$. If the plane is tilted through inclination *i*, then the amplitudes are lower limits to the true velocities.

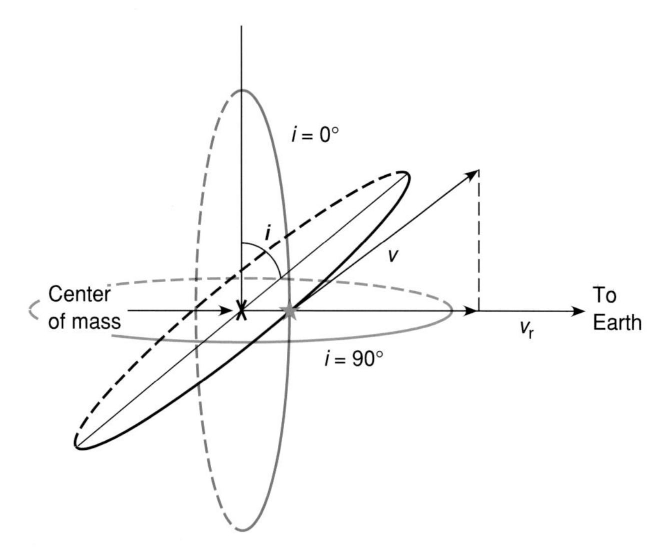

Fig. 8.9 **Orbital inclination** The circular orbit of a binary star is inclined through angle *i* relative to the plane of the sky. The observed maximum velocity *K* will be then a projection of the true velocity *v*, $K = v \sin i$. For $i = 90°$, the observed velocity is the true value. (Kaler, J.B. (1994). *Astronomy!* New York: HarperCollins, reprinted by permission of Pearson Education, Inc.)

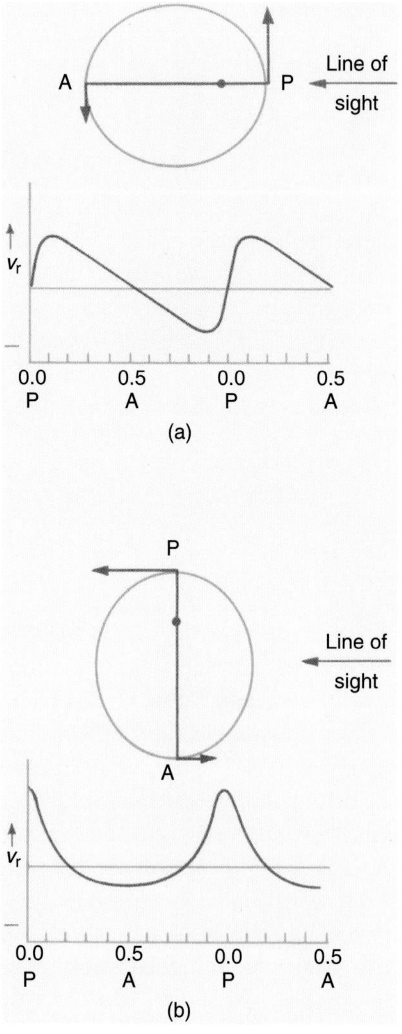

Fig. 8.10 **Eccentricity and orientation** The orbital plane of a single-lined binary with an eccentricity of 0.5 lies in the line of sight. At the top (a), the major axis also lies along the line of sight, while at bottom (b) it lies perpendicular to the line of sight. In both cases, the velocity curves are severely skewed from the simple sine wave seen in Fig. 8.8. In (a), the minimum and maximum observed velocities are the same, whereas in (b), the maximum velocity of recession takes place with the star at periastron (P) and is much larger than the maximum velocity of approach, which takes place at apastron (A). (Wyatt, S.P. *Principles of Astronomy*, Boston: Allyn and Bacon (1964)).

Given an inclination of the orbital plane to the plane of the sky, i, the projected maximum velocity for a circular orbit will be $K = v \sin i$. The orbital radius calculated from the observed projected velocity will similarly be $(a \sin i)$, which will be a lower limit to the true radius. Again, from Kepler's third law, $(M_A + M_B) = a^3/P^2$. But we know only $(a \sin i)$. Multiplying both sides by $\sin^3 i$ shows that such tilted orbits give not the sum of the masses but a lower limit, $(M_A + M_B) \sin^3 i$. However, since the individual orbits are tilted through the same angle i, the mass *ratio* from $K_B/K_A = v_B/v_A$ will still be correct. The result is useful as we can at least assume that the set of binary inclinations is randomly distributed (a tested hypothesis), which allows for the calculation of average masses for specific kinds of stars. Moreover, in many instances, lower limits are enough to constrain theories.

Now things get worse. If there is a significant magnitude difference between the two stars, the spectrum of the brighter will wash out that of the fainter. The result is that we see only one set of absorptions moving back and forth. From any such single-line spectroscopic binary, we still easily find the period and also calculate the lower limit to the semimajor axis of the orbit of the brighter component, $a_A \sin i$. However, masses, even lower limits to the masses, are impossible to derive directly. All that can be found without any assumptions is the so-called *mass function*,

$$f(m) = a_A^3 \sin^3 i / P^2 = M_B^3 \sin^3 i / (M_A + M_B)^2,$$

which by itself provides little information. Yet some assumptions still allow a little headway to be made. In some instances, M_B can rightly be assumed to be much smaller than M_A. Examples would be a red dwarf or planet [11.4.3] in orbit about a more massive star, or a black hole [14.3.3] in orbit about a much more massive supergiant. Then the known quantity $f(M) = M_B^3 \sin^3 i / M_A^2$. If we can estimate the mass of the visible primary from its temperature, luminosity, and state of evolution [13.2], we can then solve for $M_B \sin i$ and place a limit on M_B.

Moving deeper into complexity, the orbits may be elliptical, sometimes very much so, and will have some random orientation of the semimajor axis to the line of sight. The result is a deviation of the velocity curve from a simple oscillating sine wave (Fig. 8.10) in which the semi-amplitude is no longer necessarily equal to the maximum or minimum velocity. The problem is nevertheless solvable. Each combination of orbital elements gives a unique shape to the velocity curves that allows the eccentricity and ($a \sin i$) to be recovered. The result is again the ratio of the masses, $(M_A + M_B) \sin^3 i$, or, in the case of single-lined systems, the mass function.

8.4 Eclipsing binaries

If the plane of the binary orbit lies in (or close enough to) the line of sight (that is, if the inclination i is near 90°), then as the stars orbit, one can pass in front of the other, resulting in an eclipse and a periodic drop in brightness. Four such eclipsing binaries are indicated in Appendix 5. The chance of a binary's components eclipsing each other depends on the sizes and separations of the stars. Classical visual binaries are poor candidates. They are so far apart that the inclinations of the orbits would have to be almost exactly 90°, and the odds become vanishingly small. Spectroscopic binaries, whose members are close together, are far better ones.

There are almost as many combinations and types of eclipsers as there are stars themselves: the stars can be nearly equal or wildly unequal in temperature, size, and evolutionary status; the orbits can take on a variety of sizes, eccentricities, orientations, and inclinations; eclipses can be partial or total; moreover, the stars can interact, producing both distortions in otherwise straightforward changes in magnitude.

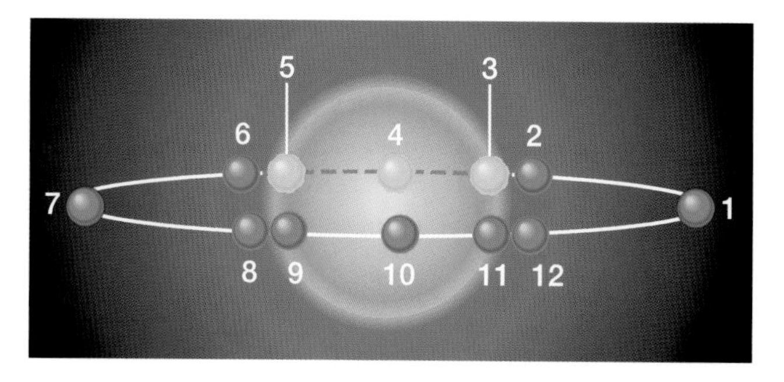

(a)

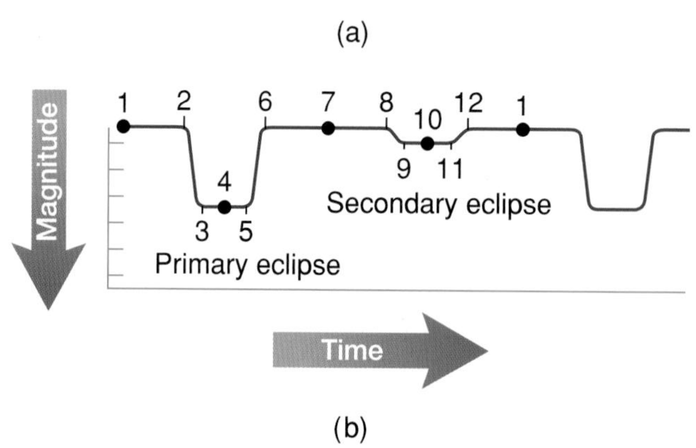

(b)

Fig. 8.11 **Classic eclipse** A small bright (primary) star orbits a larger, cool, dim (secondary) star. The orbital positions correspond to the points on the light curve below. At the primary eclipse, the bright primary star is entirely hidden, while at the secondary eclipse, the primary star cuts out only part of the light of the dimmer secondary star. (Kaler, J.B. (1994). *Astronomy!* New York: HarperCollins, reprinted by permission of Pearson Education, Inc.)

8.4.1 Light curves

A common scenario for illustration involves a bright, hot star (say, a class B dwarf) in mutual orbit with a much larger, dimmer, cooler star, typically a K giant. In Fig. 8.11, the orbits are circular and the inclination is a bit less than 90°. Once every orbital period, each star passes in front of the other, and the brightness of the combined system appears to drop (to the observer) by a factor that depends on the relative sizes and luminosities of the stars. The chain of events is recorded by the system's *light curve*, a graph of apparent magnitude (or sometimes observed luminosity) vs. time. The large star can completely hide the bright B dwarf, resulting in a deep, total, very noticeable *primary* eclipse. When the dwarf is in front of the giant, however, the *secondary* eclipse is only partial (or, in solar-eclipse terms, "annular"). Now only that part of the giant that has the projected surface area of the dwarf is hidden, and the eclipse is much shallower and in some cases even invisible.

A full light curve contains a dozen or so significant positions. Position 1 in Fig. 8.11 represents the point exactly between the primary and secondary eclipse, around which the light of the combined star is constant. At point 2 ("first contact"), the brighter primary star begins to go behind the larger secondary and the primary eclipse begins. (Within the eclipsing community, the terms "primary" and "secondary" are inconsistently used. The "primary" might be either the more luminous, the larger, or the more massive of the two stars. The reader must note the context and not make assumptions. In this example, the "primary" is the more luminous blue class B star.)

At point 3, "second contact," the primary is fully eclipsed, and the time interval 2–3 is called the "primary ingress phase." From points 3 to 4, the primary is completely hidden, and the star's brightness remains at its low level. Egress begins at point 5 ("third contact") when the primary first emerges from behind the secondary, and is over at point 6 ("fourth contact") as the star returns to normal brightness. Point 7 lies between primary and secondary rather than between secondary and primary. At point 8, the secondary star starts to cut across the primary to begin the "secondary eclipse ingress." Between 9 and 10, the secondary star remains in front of the primary and the star stays slightly dimmed. Secondary egress then runs from 11 to 12, and after a full period we are back to the original point 1.

8.4.2 Stellar properties

A great deal of information can be derived from the light curves of eclipsing binaries. Assume that the inclination of the orbit in Fig. 8.11 is 90°, that the orbit is still circular, and that the star is also a double-lined spectroscopic binary. The duration of ingress or egress during primary eclipse (T_i, the time from point 2 to 3 or from 5 to 6) is the diameter of the primary (brighter and smaller) star divided by the relative orbital velocity, v, $T_i = R_A/v$. Since $a = a_A + a_B$, $v = v_A + v_B$. Similarly, the duration of the primary eclipse (T_p, actually from points 2 to 5 or from 3 to 6) equals the diameter of the secondary (larger) star divided by the velocity, or $T_p = R_B/v$. Inverting the relations gives the physical dimensions of both stars ($R_A = T_i v$, $R_B = T_p v$)! Even if the velocities are not known, we can extract a bit of information. Since the relative velocity is $v = 2\pi a/P$, $R_A/a = 2\pi T_i/P$ and $R_B = 2\pi T_p/P$, and we can find the radii of the stars relative to the orbital semimajor axis, a. If the star is a double-lined spectroscopic binary and $i = 90°$ (such that $\sin i = 1$), we also find the stellar masses.

The inclination of the orbit complicates things. Within the set of eclipsers, there is a range of allowed inclination over which the eclipse of the smaller one will be total. If the stars are small and far apart, the inclination must be almost exactly 90° for there to be an eclipse at all. However, if they are large and/or very close, the inclination can be quite low; i well under 90°.

If the inclination is sufficiently small, the stars never eclipse. In between, for any particular stellar-size/orbital-separation combination, there will be a small range of inclinations over which the eclipses will be partial (the big star cutting off only part of the small star's light). That is the case for the most famed of all eclipsers, Algol, Beta Persei. Since the brighter primary is never completely hidden, the primary eclipse comes only to a sharp point, the system brightness returning immediately to normal; the weak secondary eclipse behaves similarly (Fig. 8.12a; see also Fig. 3.5).

The duration of ingress/egress and the length of the eclipses depends on the inclination. As long as the eclipse is total, because of the circularity of the stellar limbs, as i decreases from 90°, T_i increases and T_p decreases. Though close binaries will tend to circular orbits [8.5.2], many do have elliptical orbits and therefore varying orbital velocities. If the stars are oriented as in Fig. 8.10a, the secondary eclipse will take place midway between the primary eclipses, but the durations will again be affected. If the stars are arranged as in Fig. 8.10b, though, the secondary will be severely off-center. There is, however, a unique relation between orbital parameters (eccentricity, orientation, and inclination) and the light curve, analysis of which yields all three, and once again stellar radii and masses. No wonder astronomers love eclipsers.

Yet more information can be extracted from these stars. Outside of eclipse, the apparent luminosity is the sum of the luminosities of the two stars. But during the primary

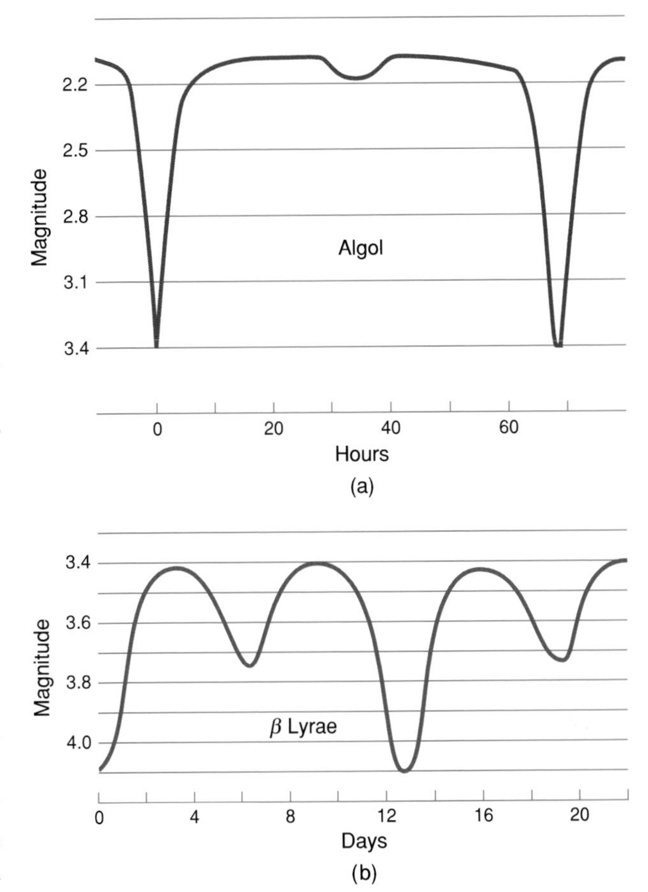

Fig. 8.12 **Distorted light curves** (a) Algol undergoes partial eclipses, so as a result, neither eclipse is flat-bottomed. The rise in brightness outside of eclipse is caused principally by the reflection effect. (b) The components of Beta Lyrae are so close and highly distorted that (including effects from flowing matter) the light curve changes continuously. (Kaler, J.B. (1994). *Astronomy!* New York: HarperCollins, reprinted by permission of Pearson Education, Inc.)

eclipse, only the bigger secondary is visible, so we can measure its magnitude in various wavelength bands and even record its spectrum. From the bolometric flux and the star's surface area, we get the effective temperature. Subtracting these measures from the combined light gives the magnitude and spectrum of the secondary and its effective temperature. Moreover, the depth of the secondary eclipse gives the amount of light that the small star cuts from the big one, allowing the derivation of the large one's surface brightness, and again the temperature. Eclipsers are the main source of data used to construct the mass–luminosity relation [12.4.1].

8.4.3 Complications

A variety of phenomena can severely distort eclipsing binaries' light curves. All stars are limb darkened [12.1.1]. Limb darkening of the primary star in Fig 8.11 will change the shape of the light curve during ingress and egress. When the primary star moves behind the secondary, the decrease in brightness will first be less than it would be with no limb darkening, but will thereafter proceed more quickly. When the primary begins to move in front of the secondary, it will cut out less light than when it is centered. As a result, the light curve of the secondary eclipse will not be flat-bottomed, but will be bowed downward.

The stars may also not be spherical. If they are close enough together, they will raise tides in each other, even to the point of filling their tidal zero-gravity surfaces [8.5.1], which makes them into ellipsoidal or even teardrop shapes. As a result, as the stars orbit, they continually present different cross sections to the observer, which not only changes the light curve during eclipse, but also outside of eclipse. For a circular orbit, the astronomer will see the maximum cross section (and brightness) when the line between the stars is presented perpendicular to the line of sight.

Asphericity, however, can be overwhelmed by the "reflection effect." As the bright star swings around in front of the larger one, its light will be reflected from the larger one's surface. (More accurately, the smaller will heat the larger's surface, causing the latter to radiate more strongly.) Maximum reflection will take place at the center of the secondary eclipse, but is masked by the eclipse itself. Maximum brightness for the system will therefore occur just before secondary ingress and just after secondary egress. These effects are beautifully seen in the light curves of Algol and Beta Lyrae in Fig. 8.12.

Finally, the stars may be rotating. Indeed, if they are close enough, rotation through tidal synchronization [13.6] will be forced upon them. If the giant is transferring mass to the small star, however, the small one's rotation can be immensely speeded up. If the star is rotating fast enough, it will be flattened at the poles, that is, further distorted. Moreover, at the pole, the temperature gradient in the atmosphere will be changed, and as a result, the star will be brightest there and darkest at the equator. The non-uniform surface brightness caused by "gravity darkening" (the "von Zeipel effect") [7.2.5, 7.3.1] will have an effect principally on the ingress and egress phases of the primary eclipse as different portions of the non-uniform small hot star disappear behind the bigger one.

All these affect the light curve in different ways, and at least in principle they can therefore all be separated from each other, even including the effects of mass transfer between the stars, which can cause the light curve to vary erratically. As a result, we can ideally determine the degree of limb darkening to find the stars' atmospheric temperature gradients, in addition to their radii, degrees of asphericity, and all-important masses. Amazingly, even something of the internal mass distribution can be learned, since if the stars do not behave as point gravitational sources, the major axis (the *line of apsides*) will rotate, which in turn causes a visible change in the orbital orientation and the light curve.

Clearly there are an almost limitless number of combinations of different kinds of stars, which gives immense variety to the eclipsing scene. Indeed, all systems are unique, no two quite alike.

8.5 Interacting binaries

If the two stars of a binary are sufficiently close together, they can interact with each other, which can profoundly change the course of their lives, produce bizarre behavior, and even cause one of the members to explode.

8.5.1 Tides and the tidal surface

Two large bodies in proximity will raise tides in each other. The lunar gravity is stronger on the side of the Earth facing the Moon, weaker on the side facing away. As a result, the solid Earth is stretched out on a line toward the Moon by about 20 centimeters. The stretching effect is much more noticeable in the Earth's flowing oceans, as anyone who has spent any time at the beach can attest. As the Earth rotates under the tidal bulge (which is actually pulled ahead of the Moon as a result of Earth's rotation), the water level at the shoreline goes up and down. A weaker tide produced by the Sun adds to the lunar at new and full Moons (when the Moon and Sun are lined up) and subtracts from the lunar tide at the lunar quarters (when the pair are at a 90-degree angle).

Dissipation of rotational energy through tidal stretching causes the Earth to slow down and the day to increase by about a thousandth of a second per century. Tides raised by the Earth in the Moon have caused the Moon to stop rotating relative to the Earth altogether and, as a result, one face of our companion points to us forever. Bring the Moon closer and the tide increases by the inverse cube of the distance. Make two bodies close enough, and one could tear the other apart: the minimum distance for a liquid body is called the *Roche limit*.

Double stars interact similarly, stellar tides producing some amazing effects. Place two stars in proximity, and around them is a three-dimensional surface where the gravitational attraction toward one of them equals the attraction toward the other. The effective gravity along this surface is therefore zero, giving it the name of *equipotential surface* (Fig. 8.13). It consists of two teardrop-shaped *Roche lobes* that are in part shaped by tides, in part by centrifugal forces caused by the stars' orbital motions. The cross section of the figure through the two stars looks rather like a "figure 8," with each star in the center of a loop. The lobes meet at the teardrops' tips, which is referred to as the *inner Lagrangian point*. A particle placed along the equipotential surface could not stay there any more than you could balance a pencil on its point. It would instead fall to one star, go to the other, or leave the pair altogether.

If the stars are very widely separated (visual doubles, for example), they sit at the cores of their Roche lobes and have nothing to do with each other except orbit. Bring them closer together (close spectroscopic or eclipsing binaries), and tides stretch the stars out, each on a line to the other. Though affecting and distorting each other, they are still *detached*. Bring them yet closer, and the larger star will meet its equipotential surface. Its mass can then leave, much of it flowing through the inner Lagrangian point where it finds itself under the gravitational control of the other star. Mass then passes from the one to the other, the large star becoming the "loser" or "donor," the smaller one the "gainer." The system is now *semi-detached*.

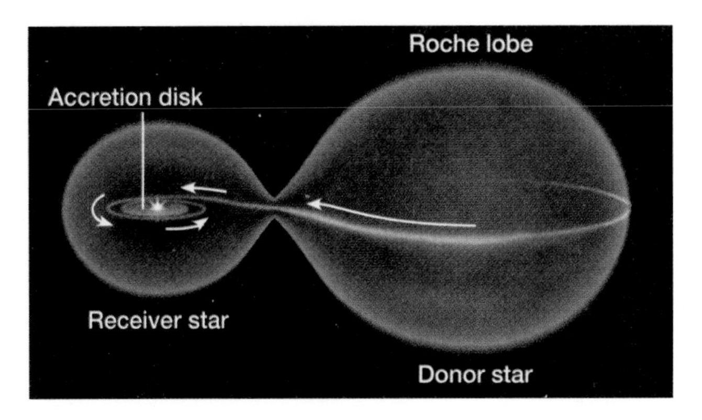

Fig. 8.13 **Roche lobes and mass transfer** Zero-gravity equipotential tidal surfaces (Roche lobes) that meet at their tips (the inner Lagrangian point) surround the members of a binary. The large star at the right fills its lobe, while the small star at left does not. The larger one transfers mass through the Lagrangian point to an accretion disk around the small star, from which it falls downward. (Kaler, J.B. (1994). *Astronomy!* New York: HarperCollins, reprinted by permission of Pearson Education, Inc.)

Though a stream of matter now flows toward the smaller star, it does not go straight in (Fig. 8.14). Because the stars are quickly orbiting each other, the outgoing mass flows out in a wide curving arc toward the secondary. There are now two possibilities. If the secondary star is large enough, the incoming matter will hit it. The point of impact will heat and radiate, as will the incoming stream. The light curve, to which is added the light of mass being transferred, now becomes messy and somewhat erratic. If the smaller star is small enough, however, the incoming flow will miss it, and will fall into a disk from which mass will eventually be accreted. Now it is the *accretion disk* around the smaller star that is hot. It can radiate so much that it competes with the stellar light. The ultimate scene comes when both stars contact their Roche lobes. They then kiss at the inner Lagrangian point and touch each other to create a *contact binary*.

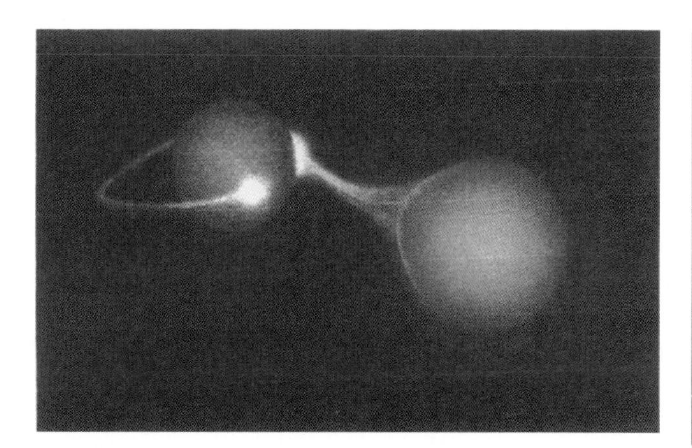

Fig. 8.14 **Algol** Mass flows from Algol's low mass K star (the loser) obliquely to the B star (the gainer). (Simpson, S. (Feb. 1990), *Sky and Telescope*, p. 128.)

8.5.2 Orbit circularization

Even detached binaries can affect each other. If a reasonably close double begins life in an elliptical orbit, highly variable tides will flex the stars and thereby dissipate energy from the system, causing the orbits to become more circular. The closest and shorter-period pairs circularize first, longer periods later. We can see the effect in clusters whose ages we know, and also in the general field. Doubles in the Hyades (which is about half a billion years old) with periods under 10 days are circularized. In the general field of solar type stars, which is a few billion years old, orbits are circularized up to periods of 14 days, and the low metal stars of the halo up to 19 days. Given good theory, the patterns make another clock with which to tell stellar ages.

8.5.3 Algol systems

Algol, a semi-detached binary, is but the prototype of a large number of similar systems. Algol's secondary star is an evolved class G or K giant or subgiant, while the unevolved one is a B8 dwarf. Mass is transferred from the dwarf directly onto the subgiant to create a hot spot (see Fig. 8.14). Stellar evolution theory predicts that stars of high mass live shorter lives than do stars of lower mass [13.1]. The K star should therefore be the more massive. It is not. The B star contains 3.7 solar masses, while the K star anomalously has but 0.8 or so. This famed *Algol paradox* is quickly explained by the observed mass transfer. The cooler K star once had to be the more massive. It has lost most of what it had to the B star, the gainer. While Algol-B (the K giant) will probably not be entirely absorbed by the primary B star, the loss of mass will probably keep the K star from fusing its hydrogen, and it will likely die as a helium white dwarf [13.5].

8.5.4 Novae

Compact secondary companions with accretion disks are common and produce a variety of phenomena related to the three classes of dead stars: white dwarfs, neutron stars, and black holes. Binaries with close white dwarfs create several classes of eruptive *cataclysmic binaries*, also known as *cataclysmic variables* (CVs). The most obvious of them are eruptive *novae* (Fig. 8.15). The term "nova", from the Latin "nova stellarum," means "new star." Though new to the observed nightly sky, novae are not new at all but old, illustrating an endemic problem in astronomy with naming something before it is understood. We observe about 10 percent of the 40 or so novae that take place in the Galaxy each year, and every couple of decades one explodes close enough to us to make first magnitude (Table 8.2).

Fig. 8.15 **Nova Cygni 1975** The surface of the white dwarf in the system now known as Nova Cygni 1975 erupted (below) to become far brighter than the system (stars plus accretion disk) that caused the mischief (top). (Copyright University of California Regents/Lick Observatory.)

A nova-to-be consists of a white dwarf [13.5] in orbit with a low-mass ordinary dwarf [6.3.1]. They are close enough that the main sequence star fills its Roche lobe and passes mass, loaded with fresh hydrogen, to the white dwarf through an accretion disk. The base of the hydrogen layer that is building up on the white dwarf becomes degenerate [Box 13.2] and mixes a bit with the carbon-rich core. When the layer becomes hot enough – 20 million K or so – it suddenly begins to fuse to helium via the carbon cycle [12.2.2]. However, being degenerate, the increase in temperature does not change the pressure or expand the layer, so the temperature goes up in a catastrophic thermonuclear runaway. The increased temperature eventually lifts the degeneracy, but the

Nova	Var. Star Name	V_{max}	$M_V(max)^a$
Persei 1901	GK Persei	+0.2	−8.5
Aquilae 1918	V 603 Aquilae	−1.1	−9.2
Pictoris 1925	RR Pictoris	+1.2	−7.4
Herculis 1934	DQ Herculis	+1.4	−5.6
Puppis 1942	CP Puppis	+0.4	−9.1
Cygni 1975	V 1500 Cygni	+1.8	−9.5
Cygni 1992[b]	V 1974 Cygni	+4.4	−7.7

[a] Corrected for dimming by interstellar dust.
[b] Famed for being the first nova observed throughout the electromagnetic spectrum.

Table 8.2 **Famous twentieth-century novae**

damage is done, as the burning layer becomes convective and dredges up part of the core. With luminosities of up to a million solar (near the star's Eddington limit [14.1.2], Nova Cygni 1975 hitting 2.5 million) the layer is blasted outward.

The spectrum of a nova at first shows strong absorption lines of a rapidly expanding, relatively cool, optically thick false photosphere. For a long time the bolometric luminosity [3.3] stays much the same, as the radiation merely shifts out of the optical and into the ultraviolet. As the shell of exploded debris grows larger and thinner we see into the hotter layers below. The spectrum now reveals emission lines and violet-shifted P-Cygni-type absorptions [7.4], from which an expansion velocity of 2000 to 3000km s^{-1} can be derived. After a few weeks, the absorptions fade, while the emissions – including nebular forbidden lines – remain. Eventually the debris shell, shredded and lumpy, grows to the point we can see it (Fig. 8.16), allowing proper motion of the expansion – and the distance [4.5] – to be derived. During the visual decline, dust develops in the expanding cloud, hiding part of the star (Fig. 8.17). Finally, decades later, the white dwarf returns to normal, the accretion disk re-establishes itself, and the binary slowly prepares to do it all over again, perhaps 100 000 years in the future.

There are several subtypes of nova that actually are just bins in a smooth continuum. Fast novae hit absolute visual magnitudes as bright as −10, but fade quickly, dropping by three visual magnitudes in a few days (Fig. 8.17). At the other end, slow novae do not get as bright, but decline through the first three magnitudes over a much longer period of time, of the order of months. The variations are caused by differences in the masses and internal temperatures (hence the states of evolution) of the white dwarfs, and in the accretion rates. The composition of the ejected debris will contain various isotopes (some rare) created by hydrogen fusion through the carbon cycle [12.2.2, Box 12.2] (and various subcycles), but will also reflect the stuff that was dredged up from the white dwarf. Most white dwarf cores are made of carbon and oxygen, but a few of the more massive ones (which pop off more frequently, giving us an anomalous view of their numbers) are rich in neon and magnesium, a by-product of carbon burning, which shows up to make "neon novae."

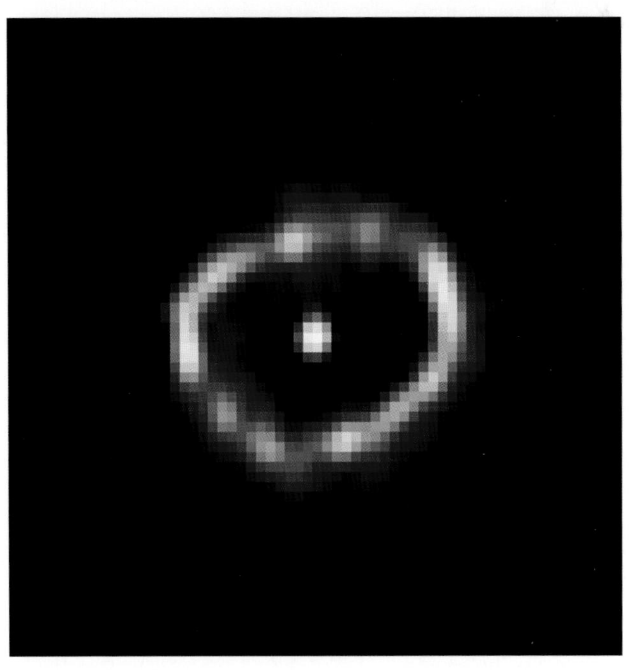

Fig. 8.16 **Nova Cygni 1992** Between February 19, 1993 (top) and May 31, 1993 (bottom), the nova shell from Nova Cygni 1992 expanded from a diameter of 800AU to just over 1000AU. (Courtesy of F. Paresce and R. Jedrzejewski (STScI), NASA/ESA.)

8.5.5 Recurrent novae

All novae are believed to repeat themselves. *Recurrent novae* (as opposed to the long-term *classical novae* of the previous section) just repeat within the span of human observation, that is, over short time periods. The best examples

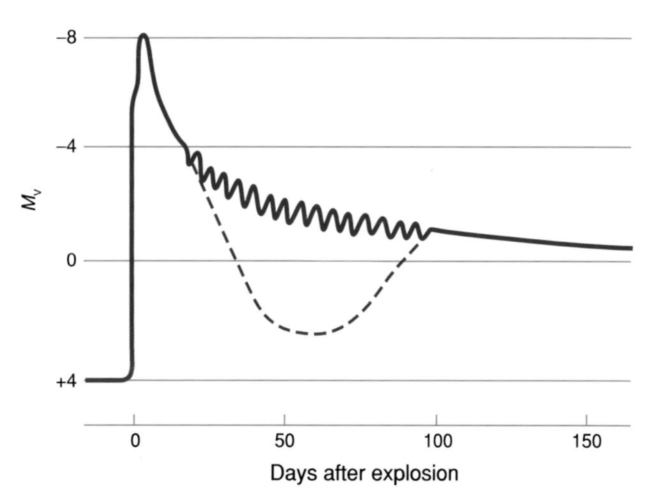

Fig. 8.17 **Nova light curve** A typical fast nova light curve climbs very rapidly and then declines by a few magnitudes over a few days. It can go into a series of oscillations or into a deep dip caused by dust that forms in the ejecta. Finally it enters into a long decline that returns the system to normal. (Kaler, J.B. (1994). *Astronomy!* New York: HarperCollins, reprinted by permission of Pearson Education, Inc.)

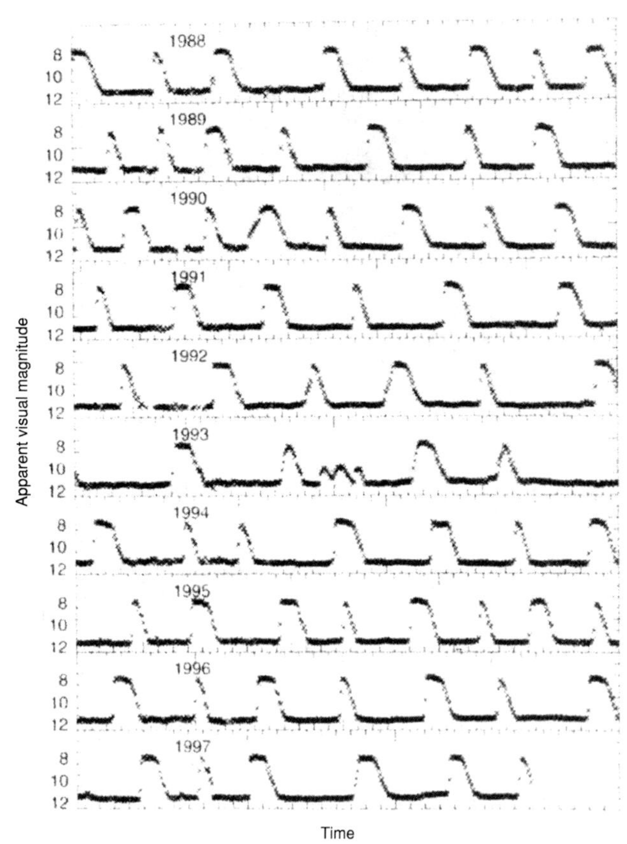

Fig. 8.18 **SS Cygni** The dwarf nova SS Cygni undergoes numerous but aperiodic bursts, then settles back to its original quiet state. (Courtesy of American Association of Variable Star Observers, Cannizzo, J. K. and Mattei, J. A. (1998). *Astrophysical Journal*, **531**, 467.)

are RS Ophiuchi (which erupted in 1898, 1933, 1958, 1967, 1985, and 2006, and reaches visual magnitude 5), T Coronae Borealis (1866, 1946, magnitude 3.6), and U Scorpii (1863, 1906, 1936, 1979, 1987, 1999, magnitude 7.5). The basic scene is the same. However, in the recurrent case, the white dwarf is massive, and near the Chandrasekhar limit [Box 13.2], as actually observed from the binary orbit for U Scorpii. Moreover, the donor stars fall into three categories: red giant (RS Oph, T CBr), subgiant (U Sco), and dwarf (as for novae). The stars with dwarf donor stars have short orbital periods (hours), as would be expected of small stars close enough to interact. Those with slightly evolved stars have longer periods (1.2 days for U Sco), while those with giants, long periods (457 days for RS Oph).

The compression on a high-mass white dwarf is higher, so the stars "go nova" more frequently, but with less vigor. Typical maximum absolute visual magnitudes are around −6.5, bright, but not as bright as they are for classical novae. Recurrent novae are candidates for real catastrophe. The eruptions are not believed to blast away all the matter that the white dwarfs accrete. As a result, the white dwarf should grow slightly in mass, despite the explosions. At some point, the white dwarf will hit the Chandrasekhar limit [13.5], at which point the whole star will burn violently, destroying itself as a Type Ia supernova [14.2.5]. With a mass of $1.37 \pm 0.01 M_\odot$, and accreting at a rate of 10^{-7} solar masses per year, U Sco could hit the limit anytime within the next 100 000 years or so, including tonight, making it the best known candidate for one of the most energetic events known.

8.5.6 Dwarf novae

Dwarf novae at first look like miniature versions of the "big event." The physical cause, however, is quite different; the name is another misnomer. Epitomized by SS Cygni and U Geminorum (the breed commonly called U Gem stars), they too consist of dwarfs that fill their Roche lobes and pass matter to white dwarfs through accretion disks. A dwarf nova will unpredictably erupt by a few magnitudes several times per year (Fig. 8.18).

Most of the light from a dwarf nova comes from a hot spot where the flow from the inner Lagrangian point smacks the accretion disk. Instead of coming from the white dwarf itself (as in the case of a nova), the outburst comes from instability in the surrounding accretion disk, when the disk suddenly changes its temperature, ionization state, opacity, and viscosity. The result is a sudden increased mass flow through the disk onto the white dwarf that is accompanied by an outburst.

There are a number of subgroups, among them the *SU Majoris stars* that display occasional "superoutbursts," and the *Z Camelopardalis stars* that can stabilize in brightness midway between the high and low states. Quite likely, dwarf novae are a subset of stars that are between true nova outbursts. None has ever been seen to explode as a nova, but the time-period between classical novae is so great that the lack of such behavior is no surprise.

8.5.7 Magnetic personalities

White dwarfs can have powerful magnetic fields that range to millions of times the strength of Earth's field [7.4]. If in a binary, the white dwarf's field is strong enough to synchronize its rotation with its orbital revolution. If the unevolved star fills its Roche lobe, the magnetic field will direct the flow from the inner Lagrangian point toward the white dwarf's magnetic pole, thus preventing the formation of an accretion disk. Without this buffer zone, the flow lands directly and forcefully onto the white dwarf, creating a shock and enough heat that the gas radiates X-rays, giving us one form of X-ray binary. Such doubles alternate between high and low X-ray states in accord with changes in the accretion rate, and are called, after the prototype, *AM Herculis* stars, or *polars* ("POL-ars"). The system that created Nova Cygni 1975, the fastest known nova on record, was an AM Her star and, once it settles down, will be again.

If the magnetic field is only moderately strong, the accretion disk can form, but will be disrupted, and we see an "intermediate polar" that radiates lower-energy X-rays. Such stars are epitomized by DQ Herculis, the eclipsing binary that remained after Nova Herculis 1934. (These eclipses led to the discovery that novae are binary systems with white dwarf companions.) The magnetism is not strong enough to synchronize the white dwarf rotation. Infalling matter hitting off-center has spun it up to a rotation period of only 71 seconds, as opposed to the orbital period of 4.65 hours. The result is an optical "flicker" with the same short 71-second period.

8.5.8 Symbiotic stars

At the fringe of the cataclysmics are the *symbiotic stars*. They were first recognized through a mixture of hot and cool spectral characteristics, most commonly those of a class M giant combined with a hot blue continuum and high-excitation emission lines. Their prototype is tenth magnitude Z Andromedae, which once lent its name to the symbiotics, and consists of an M4 or M5 giant that orbits a probable white dwarf with a period of 757 days. Every couple of decades, the system erupts by a few magnitudes and stays in the high state for several years. Beyond giant–white dwarf companionship, there is no single paradigm for symbiotics, but rather many unique individuals whose characteristics depend on the exact natures of the binary components' stars and their separations. Some systems contain K giants, others (the *yellow symbiotics*) even class G giants. The giant may fill its Roche lobe and transfer matter into an accretion disk, or the white dwarf might accrete from the giant's wind. Some symbiotics (the "d-type") are dusty, others not. The outbursts might be due to disk instability, wherein the disk suddenly dumps onto the white dwarf (CH Cygni, which produces jets during outburst and radiates X-rays), or caused by thermonuclear runaway. Some (like R Aquarii) are surrounded by very visible nebulae. Others seem to be more related to planetary nebulae. Yet others behave rather like novae. One can spend a lifetime studying them and still find new, odd, behavior.

8.5.9 Neutron stars and black holes

Interacting binaries with degenerate companions are hardly limited to white dwarfs. If the initially more massive star of a binary explodes, it may form a neutron star (perhaps a pulsar) or a black hole [14.3.2; 14.3.3]. Both are seen, creating such bizarre systems as X-ray binaries, X-ray bursters (caused by runaway helium fusion on the surfaces of neutron star companions), millisecond pulsars (spun up by infall from evaporating companions), microquasars that shoot out jets at nearly the speed of light, even gamma-ray bursters that may signal the mergers of twin neutron stars. Like the Type Ia white dwarf supernovae [14.2.5], these are best reserved for a full discussion of high-mass catastrophic evolution.

8.5.10 Chemical contamination

Binary interactions also elicit quiet, non-eruptive phenomena. The more massive component of a binary evolves first to become a giant. If the pair is close enough together, and if the giant's atmosphere had been chemically altered through nuclear processing and convective dredge-up [13.3], then matter enriched in heavy elements could pass from it to the dwarf. The giant then evolves to a white dwarf. The white dwarf's companion, whether still a dwarf or evolved to a giant, then appears strangely chemically anomalous.

CH stars and *dwarf carbon stars*, which are oddly rich in carbon, may come from such an interchange [6.4.3, 7.3.3]. It is also almost certain that such contaminated mass transfer creates the *barium stars*, which are all giants whose spectra have curiously strong absorptions of barium. All are strongly suspected of having white dwarf companions.

8.5.11 Spots

Sunspots [12.1.2] and starspots (and the activity associated with them) are caused by magnetic fields generated by convection and rotation in cool stars. Faster rotators will have more such activity than slower rotators (witness new T Tauri stars [11.2.2]). RS Canum Venaticorum consists of an F4 dwarf and a K0 subgiant in a short 4.8-day orbit, separated by only a few stellar radii. The K star recently evolved from a dwarf, just a bit more massive than the F star. Expanding, it encroached on the remaining dwarf, and the two tidally synchronized [13.6] at the same time that the K star developed a deep convection zone. The K star is therefore wildly spotted and thus variable as the dark areas swing in and out of view. Being close together, the *RS CVn stars* are also commonly eclipsers that vary erratically outside of eclipse. Growing ever larger, the subgiant might destroy and incorporate its mate,

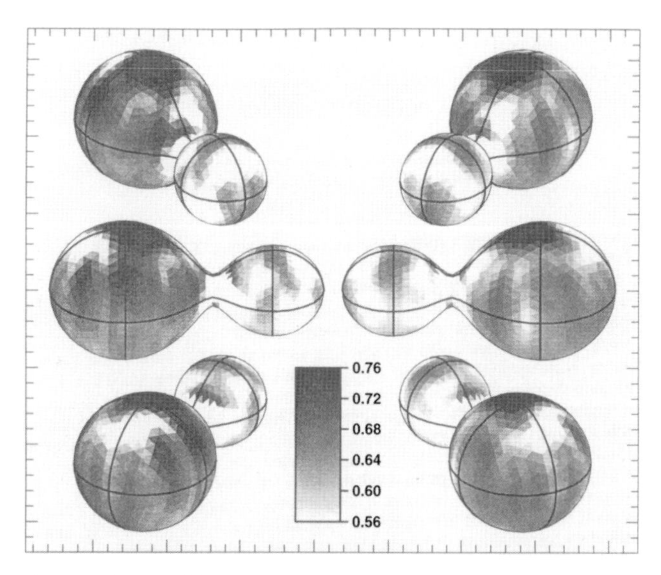

Fig. 8.19 **A contact binary** VW Cephei, a fine example of a W UMa star, consists of a pair of class F dwarfs that are in contact with each other. The rotation produces large starspots that can be tracked through Doppler imaging [7.3.3]. (Hendry, P.D. and Mochnacki, S.W. (2000). *Astrophysical Journal*, **531**, 467.)

leaving us with a single, rapidly rotating, highly spotted *FK Comae Berenices star*.

8.5.12 Contact!

An extreme in intimate stellar duplicity is the contact binary, headlined by W Ursae Majoris. The remarkably common *W UMa stars* have been brought so closely together by wind and magnetic interaction (which dissipates orbital energy) that they fill their Roche lobes and come into actual contact (Fig. 8.19). The class F dwarfs of W UMa, whose centers are a mere 1.2 solar diameters apart, orbit every eight hours and eclipse each other every four. The stars are so misshapen that the brightness of the system varies continuously. The synchronous rotation produces huge starspots. One star will probably absorb the other, perhaps creating another route to the spotted single FK Comae Berenices stars.

8.6 Multiple stars

Visual binaries are a delight to view; triples just double the fun. In the handle of the Big Dipper, class A Mizar and Alcor make what is most likely a real, though very long period, binary. Through the telescope, Mizar breaks into a much closer pair far enough apart that under low power the whole system is seen. The Castor triple is similar, consisting of a close pair of class A stars and a distant class M companion. Equally well-known, Epsilon Lyrae is a (barely) naked-eye class A double whose components are also double. Similar systems abound. Application of spectroscopic techniques reveal each of Mizar's pair to be close doubles, while the spectrum also shows all of Castor's three to be close doubles.

In the three examples, the stars form hierarchies that do not simply mill about. In a hierarchical triple, a distant star orbits a double whose components are close together (Fig. 8.20). The third star gravitationally feels the distant pair as a single unit, so in a sense the system represents two binaries, a close pair (the inner double) and a distant pair (the inner double as one plus the outer star). In principle, we can get the masses of the close binary, and then use the larger orbit of the distant star not only to get its mass, but to check on the combined mass of the close pair (which is difficult, since the orbital period of the third party is of necessity much longer than that of the inner pair). About 30 percent of doubles are in fact triple.

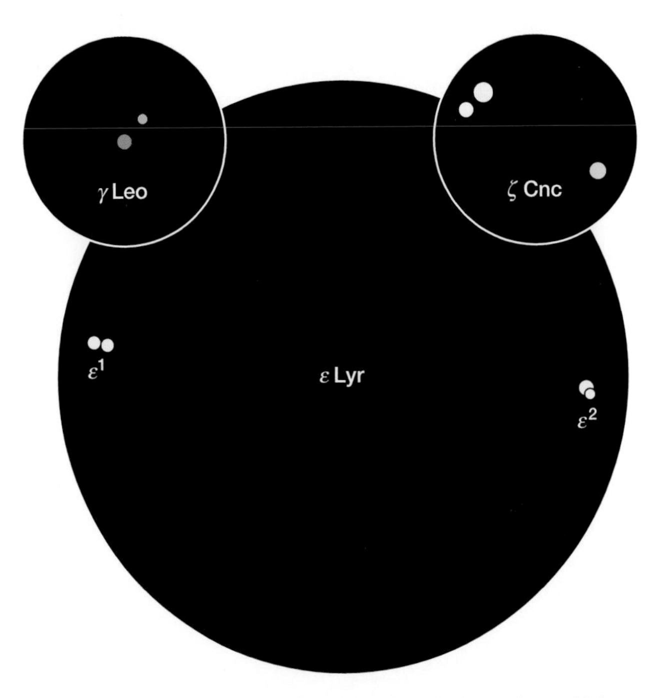

Fig. 8.20 **Hierarchical multiples** Three stars show the forms that multiples can take. At upper left is the simple double star Algieba, Gamma Leonis. The outer star of the visual triple star Zeta Cancri orbits an inner binary, much as it would a single star. The double-double Epsilon Lyrae is made of two doubles that orbit each other. (The outer star in Zeta Cancri is actually an unresolved binary, making it quadruple as well.) (Kaler, J.B. (1994). *Astronomy!* New York: HarperCollins, reprinted by permission of Pearson Education, Inc.)

In a hierarchical quadruple like Epsilon Lyrae, two distant doubles orbit each other. Each double feels the distant pair as a single star, in effect rendering the quadruple as three independent binaries. A quadruple could also consist of an inner hierarchical triple orbited by a distant fourth star. Mizar and Alcor represent a hierarchical quintuple, while Castor lies at the extreme as a hierarchical sextuple: the inner four behave like Epsilon Lyrae, while the outer M-star pair gravitationally feels the inner quadruple as one star.

Such systems are gravitationally stable, as each star or set does not disturb the other, and their orbits are all "Keplerian" [8.2.3], that is, a simple two-body solution provides a reasonable description of all of them. Though higher orders are in principle possible, six is the maximum number known. Quadruples are rarer than triples, quintuples and sextuples more so. The odds of making septuples or octuples must be therefore quite low. To have another star orbiting a sextuple like Castor, it would have to be so far out to see all six as one, and would be so loosely bound that it could not long survive the effects of passing stars or the tidal effects of the Galaxy.

If the third member of a triple is close enough to the inner pair to affect them individually, then none of the orbits will be Keplerian, or even stable. The crucial parameter is the ratio of closest approach of the third member to the inner pair's semimajor axis. If the ratio is less than about three (more if the third member is more massive than the individual stars of the pair), then the orbits become chaotic, and the triple becomes a *trapezium system*. The same arguments apply to higher orders, quadruples and so on.

Trapezium systems take their name from the four-star "Trapezium" that illuminates the Orion Nebula [11.1.3], because in a small telescope it appears as a quartet of massive stars that makes a crude trapezoid (Fig. 8.21). A larger telescope reveals a fifth star, while further analysis of the individuals show one to be itself double, one to be triple, and another to be quadruple! The five systems are all the same crude distance from each other. As a result, ordinary two-body orbits are not possible. In a trapezium triple, gravitational interaction will cause one star to be ejected, resulting in the drawing together of the other two, a triple made double. The Orion Trapezium must someday suffer the same fate, eventually breaking down to form stable hierarchical

Fig. 8.21 **The Orion Trapezium** A set of four Trapezium stars light the Orion Nebula. A closer look shows a fifth, while examination of the whole field reveals that the Trapezium is also the pinnacle of a dense open cluster. (Courtesy of Mark McCaughrean, Astrophysical Institute Potsdam/ESO.)

systems More stars will increase the total gravitational force within the system and make it more long-lived. But even these break up, the victims of a gravitational pitch-out or Galactic tides [9.2.4]. Eventually only single stars, close binaries, and hierarchical multiples will remain.

Such systems are loosely related to a large set of *runaway stars* that constitutes 20 percent of O stars and 10 percent of class B. The original example is a pair separated by some 70°, Mu Columbae and AE Aurigae, that are speeding away from each other at over 200 km s^{-1}. Tracing their Galactic orbits backward in time shows that they came from very close to the present Trapezium. The postulation is that 2.5 million years ago, two double stars violently encountered each other and for a brief interval created an unstable trapezium system. One star of each pair was kicked out in opposite directions, while the other two combined to form the spectroscopic-double Iota Orionis, whose components have an unusually eccentric orbit. Runaways are also produced by off-center supernova explosions in binaries [14.3.2].

Add more stars to a trapezium system and the result is a small open cluster [9.2]. Indeed, Orion's Trapezium is a part of one. Some contain thousands of stars, while at the pinnacle are the great globular clusters [9.4] that make some of the finest sights the Galaxy has to offer.

CHAPTER NINE

Star clusters and associations

A defining characteristic of the Galaxy lies in the gatherings of its stars, not just into binaries and multiples, but into clumps, into clusters, some so subtle as to be barely noticed, others so obvious as to be readily visible to the naked eye, yet others so grand as to be clearly seen even if we were looking from distant galaxies back toward our own. Profoundly important to our understanding of stars, clusters hold secrets that tell us how stars are born, how they live and die, indeed, how and when the Galaxy itself might have come to be.

9.1 Forms

Fig. 9.1 **Contrasting clusters** On the left is a typical open cluster, the "Jewel Box," so bright it received its own Greek letter designation, Kappa Crucis. The "open" ragged structure and relatively few stars contrast smartly with the densely-packed globular cluster M22 in Sagittarius at right. (Courtesy of NOAO/AURA/NSF.)

Like other broad categories of astronomical objects (including stars themselves), stellar assemblies possess great variety, a diversity so large that it is almost hard to believe that one type actually has anything to do with another. We can divide stellar groupings in different ways that may or may not intersect: by form and structure, HR diagram, Galactic position, number, age, orbit, chemical composition, stability and instability.

The most obvious groups are *open clusters*, epitomized by the Pleiades and Hyades of Taurus [Figs. 1.2, 4.7, Appendix 5], which are so imposing that they carry their own mythologies. Once called "Galactic clusters," the name has fallen into disuse because other, and very different, kinds of assemblies are also part of the Galaxy (though "Galactic open cluster" is commonly used to discriminate them from similar systems seen in other galaxies). The term "open cluster" is wonderfully descriptive of these generally ragged groups, many without much obvious overall design (Fig. 9.1, left). A typical open cluster may consist of anywhere from a few stars – not much more than a trapezium-type multiple system – to thousands packed into a volume a few parsecs across. Well over a thousand open clusters are known, all inhabiting the Galaxy's Population I disk [5.1], and therefore seen thronging the Milky Way [Fig. 1.1].

In striking contrast, *globular clusters* flock into a great spherical space that occupies the vastness of the Galaxy's extended halo (Fig. 9.1, right). What they lack in numbers – there are only about 150 known – they make up for in their stunning displays of up to a million stars packed into volumes not much bigger than that of an open cluster. Even at distances of thousands of parsecs, a handful are visible without the telescope. With good optical aid, they are mesmerizing. Because they, too, inhabit the Galaxy, they are often referred to as "Galactic globular clusters" (GGCs), though because of the outmoded "Galactic cluster" for "open cluster" the term leaves a trace of confusion. While there may be some instances of crossover, the divide between open and globular clusters is (at least in our own Galaxy) quite deep, and goes far beyond size and star counts. Open clusters are generally young, while globulars are ancient, indeed they are the oldest known systems, containing the oldest stars, in the Galaxy.

Clearly linked to age is stability. Unless unusually compacted, the loose open clusters cannot ordinarily survive for very long, while the dense globulars have made it through from the beginnings of things to the present day. Such stability allows us to get a glimpse of the early Galaxy, showing us that it – like the globulars themselves – were created poor in heavy elements, which begins to reveal the manner in which the Galaxy evolved [5.4]. The age difference between the two kinds of cluster shows up in greatly different HR (or color–magnitude) diagrams [6.3.1; 6.3.3], which together provide the chance to test theories of stellar evolution and to measure the age of different parts of the Galaxy, even of the Universe itself.

Back in the Milky Way, and allied with open clusters, are much looser *associations*. Though some ultra-loose open clusters may be exceptions and in the act of falling apart, the stars of globular and open clusters are bound together through the mutual gravities of their stars. Associations, however, are not so bound. Three kinds, all made of very young stars, are located in the Galactic disk. The most obvious are those that, while containing the whole stellar population, are marked by their blue O and B stars [6.2.1] and are hence called *OB associations*. So huge as to make entire constellations, they give Orion, Scorpius, Centaurus, Crux, and some others their special sparkle. Their very size renders them entirely unstable, their stars simply leaving the places of their birth among the dark molecular clouds of interstellar space [11.1.1]. Related to them are *T associations* made of unbound families of ultra-youthful lower-mass T Tauri stars, and related to them both are the lesser-known *R associations*, which are marked by flocks of reflection nebulae, the effects caused by youthful luminous stars in league with dusty interstellar blobs of cold star-birthing gas.

9.2 Open clusters

Never is the visible sky without some of its fine open clusters. Though a phenomenon of the Galactic disk, and thus of the Milky Way, several appear well off that celestial stream because they are nearby, and while still in the disk can appear at high Galactic latitudes [2.3.5]. Though constituting a single category of celestial object, they are a diverse lot, some young, others old, some sparse, others dense. To get to know them, look first to the favorites.

9.2.1 Names and numbers

Somewhere around 1200 open clusters are known in our Galaxy. Since they occupy the Milky Way, dimming by interstellar dust allows us to see only the nearby ones, or those more toward the Galactic anticenter where the dust is thinner (see the detailed set of them in Appendix 5, along with Fig. A5.1). Extrapolating from the local surroundings, there may be in excess of a hundred thousand.

The local clusters go by a variety of names, beginning with popular terms like the "Pleiades." Three even have Bayer type designations as if they were single stars [1.3.2]. The Messier catalogue is loaded with 26 open clusters, the "M" numbers in common use [1.5]. Not quite half of the known clusters are listed in the NGC and IC [1.5]. These numbers are used when there is no Messier designation, and oft-times when there is. Beyond these are myriad individual abbreviations of discoverers' names, among the most common "Mel" for P.J. Melotte (1880–1961) and "Tr" for R.J. Trumpler (1886–1956). Even "H" for Herschel is still seen. Others are too numerous to mention.

Fig. 9.2 **The Beehive** M44, in central Cancer, is surrounded by an asterism called the "Manger." There is no question about the reality of such an impressive structure. (Courtesy of Till Credner and S. Kohle, www.AlltheSky.com.)

9.2.2 Best of breed

A selection of the closest, most prominent open clusters is presented in Table 9.1, organized by right ascension. (A complementary set within a concentrated area near the Galactic anticenter is given in Appendix 5). The Pleiades and Hyades (which are represented in both sets) are the most obvious. Another, the beautiful Coma Berenices cluster, makes its own constellation. Consisting of a set of fourth magnitude and fainter stars, it sprawls like celestial lace south of Ursa Major's long tail. Much more compact, the naked-eye Praesepe, or Beehive (M44) graces the center of dim Cancer (Fig. 9.2). Also easily visible to the eye is a colorful telescopic showpiece, the "Double Cluster in Perseus," the two moving in lockstep with each other and apparently born together (Fig. 9.3). The eastern of the two is Chi Persei, the other h Persei (the Roman letters, no longer much used, continuing after the Greek). The other open cluster with a Greek letter is the lovely Jewel Box, Kappa Crucis (see Fig. 9.1), which lies just off one of the arms of the Southern Cross, due north of the black heart of the Milky Way, the famed dark Coalsack nebula.

Three great clusters almost escape notice. Deep in the southern hemisphere lies large and brilliant IC 2602. Dominated by third-magnitude Theta Carinae, and filled with hot blue stars spanning over a degree, the cluster is too far south to be appreciated by the majority of the world's population. A northern hemisphere counterpart, so big that it is hardly recognized as an open cluster, is the Alpha Persei cluster. Dominated by second-magnitude Alpha Persei, it makes the

Name	Catalogue	Const.	α (2000) h m	δ ° ′	Diam. arcmin	Dist. (pc)	Age (Myr)
...	NGC 188	Cep	00 44	+85 20	13	1500	7200
h Per[a]	NGC 869	Per	02 19	+57 07	29	2300	13
χ Per[a]	NGC 884	Per	02 22	+57 05	29	2300	13
Alpha Per	...	Per	03 24	+49 52	200	165	50
Pleiades	M45	Tau	03 47	+24 06	110	132	130
Hyades	...	Tau	04 26	+15 51	330	46.3	650
M35	NGC 2168	Gem	06 09	+24 21	28	1070	150
M41	NGC 2287	CMa	06 47	−20 44	38	690	240
M67	NGC 2682	Cnc	08 51	+11 50	18	800	4000
Praesepe	M44	Cnc	08 40	+20 01	95	175	750
...	IC 2602	Car	10 43	−64 24	70	145	30
Ursa Major	...	UMa	12	+55	1500	25	250
Coma Ber	...	CBr	12 25	+26 10	300	85	400
Kappa Cru[b]	NGC 4755	Cru	12 53	−60 18	10	2000	20
M6	NGC 6405	Sco	17 40	−32 12	14	490	95
M7	NGC 6475	Sco	17 54	−34 38	80	300	300
M11	NGC 6705	Scu	18 51	−06 17	13	1900	200
...	NGC 6791	Lyr	19 21	+37 47	8	800	8000

[a]Together, the "Double Cluster" in Perseus.
[b]The "Jewel Box."

Table 9.1 **Prominent open clusters**
[See Appendix 5 for a list near the Galactic anticenter.]

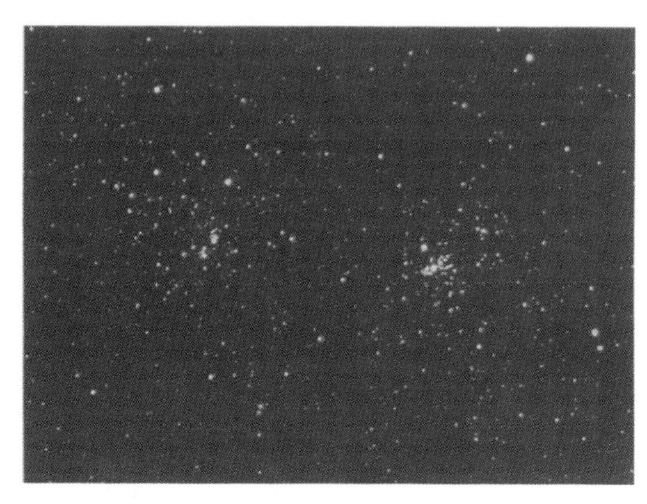

Fig. 9.3 **The Double Cluster** Two young clusters, h and Chi Persei, which together make the "Double Cluster in Perseus," travel side by side through space. (Courtesy of Mark Killion.)

heart of its bright constellation. Bigger yet, so sparse and large it is barely recognizable as an open cluster, is the core of Ursa Major, the middle five stars of the Big Dipper (the Plough) and several other fainter ones all moving through space together [Fig. 1.6].

From the heart of the Milky Way emerge another notable trio, M6 and M7 of Scorpius (the latter neighboring a magnificent star cloud) and M11 in Scutum. M11 is so compact that under low magnification it looks almost like a single star, and is so stuffed with members that it could be taken for a globular cluster. A sweep through the Milky Way with even a small telescope reveals many more. It will also show several apparent clumps that look like clusters but, like optical doubles, are mere line-of-sight coincidences (the best of these, the "Coathanger" in Vulpecula).

9.2.3 Distance, size, and structure

Distances to nearby open clusters can be found from both ground-based or Hipparcos parallaxes [4.2], the moving cluster method [4.3.1], and main sequence fitting [4.4.2] adjusted for metallicity, and unresolved binaries. Dimming by interstellar dust [3.5.1] can be calculated from a color–color fit with that of an unreddened cluster [3.4]. The end result is that the distances of open clusters are well known as long as there is a sufficient number of stars to establish a good HR diagram.

As with spectra, understanding requires classification. A standard scheme, created by Trumpler, uses three codes

Concentration	Brightness Range	Population
I. Detached[b] clusters with strong central condensation.	1. Most stars of same apparent brightness.	p. Poor clusters with less than 50 stars.
II. Detached clusters with little central condensation.	2. Medium range in apparent brightness.	m. Moderately rich clusters with 50–100 stars.
III. Detached clusters with no central condensation.	3. Both bright and faint stars; a few bright ones surrounded by many faint ones.	r. Rich clusters with more than 100 stars.
IV. Clusters not well detached, but fading into environs.		

[a]Taken with some alterations from R.J. Trumpler (1930). vol.14, number 420, *Lick Observatory Bulletin*, p. 154.
[b]Distinct from background; well separated from field stars.

Table 9.2 **Trumpler classification of open clusters[a]**

(Table 9.2). It first ranks clusters I through IV in order of decreasing central condensation, which is strongly related to their visibility against the background. Class I clusters (like M11) are obvious, whereas class IV clusters are only barely detectable as slight increases in the local stellar density. The second code ranks them according to the observed range in brightness: class 1 clusters have stars all at about the same magnitude, while class 3 clusters have a great range from bright to dim. Finally, "p," "m," and "r" respectively describe "poor," "moderately rich," and "rich" clusters. Though clearly subject to observational selection, the system at least serves to highlight cluster structure and development.

The problems of classification involve depth of observation and membership, which relates to size. There are three meaningful size measures. The first is derived from simple visual impression as to where the stars appear to merge with the background. From such observations, clusters range between approximately 1 and 10 parsecs in diameter. The *core radius*, which roughly defines where the brightness of the cluster falls to about half its central brightness (and which can also be rather well estimated by eye) is smaller, a parsec or so. Much larger is the *tidal radius*, the point at which tides [8.5.1] raised by passing interstellar clouds and by the Galaxy itself are strong enough to remove outlying stars [9.2.4]. Measured in the tens of parsecs, it is not subject to visual impression. Clusters, therefore, can be much larger than they appear.

Related, is that an open cluster is a dynamic assembly. As a cluster's stars orbit under the combined gravity of all its members, they continually encounter each other in gravitational "collisions." They do not actually hit each other, but come into significant gravitational contact, thus affecting each others' motions. In such a meeting, the more massive of the two will on the average give up energy to the lesser one. As the more massive stars slow down, they gradually migrate toward the cluster's center. As a result, the cluster becomes segregated by mass: lower-mass stars are flung out to higher orbital radii, while the more massive members settle near the middle [9.4.11].

Consider then classifying a cluster like the Pleiades. A short exposure shows a few bright blue B stars more or less uniformly spread out, and we might assign it Trumpler class III1p. A long exposure, however, reveals outliers, and vastly more stars that range down to white dwarfs, now rendering the cluster as class I3r. The nexus, though, is membership. Which stars actually belong, and which ones are just in the line of sight, is a problem that is particularly bad for genuinely poorer clusters that lie in the heart of the Milky Way. In the core, the cluster will dominate, but out toward the tidal radius, real members may be few and far between and very difficult to recognize.

To a degree, that problem can be resolved by placing all the stars within the extended angular area of a cluster on the HR diagram [4.4.2; 6.3.1]. Those in the cluster will make a coherent pattern near the main sequence and giant branch, whereas the others will scatter. Yet where do you cut them off without statistically biasing the sample? More effective is the analysis of the motions of all the candidate stars, which gives results independent of the HR diagram. Cluster members will have a common space velocity [5.3.3], while the "field stars" (those not in clusters) will stand out. Though the technique works very well, it requires great effort. As a result, far from all of the known clusters are fully analyzed – or even recognized.

Huge differences exist among open clusters. Star counts range from a few into the thousands, while masses of the common nearby open clusters range to 10 000 solar, always depending on how deep we look. There is no question that the Pleiades and M44 are rich, while there is equally no doubt about Ursa Major's lowly status. At the pinnacle are the immensely rich clusters of the Galactic center, which because of the deep intervening dust are very difficult to study and recognize (Fig. 9.4). In the intense star-forming regions of the Galaxy's core, these clusters contain great numbers of high-mass stars and top 40 000 solar masses. They are kin to the great clusters of the Large Magellanic Cloud, like R136 at the heart of the Tarantula Nebula (see Fig. 4.11), which

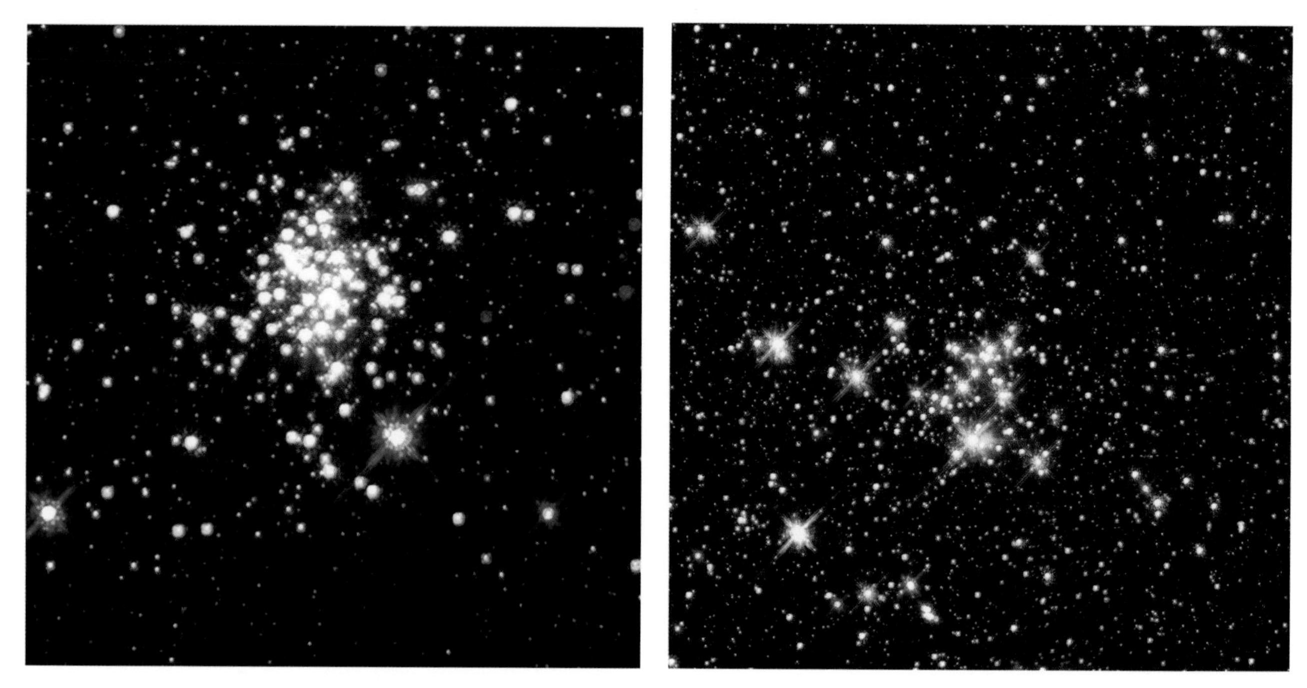

Fig. 9.4 **Quintuplet and Arches** Two of the most massive open clusters known (the "Arches," left, and the "Quintuplet," right) lie in the rich star-forming regions near the core of the Galaxy, less than 40pc from the center. Visible only at infrared wavelengths, white represents cluster stars, blue foreground stars, and red, stars that are highly reddened by dust. The Quintuplet contains the Pistol Star, a luminous blue variable similar to Eta Carinae. (Courtesy of Don Figer, STScI, and NASA.)

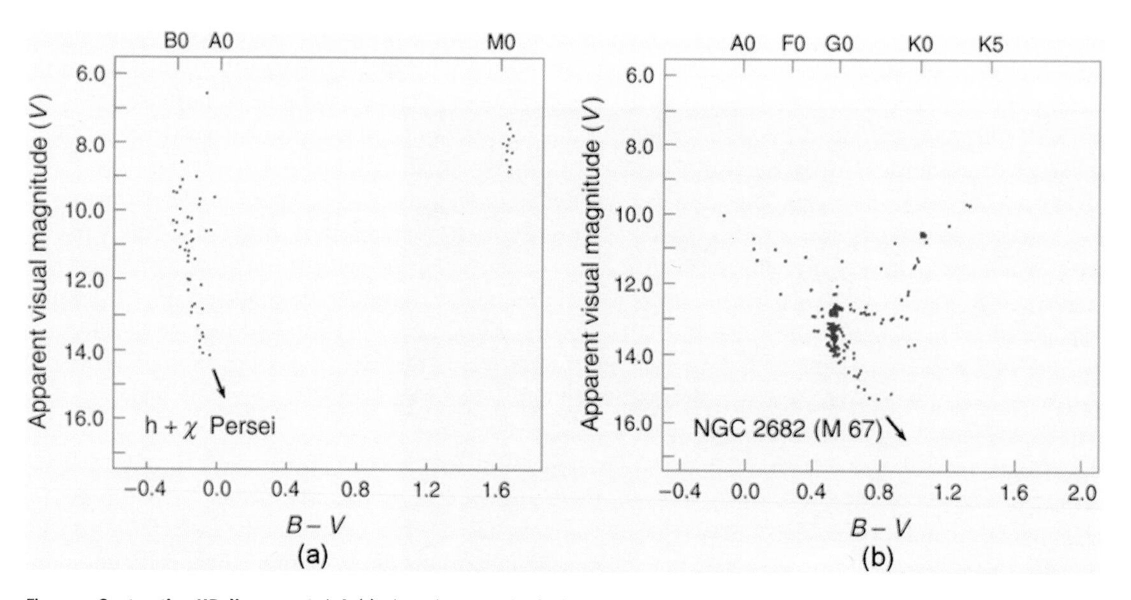

Fig. 9.5 **Contrasting HR diagrams** At left (a), the color–magnitude diagram of the very young Double Cluster displays an intact main sequence and several red supergiants. At right (b), ancient M67 has burned off all its upper main sequence, while the stars of middling mass have made a prominent red giant branch. Note the "blue stragglers" above the giant branch turnoff. (Hagen, G. (1970). *An Atlas of Open Cluster Colour–Magnitude Diagrams*, David Dunlap Obs. and Kaler, J.B. (1994). *Astronomy!* New York: HarperCollins, reprinted by permission of Pearson Education, Inc.)

is loaded with dozens of O3 stars, and vastly more on down the dwarf sequence.

9.2.4 Age and disintegration

For clusters, the HR diagram is most commonly in the form of a color–magnitude diagram (CMD), which can be established quickly from photometry alone [6.3.3]. CMDs are traditionally plotted with V magnitudes and $B - V$ colors, though infrared magnitudes and colors ($V - I$ etc.) are becoming increasingly common.

The HR diagrams of clusters are different from the diagram of the general field of stars. The "ideal" cluster is born with a full range of stars, from high mass – O stars in the tens of solar masses – to low mass, then on down to brown dwarfs [12.4.2, 13.1]. That is not to say that all clusters have massive stars. Several local star-forming regions seem to be limited to lower-mass stars, with no great luminaries present at all [13.3.3]. In any case, a young cluster will have nothing but a main sequence, and if young enough – like Orion's Trapezium cluster [Fig. 8.21] – the lower mass stars will be still evolving onto it. There will be neither giants nor supergiants.

But give the cluster a bit of time, and the highest mass O stars (if they are present to start with) will evolve [13.1] into red supergiants of the kind we see sparkling in the youthful Double Cluster (Figs. 9.3 and 9.5a). The cluster then burns its dwarfs from the top down. The result is that the turnoff point of the giants from the main sequence and the luminosities of the resulting giants that come from them slide downward. After a few billion years, the CMD will look like that of ancient M67 (Fig. 9.5b). The Pleiades and Hyades [Fig. 4.10] fall in between. Several clusters are summarized in the classic diagram, Fig. 9.6.

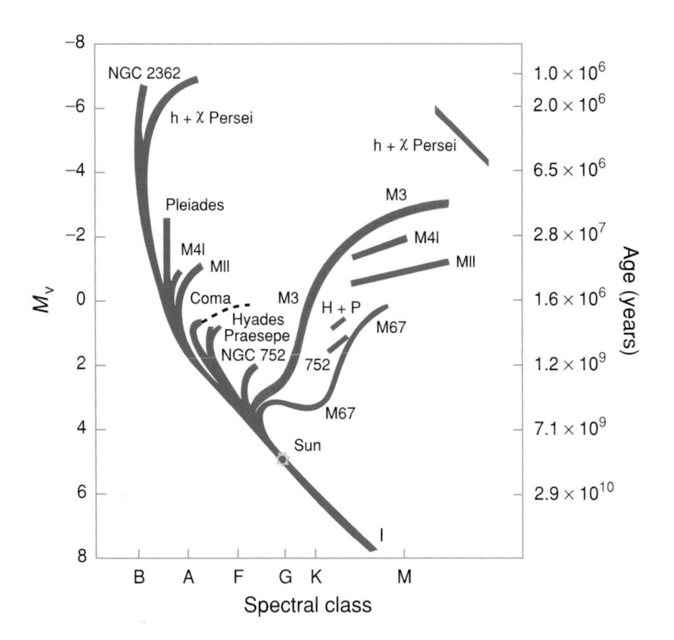

Fig. 9.6 **Comparative ages** While the ages have since been revised, this historic and classic diagram vividly shows clusters ageing: the older the cluster, the farther down the turnoff of the giant branch from the main sequence. (Sandage, A. (1957). *Astrophysical Journal*, **125**, 345, as rendered in Kaler, J.B. (1994). *Astronomy!* New York: HarperCollins, reprinted by permission of Pearson Education, Inc.)

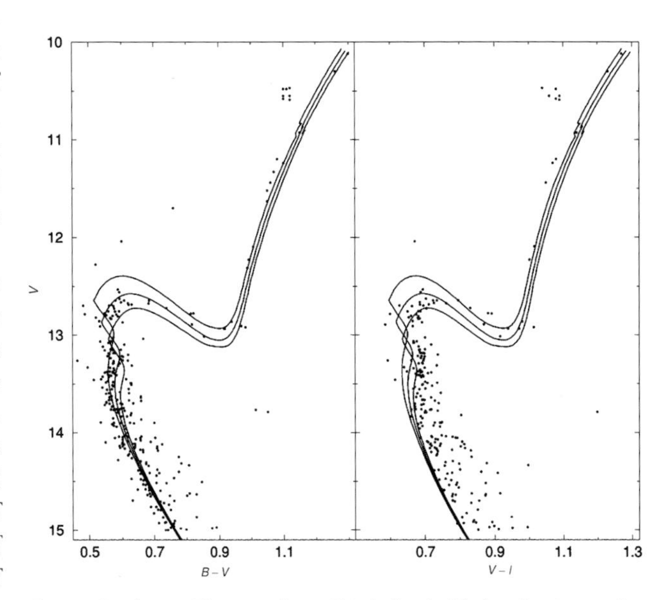

Fig. 9.7 **Isochrones** The open cluster M67 is fitted with three isochrones that represent the distribution of stars of 3, 3.5, and 4 billion years after the cluster's birth. The 3.5-billion-year isochrone fits best, and thus gives the cluster's age. (Chaboyer, B., Green, E.M., and Liebert, J. (1999). *Astronomical Journal*, **117**, 1360.)

Application of theory, which tells how long it takes to burn the main sequence down to a particular point, gives the age of the cluster [13.2]. In practice, the astronomer begins with a complete main sequence, and then calculates what the CMD should look like after various given periods of time. The resulting loci of stars, or *isochrones* (meaning "all at the same time"), are then fitted to the observed CMD to find the best match (Fig. 9.7). At the same time, the ability of an isochrone to mimic the CMD faithfully provides a test of stellar evolutionary theory, there being enough degrees of freedom to do both. We see then that standard open clusters range from 10 million years old (and only a couple million for the Trapezium cluster) into the billions of years. Isochrone fitting is labor-intensive. Fortunately, the magnitude difference between the point at which the main sequence turns off to the giant branch and the set of stars at the red giant helium-burning clump [13.2.2] (called δV) increases with increasing cluster age, and when calibrated against carefully determined isochronic ages, can be used as a *morphological age index* (MAI).

The great majority of open clusters, however, are young – well under half a billion years old. Stars that are thrown to the outside of a cluster by internal gravitational interactions are torn away from their homes by the onslaught of tidal forces [8.5.1] raised by molecular clouds [11.1.1] and the Galaxy at large, and leave forever [9.4.11]. Smaller open clusters in the thick inner portions of the Galactic disk thus rather rapidly disintegrate, their ejecta (single stars and tightly-bound

Fig. 9.8 **An ancient cluster** With a calculated age of 9 billion years, Berkeley 17, which is only barely detectable against the background of Auriga's Milky Way, is the oldest known open cluster. (Courtesy of M. Stecker.)

binaries) destined to orbit the Galaxy alone. As massive as the Galactic-center Arches and Quintuplet clusters are, they are not expected to last more than 50 million years.

In the outer reaches of the Galaxy, however, the forces that tear open clusters apart are much diminished, allowing the assemblies to last much longer, especially if they are rich and somewhat removed from the Galactic plane. Here is where we find the old open clusters, those more than about a billion years old. They are of special interest, as the oldest tell us of the conditions existing in the Galaxy in ancient times. The very oldest of them is Berkeley 17, whose isochronic age is around 9 billion years (Fig. 9.8). The next in age are well-known NGC 6791 and NGC 188, which respectively come in at 8 and 7.2 billion. Since the Galaxy's disk can be no younger than the oldest of its members, the disk must be close to 10 billion years old, which accords well with the maximum ages of ancient white dwarfs. The oldest open clusters more or less overlap in age with the youngest of the halo's globular clusters, showing that the disk had to form rapidly. While these outer old open clusters are as a group somewhat metal-poor, consistent with the Galactic metal gradient (wherein the metal abundance decreases from the center outward), they are nowhere near as metal-poor as the globulars. Ancient NGC 6791 is even metal-rich. There is in fact only the weakest relation – if any – between metal content and age, the clusters telling us that the Galactic disk's metal content built up very quickly and then remained more or less the same [5.1].

9.3 Associations

Like people, stars can associate with one another without being bound into a family from which they cannot escape. It seems probable that while maybe 10 to 50 percent of stars are born in actual clusters, the majority are created within their parent giant molecular clouds [11.1.1] in some kind of larger group that simply dissolves. Old groups spread themselves around the Galaxy in only a few hundred million or so years, and after a billion or two no trace of the original set is left.

9.3.1 Names and properties

Many unbound young groupings are brilliantly marked by their O and B stars, and hence are called *OB associations*. Observations of the relative motions of their stars show them to be moving outward from their groups' centers of mass. Living short lives, they never get far away from their birthplaces before they die. The result is that the majority of O and early B stars still belong to OB associations. These grand groups are everywhere around us, and are characterized by stars that are not only at the same rough distance, but share common space motions with only enough speed to allow them to escape from the weak gravity of the extended mob. At their cores we commonly find bound open clusters.

In spite of their vacuous natures, of the order of a hundred OB associations have been isolated, and Hipparcos observations have allowed the nearer ones to be studied rather intensively (Table 9.3). The first ones found were originally given Roman letters, followed by the constellation's genitive, for example "I Cephei," "II Cephei," in order of discovery. To make the kind of object clearer, modern names begin with the constellation name followed by "OB" and a number, for example Scorpius OB1, Sco OB2, Orion OB1. A variety of other names are in use as well. Stars within an association can be referred to by a second number, such as Cygnus OB2 #12 (one of the Galaxy's brightest stars), provided we agree on what catalogue to use.

The major problem is again that of membership. O and B stars are rare, and the vast majority are far away. Consequently, so are OB associations. Most are too distant for accurate parallax and proper motion measure [5.2.1] from the ground. Moreover, they can be angularly so large that they overlap on the celestial sphere, and must somehow be separated from one another. The problem is the same as it is for membership in open clusters, just more severe because of the associations' large dimensions. Stars that are securely listed in older catalogues are commonly rejected on the basis of their Hipparcos motions, so caution in assuming any pre-Hipparcos star is a

Name	Distance (pc)	Diameter (pc)	Age (Myr)	Prominent Members
Sco OB2[a]	135	40	5	. . .
Upper Scorpius	145	40	5	Antares; Tau, Beta Sco; Rho Oph
Upper Cen Lupus	140	70	13	Kappa Cen; Alpha, Gamma, Delta Lup
Lower Cen Crux	118	50	10	Pi, Lambda Cen
Cep OB2	615	100	5 to 10	NGC 7160
Alpha Per[b]	177	. . .	50	Alpha Per cluster
Per OB2	295	50	1.3	Zeta Persei
Cas OB5	2500	100	. . .	Rho Cas, 6 Cas
Ori OB1[a]	400	85	. . .	Orion Belt, Sword, and Nebula
NW of belt	335	. . .	10	. . .
Belt	470	. . .	2–5	. . .
Sword	500	. . .	2	. . .
Trapezium	450	. . .	<1	. . .

[a]The recognized subgroups are listed below the main entry.
[b]Surrounded by and part of the larger Cas–Tau OB association.

Table 9.3 **Prominent OB associations**

Fig. 9.9 **Sco OB1 and OB2** Scorpius's prominence is the result of its being made largely of two OB associations. (Courtesy of Akira Fujii.)

member of a particular association is advised. The Hipparcos measures include only associations within about 500pc, so one can only imagine the mess we encounter with more distant groupings.

Even when an association can be identified, we see that, though it has a certain integrity, it can be subdivided into smaller groups of different evolutionary ages as a result of sequential star formation. Such is the disposition of the two most prominent OB associations, Orion OB1 (which makes most of the constellation; see Fig. 3.2), and Scorpius OB2, which with Sco OB1 makes most of Scorpius (Fig. 9.9). Huge, Sco OB2 is made of three smaller associations, "Upper Scorpius," "Lower Scorpius–Lupus," and "Lower Centaurus–Crux" (which is notably closer than the others). Orion OB1 is made of an outer halo of sorts northwest of the Belt, the Belt star complex, the Sword, and the Orion Nebula region, all with somewhat different ages and distances (see Fig. 3.2). Ori OB1, however, maintains its integrity by belonging to the huge background Orion Molecular Cloud. We are in fact nearly surrounded by an OB association. The bound Alpha Persei cluster is part of Perseus OB3, which encompasses the cluster, and is also related to the huge (as one can see from the name) Cas–Tau association, which extends across 100 degrees of the sky from Cassiopeia to Orion!

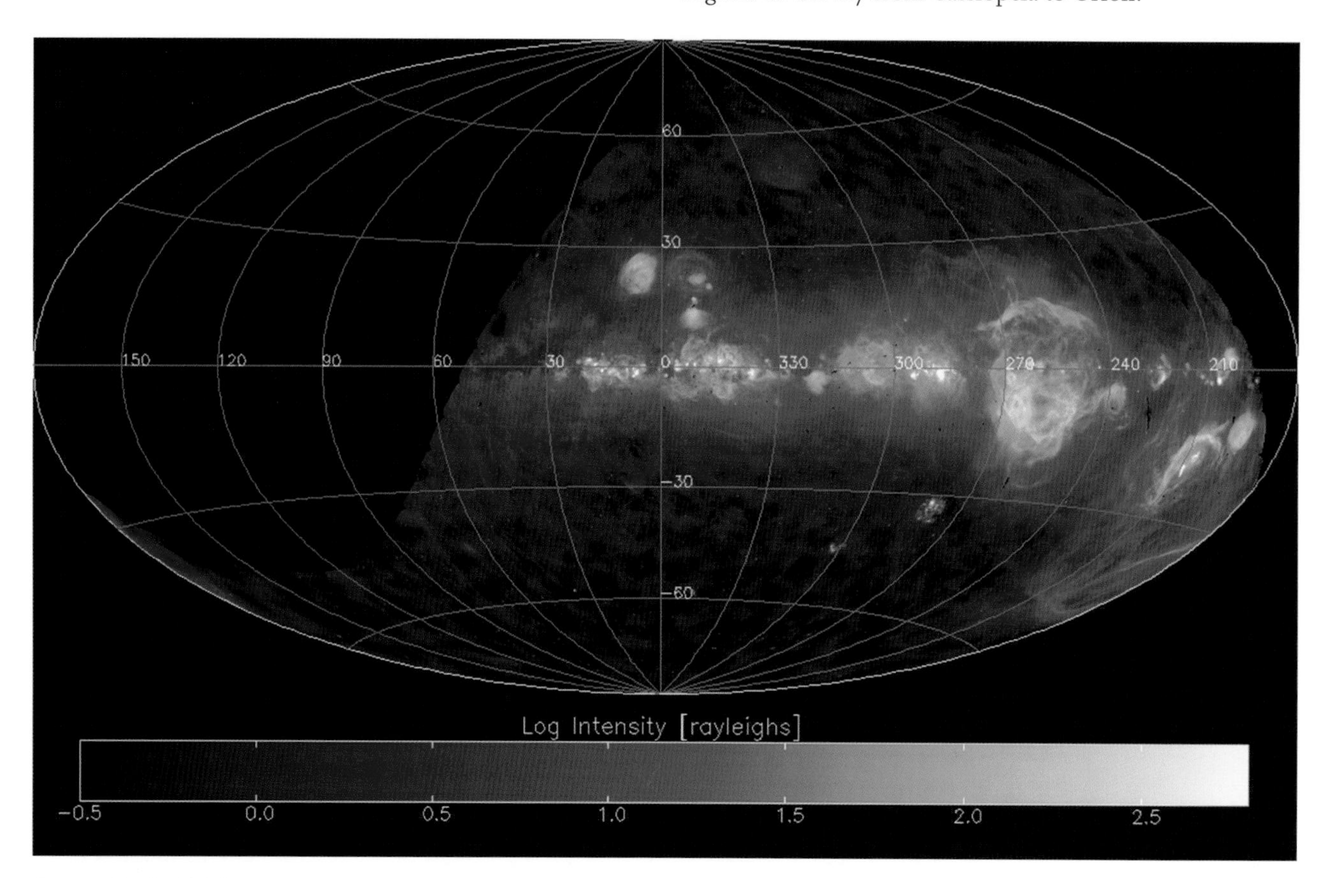

Fig. 9.10 **The Gould Belt** Distant interstellar clouds seen in H-alpha radiation hug the Galactic Center and the Milky Way, as expected. The more local clouds, however, typified by Orion (below the equator at lower right) and Scorpius (above the equator near center), form part of the tilted Gould Belt. (Courtesy of Gaustad, J.E., McCullough, P.R., Rosing, W., and Van Buren, D., "The Southern H-Alpha Sky Survey Atlas," which is supported by the National Science Foundation.)

Ages of OB associations can be found from the HR diagrams' main sequence turnoff points and from the rates of expansions of the systems. Both agree that OB associations are very young, though the two methods may not agree with each other for a specific group. Ages in Table 9.3 range from under a million years to as much as 50 million. Where O and B stars are formed, there must also be stars of the lower main sequence. The associations are so distant, however, and there are so many unrelated field stars, that the lesser lights are hard to find. Nevertheless, the closer OB associations are clearly also filled with low-mass pre-main-sequence T Tauri stars [11.2.2], relating the T associations to the OB associations. Place an OB association near reflecting dust clouds, and you get an R association. The distinctions are as much observational as physical.

9.3.2 Relations and the Gould Belt

Many of the nearby associations are connected with each other. In the nineteenth century, both John Herschel (in the southern hemisphere) and B.A. Gould (in the northern) noted that the bright blue stars near the Milky Way did not lie quite along the celestial equator, but in a circle tilted by 20 degrees to it that came to be known as the *Gould Belt*. The bright stars of Orion, for example, fall to the southern side of the Galactic equator (which runs through Monoceros and Canis Major), while the Scorpius associations fall to the north of it (Fig. 9.10; see also Fig. 1.1). The Gould Belt is caused by the tilted orientation of the whole set of local OB associations, that is, those within a few hundred parsecs. It is related to the local region of interstellar clouds and of star formation from which these OB associations have come, and seems to be centered on the huge and much older Cas–Tau association (Fig. 9.11). Moreover, not only are the stars within the local associations moving outward from a common center, but so are the associations themselves.

The concept of sequential star formation ties them all together [11.2.4]. A great burst of star formation some 50 million years ago gave rise to vast Cas–Tau. A sequence of supernova explosions produced an expansion of the immense cloud out of which Cas–Tau formed, which in turn gave rise to the birth of more associations within the expanding shell. More supernova detonations within the newly formed OB associations produced sub-associations, Sco OB2 and Ori OB1 breaking into their multiple parts, which are becoming OB associations in their own right. The local associations are thus really all part of an extended Cas–Tau: the whole structure is some 1200pc across and one of the largest "objects" in the Galaxy. The Sun, not a physical part of it, is just passing through. Similar sequential systems are seen in the Large Magellanic Cloud [Fig. 4.11], as one set of massive stars begets another. Ephemeral, the local OB associations will shortly be replaced by more, which will in their turn generate new brilliant constellations for the astronomers of the distant future.

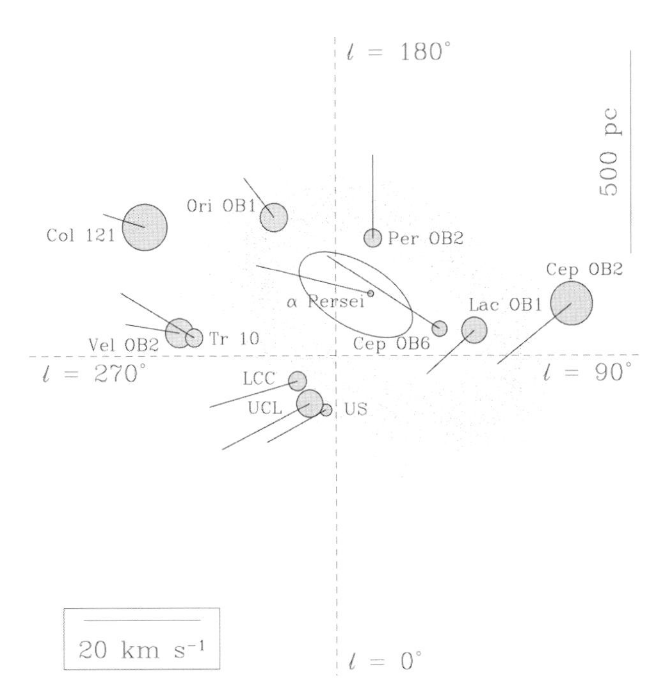

Fig. 9.11 **Local OB associations** The local OB associations, plotted according to their Galactic longitudes, create a huge tilted ellipsoidal system centered on Alpha Persei and the Cas–Tau association (the ellipse in the middle). The lines extending from them show their directions and speeds. On the whole, they are moving apart. The dots represent the Gould Belt. (de Zeeuw, P.T. *et al.* (1999). *Astronomical Journal*, **117**, 354.)

9.4 Globular clusters

The first view of a globular cluster through a reasonably good telescope will be remembered. Among the most spectacular of astronomical objects, a rich globular cluster can pack a million stars into a spherical volume only 10pc in diameter, the stars getting closer and closer toward the core, making them almost impossible to separate (Fig. 9.12). At the center, the star density can be tens of thousands of stars per cubic parsec or even more, as opposed to less than one in our sparse Galactic locale. Among the most ancient creations of the Galaxy, the globulars reveal the Galaxy's age, play a powerful role in estimating the age of the Universe, and possess clues that tell us how our Galaxy – as well as other systems – were assembled.

Fig. 9.12 **47 Tucanae** Among the most magnificent of all globular clusters, 47 Tucanae's star density increases spectacularly toward the inner core. (Courtesy of Anglo-Australian Observatory, David Malin Images.)

9.4.1 The players

Only 147 globular clusters are known in our Galaxy. Interstellar dust surely hides another 30 to 50. Though they are rare objects, their remarkable luminosities make them disproportionately populous in the nighttime sky. Integrated absolute visual magnitudes [3.2] can range to −10, with visual luminosities a million solar. Globular clusters are thus well represented in standard lists, the Messier catalogue containing 28 of them. Over 70 percent – 104 globulars – are listed in the NGC (with another 3 in the IC). Two, 47 Tucanae (see Fig. 9.12) and Omega Centauri, even carry traditional star names [1.3.2]. Most others are named after their discoverers, while two are called by their resident constellations, and a small group is yclept "Pal" for "Palomar." A selection of the best and brightest is listed in Table 9.4.

The majority of known globular clusters, 94 of them, have integrated apparent *V* magnitudes [3.1.1] brighter than 10. (Note, though, that because the clusters are extended sources they do not look as bright to the eye as does a star of the same magnitude.) Eight are brighter than magnitude 6.0, and can be seen (at least in principle) with the naked eye, while the two "named" clusters, Omega Cen and 47 Tuc, are fourth magnitude. In the northern hemisphere, M5 vies for the brightness title with M13, the "Great Cluster in Hercules," which is a bit more visible than M5 to mid northerners because it passes nearly through the zenith. Other notables are M3 in Canes Venatici, awesome M22 in Sagittarius, and M4 (the closest of them) which, with M80, lies near Antares in Scorpius. Then there is M15 in Pegasus, which is not only bright, but amazingly compact near the center, and moreover holds a rare planetary nebula [13.4], K 648 = Ps−1. M22 contains one as well. The casual stargazer, who might examine only the biggest, may succumb to the idea that if you have seen one you have seen them all. While close examination of the whole set indeed reveals similarities, it also shows both subtle and wide differences.

9.4.2 Distribution

The most distinctive characteristic of globular clusters is that even with rudimentary distances we find that the set is spread out into an enormous spherical halo that surrounds the Galactic center and encloses the disk. Globular clusters are strictly Population II, indeed they epitomize Population II [5.1]. Their distribution within the Galactic halo is much like that of stars within the clusters, in that they strongly concentrate toward the center and within the Galactic bulge. As a result, the majority are found in the environs of Sagittarius and the surrounding constellations of the Milky Way. In the other direction, they spread to some 100kpc from the center, vastly farther than do even the most extended portions of the disk.

Name	NGC	Constellation	α (2000) δ h m ° ′	V	Dist. (pc)	M_V	[Fe/H][a]	Age (Gyr)
47 Tuc	104	Tucana	00 24 −72 08	4.0	5200	−9.6	0.18	10.7
ω Cen	5139	Centaurus	13 26 −47 36	3.7	4900	−9.8	0.026	11.8
M3	5272	Canes Venatici	13 42 +28 26	6.2	10 000	−8.8	0.046	11.3
M5	5904	Serpens	15 18 +02 05	5.7	8300	−8.9	0.075	10.9
M80	6093	Scorpius	16 17 −22 59	7.3	3000	−5.1	0.033	12.4
M4	6121	Scorpius	16 24 −26 31	5.6	1720	−6.9	0.089	11.7
M13	6205	Hercules	16 42 +36 28	5.8	7800	−8.7	0.046	11.9
M92	6341	Hercules	17 17 +43 08	6.4	8900	−8.4	0.0079	12.3
...	6397	Ara	17 41 −53 40	5.7	3700	−7.1	0.017	12.1
M22	6656	Sagittarius	18 36 −23 55	5.1	5200	−8.5	0.039	12.3
...	6752	Pavo	19 11 −59 59	5.4	4600	−7.9	0.058	12.7
M15	7078	Pegasus	21 30 +12 08	6.2	10 000	−8.8	0.009	11.7
M30	7099	Capricornus	21 40 −23 11	7.2	9500	−7.7	0.012	11.9

[a]Relative to solar (Most ages and metallicities from Salaris and Weiss (2002). *A&A*, 388, 492; distances and other information from the general literature.)

Table 9.4 **Bright globular clusters**

Speeding with high radial velocities, the clusters dynamically resemble Population II field stars and the high-velocity subdwarfs [5.1, 6.4.2, 7.3.3] that are plunging through the Galaxy's disk on highly elongated elliptical orbits. The gathering of globular clusters contains recognized dynamical subsets that include those that lie in a thick disk and the distinctive clusters of the Galactic bulge.

9.4.3 Chemistry

Like subdwarfs [5.1, 6.4.2, 7.3.3], and characteristic of all Population II, the class of globular clusters is deficient in metals, as usually given by the iron-to-hydrogen ratio, Fe/H, which is typically around 0.05. The range, however, is wide, extending from well under a hundredth that of the Sun to approaching solar, showing that globulars, while an extended family, are hardly all alike. Yet, as low as these metal abundances are, they do not extend to the depths taken on by the most metal-poor halo subdwarfs, which can be 1000 times lower. The higher-metal globulars, those with Fe/H above about 0.08, are confined to the thick disk system. Bulge clusters seem to have the highest metallicities. Since the thick disk and the halo are both old, it again seems clear that metal content built up very quickly in the early days of the Galaxy.

There is also a clear relation between the "alpha elements," those built up by alpha particle capture [5.4, 13.2; 13.3] (oxygen, neon, magnesium and the like), and the Fe/H ratio. While all element abundances go down as Fe/H decreases, the alpha elements do not go down as fast. For Fe/H greater than 0.1, O/Fe is in solar proportion, while at Fe/H $= 0.01$, O/Fe is up from that by a factor of 3 (that is, it drops by about a third as much as Fe/H). The trend continues into the yet-lower-metal Population II field stars. The explanation is straightforward. The first stellar generation, Population III [5.4, 11.3.4], consisted of high-mass stars. They blew up as Type II (core collapse) supernovae [14.2.2], which ejected both alpha elements created in energy-generating processes while the stars were supergiants and iron created in the explosion, so as time proceeded, both were increased. Type Ia white dwarf supernovae [14.2.5] eject less in the way of alpha elements but relatively more iron than do Type II events. Only when lower-mass stars could evolve sufficiently to produce massive white dwarfs could the O/Fe ratio decrease. Each element in fact has its own history.

Any other correlations are weak. Some of the Galaxy's globular clusters are likely to be the products of other galaxies that have merged with ours and thus have different chemical histories, messing up whatever relations may have existed in our own early Galaxy. As evidence, M54 and three other globulars are currently being deposited in our Galaxy as the Sagittarius dwarf merges with us.

For the most part, the chemistry of the stars within a given globular is the same. Any differences are caused by stellar evolution, that includes the destruction of carbon and the increase of nitrogen during the CNO cycle, and by the normal creation of carbon stars during the second ascent of the giant branch (the "AGB" phase) [11.3]. A distinctive exception is huge Omega Centauri (Fig. 9.13), in which there is a factor of three range in the Fe/H ratio and evidence for multiple stellar generations. The cluster may be a product of mergers; indeed it may well be the leftover remains of a captured dwarf galaxy. Other clusters show vaguely similar behavior.

Fig. 9.13 **Omega Centauri** This greatest of globular clusters exhibits some chemical variations among its stars, and is not quite round as a result of some general rotation. It may be the remains of a small galaxy that merged with ours. (Courtesy of Matt BenDaniel.)

9.4.4 Color–magnitude diagrams

Low metal content renders construction of classic HR diagrams [6.3.1] impossible, because the original Harvard spectral sequence [6.2.1] is geared to the solar chemistry of disk stars. The main sequence of globular clusters is made essentially of low-metal subdwarfs [5.1, 6.4.2, 7.3.3]. Weaker metal lines render subdwarf classes too early for their temperatures. As a result, and because of ease of observation, astronomers use color–magnitude diagrams for these systems. While they share many characteristics of open cluster HR diagrams, those of globular clusters are distinctive, and have a variety of notable, even outstanding, characteristics that lead toward a full understanding of stellar evolution (Fig. 9.14).

A common trait among all globulars is a severely truncated main sequence, which stops at a $B - V$ color of around -0.6 and which, after correcting for abundances, roughly corresponds to class G8 and to masses about 0.8 to 0.9 that of the Sun. After metal-correction, the turnoff point is later (cooler) than for any open cluster, showing that the globulars are by far the older set, the simplest calculations giving ages of well over 10 billion years (the time it takes to burn the main sequence down to the observed level). The main sequence turns off to a subgiant branch and then to a standard giant branch, in which stars climb with dead helium cores and hydrogen-burning shells and then descend with helium-burning cores [13.2].

The most dramatic common trait peculiar to these clusters is the distinctive *horizontal branch* (HB) that extends the "red giant clump" of Population I core helium-burning stars [13.2.2] into the blue part of the CMD at an absolute visual magnitude of about 0.7. The horizontal branch is divided into two regions by the *RR Lyrae gap*, which is hardly a gap at all, but that part of the HB that falls into the Cepheid instability strip [10.3.1] and which (unless one is specifically studying such stars) is frequently ignored. To the right of the RR Lyrae stars [10.3.3] is the *red horizontal branch* ("RHB"), to the left, the *blue horizontal branch* ("BHB"). Extending down and to the left we sometimes see an *extended blue horizontal branch* ("EBHB"), which at the extreme might include the *subdwarf B* (sdB) *stars*.

Climbing up and to the right of the HB is a distinctive metal-poor asymptotic giant branch [13.3.1], made of double-shell (hydrogen and helium) burning stars that have dead carbon cores (the "AGB"). Far below is a white dwarf sequence [13.5] that runs parallel to the main sequence. The transition between these two states may or may not produce planetary nebulae [13.4]. If mass loss is too severe, giants might simply move on to become sdO or sdB subdwarfs [13.4.4] before passing into true white dwarfhood.

9.4.5 Distances

No globular cluster is within parallax range [4.2], so standard spectroscopic and photometric techniques must generally be used. While main sequence fitting [4.4.2] is effective, it has a problem. As a result of the low metallicity, the standard main sequence as found from open clusters is too red for the globulars. The first task is to construct a low-metal main sequence derived from parallaxes of local subdwarfs [5.1, 6.4.2, 7.3.3]. Until Hipparcos parallaxes were available, too few were known. Even now, the number is not satisfactory: main sequence position is a continuous function of metallicity, and there are still not enough local subdwarfs to cover the whole range. As a result, the globular main sequences are not yet as well calibrated as one would like.

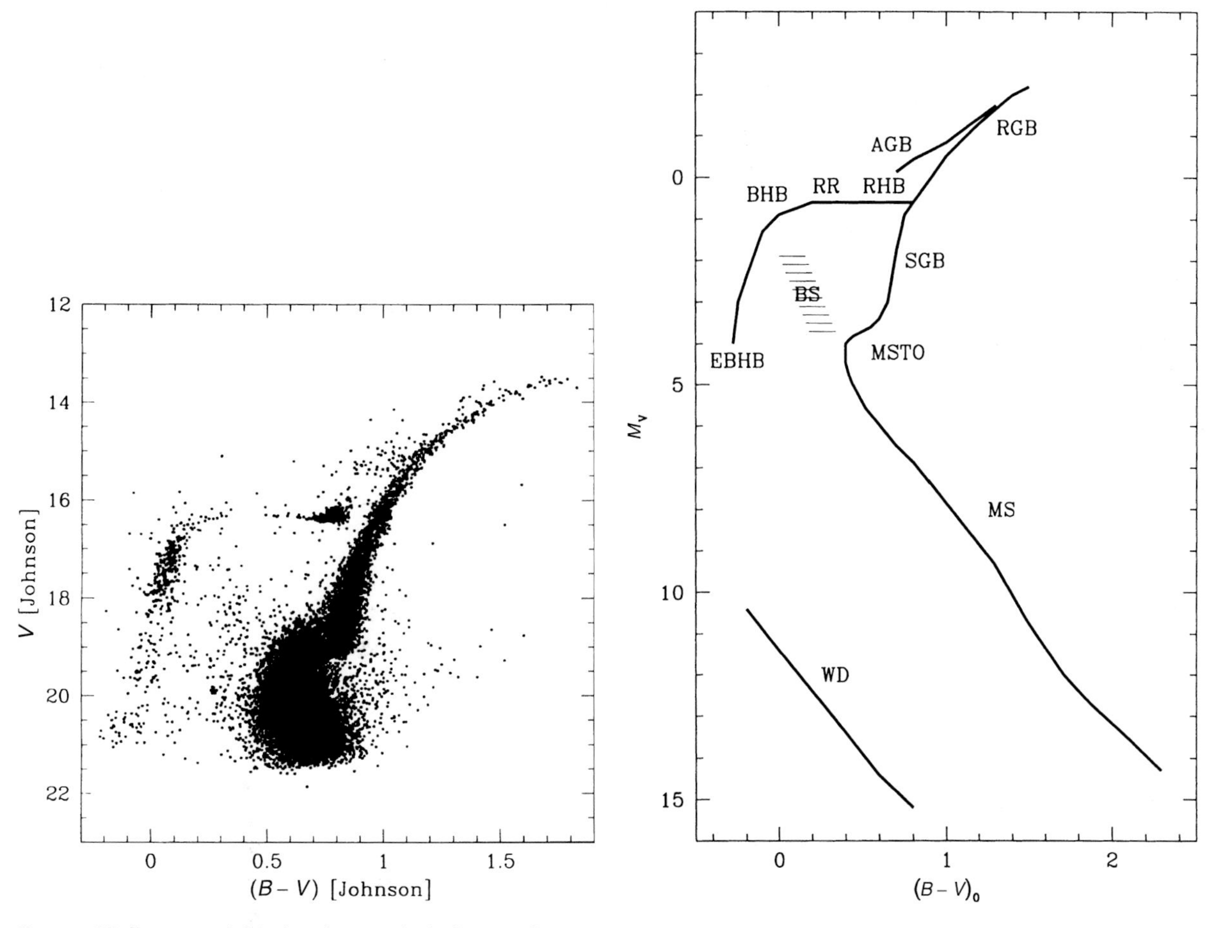

Fig. 9.14 **HR diagrams** At left is the color–magnitude diagram of NGC 2808. At right is an interpretive diagram that shows the various parts: MS = "main sequence," MSTO = "main-sequence turnoff," SGB = "subgiant branch," RGB = "red giant branch," AGB = asymptotic giant branch," RHB = "red horizontal branch," BHB = "blue horizontal branch," RR = "RR Lyrae stars," EBHB = "extended blue horizontal branch," BS = "blue stragglers," WD = "white dwarfs." ((Left) Sosin, C. *et al.* (1997). *Astrophysical Journal Letters*, **480**, L35; (Right) Adapted from King, I.R. (1999) in *Globular Clusters*, ed. Roger, C.M., Fournón, I.P., and Sánchez, F., Cambridge: Cambridge University Press.)

Equally useful in distance measure is the horizontal branch, whose absolute visual magnitude is found through the variable RR Lyrae stars [10.3.3], which are clearly identifiable HB members and shine at $M_V = 0.7$ or so (though somewhat dependent on the metal abundance). These can be calibrated by Hipparcos parallaxes of local field RR Lyraes, by their statistical parallax [4.4.3], or by the Baade–Wesselink method [Box 4.3]. We can also derive the absolute magnitudes of the RR Lyraes in the Magellanic Clouds, whose distances are known through other means. Eclipsing binaries in the globulars provide another avenue [4.6], as does the white dwarf sequence, which can be fitted with a standard absolute sequence as is done for the ordinary dwarfs. In all these methods, one must correct for the effects of interstellar extinction [3.5] – quite severe for clusters at low Galactic latitudes – which can be found from the colors of the stars or of the clusters as a whole. In the cases of some clusters near the Galactic center, variable extinction across the cluster faces enormously complicates the picture.

Better observations even allow "statistical parallaxes" for some individual objects [4.3.3]. It is now possible to observe the proper motions of individual stars within a cluster. From the whole set, we get the statistical spread, or dispersion, in proper motion. Proper motion for any star depends on the transverse velocity and distance ($v_t = 4.74\mu D$). The dispersions in proper motion and transverse velocity must thus be related. Assume the dispersion in transverse velocity to be the same as the observed dispersion in radial velocity, and derive the distance. The scheme works because there are enough stars to allow good statistics and all the stars are the same distance away.

The results of the various studies tell us that even the nearest of globular clusters is thousands of light years away. They spread outward to distances much greater than we can follow the Galactic disk, to as far as an amazing 100 kpc. The spheroidal distribution, and the reasonable assumption that the centroid of the distribution is coincident with the Galactic center, allows the distance from the Galactic center to the Sun to be found, which thus provides an absolute scale for all the Galaxy, the method first applied by Harlow Shapley (1885–1972) in 1918. While still argued, the modern value is generally taken as around 8kpc.

9.4.6 Radii

Cluster radii derive directly from distance and angular diameter. However, like open clusters, globulars have no sharp edges, but instead fade off into the background with fewer and fewer stars, so we have to be precise about the definition of radius. At any given location, an extended object has a "surface brightness" of so many magnitudes per square second, or square minute, of arc. If the energy so specified were concentrated into a point, that would be the magnitude of a star. Unlike that of an open cluster [9.2.3], the *core radius* (r_c) of a globular is based on cluster dynamics. Nevertheless, it is still reasonably close to the *half-light radius* ($r_{1/2}$), at which the surface brightness is half that at the center (and which encloses half the cluster's light). The *half-mass radius* (r_h) encloses exactly that, half the total mass of the cluster, which is larger than the half-light radius. At the important tidal radius (r_t), as for open clusters, stars are removed by tides raised in the cluster by the Galaxy. The tidal radius, which represents the absolute cutoff, is not an observable quantity and depends on local conditions. A cluster in the bulge, or one that is invading the Galactic disk, will have a smaller tidal radius than one that is far out in the halo. Typical core radii are only a parsec or so, and extend down to only a tenth of a parsec, while tidal radii tend to be around 50pc. In between, half-mass radii run of the order of 5pc.

9.4.7 Luminosities and masses

Integrated absolute visual luminosities follow directly from distances, and range from over −10 to below −3, with a broad peak around −7. Since globulars are in age and composition so similar to each other, such differences must reflect roughly equal differences in mass. Directly calculating the masses of globular clusters is particularly difficult. In open clusters, you can at least count all if not most of the different kinds of stars and add them up. Globulars are far too rich and dense.

To get the mass of any body as a whole, one applies Kepler's third law [8.2.3] to a second body that orbits it (effectively the technique for binary stars). However, nothing orbits globulars.

Fortunately, we can still use individual stars. In a loose sense, each star in a cluster orbits the cluster's center of mass under the combined gravitational influence of all the other stars. (The orbit is not a simple ellipse because of local gravitational encounters, loosely called "collisions.") If you were to increase the mass of the Sun while holding the radius of the Earth constant, the Earth's speed would increase. As a result, on the average, stellar orbital speeds must be higher in a high-mass cluster of given radius than in a low-mass cluster. Observation of the spread of velocities (the velocity dispersion) combined with dynamical calculations therefore leads to an approximate mass. From various studies, we find a simple and fairly uniform mass-to-visual-light ratio, M/L_V

Fig. 9.15 **M71** Pretty, but hardly impressive, M71 is just a millionth as luminous as Omega Cen, and made the Messier catalogue only because it is close to us. (Courtesy of NOAO/AURA/NSF.)

of around 2, where both are expressed in solar units. This ratio is quite similar to that of the Sun, and expected from evolutionary considerations. Globular clusters therefore seem to contain no dark matter [Box 5.4].

By this criterion, Omega Centauri, with an absolute visual magnitude of −9.8, has a visual luminosity 700 000 times that of the Sun and contains 1.4 million solar masses. Assuming the average mass of a star to be around half solar, Omega Centauri contains some 3 million stars. With a half-mass radius of 5pc, the inner density averages 3000 stars per cubic parsec (thousands of times local). The central densities of concentrated globulars are much higher. Imagine the view, not that any planet could survive the gravitational interactions.

Near the low end, at $M_V = -3$, masses are only a few thousand times solar, making these lesser globular lights not much more impressive than a rich open cluster (Fig. 9.15).

9.4.8 Age

Because high-mass stars burn out first, the low turnoff to the giant branch ("MSTO" in Fig. 9.14) tells of great age. Absolute ages can again be determined by fitting isochrones to the turnoff points in observed color–magnitude diagrams. The standard accepted value for the large set of oldest clusters is around 12 billion years, recent developments in the calculation of nuclear burning rates raising the figure to 13 billion, notably greater than for any of the open clusters. Because we find nothing older, 12 to 13 billion years is the age of the Galactic halo. Since (from the expansion of the Universe and the Cosmic Background Radiation) the age of the Universe is now soundly set at 13.7 billion years, it is apparent that the first stars formed very quickly after the Big Bang.

Relative ages are easier to find, yielding a general spread of about a billion years. While M92 has long been taken to be among the oldest of globular clusters, the low-to-middle metallicity clusters appear all to be about the same age. The highest-metal globulars, however, seem to spread in age down to below 10 billion years, close to that of the oldest open clusters, which reveals very rapid development of the Galactic disk. It also shows a similar rapid change in the mode of star formation, which, as the Galaxy aged, produced open clusters rather than globulars.

With the Hubble Space Telescope able to reach the dim white dwarfs in globular clusters, they too become fodder for age determinations. The faintest and dimmest yield an age of nearly 13 billion years, which, given the inevitable errors, is in strong agreement with the ages found from isochrones. Such ages are similar to those found from the abundances of radioactive elements in ancient halo field stars. Everything seems to fit, powerful (if not astonishing) proof that astronomers do indeed understand something both of cosmology and of stellar evolution.

9.4.9 The Horizontal Branch

The HB exists for the same reason that the subdwarfs are shifted blueward of the Population I main sequence: low opacity as a result of low metallicity. Differences in envelope mass then stretch the HB blueward. Before stars land on the HB, they are mass-losing red giants. Within dense cluster confines, gravitational encounters can aid in the mass loss process, and help to strip the stars toward the carbon–oxygen cores. Red HB (RHB) stars escaped the worst of it, and have thicker envelopes of a couple tenths of solar mass. On the other hand, at the extreme, the cores of the stars at the blue end of the HB can be stripped nearly bare, with envelope masses only a few hundredths solar, making them small and hot. Core masses are around $0.5M_{\odot}$ or so, as befit the remnants of $0.8M_{\odot}$ stars.

Within this general picture, globulars have an amazing variety of horizontal branch morphologies. Some display only the RHB, while others spread across the board, far into the blue HB, even to the extended blue HB. Others show just the blue HB alone. In the middle is the defining RR Lyrae [10.3.3] "gap," which will not contain the variables only if either the RHB is too short to reach their region of residence, or if an exclusive BHB is displaced too far to the left. Otherwise they will be there, whether plotted on the CMD or not.

The structure of the HB depends primarily on the globular cluster's metal abundance, as witnessed by the Fe/H ratio. Generally, the lower the metal content, the bluer the HB (as expected from lowering opacity). High-metal globulars will have an HB that more resembles the Population I clump [13.2.2], whereas those that extend to metal contents of a hundredth of so solar will have long BHBs. That is, envelope mass makes an increasingly large difference as the metal content is reduced, not stretching the HBs of metal-rich clusters, wildly stretching those of the metal-poor.

As neat as this relation seems to be, there is a major problem. Some globular clusters with low metal contents have anomalously red HBs, that is, the expected BHB is missing (while in others, the blue HB is oddly strong). There is clearly another active variable in the creation of horizontal branch morphologies, a *second parameter*, that has been driving astronomers to distraction for years. The nature of the second parameter is still not secure, although most experts now seem to agree that much of it lies in small differences in age. Older clusters have a lower main sequence turnoff and, all things equal, lower-mass HB stars. If mass loss on the RGB is constant, older clusters will therefore have lower-mass envelopes and will thus have the bluer horizontal branches. Anomalously red HBs therefore indicate relative youth. Mass stripping in dense cores may also play a role (making the stars bluer), as can increased helium content (which also makes the stars bluer), and increased CN content (which makes them redder). No one as yet claims victory over the second parameter. Indeed all sources may play roles.

At the very extreme are a few clusters in which the BHB turns downward to make the EBHB. HB stars are of very closely the same bolometric luminosity. The EBHB is caused largely by increasingly large bolometric corrections [3.3] that result from excessively high temperatures, that in turn render the stars anomalously faint at visual magnitudes.

9.4.10 Blue stragglers

Extending upward from the turnoff point of the dense lower main sequence is a smattering of high luminosity, and consequently higher mass, *blue stragglers* that have refused to become giants ("BS" in Fig. 9.14). They are present in older open clusters as well (see Fig. 9.5b), and must be of higher mass than the stars that are becoming the current giants. Blue stragglers seem at first simply to be latecomers, born after the rest of the cluster's stars. However, deferred birth seems unlikely, as the globulars are so homogeneous in all other respects. With some exceptions (like Omega Centauri), there is no reason to expect extended star formation.

More likely, blue stragglers may have had their masses increased after birth. Key evidence is that they especially like to populate the inner regions of globular clusters, where the star density is enormously greater. Here in the clusters' cores, there are so many stars per cubic parsec that binaries can form easily through gravitational interactions. Encounters with other stars can force the companions closer together, such that they merge to become single stars of greater mass. Direct collisions between single stars could do the job as well. As in so many cases, more than one explanation may be needed, especially in the case of open clusters in which the stars might actually be created over longer periods of time.

Fig. 9.16 **M15** Thousands of stars inhabit the tiny ultra compact inner region of M15, which is among the densest clusters known. (Courtesy of Hubble Heritage Team, NASA/STScI/AURA.)

9.4.11 Structure and disintegration

In spite of their unifying name, globular clusters differ considerably in mass, radius, metallicity, Galactic location, even somewhat in age. Common to them all is their extreme roundness. "Globular cluster" is as apt a name as possible. Only a few – Omega Centauri in particular – show much deviation from projected circularity, which implies that they have little if any net rotation (see Fig. 9.13).

Globular clusters also differ in the degree of central concentration, which is related to cluster mass and disintegration. To appreciate the deep differences, contrast M71 (see Fig. 9.15), which has a low luminosity ($M_V = -5.4$) and consequently low mass, and is loosely organized, with massive 47 Tuc (see Fig. 9.12), then with intensely concentrated M15 (Fig. 9.16). Clusters are traditionally ranked by concentration class, in which Roman numeral I stands for the most concentrated, XII for the least. A more modern scheme uses not discrete classifications, but a continuous variable in which the structure of the cluster is given by a "concentration parameter," the ratio of the tidal radius to the core radius, r_t/r_c. These go from around 6 in the case of loosely organized systems to just over 300, a spectacular range.

The "brightness profile" of a cluster plots surface brightness against radius. A cluster with a low value of r_t/r_c will have a flat profile through the center; that is, the star density will remain fairly constant within a radius that surrounds dead center. As the ratio climbs, this "flat" region becomes smaller and smaller, until when r_t/r_c is just over 300, the flat region essentially disappears, and the cluster's density seems to climb without bound, to a "cusp," and it appears that the cluster has "collapsed." (The cluster cannot really collapse to infinite central density; it just comes as close as possible.)

The concentration parameter roughly correlates with cluster mass, which makes sense, since giant clusters have greater gravitational energies that can drive collapse. But the scatter is huge. The greatest of all clusters, Omega Cen (Fig. 9.13) has an r_t/r_c of 34, whereas others four magnitudes fainter can be collapsed. M71 (Fig. 9.15) has a concentration parameter of 14, about right for its faint M_V of -5.4, while 47 Tuc (Fig. 9.12) climbs to 140. M15 is the prime example of a collapsed cluster which, when combined with its great luminosity ($M_V = -8.9$), makes it one of the grandest clusters in the Galaxy (Fig. 9.16).

Though much more massive than open clusters, globular clusters face the same problems of segregation and disintegration. A cluster star does not move independently, but strongly interacts with its mates in a way similar to the atoms in a gas. Though the stars do not (for the most part) physically collide (in which case they might create a blue straggler), they do gravitationally interact, which causes one star to gain energy and speed (in which case it would on the average move to the outside of the cluster) and the other to lose it, shifting it closer to the center.

An interaction between a lower-mass star and one of higher-mass will have the greater effect on the lower-mass star. Over time, the interactions are such that each star will tend to have the same amount of energy, which involves both speed and mass. More massive stars will therefore move slower, and will sink to the cluster's center, while the lower-mass stars will drift to the outside (as they will for open clusters [9.2.4]). If the lower-mass stars have enough energy and can get propelled to the tidal radius, they will leave. The result is continuous "evaporation." The effect is profoundly enhanced if the cluster crosses the Galaxy's plane, in which case it is subject to a severe "tidal shock" that can strip the cluster's outer stars away. No globular cluster can therefore be permanent. They will all disintegrate. The majority of the Galaxy's halo stars may be the remains of totally disintegrated clusters, implying that our Galaxy once held many more of these systems than it does today. What we see now are the hardy surviving remnants – as well as some that are in the last stages of disruption.

As the low-mass stars leave, the relatively higher-mass ones (those near the main sequence turnoff) descend to the core, which becomes ever more concentrated. Ultimately, the core should go into catastrophic collapse, wherein the central density (in principle) approaches infinity, which could produce a semi-massive black hole [14.3.3]. Such disaster can be rescued by binary stars.

9.4.12 Binaries

Double stars inhabit globular clusters just as they do open clusters and the general field, but with a distinct difference. Within the dense confines of the cluster, outsiders are always passing and disturbing a double's members. Several things can happen. If the members of a binary are widely enough separated, the intruder will on the average give energy to one of them and cause the pair to separate further, even to the point of disruption. If the members are close together, however, the binary will give gravitational energy to the collider, and the companions will draw closer together. The collider will speed up and move to the outside of the cluster, perhaps even to be thrown out of it. For that matter, the binary could be tossed outward too, as a result of passing its orbital energy on to energy of motion within the cluster. The results are a reduction in the number of binaries in globulars and the elimination of all wide doubles, the remaining ones separated by no more than a few AU. On the other hand, the star density within the core of a collapsing cluster is enough to create new doubles, simply by the tidal interaction of close-passing stars.

In another scenario, a relatively massive star encountering a double can replace one of the members. Neutron stars [14.3.2] and even black holes [14.3.3] left over from the early days in which the cluster contained massive stars that exploded as supernovae will preferentially attach themselves to ordinary dwarfs, thus explaining the seemingly unnatural numbers of millisecond pulsars and low-mass X-ray binaries [14.3.2] that some globular clusters are seen to contain (47 Tuc has about 100 such pulsars).

The effect of the doubles on the cluster as a whole is profound. A huge amount of energy is stored in binary orbital motion. Energy from a collapse can be banked when new doubles are formed under high density conditions. The increasing encounters between cluster stars and close binaries subsequently transfer energy back into the cluster's stars and thereby fluff the cluster, preventing a true collapse. The collapse can proceed only so far, whence the binaries make it "rebound," the bounce followed by another quasi-collapse, and so on. Only a few clusters have actually reached this state. Even then, real collapse might still be a possibility, as M15 (see Fig. 9.16) has been suspected of harboring a medium-mass black hole.

Stars by themselves give us little of their history. But when we look at their assemblies, from doubles and multiples through open clusters, to the magnificent globular clusters, indeed, to the Galaxy itself, we begin to get a sense of how our system developed, and why we are as we are today. Eventually our observational capacity may be good enough to examine in detail globular clusters in galaxies so distant that we look far enough back in time to see what they were like when we were young.

CHAPTER TEN

Variable stars

We trust the stars. That they seem constant and fixed (except for seasonal changes caused by our orbiting Earth) gives tacit stability to our lives. There is Orion, with his spectacular seven-star figure. No one wants to think that Betelgeuse or Bellatrix might be gone. Yet gone they will be. As stars move among their neighbors, all constellations will slowly be distorted. More to the point, all stars will someday die and fade, some quietly, others in awesome explosions.

We need not wait on evolutionary timescales to see stellar change, however. During a short period of each year, Mira (Omicron Ceti) makes a major impact on its dim constellation, and then for the rest of the time it disappears from view, altering the form of the Sea Monster (Fig. 10.1). Stars in Cepheus, Aquila, and elsewhere wink every few days, while Scutum sometimes harbors a reddish star, other times not. The closer we look, the more variation we see, until at the levels of greatest refinement, nothing is constant at all. Variability, one of the glories of the stars, is everywhere. Even the "constant" Sun varies, if only to a minute degree. Fascinating to watch, stellar variations are of great scientific importance, as they provide insights into stellar lives and structures, and allow new ways to test theory.

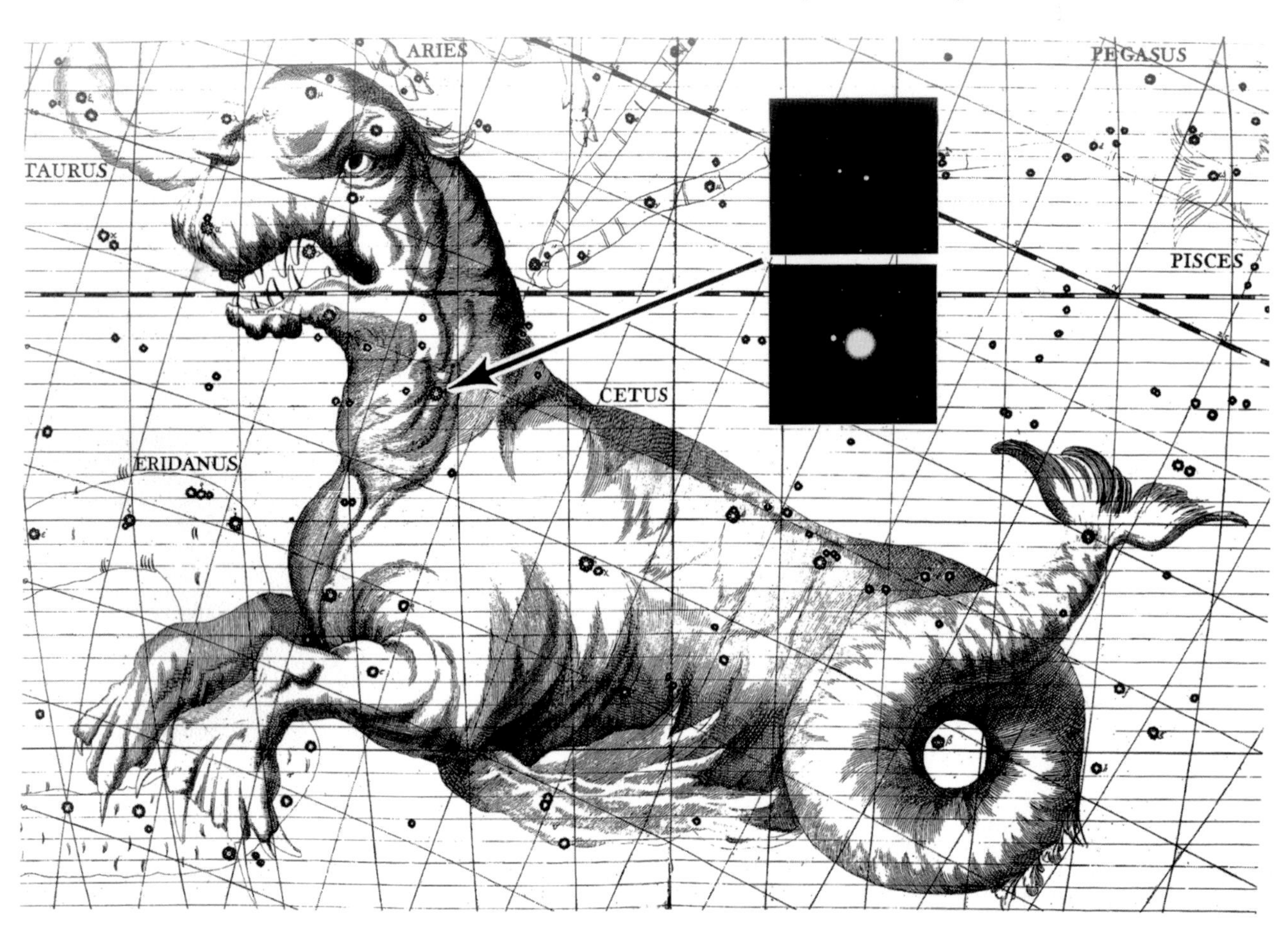

Fig. 10.1 **Cetus and Mira** Lying in the neck of the celestial Sea Monster, Mira – the classic long-period variable – goes from visual magnitude 3 or so to 10 over a period of 332 days. (John Flamsteed, *Atlas Coelestis*, edition of 1781, courtesy of the Rare Book and Special Collections Library, University of Illinois; inset photos, Lowell Observatory.)

10.1 Kinds of variation

Many are the things that can make stars vary in brightness, or at least seem to do so:

1. *Duplicity*. The most obvious reason for variation is stellar duplicity, in which two stars eclipse each other [8.4]. Interacting doubles [8.5] are responsible for effects that include gas streaming, hemispherical heating, and (most obvious), nova [8.5.4] and supernova [14.2.5] explosions.
2. *Catastrophic single-star explosions*, or Type II core-collapse supernovae [14.2.2].
3. *Eruptions*. Variations associated with sudden mass loss range from huge outbursts from supergiants such as Eta Carinae, through highly evolved stars like R Coronae Borealis, to a subset of Be ("B-emission") stars.
4. *Rotation*. Starspots, caused in part by rotation, swing in and out of view [7.2.5]. Stars with oblique magnetic axes and duplicity add to the fun.
5. *Magnetic flaring*. Rotationally induced magnetic fields can produce powerful flares in low-mass stars, and apparently in some of higher mass as well.
6. *Pulsation*. Pulsators change their brightness by altering their sizes, and accordingly, temperatures. Mira is most obvious, while Delta Cephei, Eta Aquilae, Zeta Geminorum, l Carinae, and other *Cepheids* follow close behind.
7. *Solar-type oscillations*. Stellar pulsation changes over into subtle oscillations, leading to the field of "asteroseismology" [12.3].

None of these forms of variation can entirely be separated from the others. Pulsators can be influenced by binary action, rotation, even by magnetic fields; eruptive and high-mass pre-supernova stars may pulsate and also be eclipsers. Mass-ejecting eruptive Be stars are also rapid rotators. Some of these forms are much more appropriate to other realms of study: eclipsers go to binaries, supernovae to their own division of stellar evolution. Here we take on the bulk of the others: the classic pulsators, some eruptive variables, and flare stars.

However, even a lone category like "pulsating variable" covers as much ground as "double star." There are large numbers of different kinds, as each bin of mass and evolutionary state weighs in with its own version. Subgroups and names proliferate, some kinds known by two or more different terms. Easing the way, one kind commonly moves smoothly into another, making the various terms sometimes somewhat artificial. Table 10.1 lists the principal categories, while Fig. 10.2 shows their specific locations on the HR diagram. More specifically, Fig. A5.1 in Appendix 5 displays a

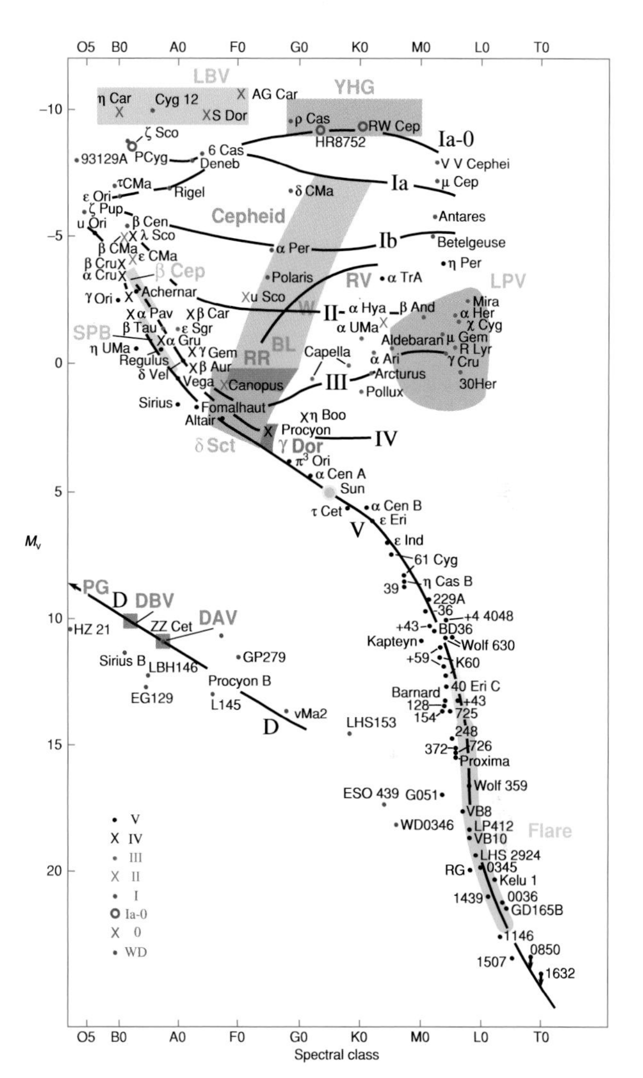

Fig. 10.2 **Variation on the HR diagram** Regions of prominent variation are drawn onto Fig. 6.12. The classical Cepheid instability strip runs through classes F and G, labelled "Cepheid." Below it lie the Population II Cepheids, which divide into BL Her, W Vir, and RV Tau stars. Other labels are LBV: luminous blue variables; YHG: yellow hypergiants; SPB: slowly pulsating B stars; LPV: long-period variables (Miras); RR: RR Lyrae stars; δ Sct: Delta Scuti stars; γ Dor: Gamma Doradus stars; DAV: DA white dwarf variables (ZZ Ceti stars); DBV: DB white dwarf variables; PG: the variable subset of the PG 1159 stars (which are up and to the left, off the diagram); Flare: dwarf K, M and L flare stars.

selection of variables that lie toward the Galactic anticenter and that, taken together, tell a story of stellar evolution from birth to near-death.

The classic listing of all variables is the *General Catalogue of Variable Stars* (GCVS) of the Sternberg State Astronomical Institute of the Moscow State University. Its various editions and

Type	Pop.	Spectrum	Amplitude (V)	Period
Classical Cepheids	I	F, G; I, II	0.5–2.0	1–50 days
s-Cepheids (for "short-period" Cepheids)				1–7 days
Pop. II Cepheids	II	F, G; II, III	0.4–2.0	1–50 days
BL Herculis stars (or AHB1 stars)				1–10 days
W Virginis stars				10–20 days
RV Tauri stars				20–50 days
RR Lyrae stars	II	A, F; III	few tenths	0.2–1 day
RRab (Bailey ab, fundamental mode)				0.5–0.6 day typical
RRc (Bailey c, first overtone)				0.3 day typical
RRd (double mode)				
Delta Scuti stars	I	A; V, IV, III	0.01–0.3	0.02–0.3 day
AI Velorum stars (dwarf Cepheids)			>0.3	
SX Phoenecis stars	II	(blue straggler version)		
Gamma Doradus stars	I	F; V, IV, III	few hundredths	0.4–3 days
roAp stars	I	Ap V	thousandths	5–15 minutes
White dwarf variables	I, II		few tenths	few minutes
ZZ Ceti stars		DA white dwarfs, 10 500–13 000K		
DB variables		DB white dwarfs, 20 000K		
PG 1159 (GW Vir) variables		white dwarfs near 130 000K		
Beta Cephei stars	I	B0.5-B2 III, IV	0.1	3–7 hours
53 Persei (SPB)	I	B3-B8		
Alpha Cygni stars	I	Be-Ae I	0.1	days/weeks
LPV (Miras)	I, II	M, R, N, S; III	5–10	100–600 days
Semi-regular	I, II	K, M; I, II, III	1–2	10–100 days
Sra		Low amplitude Miras		
SRb		Less-well-defined		
SRc		supergiant semi-regulars		
Semi-regular SRd	I	F, G, K 0-III Yellow hypergiants	0.1–4	30–1100 days
Irregular	I, II	K, M; I, II, III	1	
Lb		giant irregulars		
Lc		supergiant irregulars		
UU Her stars	II	F, G; (Post AGB)	few tenths	40–70 days
R CrB stars	I	F; III	5–10	irregular deep fadings
PV Tel	I	Bp helium	0.1	0.1–1 day; year
LBV	I	O, B; 0; Ia	3–4	years, decades

Table 10.1 **Pulsating variable stars**

[Subgroups are indented below the main class]

supplements list not only pulsators, but eclipsing binaries as well. New discoveries and discussions regarding variables are presented in the *Information Bulletin on Variable Stars* (the IBVS), produced by the Konkoly Observatory in Budapest as a publication of the International Astronomical Union. There are so many variables (the third edition of the GCVS published in 1969 listed over 20 000) that those with significant variations have long been the domain of amateur observers through such organizations as the *American Association of Variable Star Observers* (AAVSO) and the British *Astronomical Association Variable Star Section* (BAAVSS). The number of named variables as of 2004 is approaching 40 000. Huge numbers of others, discovered in wide-angle surveys but whose natures remain unknown, await cataloguing.

10.2 Names

The first variable stars discovered already had names under now-standard systems, for example Algol, Mira, Delta Cephei, and the like. However, when variables were found that had no common names (including eclipsing binaries and novae), they were considered so special that a new system had to be devised. Friedrich Argelander simply continued Bayer's scheme. Bayer first used Greek letters for stars within a constellation, then lower- and upper-case Roman letters [1.3.2]. Since Bayer stopped at Q, the first variable found within a constellation would be R, the second S, down to Z, giving us R Cygni (a Mira-type variable), S Sagittae (a Cepheid), T Coronae Borealis (a recurrent nova), U Cephei (an interacting eclipser), and so on.

No one knew of the vast numbers to follow. Z (as in Z Andromedae, a symbiotic binary [8.5.8]) was eventually reached in a number of constellations. Still unknowing about numbers, astronomers went to a double-letter scheme: RR (as in RR Lyrae, which defines a whole class), RS, RT . . . RZ, then SS, ST, SU . . . and down to ZZ. Still the discoveries flooded in. So, where to go but to AA, AB . . . AZ, BB, BC . . . BZ (leaving out J). By the time QZ arrived, there were 334 combinations, which in a rich constellation like Sagittarius were nowhere near enough. At this point, astronomers gave up and went to V335 and so on (still within a given constellation). Nova Cygni 1976 is also known as V1500 Cygni, and the last variable catalogued by 2004 in Sagittarius was V5112 Sgr! Even though the system includes a hodge-podge of forms of variables, it at least tells us immediately that the star is unsteady in its light.

10.3 Cepheid variables

Pulsating variable stars are organized in a variety of ways: by chronology of discovery, amplitude (size of variation), period, or evolutionary status. All are connected. Start with the best known and perhaps most important kind, the Cepheids, then move on to the variety of others that fall within the prime region of instability on the HR diagram. The story then moves to the other important groups (Miras and related stars), more or less following the flow of Table 10.1.

In 1784, just before John Goodricke discovered the unsteady behavior of Delta Cephei, Edward Piggot found similar variations in Eta Aquilae, forever tying these two together as the first *Cepheid variable stars*. The term used by itself is vague: the Cepheids are subdivided into "Population I" or *classical Cepheids*, *Population II Cepheids*, and *dwarf Cepheids*. To avoid confusion, the classical Cepheids are commonly called "δ Cephei stars." Population II Cepheids embrace three different kinds with different prototypical names (BL Her, W Vir, and RV Tau), while it is best to expunge "dwarf Cepheid" altogether (see Table 10.1). Here, the δ Cep stars will be referred to as "classical Cepheids," or by default, just plain "Cepheids." The other kinds will be called by their specific names.

10.3.1 Classical Cepheids

Delta Cephei, as typical as they come, varies regularly between visual magnitude 3.6 and 4.3 and back over a steady period of 5.37 days. A number of other such stars are easily visible to the naked eye (Table 10.2). The light curve of Delta Cep and its kind – the graph of magnitude against time – is characterized by a rapid one-day rise followed by a four-day fall, the variation endlessly repeating itself (Fig. 10.3). At the same time, all manner of other characteristics are changing. Temperature varies in concert with apparent magnitude, going from around 6500K at maximum to 1000K less at minimum, with the spectral class following suit, alternating between roughly early F and early G. Since luminosity is a function of both temperature and radius according to $L = 4\pi R^2 \sigma T^4$ [7.1.2], we see from the observations that the radius must change as well, by about 10 percent or so.

The pulsation and radius variation are immediately evident from the variation in radial velocity. Delta Cephei has a gross radial velocity of -17 km s^{-1}. As it pulsates, the radial velocity becomes alternately larger, then smaller, again in synchrony with the visual variations, as the surface of the star first approaches us relative to the center, then recedes during the pulsational cycle. However, the velocity curve (v_r vs. time) shows that the brightest part of the cycle is not when the star is at an extreme size, but when the contraction velocity is close to being the fastest. Similarly, the faintest part of the cycle takes place near greatest expansion velocity. As a result, the extreme radii of the star fall vaguely between the extrema of luminosity (the greatest radius occurring as the star is fading). To an approximation, the difference between

Star	α (2000) h m	δ ° ′	Period (days)	Magnitude (V)	Spectral Class
Polaris	02 32	+89 16	3.97	1.97–2.00	F5 Ia
T Vulpeculae	20 51	+28 15	4.44	5.44–6.06	F5 Ib to G0 Ib
FF Aquilae	18 58	+17 22	4.47	5.20–5.55	F5 Ia to F8 Ia
δ Cephei	22 29	+58 25	5.37	3.48–4.34	F5 Ib to G2 Ib
Y Sagitarii	18 21	−18 52	5.77	5.40–6.10	F6 I to G5 I
X Sagitarii	17 48	−27 50	7.01	4.24–4.84	F5 II to G9 II
η Aquilae	19 52	+01 00	7.18	3.50–4.30	F6 Ib to G2 Ib
W Sagitarii	18 05	−29 35	7.59	4.30–5.08	F2 II to G6 II
S Sagittae	19 56	+16 38	8.38	5.28–6.04	F6 Ib to G5 Ib
β Doradus	05 34	−62 29	9.84	3.46–4.08	F6 Ia to G2 Iab
ζ Geminorum	07 04	+20 34	10.15	3.66–4.16	F7 Ib to G3 Ib
l Carinae	09 45	−62 30	35.52	3.38–4.10	F8 Ib to K0 Ib

Table 10.2 **Bright Cepheid variables** Arranged by increasing period.

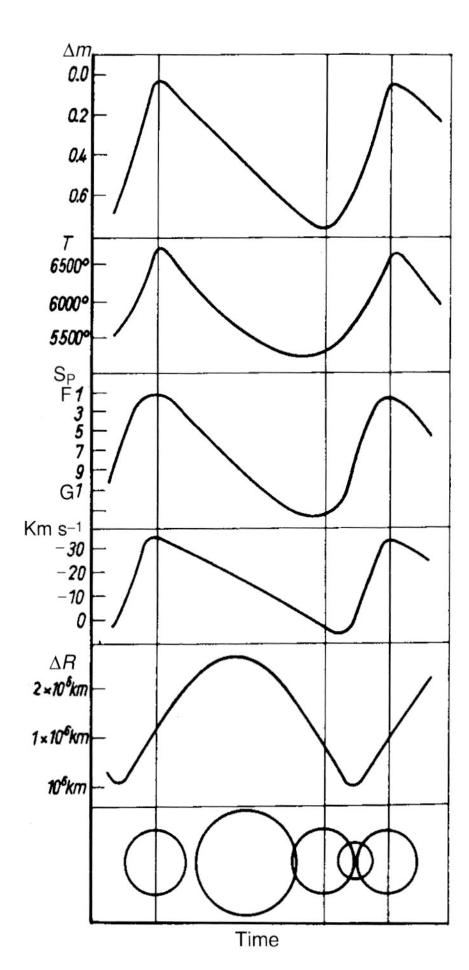

Fig. 10.3 **Variations of Delta Cephei** From top to bottom the panels show the variations of Delta Cephei: apparent magnitude, temperature, spectral class, radial velocity (the star's gross velocity at -17 kms^{-1}), radius, and an exaggerated sketch of the star's size. Time runs along the bottom axis, the period of pulsation (maximum to maximum, indicated by vertical lines) 5.37 days. (Hoffmeister, C., Richter, G., and Wenzel, W. (1985). *Variable Stars*, Berlin, Springer-Verlag, p. 35, Fig. 10, copyright Springer-Verlag.)

maximum and minimum luminosity is one of temperature, not radius.

Cepheid periods range from roughly 1 to 70 days, though most fall between 2 and 25 days. Amplitudes (the range from brightest to faintest) roughly associate with periods, half a magnitude at the low-period end, two or more at the high end, and increase dramatically into the blue, violet, and ultraviolet. However, very small amplitudes of only a few tenths or less are not unknown, as witness Polaris (see Table 10.2).

While the light curve of Delta Cephei sets a common standard, the forms of the light curves do change with period. The *Hertzsprung progression* begins at around 6 days, at which point the light curve develops a bump on the fading portion. As the period increases to 10 days, the bump travels to maximum light, and then rides down the ascending portion, disappearing at 15 days. With periods under three days are the *s-Cepheids*, whose light curves are more sinusoidal.

Classical Cepheids are nearly all F and G supergiants with absolute visual luminosities between about -2 and -6. As periods increase, the stars go to later classes, and at the extreme some step over into class K. Masses range from about 2 to 8 times solar. As short-lived high mass stars, Cepheids are therefore found only in that part of the Galaxy in which stars have recently formed, in the Galactic disk, hence their assignment to Population I [5.1]. The more massive Cepheids in fact make useful tracers of the local spiral arms. The full range of class and absolute magnitude defines part of the *Cepheid instability strip* that runs down the middle of the HR diagram (see Fig. 10.2). In principle, all stars within the strip are either Population I classical Cepheids or Population II Cepheids, though a few stars falling there refuse to behave as expected and stay stable for reasons not understood.

The most remarkable – and important – characteristic of the classical Cepheids is the strong relation between period and absolute magnitude, true whatever color (*U*, *B*, *V* and the like) is picked (see Fig. 4.12). This *period–luminosity* (or *P–L*) *relation* [4.4.3] was discovered in 1912 by Henrietta Leavitt from her observations of Cepheids in the Small Magellanic Cloud (SMC), later extended to those in the Large Magellanic Cloud (LMC) [Fig. 4.11; Box 5.1]. Since all the stars in either the LMC or SMC are at nearly the same distance, the relation is immediately evident by graphing apparent magnitude against the logarithm of the period. Once the P–L relation is calibrated against absolute magnitude, it allows the distance of any Cepheid (or of any object in which we find a Cepheid, such as a distant galaxy) to be found.

Cepheids are *radial pulsators*. Such stars pulsate as a whole, all layers moving in and out at the same time. The phenomenon, however, involves neither the whole star nor the energy emerging from the stellar core, but is instead restricted to the outer stellar layers. Delta Cephei and its kind are in a state in which these outer layers cannot find equilibrium. Like a child on a swing, the outer part of the star expands, overshooting its equilibrium radius. When it gets too large, gravity drags it back. Contracting quickly, it squeezes past equilibrium, and then when the pressure becomes too great, it expands again, the cycle repeating itself over and over. The stellar core has no idea what is going on outside. However, like the child, to keep going the star requires a driver, which is a layer far below the surface in which helium is in a state of partial ionization (Box 10.1). Only within the instability strip (and in other such areas of the HR diagram where similar drivers exist) are the conditions just right for the driver to operate.

The P–L relation makes perfect physical sense. When these massive stars cross the instability strip, either in transition to the core helium burning stage or on blue loops while burning helium [13.2.2, 13.2.3], they pulsate. The more massive stars are not only the brighter, but are also the larger, and larger stars simply take longer to pulsate. The period of a Cepheid is proportional to the inverse of the square root of the mean density, $P \propto 1/\sqrt{\rho}$. The increased radius of more massive

Box 10.1 The pulsation driver

Pulsating variable stars of all kinds are run by layers deep within them in which some atom or other is in a state of partial ionization, that is, neither fully in one state nor another. The specific driver for Cepheids is a layer that contains a mixture of singly and doubly ionized helium (that is, the region in which helium is becoming doubly ionized). Energy generated in the stellar core does not speed immediately outward from the star, but is impeded by the atoms and ions in the surrounding gas, which is partially opaque. The opacity at any wavelength (as well as the opacity averaged over all wavelengths) depends on a variety of parameters, including chemical composition, state of ionization, temperature, and density (or pressure).

Opacity generally increases with increasing density, but decreases sharply with increasing temperature. Compress a gas, and both density and temperature rise. The temperature effect is the more important, and the opacity goes down. Compress a partial ionization zone, however, and while the density increases, the energy supplies additional ionization rather than creating a higher temperature, allowing the opacity to increase rather than decrease.

When a Cepheid compresses, the radiation flowing outward through the partial ionization zone is bottled up. The trapped heat pushes the layer outward, which lowers the ionization level and the radiation is released, so the star brightens. Ionization in the affected layer then relaxes. When the star can be pushed outward no more and hits its maximum radius, it falls back. The radiation is bottled again, and the cycle starts anew. The partial ionization zone thus acts as a valve that drives the stellar oscillation. Since Greek letter kappa (κ) is the universal symbol for stellar opacity, the driver is known as the "kappa mechanism."

If the star is hot (to the left of the instability strip), the layer is too close to the surface to have much of an effect. If it is too cool, the layer is too deep. The convection that develops in cooler stars plays a role in damping the effect as well. In the traditional instability strip, the zone has just the right depth to make it all work. Similar layers run other types of stars in different regions of the HR diagram.

Fig. 10.4 **Polaris** At the end of the handle of Ursa Minor's Little Dipper (and near the upper left corner), Polaris, the north pole star, is a Cepheid that seems to be making the transition from its first overtone mode to its fundamental. (J.B. Kaler.)

Box 10.2 The strange case of Polaris

Polaris (Fig. 10.4) is the sky's brightest Cepheid variable. But do not look for its variations, as they are far below the human eye's detection limit. With a period of 3.97 days, Polaris's visual amplitude is not only small, but for over a hundred years has steadily decreased. At the beginning of the twentieth century, it was about 0.12 magnitudes. By 2000, it was down to 0.03 magnitudes. For a time, it was thought that we were seeing stellar evolution in action and that the star was exiting the instability strip. It is, however, not really that close to the edge of it.

Polaris was once thought to be a rather odd Population II W Virginis star. Instead, it is a low-amplitude example of a massive Population I s-Cepheid that is pulsating in its first overtone with a period of four days. The decreasing amplitude seems now to be part of a preparation by the star to switch over to its longer fundamental period of 5.7 days. No one knows how long this transition will take. Sometime in the distant future, however, Polaris's amplitude should increase, rendering it much like Delta Cephei is today.

stars overwhelms the increase in mass, so as a result, the average density goes down and the period P goes up.

The only hitch is in the mode of pulsation. Any vibrating body – a guitar string or an organ pipe – oscillates in a lowest-frequency fundamental mode that depends on size, giving an audio "period–radius" relation: the shorter the string, the shorter the period, and the higher the pitch. However, strings and pipes also oscillate with higher-frequency overtones that give the sound its timbre and that allow you instantly to tell the difference between a banjo and a violin [12.3]. Variable stars can do the same thing. While most Cepheids pulsate in their natural fundamental mode, the s-Cepheids seem to be pulsating in their first overtones (Box 10.2). Some stars – the "double-mode" variety – are in both states at the same time, giving us odd multiple-period light curves.

10.3.2 Population II Cepheids

Population II Cepheids, those that are found in globular clusters [9.4] and in the general Galactic halo [5.1], mimic the behavior of the classical Cepheids. The Population II variety, however, has a P–L relation that lies roughly 1.5 magnitudes below that of the classical Cepheids [Fig. 4.12]. Lower luminosities imply lower masses. Population II variables typically weigh in at $0.6M_\odot$ or so, far below the masses of classical Cepheids, and consistent with their placements in the Galactic halo, which contains no star above 0.8 solar unless, as a runaway, it has escaped the disk [8.6, 14.3.2]. Population II Cepheids are driven by a combination of partial hydrogen and helium ionization (see Box 10.1). Their instability strip (see Fig. 10.2) runs to the right of that of the Population I strip.

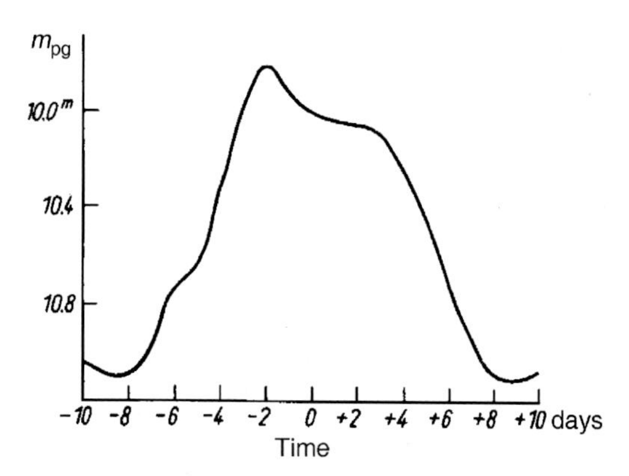

Fig. 10.5 **W Virginis** The light curve of a W Vir star (here expressed by photographic magnitude) can be recognized by its distinctive bump. (From Hoffmeister, C., Richter, G., and Wenzel, W. (1985). *Variable Stars*, Berlin, Springer-Verlag, p. 27, Fig. 5, copyright Springer-Verlag.)

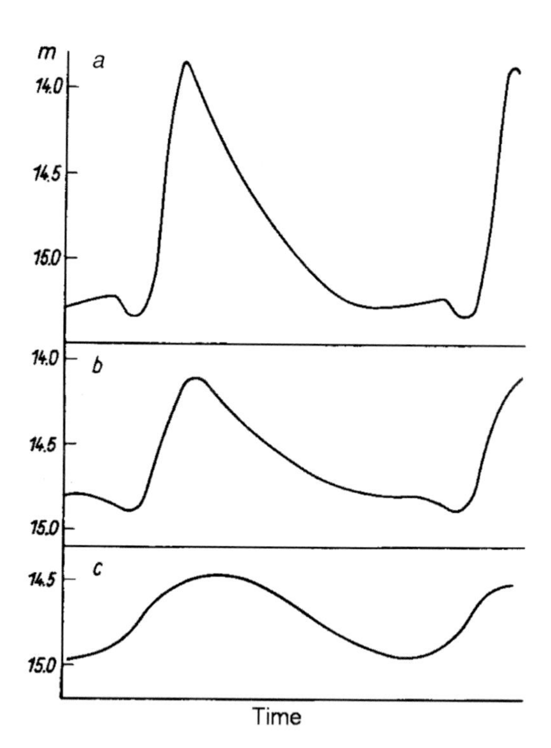

Fig. 10.6 **RR Lyrae light curves** From top to bottom are the light curves of Bailey types a, b, and c. Types a and b are nearly the same, while type c differs significantly. (Hoffmeister, C., Richter, G., and Wenzel, W. (1985). *Variable Stars*, Berlin, Springer-Verlag, p. 40, Fig. 11, copyright Springer-Verlag.)

Population II Cepheids come in three flavors (see Fig. 10.2). At the low end, with periods between about 1 and 5 days and amplitudes of a magnitude or so, are the *BL Herculis stars*, also referred to as ABH1, "above horizontal branch" [9.4.4]. Counterparts of Population I class F, they seem to be formed by stars that are leaving the blue horizontal branch [9.4.9] for the asymptotic giant branch [13.3.1] and then encounter the instability strip.

From 5 to 10 days is a void. Then between 10 and 20 days lie the *W Virginis stars*. These are the traditional counterparts of the classical Cepheids, the name once synonymous with Population II Cepheids in general. They are characterized by bumps in the ascending or descending portions of their light curves, and are probably either leaving the horizontal branch to become AGB stars or are moving to the blue away from the AGB, again passing through the Population II instability strip (Fig. 10.5).

At the top of the pile are the *RV Tauri stars*, whose periods exceed 20 to 30 days (the division is not firm). These G and K stars display very odd light curves, with alternating deep and shallow minima that themselves alternate, the shallow minima becoming the deep ones and vice versa. RVa stars continue this behavior and have constant mean luminosities, while RVb stars have longer-term wanderings. Their evolutionary state is similar to that of the W Vir stars, to which they are related. Their envelope masses are just different as they climb the AGB or leave it.

10.3.3 RR Lyrae stars

At the bottom of the Population II Cepheid category are the special *RR Lyrae stars*. These are unstable helium-burning horizontal branch stars [9.4.9] that have just the right envelope masses, hence temperatures, to place them within the lower-luminosity extension of the classical Cepheid instability strip (see Fig. 10.2). The temperatures range from 7600K at the blue edge of the strip to 6170K at the red edge, which corresponds to Population I classes A8 to F7. Typical mean absolute visual magnitudes are 0.7, while periods range between 0.2 and 0.8 days. Those of shorter period lie closer to the blue edge, as would be expected since all these stars have about the same mass (roughly 0.5 to 0.7 solar), and the warmer stars must be smaller and denser.

Though a few, like the prototype RR Lyrae, are found in the general field, the majority occupy globular clusters, which give these pulsators the alternative name *cluster variables*. Higher-metal clusters in general do not have horizontal branches that stretch far enough to the blue to contain these variables [9.4.9]. On the other hand, if a low-metal cluster's HB stars are all shifted far to the blue, it will have none either. Since the higher metals of Population I open clusters [9.2] cause the helium-burners to sit in a red clump near the giant branch, they cannot have any RR Lyrae-type stars. The near-constancy of the absolute magnitudes of RR Lyrae stars makes them wonderful standard candles with which to estimate distances to globular clusters and nearby galaxies.

It is not enough that RR Lyrae stars are already something of a subgroup of the Cepheids: like all varieties of stars, they have subgroups of their own. Most importantly, they are divided into Bailey types a, b, and c (after the American astronomer S.I. Bailey, 1854–1931) according to their light curves (Fig. 10.6). Further study showed that types a and b are basically the same, so the three original divisions collapse into RRab and RRc stars. The RRab variables brighten quickly and then dim slowly, while the RRc kind vary smoothly and more or less sinusoidally. Light curves, however, are only part of the story. The periods of RRab stars range from 0.3 to 1.2 days (averaging about half a day), while the RRc stars (which are also somewhat hotter) vary with periods between 0.2 and 0.5 days (averaging 0.3 days). Like classical Cepheids, the longer-period variety are fundamental pulsators, while RRc stars pulsate in the shorter-period first overtone mode (consistent with their sinusoidal light curves). Then there are the rare RRd stars, which pulsate in both modes at the same time.

Subtleties abound. As "cluster variables," the RR Lyrae stars divide globular clusters into two groups. In Oosterhoff (after the Dutch astronomer P.T. Oosterhoff) class I clusters, the metal abundances are higher (0.2 or greater) and the mean periods of the RRab Lyraes shorter (0.55 days) than in Oosterhoff class II (metallicity less than 0.2 and an average pulsation period of 0.65 days). Moreover, the class I clusters have far fewer RRc stars. The reasons for the differences are not understood.

The *Blazhko effect* is a long-period variation superimposed on both the amplitudes and the shapes of the light curves that for different stars range between 20 and well over 100 days, and is seen in about 20 percent of the RRab stars. Again, the reason is a mystery, though rotation and tilted magnetic fields have been suggested to be the culprits.

10.4 Near and below the main sequence

The Cepheid instability strip extends downward to the subgiants just above the main sequence, through the main sequence itself, and then on down to the lowly white dwarfs (as seen in Fig. 10.2). In each arena we find pulsating variables. Cepheid-like instability is seen in the more luminous dwarfs and in subgiants as well, in addition to other places on the HR diagram.

10.4.1 Delta Scuti and company

Nearly half the dwarfs, subgiants, and lower-luminosity giants of late A and early F that are hit by the instability strip pulsate with shorter periods of between 0.02 and 0.25 days, consistent with the stars' smaller dimensions. They take the name *Delta Scuti stars* from the fifth-magnitude prototype. The brightest is Beta Cassiopeiae. Amplitudes are small, usually far below the ability of the eye to sense, and typically range from a few thousandths of a magnitude to a few hundredths. Those that extend into the tenths go by the name *AI Velorum stars* and sometimes by the now-obsolete term "dwarf Cepheids."

Equally important, Delta Scuti stars are multi-periodic, and exhibit "non-radial" pulsational behavior [12.3]. Delta Sct itself oscillates with periods of 0.19377 and 0.11636 days, as well as with several other periods, all of which beat against each other to yield yet different periods. Non-radial oscillations require that some parts of the star move outward while others move inward at the same time, in increasingly complex patterns that depend on stellar latitude and longitude. The result is a complex pulsational pattern with so many periods going on simultaneously that they can be quite difficult to decipher. The lack of pulsation in half the candidate stars is attributed to slow rotation, which allows the ionized helium driver to settle to greater depth under the force of gravity (see Box 10.1).

There are several variations on the theme. Near the blue edge of the HR diagrams's Delta Scuti domain we find the *rapidly oscillating* (roAp) *stars*, which chatter multi-periodically by a few millimagnitudes over periods of 5 to 15 minutes. They are all slowly rotating (2 to 12-day rotational periods) magnetic Ap stars in which the rotation and magnetic fields modulate the oscillations [7.3.3]. There is also a Population II version of the Delta Scuti variables called *SX Phoenicis stars*. There should be no dwarfs or subgiants of this spectral class, since the age of the halo and Population II precludes the existence of dwarfs more massive than around 0.8 solar, which is the turnoff point of the Population II giant branch from the dwarfs [13.1]. They are believed to be blue stragglers, higher-mass stars formed by mergers of binaries or through collisions in dense globular clusters [9.4.10].

Finally, we encounter the *Gamma Doradus stars* that lie just to the red side of the Delta Scuti regime and outside the usual instability strip. These multi-periodic dwarfs and subgiants fluctuate by a tenth of a visual magnitude over periods of a half a day to three days. The may be odd "g-mode" oscillators, in which gravity rather than the usual pressure ("p-mode") provides the restoring force in the pulsation.

10.4.2 Toward higher luminosity

Upward along the main sequence lies another zone of non-radial pulsation among early B subgiants and giants. Called *Beta Cephei stars* (sometimes *Beta Canis Majoris stars*), these B0.5 to B2 giants and subgiants vary with periods of 0.1 to 0.3 days, with amplitudes measured in millimagnitudes. The brightest is Spica, a close double star whose chief member has an oscillation period of 0.17 of a day. Beta CMa has periods of 0.2500, 0.2513, and 0.2390 (plus other) days, and as in all multi-periodic stars, the different pulsation periods beat against each other producing additional patterns.

Down the line a bit, among the B3 to B8 subgiants and dwarfs, lies a more obscure set, the *slowly pulsating B stars* (SPB, or sometimes 53 *Persei stars*) with equally small variations and periods of 1 to 3 days. Further down yet, from around B7 to A3 were the purported *Maia variables*, which were supposed to have periods of hours and again low amplitudes, the prototype being one of the bright stars of the Pleiades. Alas, the class was but an artifact of observation, and does not exist.

The driver of the Beta Cephei stars was long mysterious, as these stars fall far outside the classic instability strip. Theoretical understanding of stars involves knowing the opacity of the gases through which the radiation is trying to escape, which in turn involves their ability to be heated and to re-radiate that heat. In a huge but distinctly non-glamorous study of the late 1980s, teams of astronomers re-calculated opacities with modern computers and knowledge of line absorption strengths. The Beta Cephei phenomenon is now known to be driven by zones of ionization of heavy metals, specifically iron. The jargon for the gross chemical compositions of stars is X for hydrogen, Y for helium, and Z for "metals" (everything else). The ionization and driving layer is therefore called the "z-bump." Beta Cephei and Beta Canis Majoris are "z-bump stars."

Finally, move off into the realm of the supergiants. Another of the sky's brightest stars, Deneb, is the prototype for the "Alpha Cygni variables." These multi-periodic class Be to Ae supergiants (not to be confused with Herbig Ae/Be pre-main sequence stars [11.2.4]) pulsate by a tenth of a magnitude or so, and appropriately longer periods of days to weeks, with variations also appearing in the spectra.

10.4.3 Degenerates

The instability strip dives through the main sequence down to where it cuts across the string of DA white dwarfs [6.4.6] between about 11 000 and 13 000 kelvin, akin to dwarf classes B7 to B9 (see Table 10.1 and Fig. 10.2). These *DAV stars* pulsate non-radially with multiple periods of minutes (consistent with tiny size) and amplitudes of at most a couple tenths of a magnitude. (The letter "V" confusingly refers to "variable," not the Roman numeral "V" that signifies "dwarf.") The most famous of the class is ZZ Ceti, which gives the DAV pulsators the popular alternative name *ZZ Ceti stars*. ZZ Ceti itself pulsates with two periods, 3 minutes 33 seconds and 4 minutes 34 seconds, with visual amplitudes that themselves vary from about a thousandth to a hundredth of a magnitude. Being DA white dwarfs, the driver of the ZZ Ceti stars is a layer of partially ionized hydrogen (Box 10.1).

Moving higher along the white dwarf sequence, to around 22 000 kelvin and well to the left of the classic instability strip, we encounter a tight zone in which DB white dwarfs (the *DBV stars*) similarly pulsate. Here the driver is helium ionizing for the second time.

Move upward yet again to the hottest white dwarfs between 75 000 and 140 000K, to the hot *PG1159 stars*, which are DO white dwarfs that are defined by very low hydrogen and concomitantly elevated abundances of helium, carbon, and nitrogen. About half pulsate by a few hundredths of a magnitude over periods of minutes, owing to partial oxygen and carbon ionization. They are known by the generic name of *DOV* stars (where the "O" refers to the DO white dwarfs) and also as *GW Virginis stars*, the variable-star name of PG 1159–035 itself. (Be aware: the variable class is sometimes also referred to as "PG 1159 stars," even though half of the chemically-defined PG 1159s do not vary.) PG 1159 itself oscillates with periods of 7.7 and 9.0 minutes over a range of 0.02 magnitudes. Several of these pulsators are the nuclei of planetary nebulae [13.4.4], illustrating clearly the relation between the planetaries and white dwarfs. Some of the planetary nebula pulsators are also *O VI stars*, which display unique emission lines of five times ionized oxygen radiated by a fast hot wind. The driver of the DOV stars is a layer of oxygen ionization.

10.5 The red giant realm

The set of variable red giants is dominated by the *Long Period Variables* (LPV) or (after the prototype) *Miras*. But there are other kinds with only vaguely defined boundaries, which include the *semi-regular* and *irregular* variables. As in most areas of stellar astronomy, within each of these categories lie subcategories, and within each of these yet others. All, however, are tied together under a uniform theme of stellar evolution, in which they have entered the realm of the giants, many of them losing mass at a furious rate as they prepare to convert themselves into planetary nebulae and white dwarfs [13.4, 13.5].

10.5.1 Miras

Big, red in color, with huge light variations, the Miras (the LPVs), are among the most easily discovered of variable stars (see Fig. 10.1). The typical Mira is a class M giant. Indeed, Mira (M7 IIIe) itself was not only the first of the LPVs to be found, it was the first of the regular (non-exploding) variables *ever* to be found. The initial discovery of Mira – o (Omicron) Ceti – was made by David Fabricius in 1596, before the invention of the telescope, as a star found where none had been recorded before. Four decades later, Mira was discovered to be a periodic variable, its name (from the same Latin root as comes "miracle") implying wonder. Miras – LPVs – dot the sky. While actually rare as stars go, they are luminous, and like the stars at the top of the main sequence can be seen for great distances. Hence their observed number – measured in the thousands – vastly exceeds their true space density. Many are close enough to be seen at maximum light with the naked eye (Table 10.3). The range in magnitude is such, however, that none of the classical Miras can be observed throughout the cycle without a telescope.

Mira variables typically have visual amplitudes of 8 or so magnitudes and periods of roughly a year (the synchronization to a year often making observation of behavior difficult and long-term). The range in both parameters is large, from two magnitudes to over 10, and from 80 to 1000 days. Below 80 days, the stars fall into the semi-regular class. LPV light curves, however, do not repeat as regularly as do those of the Cepheids. Indeed, no two cycles are quite alike. Mira usually hits third magnitude, but has been seen nearly at first (Fig. 10.7). Periods also tend to shift around by 10 percent or so; some Miras can change them quite suddenly, R Hydrae shortening its period by 20 percent over the last century. As to subcategories, "Alpha Miras" have rapidly rising light curves followed by slow declines, "Gammas" reverse the trend, while the "Betas" are in the middle.

Since most Mira variables are late-type M giants that fall into the M IIIe spectroscopic class, they are dominated by absorptions of metallic oxides, particularly TiO [6.2.1]. The "e" is appended because of hydrogen emission lines that arise near maximum light. However, carbon stars (C IIIe) and S stars (S IIIe) can be Miras as well [6.4.3]. While LPVs are dominantly hydrogen, the element's lines do not appear in absorption because of the very low temperatures that hover between 2400 and 3000 kelvin, making Miras the coolest of evolved stars (beaten out for the low-temperature record only

Name	α (2000) h m	δ ° ′	Period (days)	Magnitude (V)	Spectral Class
Mira = *o* Cet	02 19	−02 59	332	2.0–10.0	M5 IIIe–M9 IIIe
R Horologii	02 54	−49 53	404	4.7–14.3	M7 IIIe
R Leporis	05 00	−14 48	432	5.5–11.7	C6 IIe (class N)
U Orionis	05 56	+20 11	372	4.8–12.6	M6.5 IIIe
R Carinae	09 32	−62 47	309	3.9–10.5	M4 IIIe–M8 IIIe
R Leonis	09 48	+11 26	312	4.4–11.3	M8 IIIe
S Carinae	10 09	−61 33	149	4.5–9.9	K5 IIIe–M6 IIIe
R Hydrae	13 30	−23 17	390	4.5–9.5	M7 IIIe
χ Cygni	19 51	+32 55	407	3.3–14.2	S6 IIIe–S10 IIIe
R Aquarii	23 44	−15 17	389	5.8–12.4	M5 IIIe–M8.5 IIIe

Table 10.3 **Prominent Mira variables**

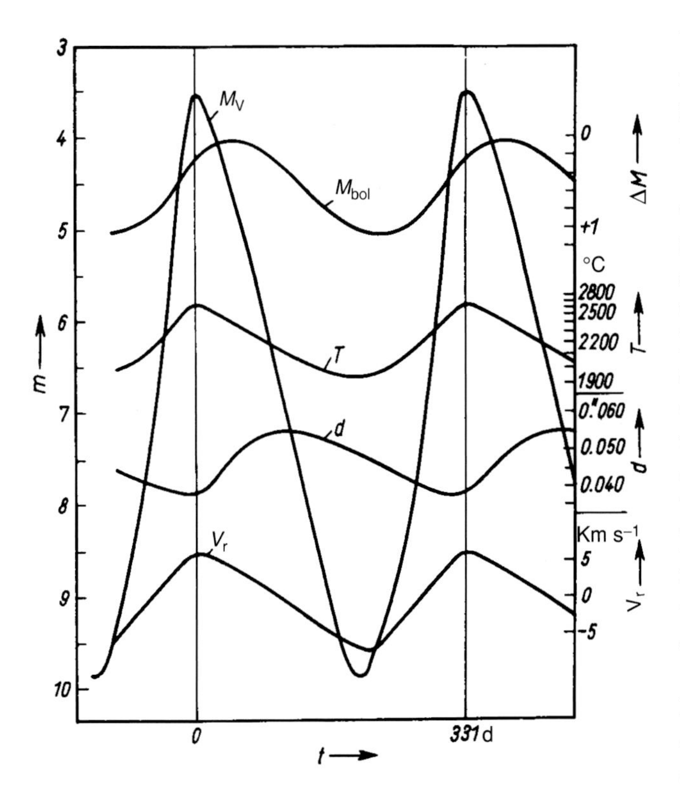

Fig. 10.7 **Variations of Mira** The superposed graphs show the visual and bolometric variations of Mira, as well as the changes in temperature, angular diameter (in seconds of arc), and radial velocity. Note the phase lag in the bolometric variation relative to the visual. (Hoffmeister, C., Richter, G., and Wenzel, W. (1985). *Variable Stars*, Berlin, Springer-Verlag, p. 65, Fig. 27, copyright Springer-Verlag.)

by L and T dwarfs, and an oddball called V838 Mon, which is discussed in Section 10.6.1). With their low temperatures and high bolometric luminosities of 1000 to 10 000 solar [3.3], Miras are also among the largest of stars, with diameters directly measured by interferometers to be in the hundreds of solar radii [7.2.2], with typical radii of over 2AU.

Radius measures, however, are fraught with difficulty. The location of the "surface" of a star depends on the opacity of the gas at the measured wavelength (or on the mean opacity averaged over the observed wavelength band). If you look at the star in a wavelength band dominated by TiO, the gas becomes extremely opaque, you cannot look as deeply into the star, and the star seems to be much larger and cooler. To obtain the "true" photospheric radius, the observer must therefore avoid strong absorption lines. The radius observed at TiO wavelengths can be double the radius to be used for luminosity calculations.

The hydrogen emission lines are odd. In the spectrum of a gaseous nebula [13.4.1], the intensities of the hydrogen lines decrease steadily from Hα toward higher-level lines, Hα nearly three times stronger than Hβ, Hβ double Hγ, and so on (the so-called "Balmer decrement"). In the set of Mira emissions, however, Hδ is dominant, which makes no physical sense. The deviation from normal Balmer behavior is caused by absorption of starlight by the titanium oxide bands, which can occur only if the emissions are produced in deeper, warmer layers that lie below those in which the molecules reign. The emissions are caused by multiple shock waves [Box 11.3] driven outward by the pulsations.

Mira variation is produced by the same mechanism that drives Cepheids (see Box 10.1), one that involves both the ionization of hydrogen and the first ionization of helium. The stars pulsate in radius some 20 to 40 percent. As they change their sizes they also change their temperatures (by several hundred K), surface radial velocities (by 10 or so kms^{-1}), and spectral classes, typically by several subclasses that can drop the stars to M9 (see Fig. 10.7). Maximum visual light takes place near maximum temperature and near minimum diameter, minimum visual light near minimum temperature but intermediate diameter. Maximum visual light also correlates closely with maximum surface approach velocity, minimum light with maximum contraction velocity. It is not clear whether Miras vary in the fundamental mode or in the first overtone; theoreticians continue to argue the issue.

The huge visual amplitudes of the Miras are deceptive. In the infrared, where most of the bolometric stellar flux emerges (because of low temperatures), the amplitudes are only about two magnitudes. The large visual variation is only partly caused by the variation in total (bolometric) luminosity. More important is the variation in temperature. As temperature drops near minimum bolometric light, the bulk of the radiation shifts dramatically into the infrared, where the human eye cannot see it. More important, declining temperature promotes the creation of the TiO molecules that absorb much of the outbound light. At minimum, the molecular opacity in the visual spectrum makes the star appear so large and the observed temperature of this "false molecular photosphere" so cool, that the stars simply become feeble visual radiators. As a result of the temperature effects, the bolometric variations are out of phase with the visual variations, the bolometric maxima taking place about 10 to 20 percent of the phase cycle after the visual maxima, and resulting from changes in both temperature and dimension.

Structurally, all Miras are asymptotic branch giant (AGB) stars near the top of their second ascent of the giant branch, and have quiet carbon–oxygen cores surrounded by helium-burning and hydrogen-burning shells that alternately switch on and off [13.3.1]. Most masses are in the neighborhood of one to two times solar, though they can be notably higher. Pulsation period correlates directly with mass. The pulsational characteristics for any star are given mostly by its mass, not its specific state of evolution.

Collectively, Miras behave oppositely to the light curves of the individuals. On the HR diagram, the cooler stars tend to be the more luminous, causing the distribution of Miras to ascend a bit up and to the right (the relation skewed by the increasing bolometric correction). Somewhat like Cepheids, Miras display a period–luminosity relation. While not as strict as the Cepheid *P–L* relation, the Miras can still be used as distance indicators. Consistently, Miras in the Galactic disk

Box 10.3 Masers and lasers

A *maser* is the microwave version of the *laser*, which stands for "light amplification by the stimulated emission of radiation." There are three ways in which an electron can make a radiative (non-collisional) transition between two energy levels [7.1.4]. Call the lower E_1 and the upper E_2. If in the lower state, the electron can absorb a photon whose energy is exactly equal to the energy difference between the levels ($\Delta E = E_2 - E_1$), and thereby jump to the higher level in the production of an absorption line. If in the upper state, it can spontaneously jump to the lower one in the production of a photon of the same wavelength, one of energy ΔE. Lesser known, but no less important, are *stimulated emissions*. If an electron in the upper state encounters a photon of energy ΔE, the passing photon can stimulate the electron to jump downward to produce a *second* photon of the same energy. The two then fly off in parallel tracks in lockstep (their directions and oscillations the same).

Within a stellar atmosphere, all three actions take place at the same time. Absorption lines are always "filled in" by both spontaneous and stimulated emissions. In a heated gas, the distribution of electrons among energy levels is controlled by the sum of collisions and radiative transitions. In general, when all the upward and downward transitions are in balance, the upper level will have a lower population of electrons (statistically, when the whole assembly of atoms is considered) than the lower level. But if you can "pump" the gas with energy and invert the population, making the upper level more populous than the lower, then stimulated emissions can dominate. One photon of energy ΔE causes a stimulated emission, each of these cause two more, and so on, until a stream of identical photons emerge from the gas to create a laser or maser. You cannot, however, get a free lunch. Energy must be supplied first to invert the levels. That energy is then turned into the laser/maser beam. Enough energy can produce an extremely powerful source of light or radio waves at a specific wavelength.

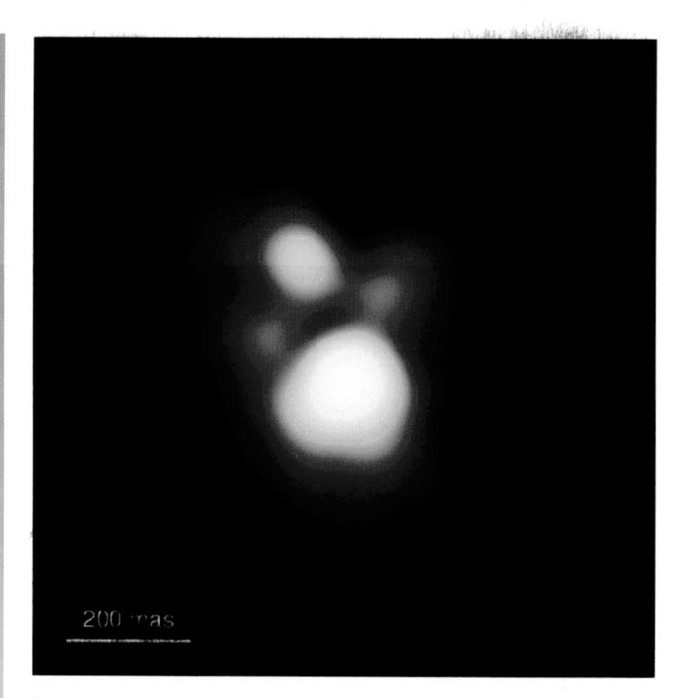

Fig. 10.8 **IRC+10 216** A carbon-rich shroud, rich in molecules, surrounds the advanced Mira carbon star CW Leonis. The scale is in thousandths of a second of arc (mas). (Osterburt, R., Balega, Y., Blöcker, T., Men'shchikov, A., and Weigelt, G. (2000). *Astronomy and Astrophysics*, **357**, p. 169, courtesy of G. Weigelt.)

tend to have longer average periods than do those of the halo, since their progenitor masses can be much higher (halo Miras coming from lower-mass solar-type stars).

10.5.2 Winds, OH/IR stars, and carbon shrouds

As the outer layers of the stars contract during brightening, new outbound pulsations drive the shock waves that produce the emission lines. The shocks lift mass from the stars, which in the cool outer layers partially condenses into dust. Radiation pressure drives the dust outward at around 15kms^{-1}. The dust entrains the gas, and the result is a potent wind [7.4] that can reach a loss rate of $10^{-5} M_{\odot}$ per year. At the extreme, a Mira can bury itself in warm dust that radiates powerfully in the infrared. The grains from oxygen-rich stars are rich in silicates, while those from carbon stars are made of amorphous carbon akin to graphite, as well as silicon carbide (SiC). From oxygen-rich dusty Miras also pours maser emission from the hydroxyl (OH) radical at a radio frequency of 1612 Megahertz, the radiation giving them the name *OH/IR stars* [Box 10.3]. Water and SiO masers are seen as well.

The OH emission comes from an expanding shell roughly 1000AU across. The emission from the front side of the shell is Doppler shifted relative to that from the rear side and, as a result, the maser emission line is split in two, allowing the measure of expansion speed of a few kilometers per second. The masers from the two sides are pumped by infrared radiation from the star, and they therefore vary in synchrony with the stellar variation. Since the front side of the shell (as viewed from Earth) is notably closer to us than the far side, because of the finite speed of light, the forward side will be seen to respond to the star before the rear side. The phase lag in the variation from one side to the other, coupled with the speed of light, gives the shell's diameter. The shells are so large that the closer ones are resolvable from Earth, allowing a measure of the angular diameter, which gives distance and therefore stellar luminosity. Moreover, the physical size divided by the expansion velocity gives the age of at least that section of the shell that produces the maser emission.

Similar extreme carbon-star [6.4.3] Miras produce circumstellar shells in which we find numerous carbon-based, organic molecules. The best known of them is IRC+10 216 (Fig. 10.8). The source is the Mira variable CW Leonis, which has a long 650-day period. Surrounding it is a dense, opaque shroud that hosts at least two dozen molecules, including hydrogen cyanide (HCN), ethylene (C_2H_4), methyl cyanide (CH_3CN), and cyanohexatryine (HC_7N), as well as table

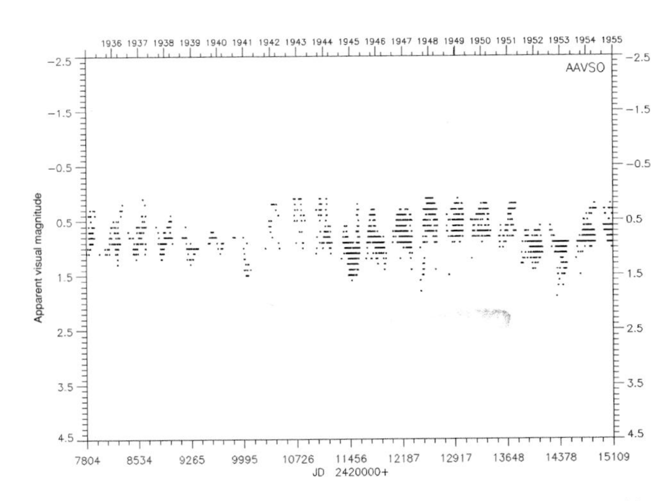

Fig. 10.9 **Variations of Betelgeuse** Betelgeuse is a semi-regular SRc variable that can fluctuate by more than a magnitude. Time on the lower axis is given by the Julian Day number, a running count since January 1, 4713 BCE. (Courtesy of American Association of Variable Star Observers.)

salt and silicon dicarbide (SiCC). Mass is ejected in bi-polar flows – jets – for no reason anyone understands.

10.5.3 Semi-regular and irregular variables

Below the level of Miras is a plethora of other types, some related to Miras, others not. The divisions among them are not always very clearly defined, and a particular star may find its way into more than one. The two broadest categories are the semi-regulars (SR) and irregulars (designated by L, not to be confused with spectroscopic class L). Both have subgroups that include both giants and supergiants.

SR stars divide into three subgroups. SRa variables have amplitudes below two magnitudes and periods that range from about a month to 1200 days, the periods varying by about 10 percent or so (as do those of Miras). SRb stars are periodic for a time, but will switch to an irregular mode, and then go back to being "periodic." The SRa stars turn out to be a rather artificial group that is a mixture of the better-defined SRb class and real Miras that just have small amplitudes. SR variables can also be broken into "blue" (T over 3200K, longer period) and "red" groups that may be in different pulsation modes and evolutionary conditions, the red group behaving more like the Miras.

SRc stars break the mold, as they are not Mira-type AGB [13.3.1] stars at all, but semi-regular, much-higher-mass, generally-red supergiants. The best known is Betelgeuse, which has irregular pulsational variations superimposed on a six-year period, with a typical amplitude of about half a magnitude (Fig. 10.9). Rotation and convective behavior may be involved as well. To these stars add a lesser-used SRd category that includes yellow giants and supergiants, and hypergiants.

Coming off the AGB as class F giants are the semi-regular *UU Herculis* (sometimes 89 *Herculis*) *stars* that vary with periods of a month or two. These will, so it is believed, turn into planetary nebulae and their central stars [13.4], leading us back to the variable worlds of the PG 1159 stars and white dwarfs [10.4.3].

Irregular variables erratically change their brightnesses with no discernable period. Lb stars are giants with amplitudes of a few tenths of a magnitude, while Lc variables are supergiants that meander by a magnitude or so. Pre-main-sequence (T Tauri and related) stars are also irregular variables, and in eruptive form become FU Orionis stars [11.2.2].

10.6 Eruptive variables

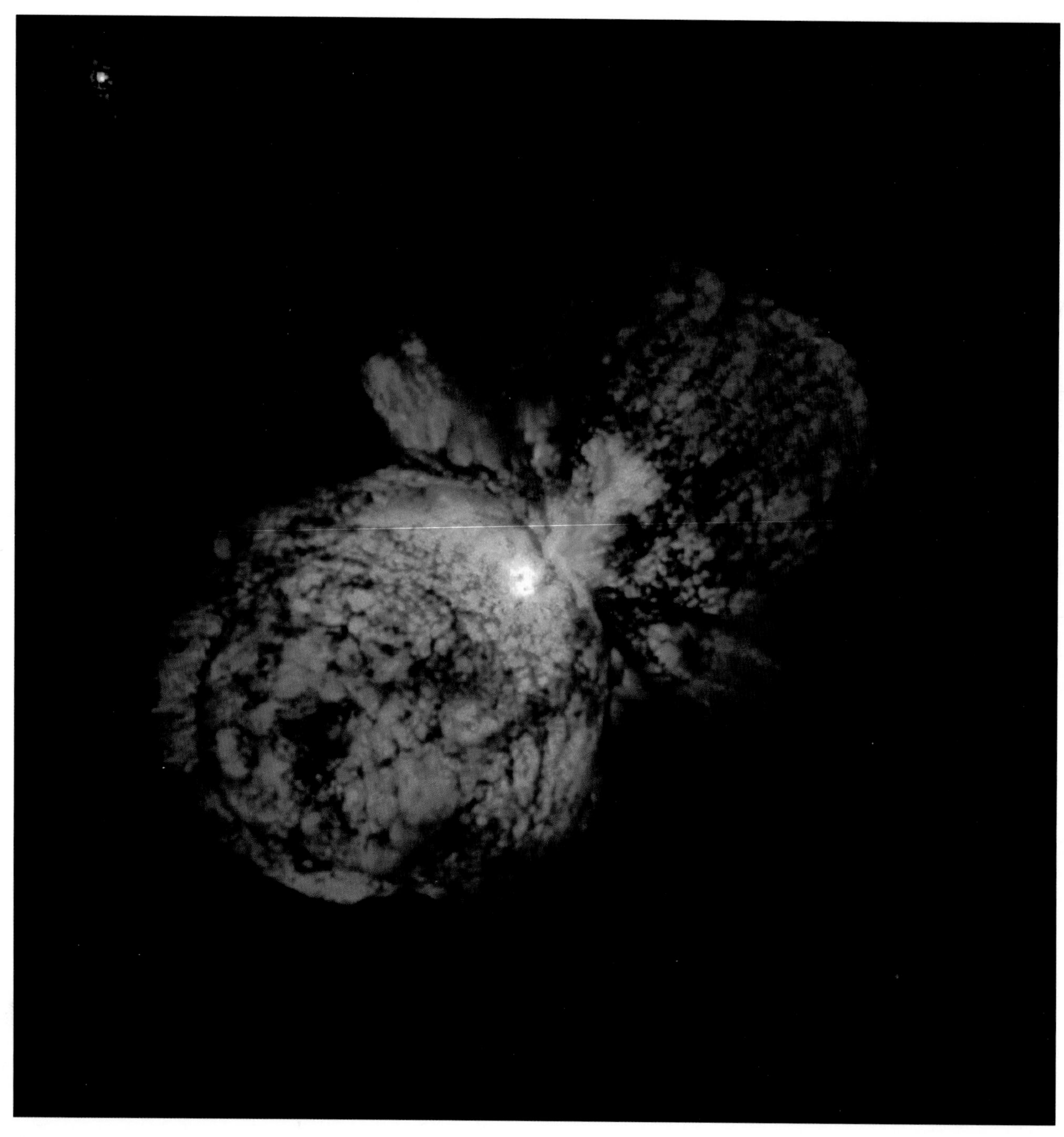

Fig. 10.10 **Eta Carinae** The Hubble Space Telescope shows a huge cloud of dusty gas – the "homunculus" – surrounding the luminous blue supergiant Eta Carinae. (Courtesy of Jon Morse, University of Colorado, and NASA.)

"Eruptive variable" is an informal catchall term for single stars whose misbehavior is related to some kind of non-explosive mass-losing event. Some brighten, while others dim as a result of dust formed within the ejecta.

10.6.1 Luminous blue variables and related supergiants

Rare and erratic, *luminous blue variables* (*LBVs*) are in the top rank of stellar brightness. They are epitomized by Eta Carinae, a B0 hypergiant that is enmeshed within the Carina Nebula of the southern hemisphere (Fig. 10.10). Radiating at a rate of some five million Suns, Eta Car is among the most luminous and massive stars in the Galaxy. Around 1846, in conjunction with a huge eruption of mass (roughly equal to that of the Sun), it rivalled Canopus and Sirius, even though 2500pc away (Fig. 10.11). The ejecta, some of which condensed to dust, then hid the star visually, and it faded away to 8th magnitude. In 1890 it underwent a smaller ejection, while also brightening a bit. It now hovers near the edge of naked-eye vision. The ejecta are seen as a dramatic, expanding cloud of dusty gas. A 5.6-year spectral variation suggests that Eta Car is a binary whose combined masses total nearly 150 Suns. Mass-losing P Cygni of "P Cygni line" fame [7.4] is also an LBV, one that erupted in the year 1600.

Such stars do not change their luminosities so much as their bolometric corrections [3.3], and to that extent have something in common with shrouded Mira stars. The surrounding dust absorbs visual and ultraviolet radiation and turns it to infrared. LBVs are apparently attempting to evolve to the red supergiant realm, but are running into the empirical Humphreys–Davidson limit [14.1.2], where luminosity-driven mass loss prevents redward excursion.

Somewhat related are the variable "yellow hypergiants" that seem to be evolving blueward after having been red supergiants and that are included in the SRd class. A fine example is Rho Cas. The star is normally something of a semi-regular (SRc) variable with multiple "periods" ranging from 350 to 820 days. In 1946, it underwent a one-magnitude dip for about a year when it ejected a cloud of dusty gas that created a "false photosphere," and changed the star's class from F to M. Rho Cas's bolometric luminosity did not change much. Instead, the energy output shifted out of the visual and into the infrared, that is, the bolometric correction increased.

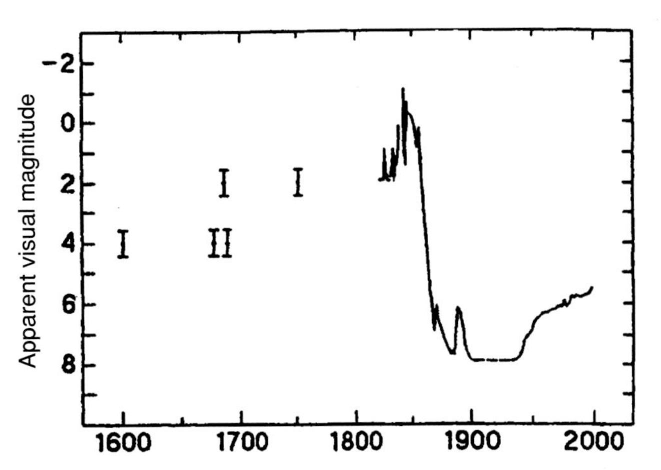

Fig. 10.11 **Eta Car light curve** Eta Carinae violently erupted in the middle of the eighteenth century, when it ejected the cloud seen in Fig. 10.10. It then quickly faded in the visual spectrum. Modern observations show a modest brightening. (Courtesy of K. Davidson.)

Nearly all large giants and supergiants are variables; their huge sizes and luminosities simply render them unstable against pulsation, whether semi-regular or completely erratic, much of it yet to be understood. Among the more mysterious are stars like V838 Monocerotis, a 12th-magnitude mid temperature star that in 2002 suddenly erupted to 6th, while turning itself first into a late M supergiant, and then of all things into one of class L with fierce water absorptions and a temperature under 2300K (and maybe as low as 1300K), making it the only known non-dwarf L star. The eruption is perhaps best known for a reflective "light echo" that dramatically illuminated a huge surrounding dust shell (Fig. 10.12). The star has since faded away.

10.6.2 R Coronae Borealis stars

While pulsating rather like Cepheids with periods of days, *R Coronae Borealis stars* are helium-rich carbon giants [6.4.3, 13.4.4] that suddenly "disappear," dropping in visual light by many magnitudes. Normally fifth magnitude, once every couple years R CrB will dim to as faint as magnitude 14, sometimes recovering quickly, other times taking months (Fig. 10.13). About 40 are known, many in the Magellanic Clouds [Fig. 4.11], from which we find luminosities of about 10 000 solar. The dimming comes from irregular clouds of carbon dust ejected by the stars, perhaps as a result of the pulsations. If a cloud is ejected toward the Earth, the R CrB star dims; if not, we notice nothing. During the dimming, chromospheric emission lines [12.1.3] are observed, showing that the dust forms close to the star. As a result of multiple ejections, R CrB and others are surrounded by extended clouds of warm dust. They may be related to the helium-carbon-rich *PV Telescopii* (PV Tel) class Bp supergiants that pulsate slightly with shorter periods of under a day as well as with year-long periods, but that do not undergo the dramatic fadings.

10.6.3 Be stars

B-emission (Be) *stars* are rapidly rotating stars of class B that are encircled by equatorial disks that radiate hydrogen emission lines. They constitute some 20 percent of B stars. The best known are Gamma Cas and Zeta Tauri. The spectrum of a Be star depends on the thickness of its disk and on the disk's inclination to the line of sight. If the disk, which is about 20 times larger than the star, is viewed more or less pole-on, the emissions are single. If the disk is tilted more into the line of sight, the lines become split by the Doppler effect. The component from the receding portion of the disk is shifted to the red of center, that from the approaching portion to the blue. If the disk lies close enough to the line of sight, metallic absorptions from the disk are superimposed on the stellar spectrum, and the Be star is called a *shell star*.

Fig. 10.12 **V 838 Monocerotis** A huge dusty circumstellar shell caused by earlier generations of mass loss is illuminated by light reflected from the wildly erupting star within. (Courtesy of NASA, ESA, and H.E. Bond (STScI).)

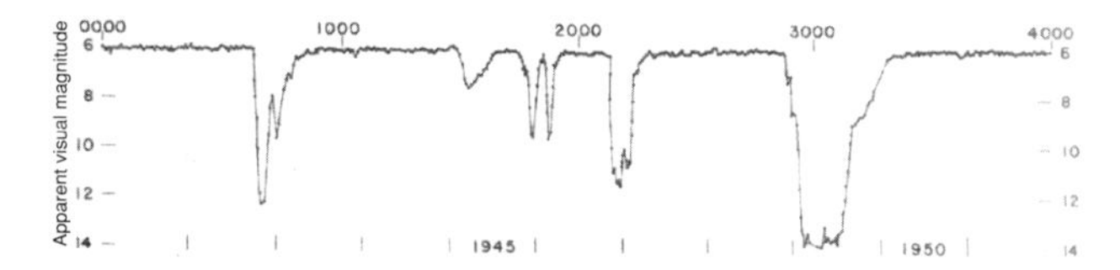

Fig. 10.13 **Variations of R Coronae Borealis** R CrB takes sudden dives as a result of erratic dust ejection. The left-hand axis gives the apparent visual magnitude. (Courtesy of American Association of Variable Star Observers.)

The intensity ratio of the blue and red components depends on the distribution of matter in the disk and changes markedly with time (commonly reversing). The spectra change as the disks empty and are refilled. A non-Be star can become one, and vice versa, as mass somehow erupts from the stellar photosphere. Delta Scorpii turned itself into a Be star in the year 2000, while at the same time brightening from its usual magnitude of 2.3 almost to first magnitude. Though always seen as a Be star, Gamma Cas similarly erupted in 1937.

The rapid rotations at first suggested that Be stars spin so fast that they are close to breaking apart, and are flinging mass from their equators to form the disks. However, the rotation speeds, even near 400 kms^{-1}, are by themselves insufficient. Non-radial pulsations or magnetic fields aided by rotation may be responsible, but no one knows.

10.7 Spots and flares

Turn now to the dim stars of the lower main sequence. The outer convection zones give these stars magnetic fields, and where there are fields there are chromospheres, coronae, and magnetic activity [12.1.2]. The fields produce starspots that cause periodic variations in the stellar luminosities as the stars rotate and the spots go in and out of view.

More dramatic are the sudden brightenings called *flares* (Fig. 10.14) that are analogous to the localized flares seen on the surface of the Sun [12.1.4], and are caused by the connection and collapse of the magnetic field lines. In dwarfs of classes K, M, and even L, flaring activity reaches such an extreme – 100 times the energetics of solar flares – that it seems to involve the whole star. A typical *flare star* is an early M dwarf that unpredictably and suddenly increases its visual brightness by a magnitude or so. The flaring can be twice the strength in the ultraviolet, and is seen across the board from the X-ray to the radio. At short wavelengths, the flares dominate the ambient radiation from the star. Intervals between flares range from minutes to days, while individual events last for several minutes. The flares are amazingly varied, even those from the same star that are close together in time. Some are near-instantaneous, followed by a few-minute fall, while the growth of others is much more gradual (and more like those seen on the Sun).

Flaring activity is clearly tied to rotation. In young clusters in which magnetic braking has not yet had much effect [13.2.1], late K dwarfs are seen to flare. As we go down the main sequence, flaring becomes more common. Between M4 and M8, a majority of stars show emission lines that indicate magnetic chromospheric heating and the potential for flaring. However, the magnetic field of a solar (or later-type) dwarf is believed to be anchored at the boundary layer between the outer convective and inner radiative envelopes that surround the core [12.4.4]. From M5 on down, stars are fully convective, which is expected to quash the magnetic fields. Yet flaring activity occurs all the way down through class M and even into early class L, where we find brown dwarfs [12.4.1, 13.1]. No one yet knows why.

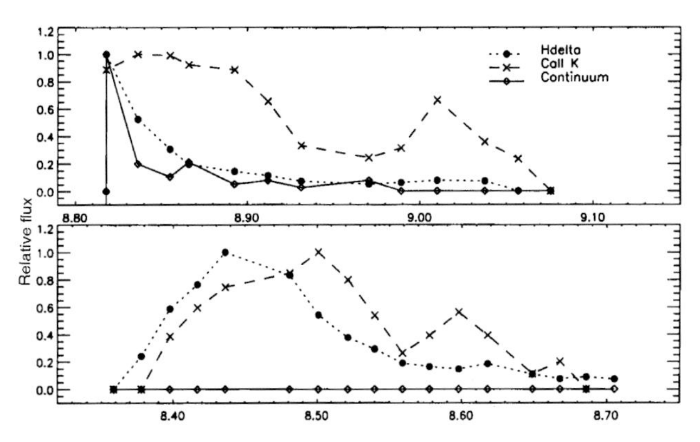

Fig. 10.14 **Stellar flares** Proxima Centauri produced two very different kinds of flares within half an hour of each other. At the top, the rise in the Hδ and Ca II K-line emissions [6.1.1], as well as in the continuum, are very sudden, while in the earlier flare (bottom), they are not only much more gradual, but there was no white-light (continuum) flare at all. The time scale is in hours. (Houdebine, E.R. (2003). *Astronomy and Astrophysics*, **319**, 1019.)

Even giants and supergiants have been known to flare, though such events are extremely rare and many are contended. Notable for such activity are Epsilon Pegasi (K2 Ib), which was observed to pop a one-magnitude flare that lasted for several minutes, Beta Eridani, and Mu Velorum. The origins of these events are a complete mystery. Data are sparse and no theory explains them.

Is any star completely "stable"? Given sufficiently refined observations, all stars vary to some degree, even our Sun, as active zones and spots go in and out of view, as flares slightly increase the luminosity, as the solar magnetic cycle changes the solar flux over not just the 11-year activity period, but over decades and centuries. Even solar oscillations [12.3] very slightly alter the luminosity over timescales of seconds and minutes. Nothing is ever perfectly stable, nothing ever perfectly still.

CHAPTER ELEVEN

Star formation

Once the varieties of stars are established and the measures of their natures are in place, we can begin to understand what they all have to do with one another. For in fact, one kind transforms itself into a different kind, sometimes with great drama. Stellar evolution – the study of the lives of the stars – begins with star birth and the processes that lead up to it.

Stars run off a finite fuel supply that begins with the hydrogen in their hot, nuclear-fusing cores. No star can therefore last forever. If stars are always dying, they must also continuously be born, else there would be few stars left for us to see. The birth process at first appears mysterious, as the breeding grounds are not immediately evident. A glance at the Milky Way, however, shows vast voids in which no stars appear. Since stellar motions would quickly fill them in, the voids must be physical entities: dark clouds. O and B associations [9.3], which must be very young (both from their expansion velocities and from evolutionary considerations of their massive stars), are clearly related to the dark clouds and the bright clouds that they illuminate. Since star birth seems mostly hidden from optical view, a logical surmise is that stars are born within the dark clouds, within the matter of interstellar space. The evidence supporting that simple, and old, hypothesis is profound.

11.1 The interstellar medium

No one has yet made perfect sense of the awesome complexity of the stuff of interstellar space, of the *interstellar medium* (the ISM, Fig. 11.1). Stars are easier to understand, as each is a unit unto itself. The ISM, however, is far from unitary, and instead consists of a profusion of shredded clouds and filaments set within a variety of uneven continua, all of different temperatures, densities, ionization states, and chemical mixes, both condensing and being ripped to pieces by the forces exerted by gravity, by internal pressures, and by stellar winds and explosions. The only thing that unifies it is that, except for some thin clouds within the halo, the medium lies almost entirely within the disk of the Galaxy. Strictly Population I, it thereby occupies the Milky Way [Fig. 1.1], most of it compressed to within 100pc of the disk's midplane. Composing some 10 percent or more of the disk's mass, the medium is not only the birthplace of the stars, it is also the repository of the stars' leavings [10.5.2, 13.3.2, 13.4, 14.2.6], and can be thought of as a "way station" for chemically enriched stellar ejecta as they wait their inclusion into new stellar generations. Stars, their births and deaths, cannot therefore be understood without reference to the ISM.

In any discussion of the interstellar medium, it is vital to remember that the ISM consists of both *gas* and *dust*, the latter the generic term for very finely divided solid silicate or carbon grains that are typically a micron or so in diameter and that constitute about one percent of the ISM's mass. Which aspect manifests itself observationally – the gas, dust, or both – depends on local conditions. The gas has the overall composition of a star (though depleted in some metals), and is 90 percent hydrogen. It can be in any of three states, neutral atomic, ionized, or molecular, depending on the local flux of radiation sufficient to cause ionization or molecular dissociation. The molecular gas can become very complex. Some molecules can apparently grow to the size at which they behave like small grains, tying all the aspects and states of matter together.

Box 11.1 Interstellar chemistry

How can chemical reactions, which to most minds are heat-driven, produce the molecules found within cold, dark molecular clouds? Energies are low, and the reactions slow, but we also have plenty of time. Two important processes reveal much about the synergy that takes place between the gas and dust in the ISM, and between these and the stars. There is insufficient energy in the gas to allow two hydrogen atoms to bang together hard enough to overcome their mutual electrical barriers and make H_2. A hydrogen ion could meet a neutral and join up, but there are no energetic stellar photons available for ionization. Hydrogen ions can be produced by speeding cosmic rays [14.3.5], but there are nowhere near enough of them to make the amounts of H_2 that constitute the clouds.

To produce hydrogen molecules in the amounts we see requires dust grains. Hydrogen atoms first stick to the grains where the individuals meet up, and – aided by the electrical forces on the grain surfaces – marry. The heat of formation, or a hit by a photon, then kicks the new H_2 molecule away, back into the gas. Cosmic-ray-driven ion chemistry can now begin, in addition to continued grain surface reactions, to make heavier molecules.

The dust, evolving within the interstellar cloud is, in its original form, the ejecta of giants and supergiants. Cosmic rays come from supernova explosions. The clouds, by virtue of their low temperatures, are also the birthplaces of the stars. The molecules – or similar processes in circumstellar disks – may grow into the seeds of life itself. Everything seems to depend on everything else.

11.1.1 Dark nebulae

The most obvious features of the interstellar medium, their effects easily "visible" to the naked eye, are dark interstellar dust clouds (Fig. 11.1) first carefully catalogued by E.E. Barnard (1857–1923) of Yerkes Observatory, many carrying his name. Unilluminated, the clouds appear in relief, blocking the light of background stars, the effect vividly seen in Fig. A5.1 in Appendix 5 where the Taurus–Auriga dark cloud is outlined by distant clusters and variable stars. The clouds range in size from the smallest of *Bok globules* (after Bart Bok, 1906–1983) that contain no more than a few solar masses and are a parsec or so across, to huge lumpy complexes that can contain 100 000 solar masses or more. In total, the dark clouds constitute 25 to 50 percent of the ISM (the remainder in cool neutral clouds and a complex intercloud medium). A view of the Milky Way on a dark night shows it divided in two, split by the blending of the nearby dark clouds (see Fig. 1.1).

The dust absorbs starlight that would otherwise heat the gas, making the gas naturally cold, temperatures falling to only a few degrees kelvin. At such conditions, the atoms of the gas combine into molecules, the dark clouds thus becoming *molecular clouds*, the biggest ones *giant molecular clouds* or GMCs. While the dominant molecule is molecular hydrogen (H_2), over 100 others have been discovered via their low-energy radio radiation, including carbon monoxide (a prime tracer for hard-to-observe H_2), formaldehyde, alcohols, various hydrocarbons, and long-chain carbon molecules. The densities of the clouds are so low – 10^5 to 10^6 atoms per cubic centimeter at most – that even a variety of unstable radicals like hydroxyl (OH), CN, and CH can survive intact (Box 11.1). There is no known limit to molecular complexity.

Fig. 11.1 **The interstellar medium** Masses of stars in the Milky Way in Sagittarius intermingle with the interstellar medium. Pervading the picture are sheets of warm, reddish nebulae ionized by O stars. Stars and bright nebulae provide backlighting for dark dust clouds that range from tiny Bok globules to vast complexes that mark the sites of great molecular clouds. (Courtesy of Matt BenDaniel.)

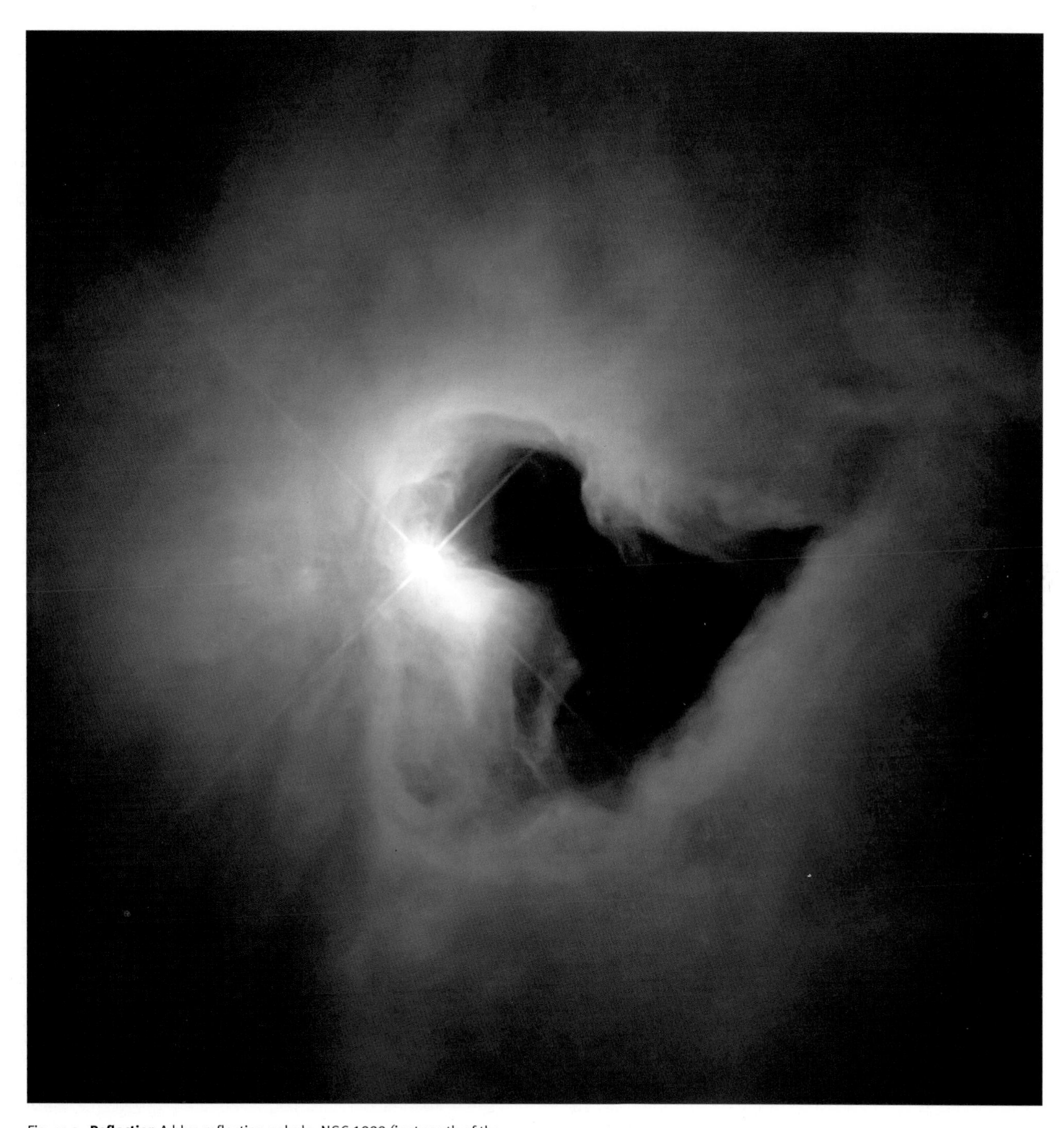

Fig. 11.2 **Reflection** A blue reflection nebula, NGC 1999 (just south of the Orion Nebula: see Figs. 8.21 and 11.4), is formed by starlight scattered from the fringes of the associated dark globule. (Courtesy of NASA and the Hubble Heritage Team (STScI).)

Researchers have uncovered linked benzene rings, known as "polycyclic aromatic hydrocarbons," and "fullerenes" (ball-like structures of carbon atoms), and are searching for heavy molecules that may form the basis of life.

The dust grains, which aid in cloud chemistry, are affected by their environments. Ejected largely by evolving stars (silicates from oxygen-rich stars, graphite-like carbon from carbon stars [6.4.3, 13.3.2]), they absorb molecules and free metal atoms from the gas, leaving the gas metal-poor. These are the same grains that find their way into the disks that surround new stars and that, in turn, create the planets.

11.1.2 Reflection nebulae

The Pleiades cluster (see Fig. 1.2) is currently passing through a low-density interstellar cloud, one not thick enough to block much starlight, and is thereby enmeshed in a blue haze. The grains scatter stellar radiation with an efficiency in inverse proportion to wavelength, one of the reasons for the interstellar absorption and reddening of starlight [3.5.1]. Since blue and violet light are preferentially scattered from the grains, and since the stars are rather blue to start with, the cloud (at least most such clouds) turns into a blue *reflection nebula* whose spectrum is a reproduction of that of the illuminating stars.

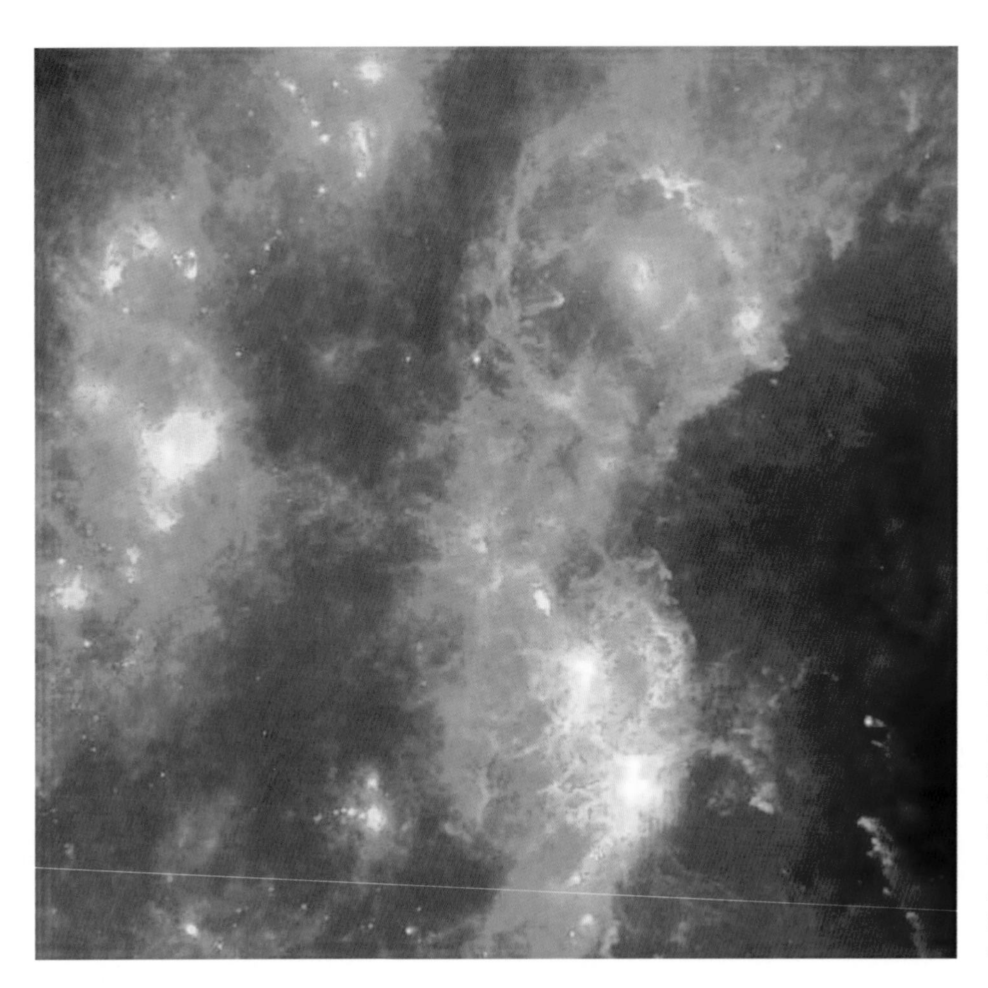

Fig. 11.3 **Dust emission** What appears to be dark to the eye glows brilliantly in the infrared, compliments of heated dust. The Orion Belt and Nebula complex are near the bottom, while a huge ring (created by stellar winds and supernovae) surrounds Lambda Orionis at the top. Infrared "cirrus" shines throughout the whole picture. (Courtesy of Infrared Processing and Analysis Center, Caltech and JPL. IPAC is NASA's Infrared Astrophysics Data Center.)

Reflection nebulae abound at the edges of molecular clouds where there are luminous class B stars to create them (Fig. 11.2), giving rise to R associations [9.3.1]. Just look at any color photo of the Milky Way for the blue light.

Dark dust clouds and reflection nebulae do glow on their own. Heated to some degree by the absorption starlight (the other reason for interstellar absorption), the dust grains cool by emitting infrared radiation whose quality depends on temperature. Such infrared-glowing clouds fill the Milky Way (Fig. 11.3).

11.1.3 Diffuse nebulae

Second only to the dark clouds as obvious manifestations of the interstellar medium are the *diffuse nebulae*, epitomized by the magnificent Orion Nebula at the heart of Orion's Sword (Fig. 11.4). Seeming to surround the Trapezium [8.6; Fig. 8.21], this shining cloud is so large, some 8pc across (depending on where you want to draw the boundaries), that even though 450pc away it still appears a degree or so wide in our sky. Great numbers of similar bright clouds of differing sizes lie within the plane of the Galaxy, and therefore from our point of view within the Milky Way, confined to the same thin plane as the dark clouds. The gas in a reflection nebula is practically invisible; all the action is produced by the dust. Diffuse nebulae reverse the roles, their glow coming from emission line spectra radiated by ionized gas (Fig. 11.5). Because the dominant hydrogen is nearly fully ionized, diffuse nebulae are also commonly referred to as "H II regions," as opposed to neutral "H I" clouds.

The electron of a hydrogen atom in the ground state is bound to the proton with an electrical energy of 13.6 electron volts (eV) [7.1.1]. If the bound electron can absorb at least that much energy by absorption of a photon, it can be ripped away, and the atom ionized. Electrons recaptured on upper energy levels skip downward until they get to the ground state, in the process radiating hydrogen emission lines. Such *recombination lines* are seen from helium, oxygen, neon, and other elements. Collisions between energetic free electrons and heavier ions create the so-called "forbidden lines" that are so prominent in nebular spectra [13.4.1].

The ionization energy of hydrogen corresponds to a photon with a far ultraviolet wavelength of 912Å (the Lyman limit), which is at the high-energy end of the Lyman series of hydrogen lines, those that connect directly to the ground state [7.1.4]. The hottest stars of the Pleiades (class B6 Electra, Merope, and Taygeta) have temperatures of about 14 000 kelvin, which is not hot enough to radiate ionizing radiation. Not until a star hits 25 000 kelvin does the blackbody curve (or the stellar equivalent) sneak past the limit, which corresponds to spectral class B1. From there and earlier, the stars make diffuse nebulae, later than that, reflection nebulae. The Orion Nebula is lit largely by the O6 (40 000K) dwarf Theta-1 Ori C.

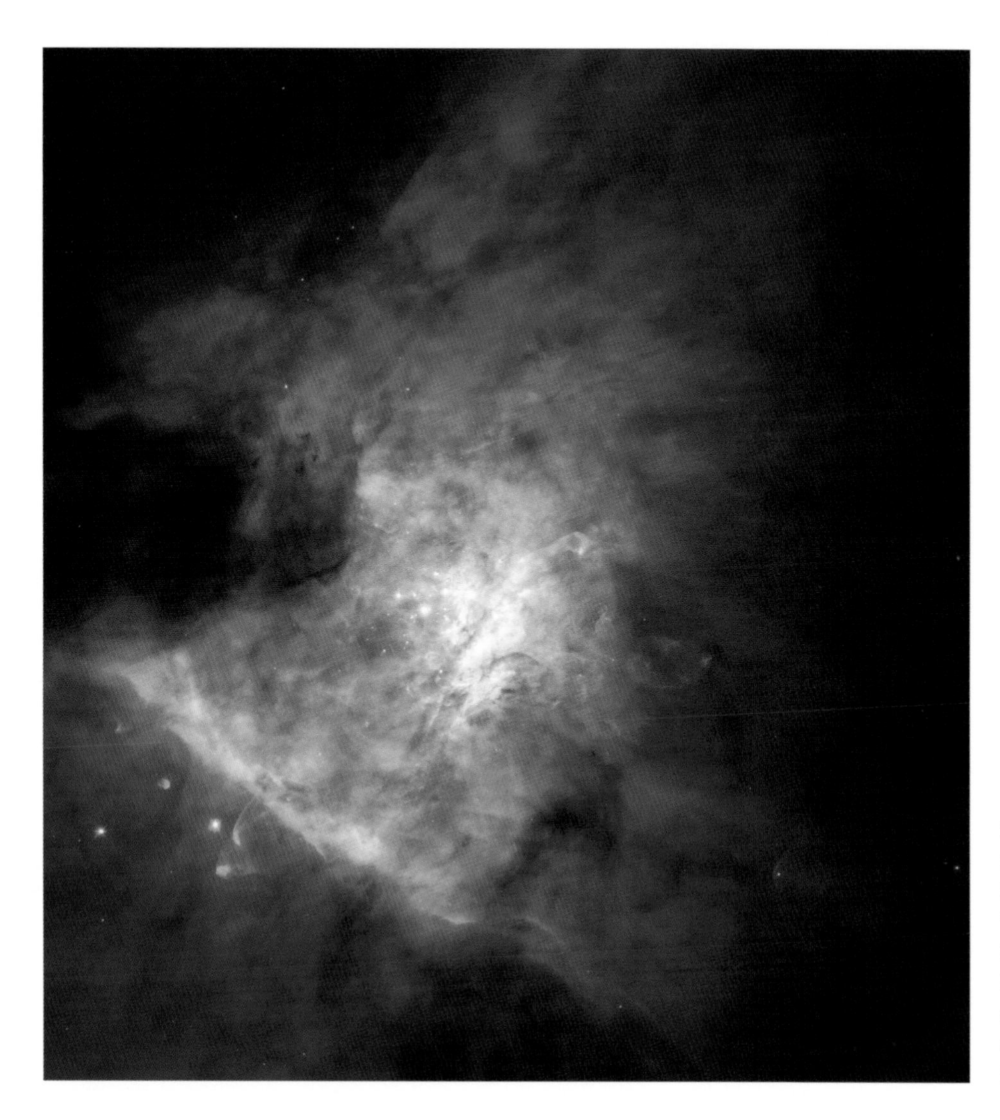

Fig. 11.4 **The Orion Nebula** The core of the Orion Nebula, seen by Hubble, lies in back of the illuminating Trapezium, the extended nebula a blister on the face of the dark Orion Molecular Cloud. Compare with Fig. 8.21. Foreground dust clouds add spice. (Courtesy of C.R. O'Dell and S.K. Wong (Rice University), NASA.)

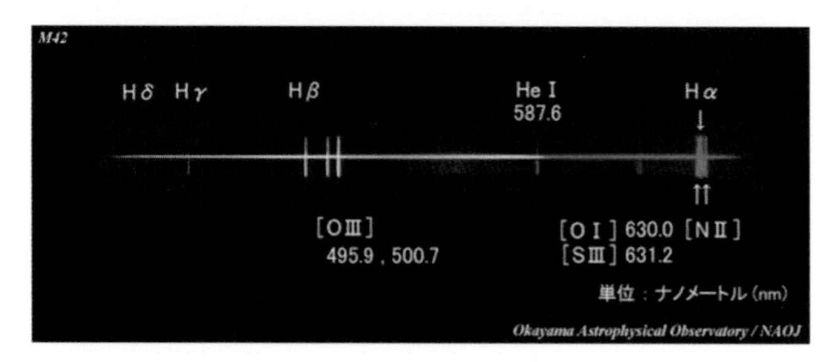

Fig. 11.5 **Orion spectrum** Emission lines fill the spectrum of the Orion Nebula. Square brackets indicate forbidden lines [13.4.1]. Wavelengths are in nanometers. The continuous spectrum of one of the Trapezium stars runs down the center. (Courtesy of Okayama Astrophysical Observatory/NAOJ.)

Diffuse nebulae therefore belong to the realm of the rare massive O stars, into which we toss classes B0 and B1. Class B1 roughly corresponds to a mass of 10 solar masses, which is near the dividing line between stars that produce giants and white dwarfs, and those that produce supergiants and supernovae [13.1]. To a good approximation, the diffuse nebulae tell us immediately where we can find stars that are going to blow up.

Since diffuse nebulae are ionized regions of ambient interstellar matter, they must associate with dusty dark clouds that range from globules to GMCs. Diffuse nebulae are surrounded by dark protrusions, and provide excellent backlighting for foreground clouds. Conversely, many diffuse nebulae – foremost the Orion Nebula – are blisters raised on the Earth-facing sides of background molecular clouds that are so huge as to be unnoticeable to the eye. Other diffuse nebulae are so buried within their molecular clouds that we know of them only by their radio radiation.

Nebulae of any kind – diffuse, planetary [13.4], supernova [14.3.4] – are not blackbodies, so temperature does not give bulk radiative properties: the Wien and Stefan–Boltzmann laws do not apply, any more than they do to the solar corona [7.1.2]. Nebular parameters are derived from the strengths of the emission lines and theories of line formation. "Temperature" for such gas clouds refers only to the velocities of particles in the gas, specifically to the electrons, and is therefore properly called the "electron temperature," T_e, that appropriate to the Maxwellian velocity distribution [Box 7.3]. Typical values of T_e range from 8000 to 10 000K, while densities run from under 10 atoms per cubic centimeter for barely visible sheets of gas, to over 10 000 for objects like the Orion

Box 11.2 The 21-centimeter line

Neutral H I clouds radiate the powerful 21-cm line. Electrons and protons have a property called "spin," a quantum version of classical rotation. Since spin involves "movement" of an electrically charged particle, it generates a magnetic field. In a neutral hydrogen atom, the magnetic fields of the proton and orbiting electron link through their magnetic fields. "Spin" is quantized: the electron's spin can be aligned either parallel to that of the proton or oppositely. If the first case, the total energy of the ground state is slightly higher, in the second, slightly lower; that is, the ground state of hydrogen is divided in two (a phenomenon called "hyperfine splitting"). Collisions between neutral hydrogen atoms can raise an electron from the bottom state into the higher-energy "parallel" state. Given enough time, the electron will spontaneously reverse its spin and drop back to the anti-parallel state with the radiation of a photon at 21cm.

Nebula (which is still quite vacuous). Chemical compositions are roughly solar. Dimensions range from a tenth of a parsec for dense, buried *ultracompact H II regions*, to well over 100pc for *giant H II regions* like 30 Doradus (the Tarantula Nebula) in the Large Magellanic Cloud, which is lit by a whole dense cluster of early O stars [Fig. 4.11]. Beyond these, we find much of the Galactic plane to be aglow with vast billows of radiating gas that have no specific names, as in Fig. 11.1.

11.1.4 Neutral clouds

Cool neutral clouds litter the Galactic plane. Not lit by O stars, nor so dense that they block the light of the background, they escape ready detection. Nevertheless, up to half the ISM is tied up in them. They are detectable optically when they superimpose the spectra of their metals onto those of background stars. More telling is that, depending on the background, they either radiate or absorb the 21-cm line of neutral hydrogen, from which we find that they have typical temperatures of 100 kelvin or so and densities around 10–100 atoms per cubic centimeter (Box 11.2). Along any line of sight there is about one cloud every 100pc. It is from such 21-cm emitters that we can map the rotation and structure of the Galaxy [5.3.3].

Calling them "clouds" is a stretch, as they are more like shredded filaments of extraordinary complexity. They are set into a warm neutral medium of 8000K and a density under 1 atom per cubic centimeter that also radiates at 21 cm, a warm ionized medium (that relates to extended diffuse nebulosity), and an ultra-low-density hot ionized medium with a temperature over 100 000K that is produced by the passage of shock waves (Box 11.3) sent out by supernova explosions [14.3.4].

Here, in all its complexity, is the medium that fosters the births of stars, stars that begin their lives as no more than density fluctuations within dark interstellar clouds.

Box 11.3 Sound and shock

Common wisdom has it that "there is no sound in space." Untrue. There most certainly is. Sound is a natural transmitter of energy via compression waves through a medium, and nowhere in "space" is it empty. Thin as it is, the interstellar medium will transmit sound just as much as will the air we breathe. No matter that you could not hear it; the sound is still there.

The speed of sound (v_s) in a medium depends primarily on the square root of the temperature (T) divided by the "mean molecular weight" (μ), the average mass of the mixture of the atomic and/or molecular particles relative to the particle mass of hydrogen, or

$$v_s\ (\mathrm{km\,s^{-1}}) = \sqrt{(0.0083 \times C\ T/\mu)}.$$

The constant C depends on the ability of the medium to absorb heat, and for simplicity can be taken as 1. For air near sea level, T = 293K, and for a composition 80 percent molecular nitrogen and 20 percent oxygen, $\mu = 29$, giving $v_s = 0.29\ \mathrm{km\,s^{-1}} = 1040\mathrm{kmhr^{-1}}$ (645 miles per hour), close to the familiar measured value. In a diffuse nebula that consists of 90 percent atomic hydrogen and 10 percent helium, μ is down to 1.3, and T is up to 10 000K, yielding a much higher value of $v_s = 10\mathrm{km\,s^{-1}}$. In a cold molecular cloud, v_s is a creepingly slow 0.2cm $\mathrm{s^{-1}}$.

Move a body through a medium at a speed greater than that of its natural speed of sound, and the atoms and/or molecules within it cannot get out of the way fast enough. As a result, they pile up to produce a wall of moving particles, an "overpressure" called a *shock wave* that can smack hard against a barrier. The most familiar example is the steep wall of water that develops off the bow of a boat moving faster than the natural wave motion (though here we deal with transverse water waves rather than compression waves). Such shocks are therefore called "bow shocks." Airplanes flying overhead faster than sound produce the same thing, and a devastating "thud" – a *sonic boom* – when the shock hits you. (The boom is not produced by the aircraft "breaking" the sound barrier, but is continuously produced off the bow of the craft as long as it is flying supersonically.) The crack of a whip, the crack of a rifle bullet, are caused by shock waves induced by supersonic motion.

Shocks are everywhere in astronomy. The solar wind forms a bow shock when it encounters the magnetic fields of the Earth and other planets. Winds from binary stars hit each other and create shocks, as do supernovae within the ambient medium. As a shock passes through a medium, it deposits energy and heats the fluid left in its wake. The result is the 100 000K hot ionized medium of interstellar space and the more modestly heated gas that makes the Herbig–Haro objects as jets from T Tauri stars plow forward at speeds greater than sound.

11.2 Evidence for star birth

The first step in the comprehension of star birth is to look at our Sun and Solar System to see if they can provide us with clues from the distant past as to how we all came together. Then we need find true stellar youth, to see what the various stages of star birth look like. Finally, we let the theoreticians loose to explain it all. The stars tell us not only how they came to be, but return us back to the Earth and our Sun to see how we came to be as well.

11.2.1 The Solar System

The central body of the Solar System is an ordinary class G2 main sequence dwarf that fuses hydrogen into helium in its core [12.1]. Orbiting around it are various sets of satellites, chief of which are the 8, 9, or 10 bodies (depending on your point of view) called *planets* (Fig. 11.6). The planets fall into four distinct categories. The inner four – Mercury, Venus, Earth, and Mars – are similar small balls of rock that surround iron cores and orbit within 1.5AU of the Sun. Our Moon is much the same, though smaller, and could be considered the fifth of these *terrestrial planets*. Farther out, at 5 and 10AU, are giants Jupiter and Saturn (roughly 10 times Earth's diameter, Jupiter at 300 Earth masses, Saturn at 100), which have hydrogen and helium compositions similar to those of the Sun. Farther yet, at 19 and 30AU, are intermediate-size (15 Earth radii, 4 Earth masses) Uranus and Neptune that, while loaded with H and He, contain much more heavy stuff. And at the end, averaging 40AU distant, is icy Pluto, which because of its small size and its highly eccentric and tilted orbit is sometimes dropped off the planetary list.

Orbiting between Mars and Jupiter are countless "minor planets," *asteroids*. Though tens of thousands are known, they would not accumulate even to an Earth. We have direct contact with them, as Jupiter's gravity hurls a few into the inner Solar System, where they hit us as *meteorites*. Most are stones of various kinds, but a fraction are made of iron, suggesting that we are looking at the collisional remains of small planets constructed like the Earth, with rocky exteriors and iron cores.

From Neptune on out to 50 or more AU, in the *Kuiper Belt*, is another set of small bodies, hundreds of which have been observed. They are known to be made largely of dusty ices, as the gravity of the outer planets sends a few toward the Earth where they are seen as *comets* with short periods, under 200 years. As a comet gets close to the Sun, the ices sublime to gas that is heated and ionized by solar radiation, and then pushed backward by the solar wind to create a long bright ion tail. Melting ice releases the dust and rocks that are pushed back by the solar wind to make a dust tail that shines by reflected sunlight. Pluto orbits in the inner Kuiper Belt, and is clearly a member, though it is not a comet but a more evolved body. Finally, from huge distances away arrive the long-period comets, some with orbital periods that can range into the hundreds of thousands of years. Trillions of these cold bodies inhabit the *Oort Comet Cloud*, which extends tens of thousands of AU away, and perhaps halfway to the nearest star.

The planets all orbit the Sun in very nearly the same plane, that of the ecliptic, and all in the same direction, that of solar rotation, as do the bodies of the Kuiper Belt. Most spin in the same direction too. The Sun and planets seem to have been born simultaneously, the planets within a disk of gas and dust – the *solar nebula* – around the early Sun. The radioactive record tells how long ago. A radioactive isotope decays exponentially

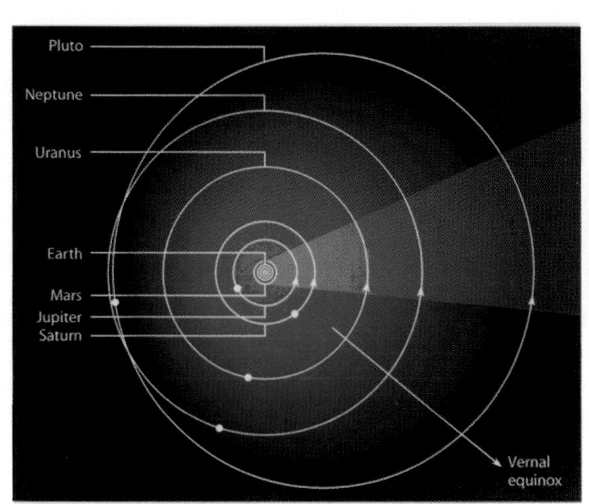

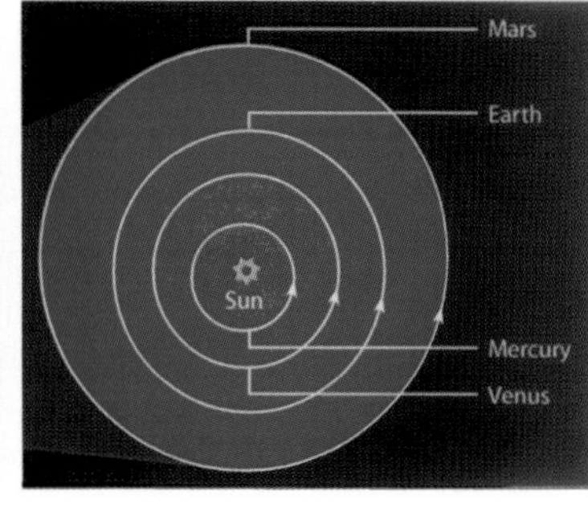

Fig. 11.6 **The Solar System** The planets orbit the Sun in a tightly restricted plane, all in the same direction, instantly suggestive of their formation from a prior "solar nebula." The asteroid belt lies between Mars and Jupiter, the Kuiper Belt of small debris bodies lies outside the orbit of Neptune, while the Oort Comet Cloud surrounds the whole thing. (Kaler, J.B. (1994). *Astronomy!* New York: HarperCollins, reprinted by permission of Pearson Education, Inc.)

to a final stable daughter product over a period of time called its "half life" [Box 7.4]. A comparison of the amount of a daughter product in a rock with the amount of its parent, and a knowledge of the half life, allows the determination of the age of the rock since solidification. The oldest rocks of Earth stretch back to over 4 billion years, those of the Moon back to 4.5 billion, and the oldest meteorites to nearly 4.6 billion, which provides the upper limit and therefore the Solar System's age. Solar oscillation theory [12.3] gives a similar answer. Can we see ourselves as we were 4.6 billion years ago? To find "us" in the stars of today, look for the disks.

11.2.2 T Tauri stars

The dark clouds are our best candidates for locating stellar youth. In and around them we find odd variable stars, the first of which, discovered in 1945, was T Tauri [Appendix 5]. *T Tauri stars* tend to come in loose expanding groups, in "T Associations," that seem like lower-luminosity, low-mass versions of OB associations [9.3.1]. Like the cooler B stars, they are also associated with reflection nebulae [11.1.2]. T Tauri stars fall into classes F through M. Their distances can be found from those of the interstellar clouds with which they are linked. These are derivable from normal stars that are associated with the clouds, or from star counts, which involve the statistical determination of the number of stars of given apparent magnitude that lie in front of the cloud. The resulting absolute magnitudes show that the stars lie on the HR diagram above and to the right of the main sequence [6.3.1], that is, they are optically over-luminous for their classes. They are even more over-luminous in both the infrared and ultraviolet.

T Tauri spectra exhibit hydrogen and forbidden sulphur emissions [13.4.1] as well as strong absorption lines of lithium. Lithium is easily destroyed in nuclear reactions at modest stellar temperatures far below that required for nuclear fusion [13.2.4, 13.2.5]. Recycling of stellar envelopes through convection thus depletes lithium in older stars. Solar lithium, for example, is down from that in primitive meteorites by a factor of over 100. But T Tauri stars have the full complement: they must be very young indeed. The infrared radiation has the spectrum of heated dust, so much of it that the star should be hidden optically, which it very much is not. A reasonable explanation is that the dust is in the form of a tilted disk, allowing us to see the star over the edge. The emission lines reveal low-density circumstellar gas. The ultraviolet continuum can be explained by matter that heats via violent shock waves as it falls to the stellar surface from the surrounding disk, controlled and channelled by strong magnetic fields. The stars thus grow apace, giving us suns in the making. Much of the accretion may come during an especially violent phase that produces *FU Orionis stars*. In 1937, the prototype of these erupting variables went from 16th to tenth magnitude in under a year. One of the FU crew was a known T Tauri star, and probably they all were.

The spectra of T Tauri stars exhibit a peculiar phenomenon: the hydrogen emissions commonly have P Cygni profiles [7.4] (or blue-shifted absorptions), indicating that they are *losing* mass rather than gaining it. In contrast, some T Tauri stars do have the reverse, *inverse* P Cygni profiles and red-shifted absorptions, showing accretion. Apparently, these nascent stars can accrete and lose mass at the same time, and they can grow as long as the former outweighs the latter.

"Classical T Tauri stars" (the CTTS) have all the spectral signatures listed earlier. They have as counterparts the "weak-lined T Tauri stars" (WTTS), whose spectra display chromospheric emission lines and sometimes a bit of infrared dust radiation, but no ultraviolet. A subset, the "naked T Tauri stars" (NTTS), eliminate even the dusty disk. WTTS tend to be fainter, and appear to be a subsequent stage of the CTTS in which accretion has largely shut down. Here is something of an evolutionary sequence. These and their predecessors are commonly termed YSOs for "young stellar objects" (Fig. 11.7).

YSOs can be classed according to their infrared spectra. Class I objects are deeply embedded within their clouds. In the near infrared they behave like blackbodies, but in the far infrared they have huge excesses from the surrounding heated dust. In class II, the far infrared, while still strong, is suppressed (think T Tauri), while class III objects have lost their IR signatures and are identified with weak-lined T Tauri stars or their successors.

11.2.3 Disks, flows, and cores

The partial solution to the mystery of the mass loss in T Tauri stars began over half a century ago, when George Herbig and Guilerrmo Haro (1913–1988) independently discovered odd glowing gas clouds. Though radiating emission line spectra, the clouds seemed to have no obvious sources of illumination. There were no embedded or neighboring stars to light them up, as there are for diffuse and reflection nebulae (or for that matter, planetary nebulae). Some were seen to change form and structure, new blobs developing internally over only a few years. For a time, these *Herbig–Haro* (HH) *objects* were thought to be stars in the making (see Fig. 11.7). Not quite, but close.

A wealth of optical, infrared, and radio data established the overall ideal picture. HH objects are found in pairs separated on the order of parsecs. Between them is a T Tauri or related star (or a binary or multiple). Emerging from the star is a pair of jets – a *bipolar flow* – lined right up with, and plowing into, the HH objects. The flows are the origins of the T Tauri forbidden lines. In individual cases, dust clouds (or circumstellar dusty disks) may hide one or more of the phenomena, including the T Tauri stars. A flow might be tilted so that all we see is the jet that comes toward us, while the other is obscured. Perhaps all we will see is a jet and its aligned HH object at the edge of a dark cloud, and maybe only the HH object itself. No matter: all must be present.

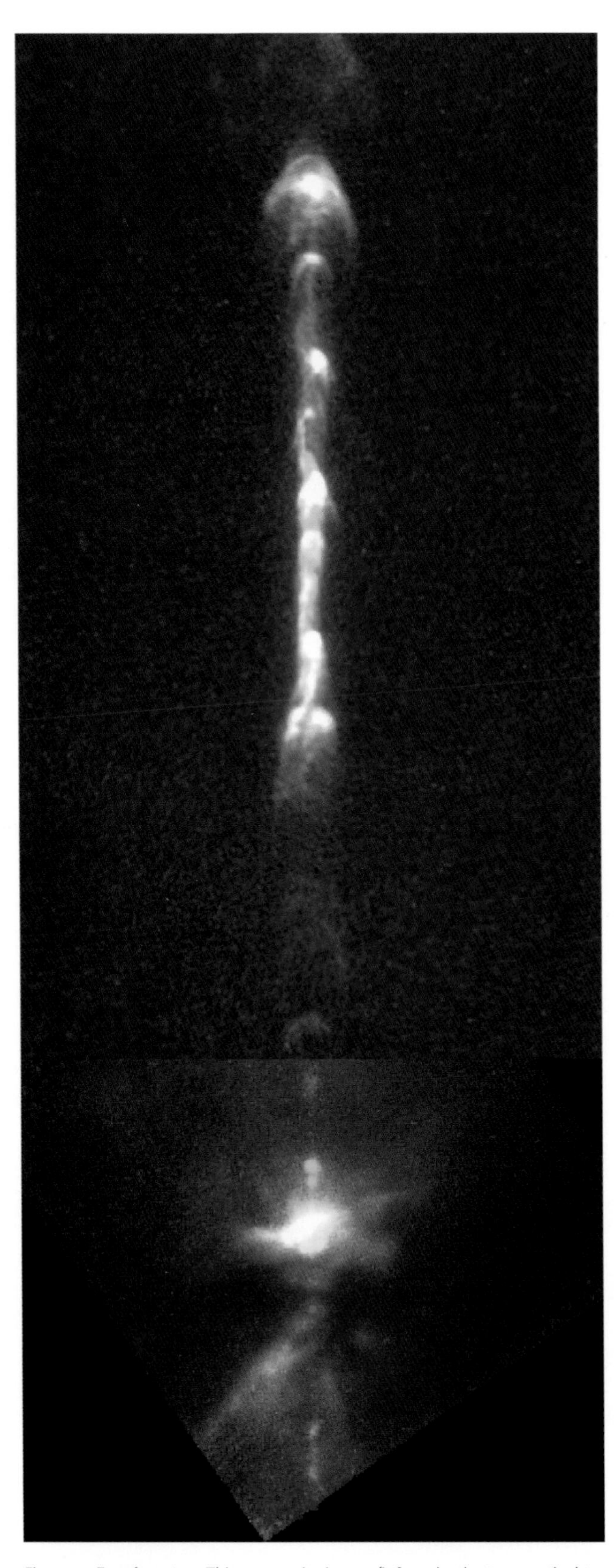

Fig. 11.7 **Forming stars** This composite image (infrared at bottom, optical to the top, both from Hubble) shows a trio of YSOs near the circumstellar disk at the lower edge, from which streams one half of a bipolar jet filled with shock waves and ends in a classical HH object. (Courtesy of NASA and B. Reipurth (CASA, University of Colorado).)

Radial velocity and proper motion observations of knots that travel along the jets show that they speed out of the central star from 20 to 500 km s^{-1}, above the ambient speed of sound both in the jet and in the surrounding medium. As a result, they sweep up interstellar matter into great bow shocks; the arc-like structures of typical HH objects look like they were made by a speedboat plowing through water [Box 11.3]. In return, the ambient medium sends a shock into the jet, giving it a sort of doubled structure. The jets are knotty, telling of sporadic ejecta from the star, and are the bright cores of broader, slower molecular flows. Some of the knots have the forms of bow shocks themselves, as faster-flying bullets from the star catch up and shock the slower-moving stuff. The very existence of bipolar flows suggest some kind of mechanism that focuses mass loss. The jets are most likely emerging through the poles of a dusty circumstellar disk that surrounds the inner star (or stars) and creates its infrared emission. High-resolution observations with both radio and optical telescopes reveal the disks themselves (Fig. 11.8). Others are seen projected against the background of the Orion Nebula. Called "propylids," they are being ionized out of existence by the intense ultraviolet radiation from the Trapezium (Fig. 11.9).

Looking more deeply into dense dusty molecular clouds with radio telescopes, we find numerous small clumps that are identified by emission from the ammonia molecule, which preferentially radiates from regions of high density (as do a few other molecules). These *dense cores* have a large range of properties, but typically are a tenth of a parsec across, have densities of a few tens of thousands of molecules per cubic centimeter, and contain stellar-sized masses cocooned within much more massive, lower-density shells. The linkage between dense cores and T Tauri stars is strong. Not only are the two kinds of objects found in the same neighborhoods, but dense cores are strongly associated with bipolar molecular flows beautifully outlined by emission from carbon monoxide. Some dense cores even contain YSOs. It seems clear that dense cores are the first actual stages of stars in formation, and are the predecessors of T Tauri and even higher mass stars. While some dense cores are isolated, many are also found in groups as they seemingly prepare to make T associations. We thereby look back in time to see what our Sun once was, how it began its life.

11.2.4 Higher mass

Related to T Tauri stars are higher luminosity (and higher mass) versions that fall in some semblance of spectral classes A and B and have emission lines in their spectra. Discovered by Herbig, they are called *Herbig Ae/Be* (or just Ae/Be, even HAEBE) *stars*. (Do not confuse them with "Be stars," which are rapidly rotating, mature main sequence dwarfs, subgiants, and giants that have emission lines in their spectra as a result of circumstellar disks [10.6.3], or with highly evolved B[e] stars that are related to Wolf–Rayet stars [6.4.4, 14.1.2].)

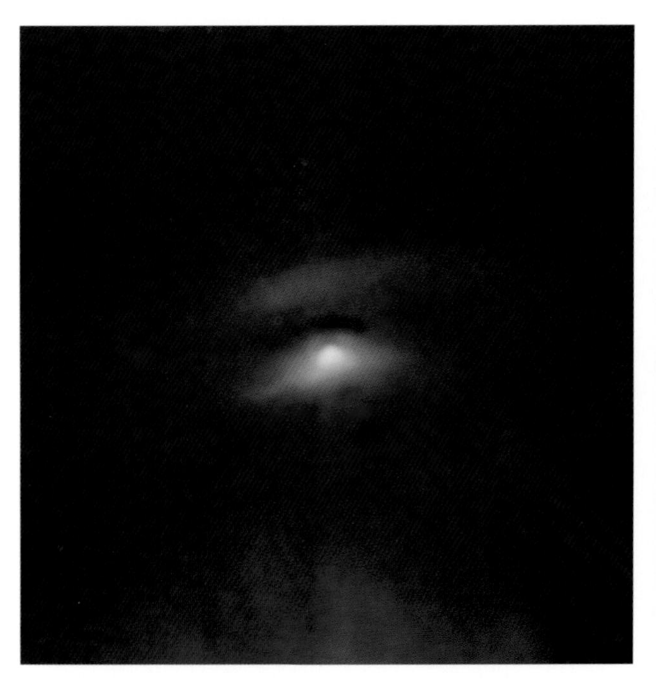

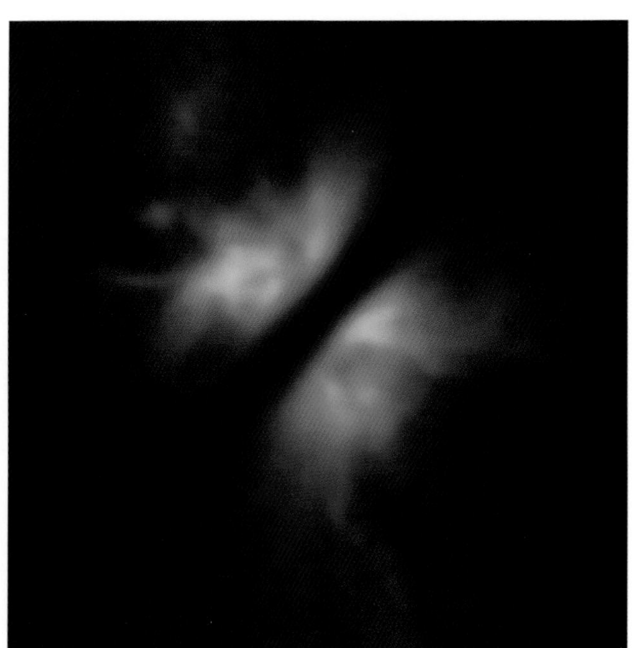

Fig. 11.8 **Disks** Circumstellar disks, observed with the Hubble Space Telescope, surround Haro 6-5B (left) and IRAS 04302 (right). Are planets forming within? (Courtesy of D. Padgett (IPAC/Caltech), W. Brandner (IPAC), K. Stapelfeldt (JPL) and NASA.)

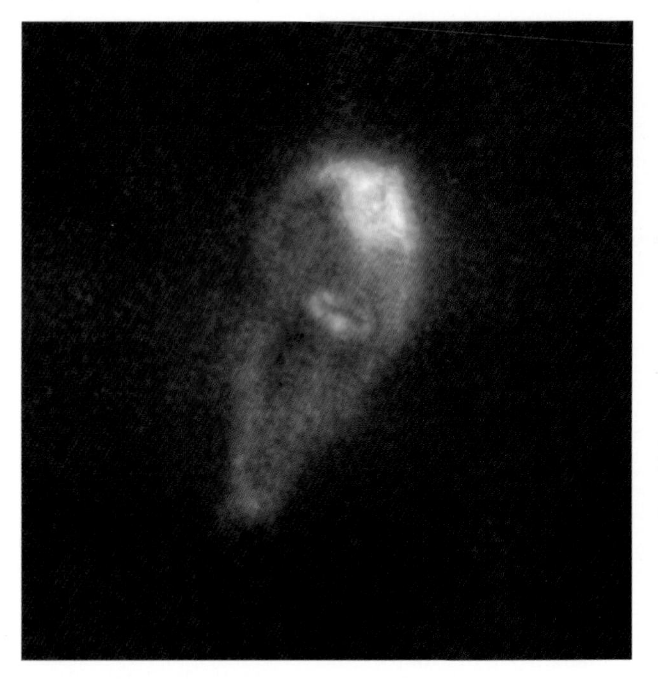

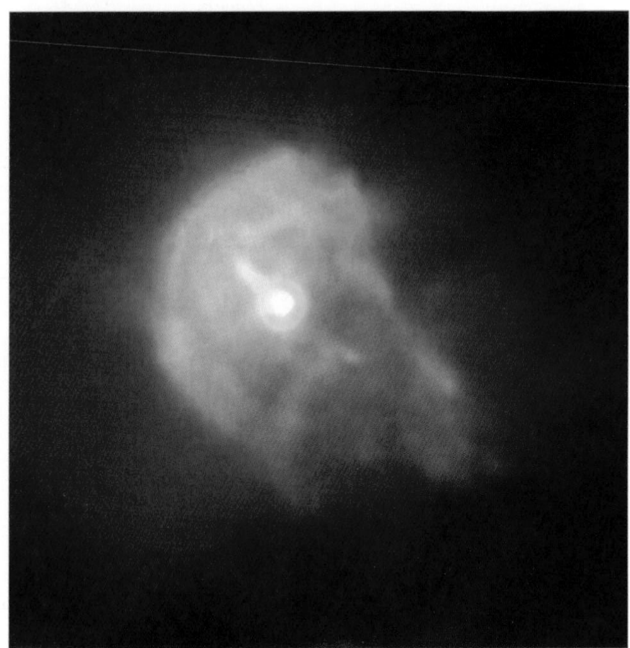

Fig. 11.9 **Orion's "propylids"** Disks surrounding young stars in the Orion Nebula cluster are being ionized out of existence by the harsh ultraviolet light of the Trapezium. (Courtesy of NASA, J. Bally (University of Colorado), H. Throop (SWRI), and C.R. O'Dell (Vanderbilt University).)

While Ae/Be stars also possess disks, theirs – unlike those of the classic Be stars – are very dusty. Moreover, Ae/Be stars are closely allied with dusty molecular clouds and bipolar flows, as are the T Tauri stars.

A prime example is R Monocerotis, an Ae/Be YSO that lies at the tip of Hubble's Variable Nebula (NGC 2261), a fan-shaped dusty reflection nebula that extends half a parsec to the north of the star; a faint counterfan projects in the other direction (Fig. 11.10). Dimmed in the optical by 13 magnitudes, the star itself can be seen only in the infrared. Emerging from it is a 100 km s^{-1} bipolar flow that runs perpendicular to a thick inner disk of molecular hydrogen,

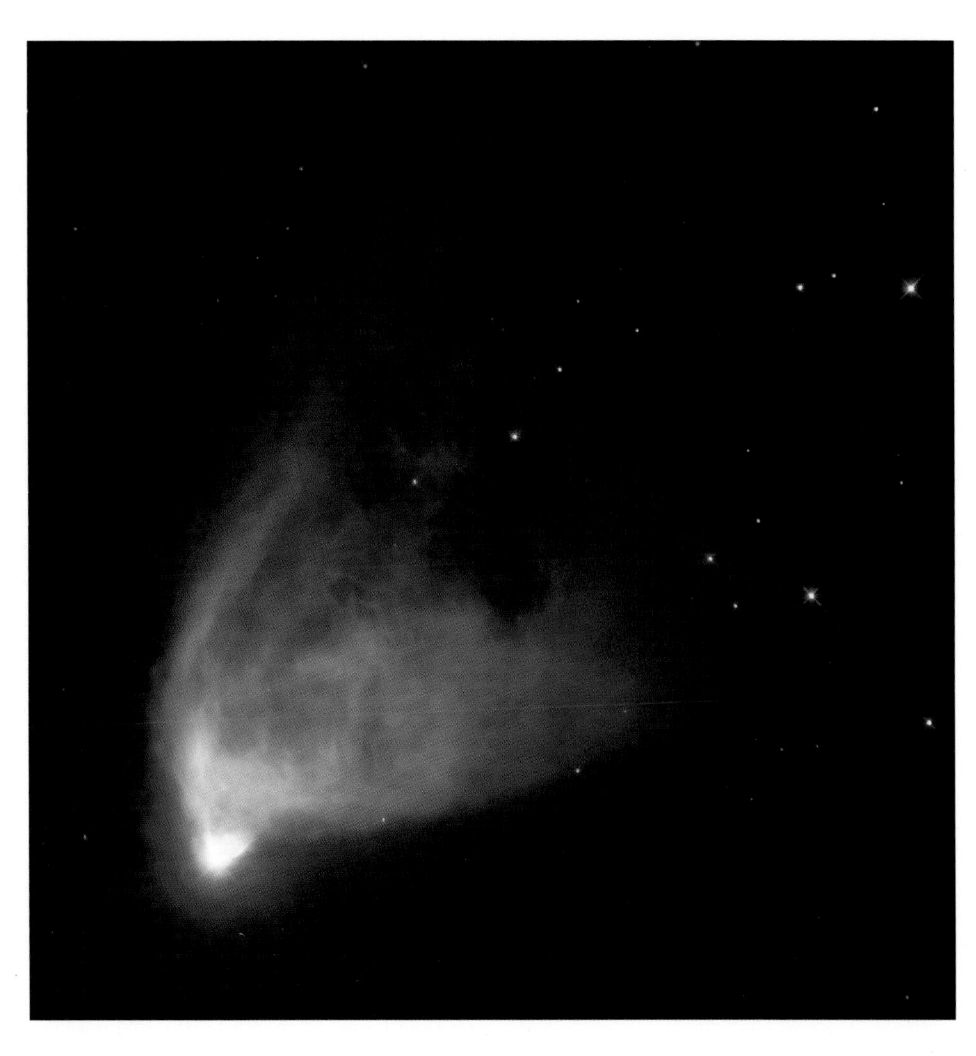

Fig. 11.10 **R Monocerotis** The fan-shaped reflection nebula is illuminated by a Herbig Ae/Be protostar at the center, whose light shines through the pole of a circumstellar disk. (Courtesy of NASA and the Hubble Heritage Team (AURA/STScI/NASA).)

through which starlight pours to illuminate the reflection nebula. Heliacal filaments within it suggest a strong magnetic field from either the disk or the star. The variation in the structure and brightness of the nebula seems to be caused by variable shadows projected by moving knots of dust that lie near the star.

At the top end of the mass scale are deeply embedded *ultracompact H II regions*, in which a massive star has sufficient ultraviolet power to ionize its dense surroundings. Quite invisible in the optical because of the thick dust, they are observed via radio. Some are associated with poorly collimated flows. Their predecessors are warmer versions of dense cores called *hot molecular cores*, in which temperatures can reach 100–300K. These are the stars that eventually explode as supernovae, which in turn provide the shock waves that begin the process of star formation in other birthclouds, and lead to sequential star formation and generations of OB associations [9.3.2].

11.3 The process

The birth of a star is among the grandest examples of synergy in nature. Stars are not created in isolation. They require the action not only of other stars, but of the Galaxy itself to bring them to fruition. Though we know that the action takes place through the gravitational collapse of dense knots of matter within the molecular clouds, there is no certainty as to the exact mechanisms involved, forcing the invocation of partial theory and speculation in stringing together the various kinds of protostars that are observed.

11.3.1 Collapse

In the beginning are the molecular clouds, which seem to be related to compression caused by the density waves that make the Galaxy's spiral arms [5.1]. Random instabilities and shock waves from supernovae help shred a cloud into large-scale clumps and these into dense cores. Fragmentation on smaller and smaller scales seems to be an important part of the process that produces gravitationally unbound T and OB associations [9.3] (which together contain the majority of all new stars), gravitationally bound clusters [9.2] both within and outside the associations (which contain between 10 and 50 percent of the stars), then multiples [8.6], doubles, and finally single stars (which may also develop from collapsing isolated globules).

A basic concept is the *Jeans mass* (after Sir James Jeans, 1877–1946), M_J, which is the minimum cloud mass that for a given density, temperature, and chemical composition will gravitationally collapse (whether as a single star or a nascent cluster). $M_J = K\,(T/\mu)^{3/2}/\sqrt{\rho}$, where K, T, μ, and ρ are respectively a constant, temperature, mean molecular weight (see Box 11.3), and density. M_J is so large for warm thin clouds that they are nowhere near massive enough to collapse. But drop the temperature to near zero and increase the density to that in dense molecular clouds, and the fragments within them should contract quickly to form stars and star clusters.

Except that they do not. The low star formation rate (2–5 new stars in the Galaxy per year) is nowhere near what it apparently should be. The clouds are in some way supported against quick collapse. Moreover, the larger clumps and smaller cores that will form new stars must be rotating; indeed, Bok globules are found to be spinning, albeit slowly. Through the conservation of angular momentum [Box 8.3], a collapse to form a star would spin up the rotation speed to the point where the cloud would fly apart so as to keep the process from completion. Something must also first slow the cores down.

Turbulent mass motions within a cloud (perhaps provided by the winds and jets from new stars) provide an outward pressure that, along with gas pressure, slows or even prevents, contraction. Magnetic fields help. The rotation of the Galaxy, filled with an ionized interstellar gas, generates a weak magnetic field about a tenth of a millionth the field strength of Earth's. Though almost entirely in the neutral state, a cold cloud will contain at least a few ions produced by the passage of energetic cosmic rays (high speed particles) from supernovae [14.3.5]. These ions grab onto the magnetic field, the process creating an additional outward pressure to retard the collapse. As a result, the star formation efficiency is low before the parent cloud dissipates. Since the magnetic fields are attached to the Galaxy, and the ions are constantly interacting with the neutral matter in the cloud, the field acts as a brake that slows rotation, preventing the stars-to-be from spinning themselves apart before they can be formed.

11.3.2 Birth

In spite of these outward pressures, in a larger clump that may form a cluster, or in a dense core, gravity can win out (Fig. 11.11). As the core contracts, the neutral molecules and atoms slip past the ions that are held up by magnetic forces, causing the magnetic field to lose its grip, a process called *ambipolar diffusion*. Theory suggests that the blob of matter then collapses from the inside out, mass falling from the surrounding cloud into the growing star in the middle. A shock wave from the collapsing mass provides the energy to heat surrounding dust and to create a large "false photosphere" that radiates in the infrared, announcing that a star is about to be born.

Though magnetic braking has slowed the collapsing mass's spin, it can hardly stop the entire rotation. As a result of the conservation of angular momentum [Box 8.3], it rotates faster and faster. The stuff that is not falling into the new star then spins out to form a disk. Still-mysterious forces, perhaps related to magnetic fields generated by the circumstellar disk, produce orthogonal bipolar jets that carry away further angular momentum. The jets might be directed by the forces in the disk, or they might just naturally flow along a path of least resistance within otherwise empty channels formed by the collapse process.

As the nascent "protostar" in the middle grows, its core heats and radiates by gravitational compression, though still quite buried in its dusty glowing birthcloud. At a central temperature of about a million K, it begins to fuse its natural deuterium into helium [12.2.2]. Increased opacity in

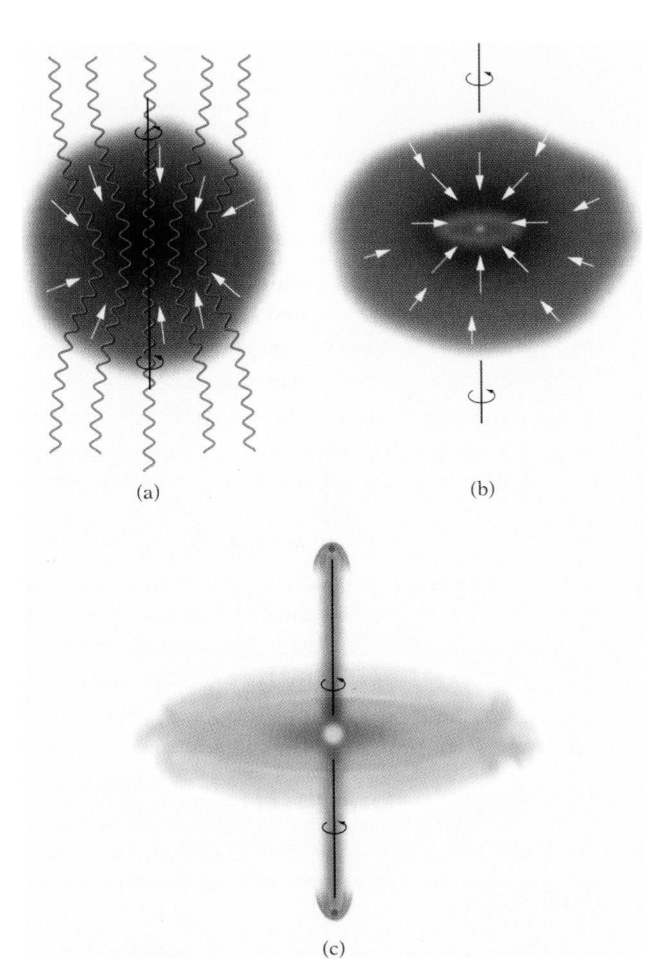

Fig. 11.11 **Star formation** (a) The Galaxy's magnetic field helps support a contracting blob of interstellar gas and dust and slows its rotation. (b) The core contracts as the magnetic field loses its grip, while conservation of angular momentum creates a disk around the new star, from which emerges a bipolar flow (c). (Kaler, J.B. (1997). *Cosmic Clouds*, Scientific American Library, New York: Freeman.)

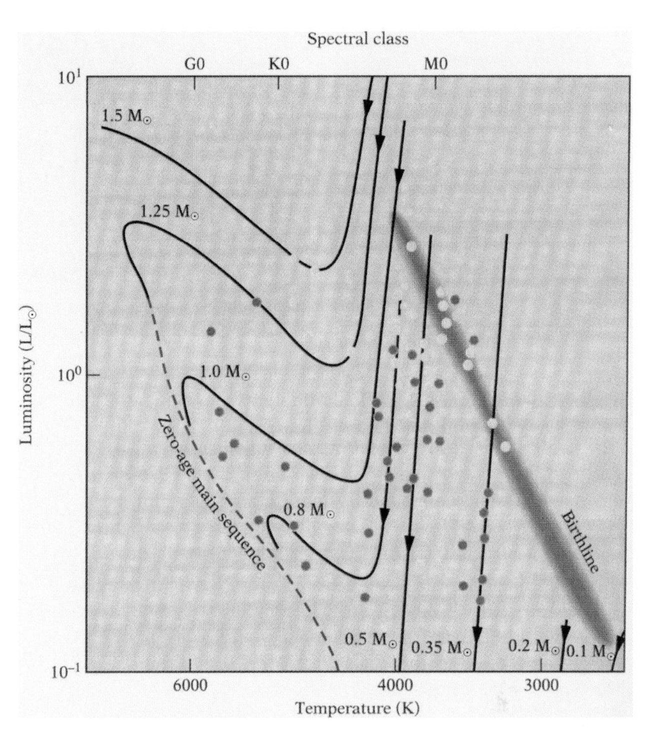

Fig. 11.12 **Pre-main-sequence evolution** Convective, deuterium-fusing stars descend on the HR diagram from the birthline, and then move onto the zero-age main sequence where they run full hydrogen fusion. Classical T Tauri stars are shown in yellow, older weak-line T Tauri stars are in red. (Kaler, J.B. (1997). *Cosmic Clouds*, Scientific American Library, New York: Freeman; and Stahler, S. *Astrophysical Journal*, **332**, tracks by I. Iben, Jr.)

the surface layers, as a result of the creation of the negative hydrogen ion (which dominates the opacity of the solar photosphere [12.1.1]), tips the temperature gradient (the run of temperature against stellar radius) so high that the stellar gases enter into a state of convection [Box 12.1]. This process brings fresh deuterium into the nuclear burning core, both from the outer parts of the star and from the accreting material. The rotation and convection in turn generate a magnetic field that makes the star active and creates a stellar wind. Here is the divide between collapsing protostars and real *pre-main-sequence* (PMS) *stars* (T Tauri and related stars), as the process for a time stabilizes the new stars at the *birthline*, which on the HR diagram lies above and roughly parallel to the main sequence (higher-mass stars lying at greater luminosities) all the way to the Ae/Be stars (Fig. 11.12).

Heavily accreting matter from their disks, the new T Tauri stars increase their masses until the disks dissipate, or at least until their inner edges can no longer reach the stellar surfaces. Accretion at this stage can be violently random. Huge mass infalls from unstable disks may produce sudden brightenings and the set of FU Orionis variables, during which stage the stars may accrete a good portion of their final masses. At the same time, the stars separate from their dissipating birthclouds to become more easily visible to us.

As accretion slows, the stars dim on the HR diagram at constant temperature along convective *Hyashi tracks* (after the Japanese astronomer C. Hayashi). Dropping accretion turns classical T Tau stars into their weak-line versions. The greater the mass, the smaller the degree of dimming before decreased opacity near the core squelches the convection. Near this point, nuclear reactions through the proton–proton chain or CN cycle [12.2.2] start up in earnest, while the star now heats at roughly constant luminosity until it stabilizes at the *zero-age main sequence* (the "ZAMS"), from which we count the time of all further evolutionary changes (see Fig. 11.12). Classical T Tauri stars turn into main sequence stars from roughly mid class B on down, while Ae/Be stars become dwarfs farther up the line, above about $5M_{\odot}$. The lowest masses, below 0.075 solar, become brown dwarfs [6.2, 12.4.2, 13.1] that are so light that they cannot run the full proton–proton chain [12.2.2]. The whole process takes only a few million years.

High-mass (early O) star birth is shrouded in mystery. These stars develop so fast that the stars are created even before they clear out their surrounding clouds. Moreover, they begin the fusion process and become "stars" even before they have accreted their final masses. As a result, the formation

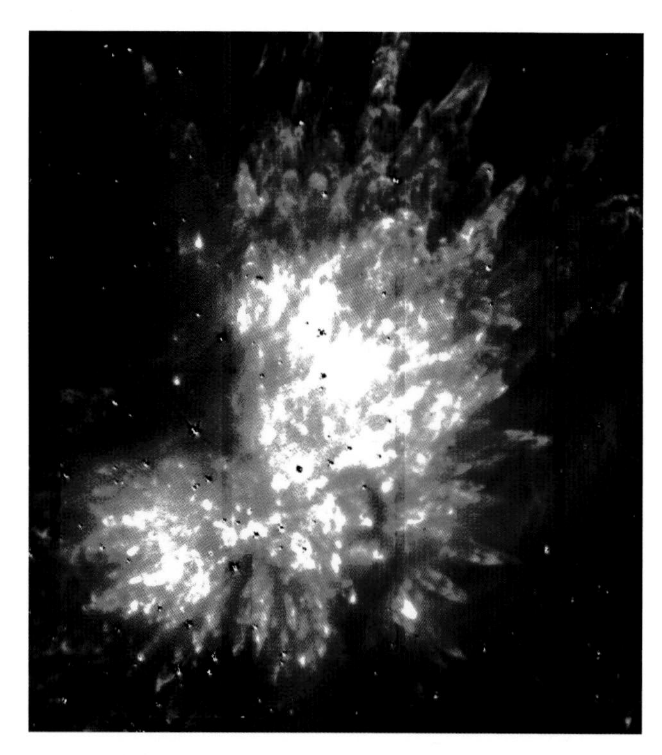

Fig. 11.13 **High-mass star formation.** Buried in the cloud behind the visible Orion Nebula is the K–L nebula (here imaged in infrared radiation from the H_2 molecule), where a huge, poorly collimated bipolar flow with dozens of HH objects signals the birth of a 20-solar-mass star. (Copyright © Subaru Telescope, NAOJ. All rights reserved.)

process is difficult to observe. There is yet only small evidence for circumstellar disks, though some new massive protostars are associated with bipolar flows. Radiation and winds from very high-mass stars would seem to destroy the surrounding clouds and prevent accretion. An alternative is that high-mass stars form by collisional accretion of low-mass stars within the cores of dense clusters (Fig. 11.13).

Box 11.4 The IMF

A fundamental constraint in the theory of star formation is the *initial mass function* (IMF). From the luminosity function of the main sequence (the number of stars per unit luminosity or absolute magnitude) combined with main sequence ages, we can derive the number of new stars born within a specific mass interval. Following the luminosity function, the birth of a high-mass star is a rare event, while low-mass ones come into being at a fierce rate. A final object of theory is to replicate what we see and to explain why nature likes making lower-mass stars. Given that theory is still unable to explain fully how even single stars are born, the explanation of the IMF is as yet out of reach, much less an explanation of why the IMFs of some clouds are different from those of others. The distribution of mass ratios in binaries for new stars and for those of mature field stars represents a similar unsolved problem. The high-mass end of star formation most likely lies around 100 to 120 solar masses, though the number cannot be checked with binary stars, as there are too few such massive bodies. The low-mass limit is unknown. The masses of brown dwarfs could extend downward into the range of the giant planets.

11.3.3 Collections

Stars are not so much formed singly, but in hierarchical collections, from doubles to multiples to bound clusters to unbound T and OB associations. What we get seems more than anything to be related to the kind of molecular cloud in which birth takes place. Star formation appears to be bimodal, in the sense that some molecular clouds (nearby Taurus–Auriga for example [Appendix 5]), generate only lower mass stars and T associations, while others (like the Orion molecular clouds) form stars from high mass on down that make the great OB associations, of which lower-mass T associations will be a part (Box 11.4). Fully three-fourths of the Galaxy's new stars seem to have been born in OB associations, whose central regions can be incredibly densely packed. The Orion cluster, which centers on the Trapezium, contains 10 000 stars per cubic parsec. Clusters themselves, whether parts of OB associations or not, hold anywhere from 10 to 50 percent of the pre-main-sequence stars. All but the most tightly bound of these structures are doomed to destruction as a result of dynamical "evaporation" of stars and of tides raised by the Galaxy, the stars thereafter – like the Sun – running free [9.2.4].

Binaries present a particular problem. To ignore binary formation is almost to ignore star formation altogether, as there are so many of them. Proto-binaries seem to have most of the same characteristics as main sequence (mature star) binaries. Proto-binary nature is found as far back along the creation chain as we can reasonably observe, implying that doubling is integral to the birth process. We see close doubles and wide doubles, eccentricities of all varieties, and both circumstellar and circumbinary disks.

There are several ways in which binaries may form, and all may operate under different conditions. The original idea is fission, in which a star rotates increasingly faster as it contracts until it divides. The concept is now largely discounted, as theory shows that a protostar spinning near its breakup velocity will cast out rings rather than double up. Nevertheless, fission still has its adherents.

Capture is a possibility. Two well-separated stars would act like point sources of mass, and cannot capture each other gravitationally, as they would approach on hyperbolic orbits, then leave the same way. However, in a near miss, they can raise tides in each other that can dissipate energy and allow them to join in elliptical orbits. More likely, approaching protostars might interact with each others' extended accretion disks, slow down, and thereby join into mutual orbit. Three-body interactions could also do the job, as a third star could absorb energy through gravitational "collision" from two others, allowing a pair to go into orbit. Collisional formation of binaries in globular clusters, where the star density is extremely high, clearly takes place [9.4.12]. High star density is required, however, so collisional binary creation probably occurs only in very dense OB-association cores.

The most likely routes seem to be fragmentation of a collapsing core that gets the process going before the stars actually form, and the formation of a companion from instabilities that emerge within a protostellar disk. The former may operate for wide binaries, the latter for close ones. Whatever the process, the global environment seems to make a difference. The fraction of binaries in Taurus–Auriga [Appendix 5], which is close enough to us to allow decent statistics, is near 100 percent, roughly double the binary frequency of mature field stars. The disintegration of these binaries to produce the field star binary population is not an option. The binary frequency in high-mass clouds such as Orion (which produce by far the most stars) is lower, and closer to what is found in the field. Yet binaries are ubiquitous among very high-mass early-type stars, as witness the rich binary nature of the Trapezium members.

11.3.4 Population III

Where do these theories leave us with zero-metal Population III [5.4, 14.2.6]? With no dust to act as coolant, how could they have formed from the original "interstellar" gas? Hydrogen alone apparently can still do the job. Theory suggests that chaotic blobs of hydrogen gas with a Jeans mass of $1000 M_\odot$ were created in the aftermath of the Big Bang. Relatively warm temperatures allowed the formation of hydrogen molecules directly from the gaseous state. Collisions among the H_2 molecules would raise their electrons to higher orbital levels, the energy coming from the velocities of the colliders, slowing them down. Downward transitions then radiated the energy away, cooling the gas much as do the forbidden lines in a planetary nebula [13.4.1]. The gas could then condense into a star.

The sizes of the condensations indicate that a zero-metal gas would most likely form very massive stars, perhaps in excess of $100 M_\odot$, which quickly exploded as supernovae, thereby creating the first metals and the dust for the next stellar generation. There are therefore no longer any Population III stars left to find.

11.4 Planets

The Sun's circumstellar disk produced planets. Over 100 other planetary systems are now known, and the number is rising quickly with increased observation time and more and more sophisticated techniques.

11.4.1 Formation of the Solar System

The early Sun must have been a T Tauri star surrounded by a dusty circumstellar disk. Near the Sun the disk was hot, some 1400K, while at the fringe, where we now find Pluto, it dropped to only 150K or less. The dust, from the interstellar cloud in which the Sun was born, was in chemical equilibrium with the gas in which it was embedded. In the cold outer disk, it accumulated ices and other volatile substances, while in the hot inner disk, only refractory (high melting point) elements and compounds could condense. Beyond 4 or 5AU (the *snowline*), the grains were "wet," inside that distance, dry.

The early microscopic grains were in approximate Keplerian orbits around the forming Sun. Drag induced by the gas, however, made them slowly cross each other's paths. Individual grains collided and stuck together, larger grains growing from smaller ones. The process continued until they reached visible size, then a millimeter, a centimeter, and beyond. Because the orbits were constantly changing, there was always a new supply of material for larger bodies to accumulate.

In the kilometer range, the increasing gravity of the countless new *planetesimals* allowed the bigger ones to grow at the expense of the smaller, leading to the formation of several Earth-size or larger rocky-metallic bodies. In the inner Solar System, that is where they stopped, as it was too hot to accumulate the most common of all elements, hydrogen, which the Sun was clearing out through its wind anyway. The result was the set of small terrestrial planets. Farther away, though, in a colder environment, nascent Jupiter and Saturn (whose cores grew to 10 Earth masses or more) could gather huge amounts of hydrogen and helium and grow to 100 or more times the mass of Earth (Fig. 11.14). Yet farther away, the density of the disk dropped off. Uranus and Neptune did not have the time to gather in much volatile hydrogen and helium, and thus came out smaller and denser. At the fringe, the disk's density became so small that planets could no longer grow. Lack of accumulation and gravitational migration caused by the larger planets (which may have migrated themselves) resulted in a huge number of small icy planetesimals that remain in the distant Kuiper Belt. An alternative picture has the giant planets accumulating directly from the gas through instabilities in the disk, much as a binary companion might.

The last collisions that created the inner terrestrial planets were whoppers, and deposited so much energy into the bodies that they at least partially melted, which sent the heavy metals – mostly iron and nickel – to the centers, while the light stuff (silicates and so on) floated to the outside. The debris from a last giant collision between Earth and a competing Mars-sized body is believed to have consolidated into the Moon (Fig. 11.15). A planet would have formed outside the orbit of Mars, but Jupiter kept the planetesimals so stirred up that collisions from swift orbital crossings broke them back down again, resulting in the current asteroid belt. Satellites around the giant planets might have formed in some similar manner. The whole process should have taken no more than 100 million years.

For about half a billion years, the thick debris of the new system, stirred up by the giant planets, rained down on the now-cooled inner smaller planets (and outer moons), cratering their surfaces with violent impacts. Earth wiped away the record of this "heavy bombardment" through erosion and tectonic processes (that can move whole continents), but the smaller planets and the planetary satellites, with neither atmospheres nor crustal motion, did not, resulting in the still-battered surfaces of the Moon, Mercury, and many other bodies. The icy planetesimals that crashed into the Earth brought our water, which could not have been originally incorporated into our early hot planet. Trillions of icy planetesimals were ejected by the giant planets far beyond the local system, to distances of tens of thousands of AU, to populate the Oort Comet Cloud.

Small asteroids – some carrying pre-solar dust grains – are still tossed into the inner Solar System, where they become meteorites. At the same time, passing stars or molecular clouds throw the occasional icy planetesimal from the Oort Cloud toward us, and a long-period comet graces the sky, the short-period ones the remains from the Kuiper Belt (Fig. 11.16).

11.4.2 Mature disks

Though the original solar nebula quickly dissipated as a result of planet formation and clearing by the early solar wind, a residual dusty disk remains behind even today. Consisting of particles that flake from short-period comets and of the debris of smashed asteroids and Kuiper Belt objects, the disk reflects sunlight and is seen by us as the *zodiacal light*, which appears as a conical illuminated band that stretches up from the horizon after the end of evening twilight or before dawn along the ecliptic. Heated by sunlight, the band – stretching

Fig. 11.14 **Jupiter** With its swirling ammonia cloud belts, the giant of the Solar System is seen here by the Cassini spacecraft, half of it in daylight, half in night. Made mostly of hydrogen and helium, the planet's diameter and mass are respectively 11 and 300 times that of the Earth. Such planets are found orbiting other stars as well. (Courtesy of NASA.)

Fig. 11.15 **Earth and Moon** Two rocky bodies of the inner Solar System, Earth and its Moon, provide a stunning contrast, the Moon beaten by the colliding debris of the early Solar System, the Earth, covered with water and clouds, sustaining life. (Courtesy of NASA.)

Fig. 11.16 **Comet Hale–Bopp** This icy body, a leftover of planetary creation and seen against the background of the constellation Perseus, makes a visit to the inner Solar System from the Oort Comet Cloud. (The bright star to lower left is Algol, Beta Persei.) (J.B. Kaler.)

50 or more AU away from the Sun – radiates weakly in the infrared and could conceivably be detected from a nearby star.

In the early 1980s, the *Infrared Astronomical Satellite* (IRAS) detected suspiciously similar dusty circumstellar disks around the class A stars Vega, Fomalhaut, and others. More such stars have been found with Earth-based instruments, including the best example, Beta Pictoris, whose edge-on thick disk is over 10 times the diameter of ours (Fig. 11.17). Several of these disks are warped or emptied in the middle, suggesting the formation of planetary systems. And the planets themselves are starting to turn up.

11.4.3 Other planets

In our initial parochial view, all other planetary systems were expected to look like our own. What a surprise it was to find that many are nothing like ours at all. There are several ways of detecting other planets and/or possible planetary systems.

1. *Visual sighting*. We might see a planet in orbit around its parent star. Given the distances to the stars, the closeness of the planets to their parents, the faintness of the reflected light compared with stellar brightness, such detection is currently impossible.
2. *Radio signals*. Signals from an intelligent civilization would instantly give away a planet. The ongoing "Search for Extraterrestrial Intelligence" (SETI) has so far yielded nothing. Perhaps more likely are radio bursts from the magnetospheres of extraterrestrial Jupiters.
3. *Variations in proper motion* [5.2.1]. Astrometric shifts produced by planetary bodies are generally too small to be readily measured with current techniques. The technique has, however, been used successfully, especially to place upper limits on planetary masses.
4. *Transits* that depress the light of a star.
5. *Gravitational lensing* of a distant star.
6. *Radial velocity variations* such as we see for single-lined spectroscopic binaries. This method is responsible for the detection of well over 100 planets.

The Sun moves around the center of mass of the Solar System (controlled largely by Jupiter) at a speed of around 3 km s^{-1}, which is greater than the current capability of

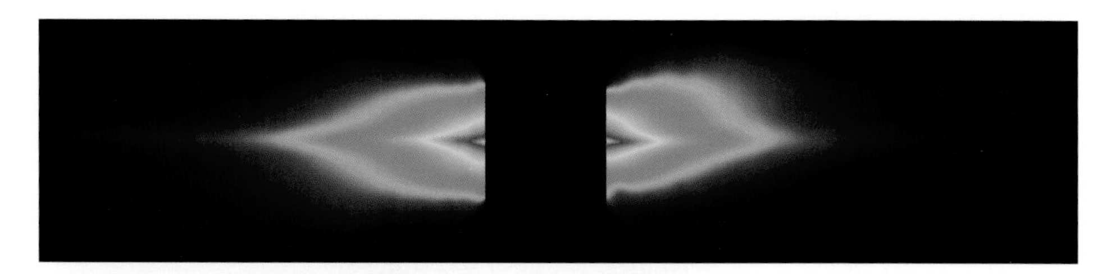

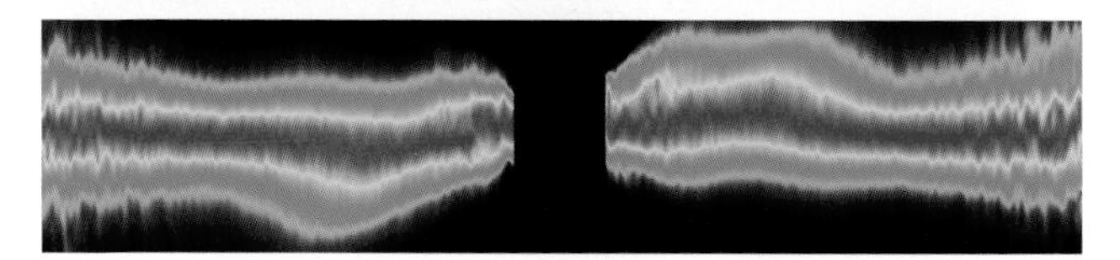

Fig. 11.17 **Beta Pictoris** A dusty disk (above) stretches 400AU away from Beta Pictoris. The warping of the inner 15 percent (below) suggests planets. (Courtesy of A. Schultz (CSC), S. Heap (NASA GSFC), and NASA.)

modern radial velocity measures [5.2.2]. A larger planet, or one placed closer to its sunlike star, would produce a greater variation (Fig.11.18). A single-lined spectroscopic binary yields little information. But a planet provides a great advantage in that we know it to be much less massive than the star it orbits. The result is the semimajor axis of the planetary orbit and a lower limit to its mass (Box 11.5).

Extrasolar planetary astronomy exploded in 1995 with the discovery of a Jupiter-sized planet in orbit about the G2.5 subgiant 51 Pegasi. With a period of only 4.23 days and a mass of at least 0.46 that of Jupiter (M_j, not to be confused with the Jeans mass, M_J), the planet has a semimajor axis of but 0.051AU. Dozens more such discoveries flooded in, mostly involving F, G, and K dwarfs and subgiants. Values of $m_p \sin i$ (see Box 11.5) range from well under that of Uranus to greater than $10M_j$ and into the realm of deuterium-fusing brown dwarfs [6.2.1, 12.4.2, 13.1] above about $13M_j$. Planetary masses above $10M_j$ are rare, however, leading to the concept of a "brown dwarf desert," that there are few close-in brown dwarf companions to G stars. To be sure, the planets' masses are all larger than given, and some could indeed be brown dwarfs. But so many observations are now available, that too many more would be statistically improbable. The majority of the bodies are indeed of planetary mass. One of the stars, Epsilon Eridani, was already known to possess a circumstellar disk, giving credence to the idea that all stars with such disks have planets as well.

Most of the early discoveries, and about a third of the planets so far found, are close-in "hot jupiters" that revolve closer than Mercury does to the Sun, which is explained by observational selection, as close-in massive planets will produce the shortest periods and the greatest velocity amplitudes. As the techniques improved and the time baseline lengthened, astronomers discovered planets farther and farther out. Roughly half of them are now over one 1AU away from their stars. Moreover, more than 10 percent are real systems, with more than one planet.

Box 11.5 Extrasolar planet analysis

A planet orbits another star. For simplicity, assume a circular orbit that lies in the line of sight. Kepler's third law [8.2.3] in solar units is

$$P^2 = a^3/(M_* + M_p),$$

where M_* and M_p are the respective masses of the star and planet, and P and a are the orbital period and semimajor axis in years and AU. But since M_p is much less than M_*,

$$P^2 = a^3/M_*, \quad \text{and} \quad a^3 = M_* P^2.$$

The mass of the star is known accurately from its temperature, luminosity, and evolutionary considerations; that is, from its placement on the HR diagram. From the orbital period, we thus know a. The semimajor axis is actually $a = a_* + a_p$ (where these are respectively the distance of the star and planet from the center of mass), but since the star is so massive relative to the planet, $a = a_p$. We now know the planet's orbital radius.

Since orbits are mutual,

$$M_p/M_* = a_*/a_p,$$

so

$$M_p = M_* a_*/a_p.$$

The semimajor axis of the stellar orbit, a_*, is just $Pv_*/2\pi$ (where Pv_* is the orbital circumference) and where v_* is known from radial velocity observations (see Fig. 11.18). With a_p, a_*, and M_* known, we find M_p, the mass of the planet. Elliptical orbits add some spice, but the radial velocity curve quickly yields the eccentricity, plus the planetary mass and semimajor axis. The problem, as with any spectroscopic binary analysis [8.3], is the inclination of the orbit, which is unknowable without further information. The velocity curve provides only a lower limit to the radial velocity, which in turn yields only a lower limit to the planetary mass (rather it gives $v \sin i$, thus $m_p \sin i$, where i is the orbital inclination to the plane of the sky).

"Jupiters" should be formed far out, where it is cold enough to accumulate molecular hydrogen from the swirling birth cloud. If, however, after the planet is formed, the birthing disk is still thick, viscous friction can make the planet spiral inward. The giant planet could in fact fall into its star and be destroyed. If the disk dissipates at just the right moment, the planet then gets stuck close to the star without

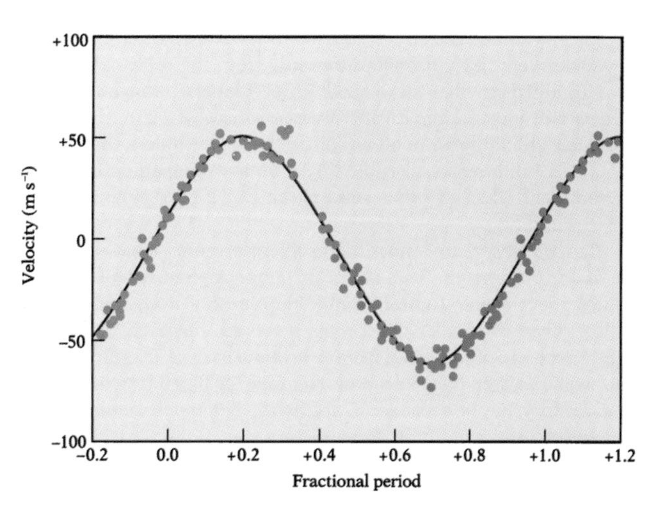

Fig. 11.18 **51 Pegasi** The velocity curve of the G subgiant 51 Pegasi revealed a jupiter-like planet. The red points show recession and redshift, the blue approach and blue shift. (Kaler, J.B. (1997). *Cosmic Clouds*. Scientific American Library, New York: Freeman; and courtesy of G.W. Marcy and P. Butler, California and Carnegie Planet Search.)

falling in. Theory shows that once a jupiter is made, it can survive, at least for a time, as a hot body near its star. Systems more like ours (with multiple planets) are turning up, and we expect to find more and more of them. And someday, we will see even the terrestrial planets that may co-exist, as they do in our system.

The search continues photometrically in trying to find evidence of planetary transits. One of the hot jupiters detected by the radial velocity technique was finally seen to cross in front of its star, the "eclipses" repeated right on time. (Another was found by photometric monitoring.) Application of eclipsing binary rules demonstrates that it is a jupiter-sized body. We really have found planets. Spectra taken of the star during transit provide evidence for mass boiling away in the stellar heat (Fig. 11.19), and even allow for marginal "detection."

No one yet knows the range of planetary masses. In our own system they start out as low as Pluto, the Moon, perhaps Mercury, depending on your definition of a planet. Planets of other systems may be created into the brown dwarf range, and some – if they can reach $13M_J$ – may even fuse deuterium. On the other hand, brown dwarfs may exist below this limit into the range of planets. The distinction between the two is that stars, including brown dwarfs, are built from the top down – by fragmentation and condensation of a birth cloud – while planets are built from the bottom up, by accumulation of dust and then (for the jupiters) gas. However, one theory for the formation of binary companions involves instability in circumstellar disks, and so does one theory for planet formation! At the end, there may be little difference except semantic. On the other hand, if the brown dwarf desert continues to hold up with time, it may provide the needed separation between the two.

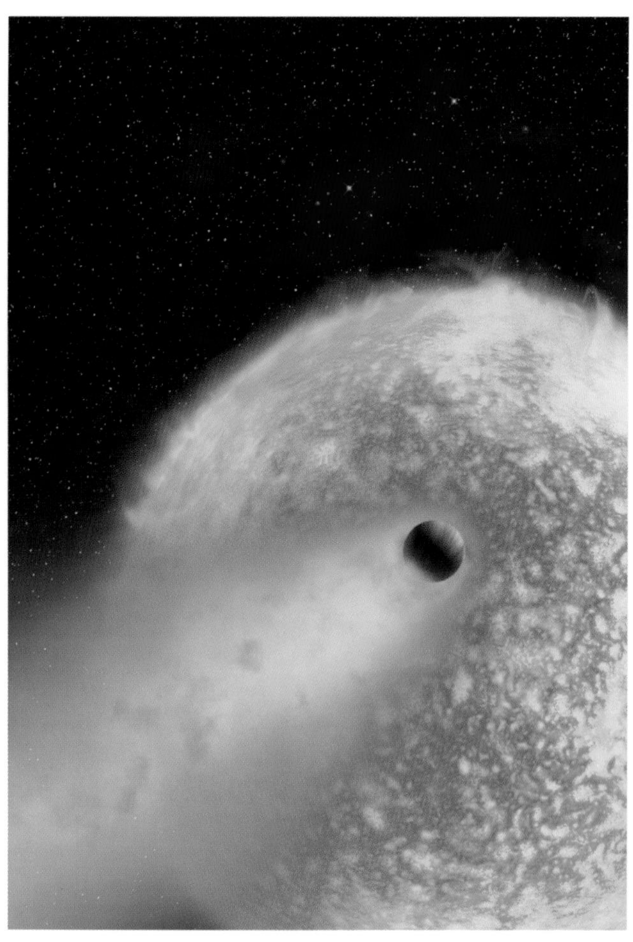

Fig. 11.19 **Transiting planet** In this artist's conception, the "hot jupiter" belonging to HD 209458 and trailing gases behind it, cuts in front of its star, producing a detectable "eclipse." (Courtesy of ESAS, A. Vidal-Madjer (Institut d'Astrophysique de Paris, CNRS, France) and NASA.)

The number of candidate stars equipped with planets hovers at around 10 percent, though allowance for selection effects can make the number much higher. For some stars, the disks are so thick that planets spiral in and are lost. Others are too close to O stars in their birthing associations, where ultraviolet light can cook away a disk before planets can even be born [Fig. 11.9]. More, there is a strong propensity for stars with planets to be metal-rich compared with the Sun, though it is not an absolute requirement. Either high metallicity is needed to make planets, or the stellar atmospheres become polluted from planets that spiral inward, most astronomers leaning toward the former idea. Planets therefore come along relatively late in the Galaxy. Maybe life and civilizations do too.

Chapter Twelve

Sun and main sequence

The first station on the trainride to stellar destiny is the main sequence. Here is the greatest zone of stability before death, where stars of all masses reside in some semblance of peace while they fuse the available hydrogen fuel in their cores into helium. The classical main sequence runs from O2 stars and 100-plus solar masses at the high end, down to the 0.073 or so solar at the low end in late class M and early L, below which full hydrogen fusion (the conversion of hydrogen into helium with the production of energy, which powers the Sun and most stars) no longer takes place. At this point are appended the great number of brown dwarfs of classes L and T being discovered at ever lower masses and luminosities, perhaps right down into the realm of the planets. No one yet knows where the full star/brown dwarf sequence ends.

However the main sequence is defined, in the middle of the range is that most familiar of stars, the Sun, which serves as the paradigm of normal dwarfs. Here we can see features and events that we know must exist on other stars. To understand other stars therefore, we must begin with the Sun.

12.1 The Sun

High in the sky, our star shines down upon us, heating the Earth (Fig. 12.1). So bright we dare not look at it without protection, it is no wonder our forebears considered it a god. Seen through clouds, the Sun looks like a perfect, unflawed circle of white light. Ancient Chinese astronomers, however, noted occasional dark patches, which were rediscovered by Galileo with his first glance through his primitive telescope. The Sun's seeming imperfection, at odds with the dogma of the day, helped get him into trouble with the Church, which eventually stopped his astronomical endeavors.

Far from just having a few spots, the Sun is a complex layered body made of heaving, rotating gases filled with magnetic activity that extends from deep inside to the far reaches of the Solar System. Unlike the distant stars, the Sun is so close that it provides us with an overwhelming amount of detail, much of which is still incomprehensible.

12.1.1 The photosphere

The bulk properties of the Sun are defined largely through its apparent "surface," its *photosphere* (from the Greek "sphere of light"), the 600-km-thick external layer from which the solar radiation escapes freely toward Earth (see Fig. 12.1) and which is also called the *solar atmosphere*. The great opacity, the result of absorption of light by the negative hydrogen ion [7.1.7], gives the photosphere its razor-sharp edge, allowing us to define a precise radius (taken as where the optical depth [7.2.2] equals unity) of 6.96×10^5km, equal to 0.0465AU or 54 Earths. Measurement of the energy of sunlight falling on Earth per unit time, the *solar constant* of 1368 watts per square meter, multiplied by the surface area of a sphere with a radius of 1AU = 1.496×10^{11} meters, gives the solar luminosity of 3.85×10^{26} watts. Dividing by the solar surface area (for a sphere of radius 6.96×10^8km) yields a surface flux (radiation per unit solar area) of 6.32×10^7 watts per square meter, whereupon the Stefan–Boltzmann [7.1.2] law gives an effective blackbody temperature of $T_{eff} = 5777$K. At this temperature, the Wien law [7.1.2] gives a peak in the blackbody spectrum of 5010Å, which falls in the green part of the spectrum, the combination of colors making the Sun appear a bit on the yellow side of white. The temperature is consistent with a G2 spectral class [6.2.1] and a $B - V$ color index [3.4] of 0.65.

The Earth's orbital period and semimajor axis entered into Kepler's third law [8.2.3] provides the solar mass of 1.99×10^{33}g. The mass divided by the volume (computed from the radius) shows the Sun to have an average density of 1.4gcm^{-3}, not far above that of water and similar to Jupiter's 1.2gcm^{-3} (ignoring the fact that the photospheric density is vastly less and the central density vastly more). Jupiter is largely in the liquid state as a result of high internal pressure. Add mass to it, and the radius would not increase so much as the density. The low mean density for such a large body as the Sun shows it to be gaseous throughout, as are all stars but neutron stars and cool white dwarfs [13.5, 14.3.1].

Other than its gross properties, the photosphere has two outstanding characteristics. Even quick observation shows that the solar disk is darker at the edge (the *limb*) than it is toward the middle (see Fig. 12.1). *Limb darkening*, which afflicts all stars, is the result of a gaseous photosphere, a curved surface, a *temperature gradient* (change of temperature with radial distance from actual center of the three-dimensional Sun) that increases inward, and the Stefan–Boltzmann law. The photosphere does not behave like a "solid" blackbody, as the radiation we see is actually the sum of that from different layers at different depths, wherein the radiation from each layer is attenuated by the opacity of the layers above it [7.2.3]. When you look into the center of the apparent solar disk, your line of sight is along a path perpendicular to the surface to a particular depth and you receive radiation from all along that line of sight. Now look to the limb. The line of sight penetrates to roughly the same physical distance, but since it is along an angle to the solar surface, not to the same perpendicular depth. On the average, the radiation from the limb therefore comes from cooler – and therefore dimmer – layers than does the radiation from the center. Cooler layers also mean that the limb is redder than the solar center.

Limb darkening/reddening is a double-edged sword. On the one hand, it allows direct measurement of the photospheric temperature gradient, which is a prelude to the general temperature gradient that takes us all the way to the solar core. The degrees of limb darkening for other stars can be derived from the light curves of eclipsing binaries [8.4.3], allowing the components' upper temperature gradients to be found as well. However, the phenomenon clouds the measures of the radii of stars as found from interferometry and lunar occultations [7.2.2], for which theoretical corrections must be applied.

The other characteristic is the heavy *granulation* of the solar surface (Fig. 12.2). Far from being smooth, the photosphere is actually composed of the order of a million bright "granules" that are separated by darker, cooler lanes. Doppler [5.2.2] observations reveal that the bright grains are columns of gas rising at speeds of several tenths of a kilometer per second, whereas the cool lanes are falling, implying that the bright patches are the tops of convection cells. The cells

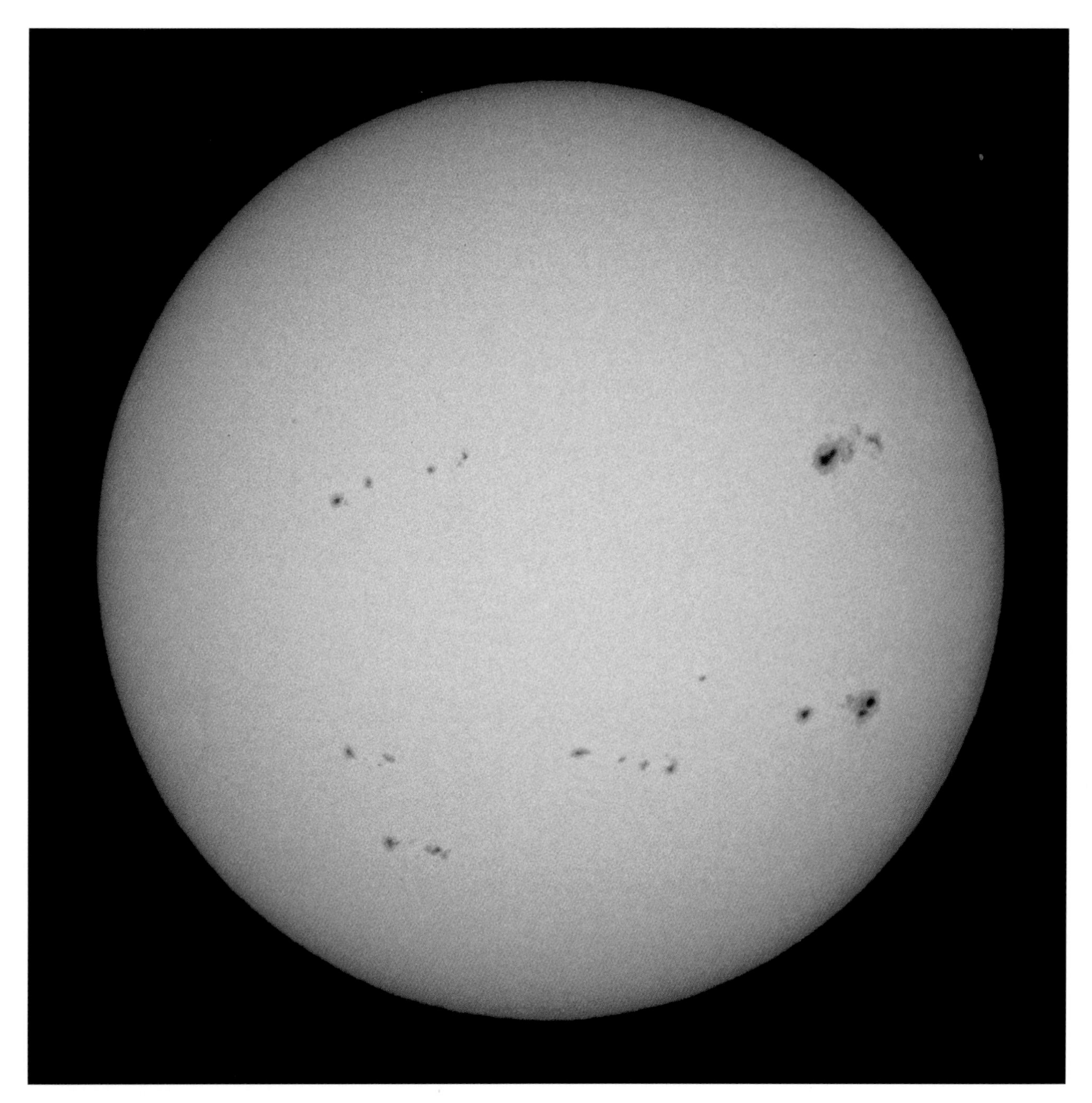

Fig. 12.1 **The Sun** A white light view of the Sun shows the darkening and reddening of the solar limb, and several magnetically active regions marked by sunspots. (Courtesy of Matt BenDaniel.)

continuously form and disappear, much as clouds on a summer day (which also mark the tops of convection cells). Typically 1000 or so kilometers in diameter, a granule will dissipate its heat and disappear over a timescale of no more than five to ten minutes, making the outer part of the Sun look like a pot of boiling oatmeal. Superimposed on this obvious structure is a "supergranule" pattern tens of times the scale of the granules (Box 12.1).

Given limb darkening and convection, the Sun (or any star) cannot really be characterized by any single temperature. At the reference level of the photosphere, at an optical depth [7.2.2] of 1 at 5000Å, the temperature is 6520K. At an "altitude" of 500km above this level, it drops to a minimum of 4400K. The effective blackbody temperature is only a kind of average over all the various layers through which we look [7.2.3], and of the granule–intergranule pattern, sunspots adding to the brew.

12.1.2 Sunspots and the magnetic activity cycle

The photosphere exhibits a variety of dark spots, *sunspots*. At times there are almost none at all, while at other times they appear nearly everywhere. They range in size from barely visible motes between granules to monsters much bigger than Earth. A sunspot consists of a dark, central, depressed *umbra*

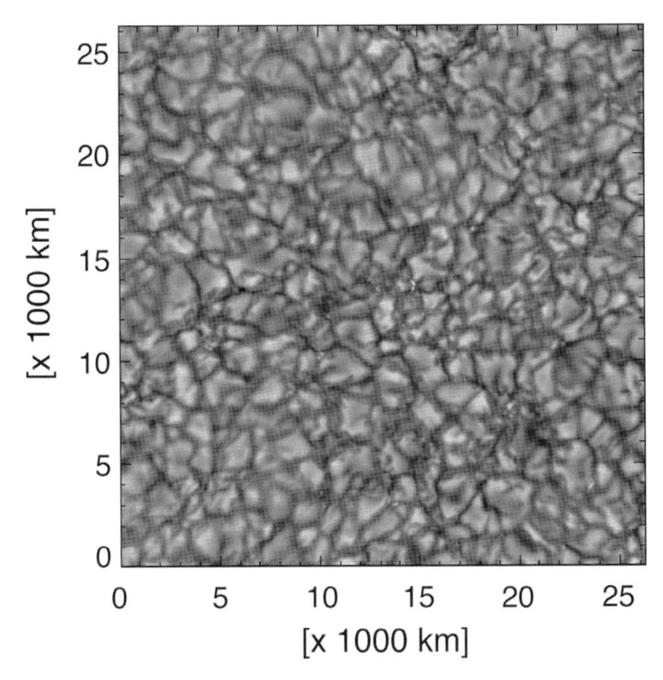

Fig. 12.2 **Solar granulation** The apparently smooth solar surface breaks up into myriads of granules, the tops of convection cells that are typically 1000km or so across. (Courtesy of T. Rimmele/NSO/AURA/NSF.)

that is connected to the bright photosphere by an upward-sloping, reddish, striated *penumbra* (Fig. 12.3). With temperatures of about 4500 kelvin, the umbrae are hardly black: they just look dark in contrast to the unspotted surface. Single spots are rather rare. Instead, they tend to come in pairs, and in larger groups within well-defined *centers of activity*. The Zeeman effect, the splitting of spectrum lines created in a magnetic field [7.1.6], shows the sunspots to be highly magnetized (field strengths 5000 times that of Earth's), with one spot of a pair having one polarity and the other the opposite polarity, indicative of a magnetic loop that has popped up through the photosphere. Within sunspot groups, the connections are awesomely complex. The intense magnetic fields inhibit the upward convection, and thereby refrigerate their surroundings. The fields are unconfined and dynamic, so all spots are doomed to extinction, some lasting a day or two, others for a month or more.

As we watch, the spots march across the solar surface, showing the Sun to be rotating, the phenomenon confirmed by Doppler measures at the limb. Because the Sun is gaseous, it does not rotate as a solid body, but *differentially*. At the equator, sunspot groups and the solar magnetic field give a sidereal rotation period of 25.0 days and an equatorial speed of 2.0km s^{-1}. (Because the Earth orbits in the direction of solar rotation at a degree per day, the "synodic" equatorial period relative to us is 26.8 days.) Toward the poles, the rotation period lengthens, to 26.4 days at 30° latitude, and to 31 days at 60°. Nearly all stars rotate this way, with, of course, very different periods [7.2.5]. The combination of convection and rotation are in turn responsible for the spot-producing fields. Together, the moving, ionized gases create a natural dynamo

Box 12.1 Convection

Energy can be transferred from one place to another in a variety of ways that include acoustic, shock, and magnetic waves, radiation (photons), conduction (atom-to-atom heat kinetic motion transfer), or convection. Convection transfers heat by mass motions. Heat a portion of a fluid (gas or liquid, or even a plastic) and it expands, becomes less dense relative to its surroundings, and like a helium balloon, floats; that is, rises upward. As the bubble rises, it also radiates heat to its cooler surroundings, thereby cooling itself, and finally becomes so dense that it sinks back downward, to start the process anew. A convective fluid is one that is in a state of constant circulation. Convection appears everywhere, from a pot on the stove, to the air in a room, to the semi-plastic mantle of the Earth (whose convection drives continental drift), to layers within stars.

Heat flows from higher temperatures to lower. The temperature in a star increases with depth, from the surface to the core. Whether any layer "convects" depends on whether a given blob of gas, when moved upward, will continue to climb or will drop back down to where it was to start with. That behavior in turn depends on how the density of the upwardly forced blob compares with its surroundings, and that in turn ultimately depends on the temperature gradient. If the gradient is sufficiently low (that is, if temperature changes only slowly with depth), convection will not set in, and energy is transferred by radiation. However, if the gradient increases beyond a critical value, a rising blob is continually forced upward, and the gas churns over. Though the process sounds simple, the conditions for convection and its extent is one of the great unsolved problems in astrophysics.

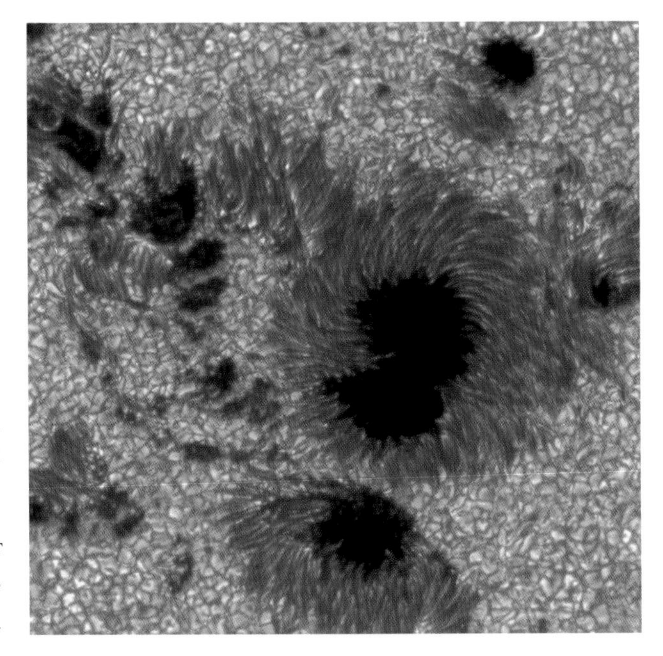

Fig. 12.3 **Sunspots** An extraordinarily detailed view of a sunspot group shows the spots to have dark central (depressed) umbrae surrounded by swirling (sloping) penumbrae, whose striations seem to follow magnetic field lines. Though spots are magnetically paired, it is difficult to tell which is coupled with which. Note the suppression of the convection. (Courtesy of Institute for Solar Physics, Royal Swedish Academy of Sciences.)

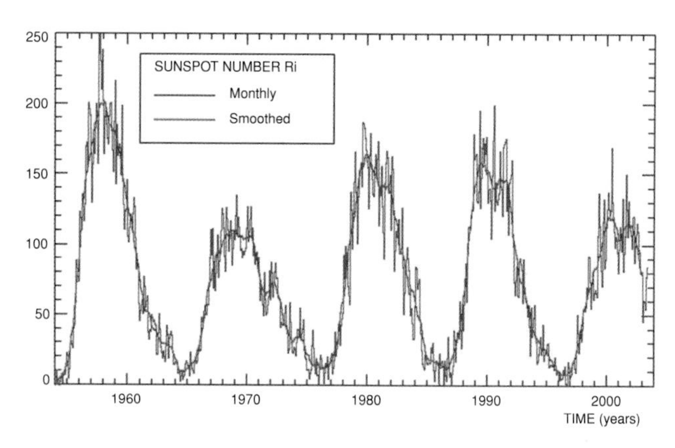

Fig. 12.4 **Sunspot cycle** Sunspot numbers wax and wane in an 11-year cycle, as do a variety of other related magnetic phenomena. Note the variation from one cycle to the next. The 1960 peak was the greatest on record, and was accompanied by spectacular auroral displays. (Courtesy of SIDC, RWC Belgium, World Data Center for the Sunspot Index, Royal Observatory of Belgium, 1954–2004.)

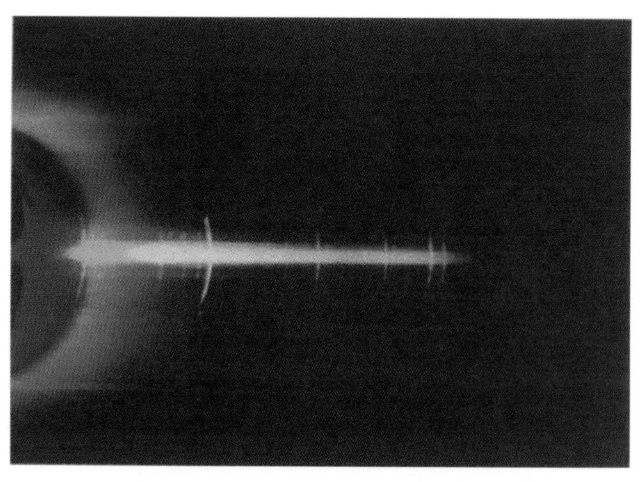

Fig. 12.5 **Flash spectrum** The emission spectrum of the chromosphere emerges at the beginning of a solar eclipse. (Courtesy of Richard Berry.)

that produces an averaged magnetic field similar in overall strength to Earth's, but then concentrates and twists it into powerful subsurface ropes. The ionized gas trapped by the magnetic ropes is buoyant, and it and the field are lofted to the surface. Where the field loops through the photosphere, we find the spots, one magnetically positive, the other negative.

The number of spots on the Sun at any one time varies dramatically over an interval that averages 11 years (Fig. 12.4). At the beginning of a cycle, the spots first appear in mid latitudes, and then as they increase in number, their mean position marches to the equator. Half a dozen years into the cycle, the number falls, and then after 11 years goes to near zero while at the same time the new cycle begins. In each hemisphere, over the course of a cycle, the spot of a pair leading in the direction of rotation will always have the same magnetic polarity as that of the hemisphere's pole. As a result, the polarities in one hemisphere are opposite those in the other. In the next cycle, all the polarities will be reversed, the full *magnetic cycle* taking 22 years rather than 11.

The cycle's origin seems to be the differential rotation. As time proceeds, the fields become ever more wrapped up, producing more and more spots. The sunspot fields do not really disappear, they just expand and scatter into the photosphere. At maximum complexity, the scattered remnants drift to their opposite polarities at the poles, and the whole thing re-orders itself to start anew. Other stars with outer convection zones (from class F on down) show much the same thing, allowing the measure of both their equatorial rotation periods and of the lengths of their spot cycles [7.4].

12.1.3 The chromosphere

At the low temperature point near the top of the photosphere, the solar gases have lost much of their opacity and become transparent. While continuously dropping in density with height above the photosphere, the temperature of the gas now remarkably begins to increase, from 4400 kelvin up to around 10 000 kelvin. As a hot, thin gas, this layer, about 2000km thick, is subject to Kirchhoff's third law [7.1.5], and thereby radiates an emission spectrum. Practically invisible against the photosphere, the warming layer becomes visible as a thin red band at the edge of the Sun during the progress of a solar eclipse, compliments of the strong Hα line in emission [7.1.4]. The vivid color gives the layer the name of *chromosphere* (from Greek "chromos," for "color"). As the photosphere is covered in the eclipse, a spectrograph reveals myriad emissions which, since they are seen only briefly before the Moon covers the layer, is called the *flash spectrum* (Fig. 12.5). Since the flash spectrum switches photospheric absorptions for emissions, the chromosphere is sometimes called the "reversing layer." In 1868, one emission at 5876Å with no photospheric counterpart (labelled D_3 for its proximity to the sodium D lines [6.1.1]) could not be identified with any element then known, so it was ascribed to a "new" element, helium, named after the Sun itself (the element not discovered on Earth until 1895). Among the best-known chromospheric spectral features is the complex K line of ionized calcium, which displays an absorption (K3) in the middle of the emission (K2), that lies inside the photospheric absorption (K1) [4.4.6].

The chromosphere can be seen directly by taking an image of the Sun in the light of one of the flash spectrum's emissions (or in a strong photospheric absorption in which the high opacity does not allow our vision to penetrate to the photosphere). The result is called a *spectroheliogram* (Fig. 12.6). The chromospheric "surface" is composed of turbulent "spicules" that are apparently related to the Sun's five-minute oscillation pattern [12.3], which injects hot gases upward through existing magnetic tubes. The higher temperature of the chromosphere appears to be a distinct violation of physical laws, in that a cooler body (the photosphere) cannot heat a hotter one. Photospheric radiation can therefore have nothing to do with the inverted temperature gradient.

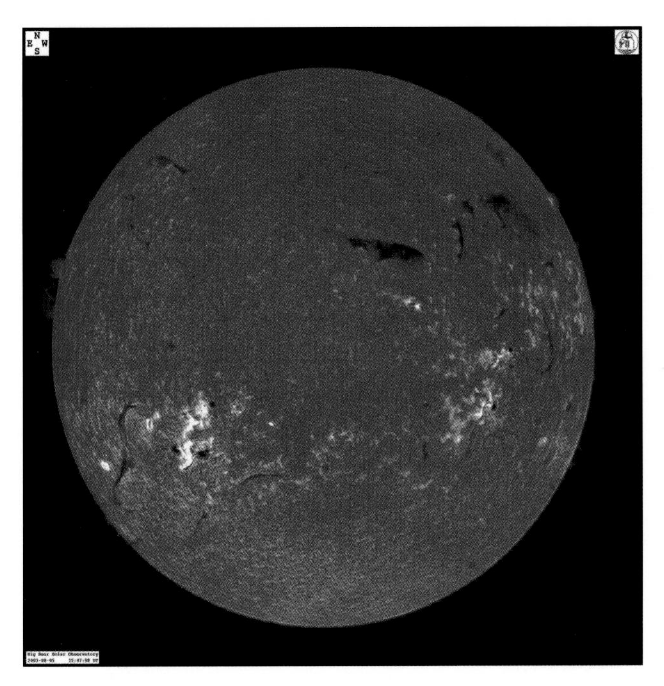

Fig. 12.6 **The solar chromosphere** A spectroheliogram of the chromosphere, taken in the light of the K line of Ca II, shows bright magnetic *plages* hovering above centers of activity, dark cool magnetically-supported *filaments* and *prominences*, which are filaments seen against the sky, where they appear bright. (Courtesy of Big Bear Solar Observatory, New Jersey Institute of Technology.)

Fig. 12.7 **The solar corona** Magnetically confined coronal streamers, made visible by a total eclipse of the Sun, fade out into space. The corona does not really end, but through the solar wind extends past the Earth to beyond the planetary system. (Copyright 2004 by Fred Espenak, www.MrEclipse.com.)

While the heating process remains mysterious, it probably involves energy deposited by the magnetic fields that produce the sunspots, as well as by shock waves set up by the photospheric convection.

Though other stars are too far away to allow resolution of their chromospheres (providing they have any), emission lines, particularly of H and K of Ca II [6.1.1] and of h and k of Mg II (deeper in the ultraviolet), give them away, allowing us to follow their magnetic cycles, and even to measure their distances through the Wilson–Bappu effect [4.4.6].

12.1.4 The *corona*

A total eclipse of the Sun reveals a stunning white pearly "crown" around the Sun, the solar *corona*, which extends ever-fainter outward for several solar radii before fading to invisibility (Fig. 12.7). The optical spectrum of the corona consists of emission lines superimposed upon a continuum. The outer observed coronal continuum, which contains solar absorption lines, is caused by the scattering of sunlight from interplanetary dust grains, and is of little consequence. The inner corona, however, has no absorptions. It is created by sunlight scattering from free electrons that must be moving so fast that they Doppler-broaden [7.1.6] the solar absorptions right out of visibility. The emission lines were initially unidentified, and were surmised to come from a "new element" called "coronium." In the mid twentieth century, they were finally found to come from highly ionized metals (particularly iron and nickel) that have half or more of their electrons stripped away. To achieve such conditions, the average temperature must be of the order of two million kelvin. Powerful X-rays confirm the deduction (Fig. 12.8). Given its optical faintness, the corona is far from being a blackbody. (X-rays from other convective stars show that they too can have coronae similar to the Sun's.) The corona is connected to the lower chromosphere by a narrow *transition layer* in which the temperature suddenly climbs through the tens, then hundreds of thousands, of kelvin.

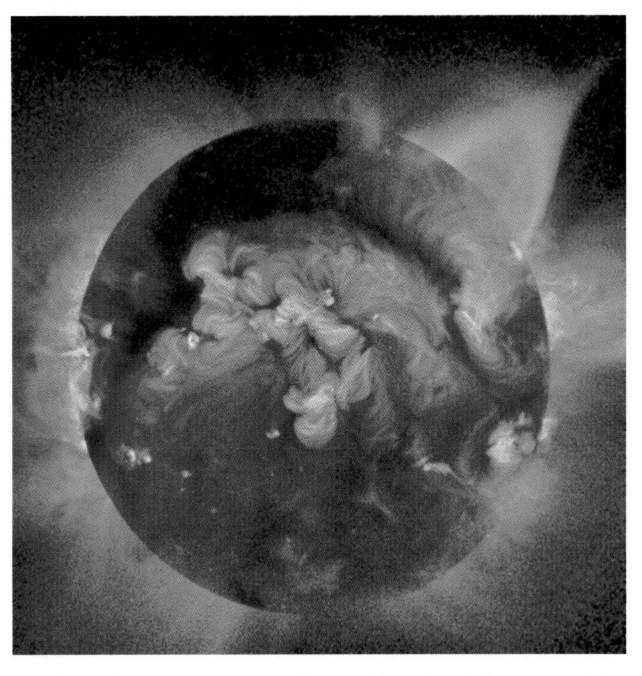

Fig. 12.8 **The X-ray corona** X-rays allow a full-face view of the corona, while the photosphere is not visible at all. Obvious magnetic loops confine the two-million-kelvin solar gas; the solar wind blows through the holes. (Courtesy of Yohkoh X-ray Telescope, Lockheed-Martin, National Observatory of Japan, NASA/ISAS.)

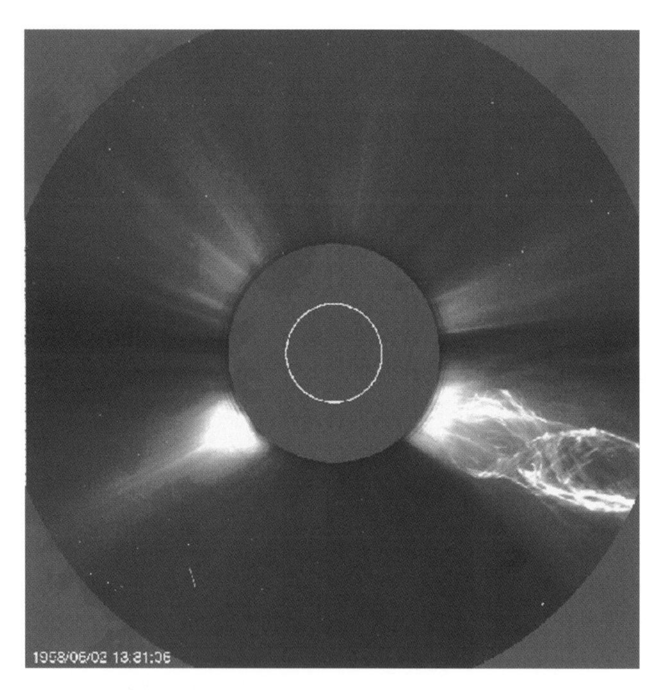

Fig. 12.9 **Coronal mass ejection** A spectacular coronal mass ejection (CME) blasts off the Sun as the result of violently relaxed magnetic fields. A central disk in the SOHO spacecraft hides the bright Sun. (Courtesy of SOHO LASCO consortium, ESA and NASA.)

Neither photon absorption nor acoustic waves can heat the corona. The only viable mechanism is the deposition of solar magnetic energy. In confirmation, the inner corona is made of a fantastic number of magnetically confined loops that relate to the sunspot loops and to wider loops that connect centers of activity. Though partial theories abound, no one yet really knows how the mechanism actually works.

Where the corona is not confined by magnetism, the hot coronal gases expand from *coronal holes* to create the *solar wind*. This thin gas blows past the Earth at a temperature of 150 000 kelvin with speeds between 200 and 700 km s^{-1}, filling the Earth's magnetic field with particles and causing the sky to glow faintly. As a consequence, we and the other planets orbit within the extended solar corona, the wind responsible for blowing the gas tails of comets backwards [11.4.1; Fig. 11.16]. Through the wind, the Sun loses mass at a rate of 10^{-14} solar masses per year, about one Earth mass in a billion years (making it of little consequence to early solar evolution).

The magnetic fields that heat the corona can interconnect and neutralize each other. As they collapse, they accelerate electrons both upward and downward. The latter collide violently with the chromosphere to create a sudden brightening, a *solar flare*. The disappearance of the field releases the confined coronal gases into an outward-bound *coronal mass ejection*, or CME (Fig. 12.9). If it hits the Earth, the disturbance of our own magnetic field lights the sky in an *aurora* (*borealis* in the north, *australis* in the south). Not only can a CME wipe out the electronics aboard an Earth satellite, but the electrical disturbances can trip circuit breakers in power grids. "Space weather" is therefore continually monitored to provide warnings of such events.

All solar activity, including the appearance and intensity of the corona, depends on the 11-year solar magnetic cycle. In ways yet unknown, magnetic activity is funnelled to, and affects, the Earth. During peak magnetic activity, the atmosphere expands, to the detriment of orbiting satellites. Moreover, the average global temperature depends on the length between cycles; the shorter they are the warmer we are. Every few hundred years, the cycle apparently shuts down (as it did in the *Maunder Minimum* between 1650 and 1715), and the Earth chills in a "mini ice age." The Sun affects us not just by its great luminosity, but through its rotation as well.

12.2 Inside the Sun

The surface features of the Sun and stars are driven by the processes and events that take place deep inside, within the regions that stellar photospheres seem to hide from view and that extend all the way to the stellar centers (Fig. 12.10). There are several ways, however, to "see" into stars, to assess their internal natures. First is by theory, in which the astrophysicist uses known physical laws to construct a model of pressure, density, temperature, energy generation, internal chemical composition, and more, as a function of depth within the star. To varying degrees, theory is applicable to all stars.

Second is to use natural oscillations of stellar surfaces to probe the interiors. Oscillations have revolutionized our view of the Sun, and are becoming ever more useful in probing stars. Third is the direct observation of neutrinos, minute atomic particles generated by the nuclear reactions that power the stars and that immediately escape the stellar interiors [12.2.2]. Neutrino emission is vital in checking solar theory, and has had limited, though powerful, application to other stars. To these, add stellar shapes caused by rotation [7.2.5] and the rotations of the semimajor axes of close binaries [8.4.3], both of which depend upon internal constructions.

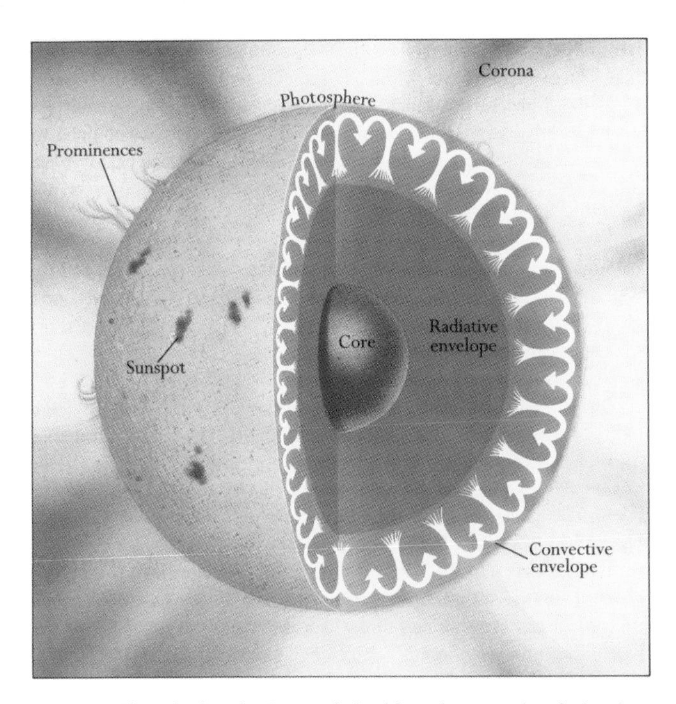

Fig. 12.10 **The solar interior** Energy, derived from thermonuclear fusion, is generated only in the solar core, which extends about 25 percent of the way to the solar surface. Surrounding it is a radiative envelope, which is encased in a convective envelope that begins 71 percent of the way out. Magnetically activated spots appear at the surface, while all is surrounded by a hot corona. (Kaler, J.B. (1992). *Stars*, Scientific American Library, New York: Freeman.)

12.2.1 Solar/stellar structure

The calculation of the interior construction of a star requires the mathematics of calculus. Though not generally used here, we can easily approach it by looking at simple differences in conditions as we probe into the solar interior. In their simplest forms, theoretical stellar models are derived from four equations.

1. The first is that of *hydrostatic equilibrium*. Think of a layer of gas within a star of thickness Δr (in physics, "Δ" stands both for "uncertainty" and "spread"). A unit area cross section of the layer has a mass $\rho\Delta r$, where ρ is density. From Newton's law of gravity [8.2.3], the section's weight is $G\rho\Delta r M_r/r^2$, where r is the distance of the layer from the center of the star, and M_r is the mass interior to distance r (which is all that acts on the cross section; the outer mass is irrelevant). For this layer to keep its position, the inward pull of gravity must be balanced by the outward change in pressure, ΔP, equal to the pressure at the bottom of the layer minus that at the top (where P is proportional to the number density of atomic particles and the temperature [13.2.1]). As a result,

$$\Delta P = -G\rho\Delta r M_r/r^2,$$

(the minus sign coming from the fact that pressure increases with decreasing r), which can be written as

$$\Delta P/\Delta r = -G\rho M_r/r^2.$$

The pressure, however, does not change in discrete jumps, nor is the density constant, even within Δr. If, however, you make Δr, and consequently $\Delta\rho$, ever smaller and smaller, and proceed to the infinitesimal limit (which in mathematical parlance converts "Δ" to "d"), you have a perfectly continuous function, now written as

$$dP/dr = -G\rho M_r/r^2,$$

which is a *differential equation* written in the language of calculus. Such was the inventive genius of Newton and Leibniz, working on the theory of limits first derived by Descartes.

The outward push of gas pressure is supplemented by the pressure produced by radiation being absorbed by the gas, which depends on the intensity of the radiation and the gaseous opacity. Radiation pressure is unimportant in the Sun, but must be taken into account among the more luminous stars. It is crucial in limiting the sizes of massive stars that approach the "Eddington limit," at which point the radiation pressure produced by the scattering of radiation from electrons (an important opacity source) equals the inward pull of gravity, the two just balancing each other.

2. Next is the *mass equation*. The stellar layer of thickness Δr is actually a spherical shell of radius r, which has a surface area of $4\pi r^2$ and consequently a volume of $4\pi r^2 \Delta r$. The mass of this layer is thus $\Delta M = 4\pi r^2 \rho \Delta r$, or (squeezing the deltas to the infinitely small),

$$dM/dr = 4\pi r^2 \rho.$$

3. The *luminosity equation* gives the rate at which energy is generated (which involves the rates of the internal thermonuclear fusion reactions generated from the conversion of hydrogen into helium [12.2.2]). It is similar to the mass equation, except that luminosity L is substituted for mass, and the amount of energy produced per unit volume (ϵ) is substituted for density, hence

$$dL/dr = 4\pi r^2 \epsilon.$$

The rate ϵ has a finite value only within the solar core, and is zero outside of it where the temperature is too low to allow fusion.

4. Finally we look at the *temperature gradient*, ΔT, across a layer of thickness Δr (which converts to $\Delta T/\Delta r$, hence to dT/dr). Assume that energy is transferred by radiation, and not convection (which introduces much greater complications). The temperature gradient depends on the opacity of the gas averaged over wavelength, expressed by Greek kappa (κ); the greater the opacity (and the greater the density, ρ), the greater the temperature gradient needed to nudge the radiation along. At the same time, the greater the luminosity, the more radiation there is to get out, which also requires a greater gradient. At the end,

$$dT/dr = -(\text{constant})\kappa \rho L_r / T^3 r^2.$$

If the gradient gets too steep, convection sets in, whence

$$dT/dr = -(\text{approximate constant})\, \mu M_r / r^2,$$

where μ is the mean molecular weight [Box 11.3].

Recalling the basic problem in algebra of two equations in two unknowns, here are four *differential* equations in "four unknowns" (as functions) to be solved simultaneously so that they also fit or satisfy various observed constraints, such as mass, luminosity, chemical composition, surface temperature, how the composition might change as a result of energy generation, and so on. The result is a *solar* (or *stellar*) *model* that gives the run of temperature, pressure, density, energy-generation rate, and so on, from the core's center to the photosphere (Figs. 12.10 and 12.11).

12.2.2 Energy generation and the final model

The key is the energy generation mechanism. Chemical combustion is woefully inadequate to produce the solar luminosity. The most obvious energy source is gravity through *Kelvin–Helmholtz* (after Lord Kelvin and Hermann von Helmholtz, 1821–1894) *contraction*. As the Sun compresses under the force of the consolidated gravity of its entire mass (which is what makes the Sun a sphere), it heats inside, the deeper the hotter. Each layer acts like a blackbody that radiates its energy toward the outside, the direction in which the temperature decreases. The removal of energy would cool the gas and drop the pressure, so to compensate and to maintain the equilibrium, the Sun (or any star) must therefore shrink as it radiates.

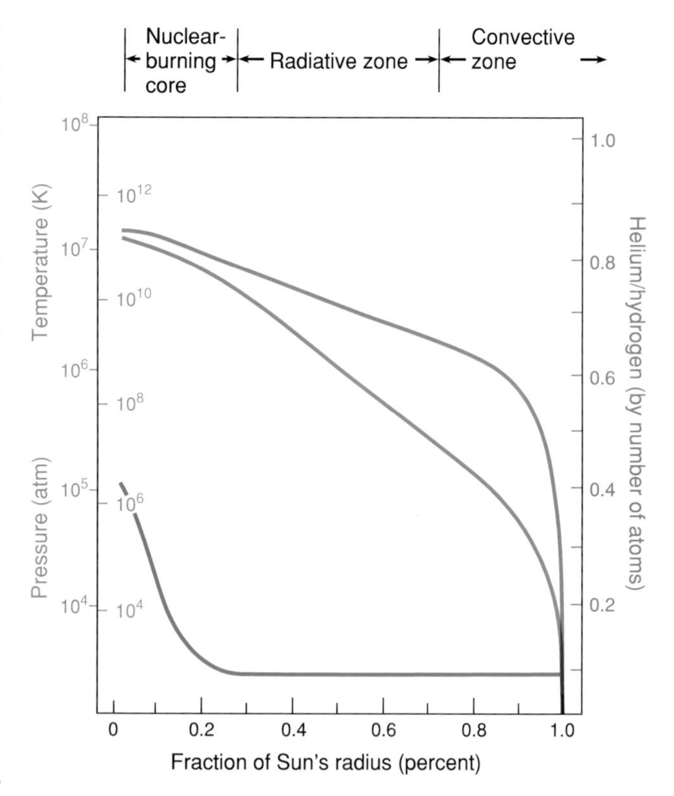

Fig. 12.11 **A solar model** A complete model of the Sun gives physical parameters as a function of radius from the center to the surface. (Adapted from Robert Noyes in Beatty, J.K. and Chaikin, A. eds. (1990). *The New Solar System*, third eds., Cambridge University Press and Sky Publishing Corp, as rendered in Kaler, J.B. (1994). *Astronomy!* New York: HarperCollins, reprinted by permission of Pearson Education, Inc.)

Although Kelvin–Helmholtz contraction would work to produce the solar luminosity, it cannot produce it for a long enough period of time. To generate 4×10^{26} watts, the Sun would need to shrink by about 20 meters per year, and would therefore last just a few million years. But the geological record shows the Solar System, hence the Sun, to be 4.6 billion years old. Kelvin–Helmholtz contraction will not work except to make the core of the Sun hot enough in the first place to run the real mechanism, *thermonuclear fusion*, in which one element is converted to a heavier one: in the case of the Sun, hydrogen into helium.

The discovery of fusion power required many smaller discoveries that would come together. First, that four atoms of hydrogen are 0.7 percent less massive than one atom of helium. If nature can make helium from hydrogen, mass is lost. Second, that mass and energy are equivalent through Einstein's $E = Mc^2$, where E is energy, M is mass, and c is the speed of light. Third, from the analysis of the solar spectrum that the Sun is made mostly of hydrogen (as are most other stars), so that there is plenty of fuel. Fourth, the knowledge of the different atomic particles and how they work, and of the forces that hold them together.

Nuclear fusion (*nuclear burning* in the vernacular) – can take place only at the extremely high temperatures of the solar core, such that protons (and other nuclei) can move sufficiently fast to overcome their electrical repulsion and get close enough to stick via the strong force [Box 7.4] (rendering the process *thermonuclear* fusion). Only the extreme tail of the Maxwellian velocity distribution [Box 7.3] can do the job, and even then only by means of quantum mechanical "tunneling," in which the wave nature of the proton [Box 7.4] allows it to jump through electrical barriers rather than having to force itself over them.

Numerous reactions are involved, some more important than others. The basic nuclear process that runs the Sun, producing 93 percent of its luminosity, is the *proton–proton chain* (the p–p chain), which requires a minimum temperature of about eight million kelvin to run to full completion. It begins with the collision of two protons that come together so violently that they can momentarily be held by the strong force. But even that cannot keep them bound. Instantaneously, one of them releases its positive charge via the weak force as a *positron*, a positive electron (an example of antimatter, matter with ordinary properties reversed), which turns the proton into a neutron with the release of a nearly massless particle called the *neutrino*. Since the neutron also carries the strong force, the two become locked to make deuterium, ^{2}H. (A bit, less than one percent, of the deuterium is also made by a three-body collision between two protons and an electron, with the ejection of a neutrino).

Protons and neutrons are held together under the strong force with powerful negative binding energies [Box 7.4]. Upon fusion, the energy released to bind the protons goes into giving the combined nucleus a huge kick, a higher speed, a greater "kinetic energy" that in turn goes into heating the surrounding gas. All the energy-generating fusion reactions act the same way. Add to that the effect of the positron. Matter and antimatter annihilate each other with the production of energy. The positron does not get far before encountering an electron, and the two disappear in a burst of two gamma rays. The deuterium nucleus quickly picks up another proton to make ^{3}He with the creation of another gamma ray. The gamma rays do not get far before absorption by an ion, and thereby add to the deposition of heat into the stellar core.

Now things get tricky, as there are four routes to full helium (^{4}He) glory (and more energy). Eighty-five percent of the total solar p–p energy comes from "Branch 1," the fusion of two ^{3}He nuclei to form ^{4}He (with the deposition of more kinetic energy) with a pair of protons left over, which is where elementary textbooks usually stop. A significant fraction, however, 15%, comes from a more complicated "Branch 2," in which ^{3}He and ^{4}He combine to make beryllium as ^{7}Be and another gamma ray. ^{7}Be is very unstable, its half-life [Box 7.4] on Earth a mere 53 days. Before it decays, it absorbs an electron, which turns one of its protons into a neutron, converting it to the next *lower* atom, lithium as ^{7}Li (which is stable) and a neutrino. The ^{7}Li then absorbs a proton, which would make it into ^{8}Be, but which is so wildly unstable (half-life 10^{-17} seconds) that it immediately becomes a pair of ^{4}He nuclei. In "Branch 3" (0.02 percent of the p–p energy), the ^{7}Be of Branch 2 nails a proton instead of an electron to become boron (^{8}B) and a gamma ray, which then decays into a ^{8}Be nucleus with a gamma-ray-producing positron and a neutrino, which again instantly splits into two ^{4}He atoms. An even-rarer "Branch 4" reaction (0.00002%) has ^{3}He (made in the second step of the p–p chain) absorbing a proton to make ^{4}He directly, along with a positron and again a neutrino. All these are summarized in Box 12.2.

Through successive absorption and re-emission of ordinary blackbody radiation, the heat slowly works its way out of the Sun by radiation along the gentle downward slope of the temperature gradient (aided by outer-layer convection). Each shell within the star outside the core must transfer the same amount of energy. But on the average, each energy release is at a lower temperature, so as the energy moves outward, the star passes more and more of lower and lower energy photons. After hundreds of thousands of years to make it to the cooler surface, the energy is released as a spray of relatively low-energy optical photons. As near-massless particles moving near the speed of light, the neutrinos carry energy away as well, but since they are part of the weak force, they do not much interact with atomic particles, and escape the Sun immediately, without causing any significant heating of the solar gases.

Still missing is seven percent of the solar energy. It comes from the *carbon cycle* (or CN cycle or CNO cycle), which uses carbon as a nuclear catalyst to make helium from hydrogen. Ordinary ^{12}C first picks up a proton, which changes it into ^{13}N with the ejection of a gamma ray. Within a few minutes, the highly unstable nitrogen ejects a positron and a neutrino to become stable ^{13}C, which absorbs first one proton to make common ^{14}N (plus a gamma ray) and then another to make ^{15}O (plus a gamma), which quickly decays with the ejection of another positron and neutrino into stable ^{15}N. At this point, the cycle branches. The more common route is for the ^{15}N to collide with a proton, but instead of making ^{16}O, it divides to re-create the "original" ^{12}C nucleus plus ^{4}He. Summing it all up, four hydrogens were added to the original carbon to make helium with the return of the carbon. The second, more complex, branch goes through ^{16}O, fluorine (^{17}F), ^{17}O, and then back to ^{14}N and a ^{4}He nucleus with the usual neutrinos and gamma rays (spelled out in Box 12.2). All these various reactions (in both the p–p chain and the CN cycle) have very different sensitivities to temperature, so the percentages above are averages over the solar interior. The relative rates also change as solar ageing causes the chemical composition to change, and are quite different in stars of different masses.

Why does the Sun not explode like a hydrogen bomb? The first p–p reaction, which starts all the branches, is exceedingly slow. Under current solar conditions, the "half-life" of a proton to merge with another proton is 14 billion years! The second reaction then takes only 6 seconds. The first maintains tight control, preventing any kind of runaway reaction. The proton-absorbing steps of the carbon cycle serve the same purpose, as none of them goes very fast. The Sun cannot explode, nor can similar stars.

From the final solar model, which incorporates the energy sources, the Sun hits the key fusion temperature of 8 million kelvin at a radius about 25 percent that of the photosphere, which occupies only about 1.5 percent of the solar volume, but because of the huge internal density, includes 45 percent of the solar mass. When the proper energy-generating mechanism is used, the Sun is seen to have a central temperature of 15.6 million kelvin and a density of 160 grams per cubic centimeter, 11 times that of lead – and it is still a gas. This nuclear-burning core, with some mass from its outer surroundings added to it, will someday become the final product, a white dwarf [13.5].

Surrounding the energetic core is a quiet envelope that passes energy along the temperature gradient purely by radiation. At a radius 70 percent that of the photosphere, the solar gases suddenly become convective (see Box 12.1), the circulation ultimately seen as photospheric granulation (Figs. 12.10 and 12.11).

12.2.3 Neutrinos

Among the longest standing major problems in astrophysics was that involving the lowly neutrino. Neutrino detection allows a direct view into the solar core. But as weak force particles, neutrinos only barely interact with matter: it would

Box 12.2 Hydrogen-burning reactions

[e^- = electron; e^+ = positron; γ = gamma ray; ν = neutrino]

Proton–Proton Chain

(93% solar energy)

(0.996) $^1H + {}^1H \rightarrow {}^2H + e^+ + \nu$

$e^+ + e^- \rightarrow 2\gamma$

(0.004) $^1H + e^- + {}^1H \rightarrow {}^2H + \nu$

$^2H + {}^1H \rightarrow {}^3He + \gamma$

Branch 1 (85%) $^3He + {}^3He \rightarrow {}^4He + 2\,{}^1H$

Branch 2 (15%) $^3He + {}^4He \rightarrow {}^7Be + \gamma$

$^7Be + e^- \rightarrow {}^7Li + \nu$

$^7Li + {}^1H \rightarrow 2\,{}^4He$

Branch 3 (0.02%) $^3He + {}^4He \rightarrow {}^7Be + \gamma$

$^7Be + {}^1H \rightarrow {}^8B + \gamma$

$^8B \rightarrow {}^8Be + e^+ + \nu$

$e^+ + e^- \rightarrow 2\gamma$

$^8Be \rightarrow 2\,{}^4He$

(Branch 4 (0.00002%) $^3He + {}^1H \rightarrow {}^4He + e^+ + \nu$

$e^+ + e^- \rightarrow 2\gamma$

Carbon Cycle (CNO Cycle)

(7% solar energy)

$^{12}C + {}^1H \rightarrow {}^{13}N + \gamma$

$^{13}N \rightarrow {}^{13}C + e^+ + \nu$

$e^+ + e^- \rightarrow 2\gamma$

$^{13}C + {}^1H \rightarrow {}^{14}N + \gamma$

$^{14}N + {}^1H \rightarrow {}^{15}O + \gamma$

$^{15}O \rightarrow {}^{15}N + e^+ + \nu$

Branch 1 (99.96%) $^{15}N + {}^1H \rightarrow {}^{12}C + {}^4He$

Branch 2 (0.04%) $^{15}N + {}^1H \rightarrow {}^{16}O + \gamma$

$^{16}O + {}^1H \rightarrow {}^{17}F + \gamma$

$^{17}F \rightarrow {}^{17}O + e^+ + \nu$

$e^+ + e^- \rightarrow 2\gamma$

$^{17}O + {}^1H \rightarrow {}^{14}N + {}^4He$

Box 12.3 Astonishing numbers

An individual fusion reaction produces very little energy. The luminosity of the Sun therefore requires huge numbers of such reactions. For simplicity, assume that all the solar energy is run by Branch 1 of the p–p chain. The making of one helium nucleus requires the melding of 4 hydrogen atoms and the loss of 0.007 of their combined masses, which is converted to energy. Given the mass of the proton (1.67×10^{-27}kg), for each helium nucleus created, 4.2×10^{-12} joules of energy is liberated. (The familiar watt, a unit of power, is one joule released per second.) The solar luminosity of 4×10^{26} watts requires $4 \times 10^{26}/4.2 \times 10^{-12} = 9.5 \times 10^{37}$ complete reactions per second, and the total consumption of 4×10^{6} metric tons of mass per second.

Each helium nucleus requires the creation of two neutrinos, so 2×10^{38} of the tiny particles are released per second. At the distance of the Earth, that number must pass through a sphere with a radius R of 1 AU $= 1.5 \times 10^{13}$cm. The number of neutrinos passing through one square centimeter of the Earth per second – and of you – is therefore $2 \times 10^{38}/4\pi R^2 = 7 \times 10^{10} = 70$ billion. And you are completely unaware of them.

take the order of a light year of lead to provide significant "optical depth" to them. The vast majority pass directly out of the Sun then go right through the Earth without even slowing down. Yet, since the weak force exists, neutrinos *must* interact at some level. And because of their huge numbers, a flux of 70 billion per square centimeter per second at the Earth (Box 12.3), even an exceedingly low reaction rate has to produce some effect.

The scientific return of neutrino detection is so important that a considerable variety of "neutrino telescopes" have been built around the world. The first was Ray Davis's 100 000 gallon vat of chlorine cleaning fluid, buried a mile deep in Lead, South Dakota. The common isotope of chlorine, ^{37}Cl, has a particular affinity for neutrinos. A rare hit with one that is above a particular energy threshold converts the atom into ^{37}Ar ($^{37}\text{Cl} + \nu = {}^{37}\text{Ar} + e^-$). The argon is radioactive, and decays back to ^{37}Cl at a known rate with the emission of an electron, which then allows the number of impacting solar neutrinos to be counted. Factoring in the probability of capture allows the neutrino flux above the critical reaction energy to be counted. Unfortunately, only p–p Branches 2 and 3 (which run through beryllium and boron) produce neutrinos with sufficient energy, as given by the neutrinos' wavelengths. Since these branches generate only a small portion of the solar energy, a count of the total number of neutrinos of all energies requires multiplication of the observed number by a large correction factor.

The chlorine experiment consistently counted only a third the number expected. Later detectors that used other isotopes, and the count rate of collisions between neutrinos and electrons in pure water (which causes light to be released), told the same story. There were two alternatives. Either the core of the Sun is cooler than expected (which no model could account for) or something was wrong the standard picture of neutrinos.

The original theory had neutrinos as massless. There are three kinds of neutrinos, however, each of which is associated with a different "family" of matter. Among the zoo of atomic particles, electrons have two (so far as we know) heavier, unstable analogues, the "muon" and the "tau particle", each of which has its own kind of neutrino. Those made in the Sun (and captured by chlorine or ordinary water) are of the electron variety; the others are undetectable. If neutrinos have mass, theory allows them to switch from one kind to another in their eight-minute flight to Earth, which would account for their deficit. Finally, a heavy water (D_2O) neutrino detector that could count all the different kinds revealed the predicted number emerging from the Sun, clearly demonstrating both that our theory of the solar interior is correct, and that neutrinos indeed have mass, a prime example of the results from one branch of science feeding into another. The Sun works just as we thought; neutrinos did not.

12.3 Oscillations

Pluck a guitar string, and it vibrates with a lowest frequency that depends on its thickness and length. Fastened at the top and bottom, at forced "nodes" where it cannot move, the string will wiggle up and down in a "standing wave" that gives the fundamental wavelength (or frequency), or the audio note on the scale. It can also vibrate at half the string's length with an additional node in the middle, which gives the first overtone, and which will have a lower volume such that the fundamental is still nicely heard. In the first overtone, one half of the string moves in one direction while the other moves oppositely, it and the fundamental piling atop each other. More nodes mean more overtones. The gathering of overtones and how they interact with each other tells the listener the nature of the instrument. The same vibrational phenomenon occurs in the air within an organ pipe or in the tube of a trumpet.

Stars vibrate much the same way. A star is not a one-dimensional string, however, but a three-dimensional sphere, and therefore can "vibrate" in three dimensions as well. The fundamental "tone" corresponds to the whole star breathing in and out along all its radii at the same time, that is, it pulsates *radially* with no dependence on latitude and longitude. Overtones, however, are different. In an overtone, one part of the star will be moving out while another will be moving inward, as happens for the guitar string; that is, it will be pulsating *non-radially*. Non-radial surface pulsations can be divided into those whose nodes lie parallel to the rotational equator (along parallels of latitude), and those that lie perpendicular to it along meridians of longitude. They are independent of each other: there can be multiple nodes in one direction and none in the other, or for that matter any combination at all. With high numbers of nodes in each direction, the star can take on a spectacularly pulsing appearance.

At the pinnacle of pulsating stars are the Cepheid [10.3.1] and Mira [10.5.1] variables that may be pulsating in their fundamental radial "tones" (periods), their first overtones, or both at the same time. Delta Scuti, Beta Cephei, and ZZ Ceti stars [10.4.1, 10.4.2, 10.4.3] sing much more complicated songs, with multiple periods and complicated non-radial pulsations. The closer we look, the more complex such stellar vibrations become, until with the finest view – that of the Sun – we see millions, all piled atop each other.

The pluck of a finger, or the friction of bowing, sets a musical string into vibration. In the Sun, the action is caused by the mass motions of convection that send columns of hot gas climbing into the photosphere, at speeds of a good fraction of a kilometer per second, to produce the ubiquitous granulation. Sound waves with millions of frequencies thereby rock the Sun in a cradle of non-radial pulsation. The result is an intricate rolling action with an amplitude that can reach 10km and velocities in the hundreds of meters per second. The sound waves conspire to give the Sun a dominant five-minute oscillation (the first discovered), upon which are superimposed countless others. Even if you could approach the Sun through some undreamed-of technology, you could never hear them, as the frequencies are far below those that can be detected by the human ear (Fig. 12.12).

The path of a wave (whether light or sound) through a substance is not necessarily a straight line, but is bent or refracted by variations and sudden discontinuities in density, temperature, and chemical composition (molecular weight [11.3]). Waves in the Earth, produced by earthquakes, rumble through the entire planet. They change direction in response to internal structure, and even disappear. The progression, speeds, and strengths of these waves, as monitored by motion detectors ("seismographs") spread around the planet, form the basis of the science of *seismology*, and allow a "tomography" of the terrestrial interior and a highly refined knowledge of its construction.

The sound waves in the Sun behave similarly, leading to the analogous science of *helioseismology*. Refracting within the solar gases and at discontinuities, bouncing off the lid of the photosphere, some waves penetrate through to the core, while others stay near the surface. Their presence is made known by photospheric velocity changes observed through the Doppler effect, and by subtle changes in point-to-point solar brightness. Helioseismologists then use them to "look" deeply into the Sun to build a picture of its internal construction. From them they find the convection layer begins at a radius of 71.3 percent that of the photosphere, find the helium-to-hydrogen ratio, and determine a temperature for the core of 15.6 million degrees kelvin, quite in accord with that found from theoretical models. Even the interior rotation falls to solar seismology. Sound waves running around in the direction of rotation have a different vibrational frequency than those running oppositely, such that an oscillation is "split." Rotational splitting allows the discovery that the differential rotation extends down through the convection zone to the radiative zone, to the *tacholine* where it becomes more like solid body rotation. This discontinuity is thought to be the site of the dynamo where the powerful solar magnetic field is generated. (But tell that to low-mass magnetic red dwarfs where convection extends all the way through and there is no tacholine [10.7].)

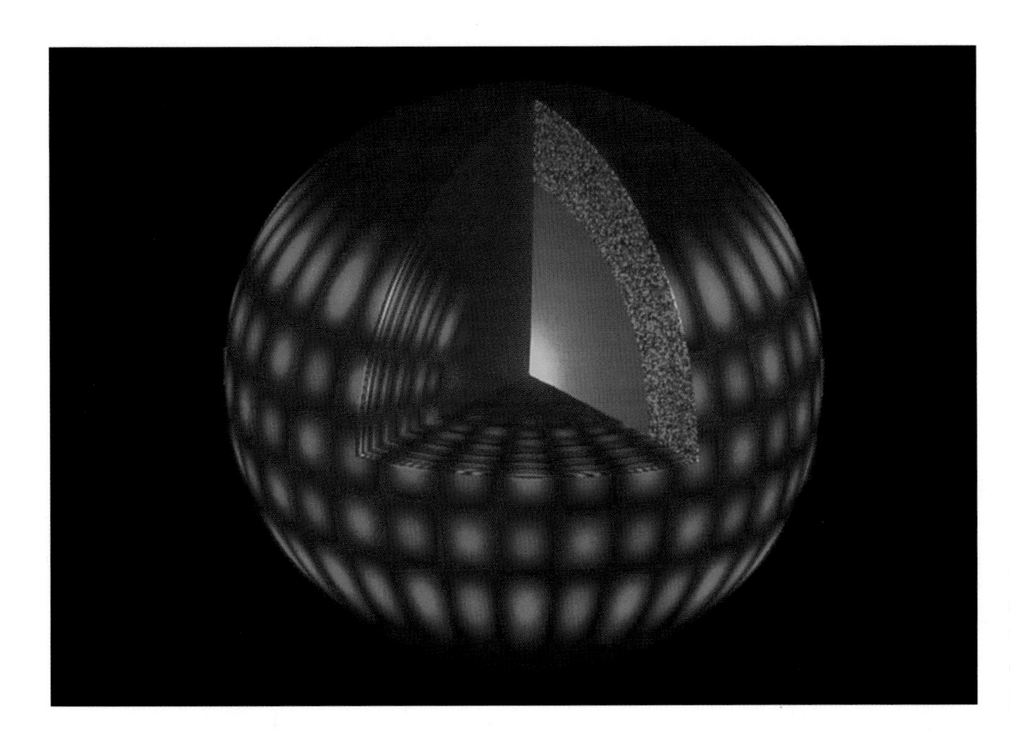

Fig. 12.12 **Solar oscillations** Solar oscillations cause the Sun to vibrate in and out in a complex pattern that can extend all the way to the core. (Courtesy of NSO/AURA/NSF.)

In principle, we can apply a similar study to all stars in the broader subject of *asteroseismology*. While the field is developing quickly, asteroseismologists are still limited to brightness and gross velocity variations. Even so, such observations – from the simplest Cepheid radial pulsations to the most complex observed for Delta Scuti, ZZ Ceti, and even for "normal" stars – can be used to test theories of stellar structure.

12.4 The main sequence

Starting at the upper left of the HR diagram [6.3.1; Fig. 6.12] and plunging to lower right, the main sequence is by far the diagram's most significant feature. It is a mass sequence, high at the top, low at the bottom. For stars of solar chemical composition, it spreads to the right (toward lower temperatures) of the zero-age main sequence [11.3.2], where the spread is the result of the ageing process [Chapters 13 and 14]. Low metal subdwarfs [5.1, 5.3.2, 6.4.2, 7.3.3], which for the same mass are hotter than solar-mix stars, add to the broadening.

The Sun, which occupies the middle of the observed main sequence, runs off hydrogen fusion. It logically follows that the entire main sequence – the whole array of dwarfs from class O through some of early class L – works by the same process. Since we know that our solar model works well for the Sun, similar models should provide the internal constructions of other stars as well.

12.4.1 Masses

The study of binary stars, from which masses can be derived [8.2.3], shows that on the main sequence [6.3.1] masses tightly correlate with luminosity (Fig. 12.13). Taken as a whole, luminosity is roughly proportional to the 3.5 power of the mass ($L \approx M^{3.5}$). A closer look at this *mass–luminosity (M–L) relation* shows the exponent to be a continuous function of the luminosity, varying from around 2.7 for masses significantly less than solar, to around 4 for solar-type stars, and back to 3 or so for significantly more massive stars.

The calculations of stellar models from the equations of stellar structure for various masses along the main sequence, which includes differences in the exact mechanisms that generate energy, replicate the mass–luminosity relation very nicely. The basic concept is that the greater the mass of the star, the higher the internal compression and temperature at any given radius. Since each layer radiates as a blackbody, higher-mass stars therefore have greater luminosities. The close agreement demonstrates that the structure calculations and thermonuclear reaction rates are correct, and that the main sequence is indeed a hydrogen-burning sequence. The luminosities of evolved stars are controlled by a variety of other processes, so that they do not fit the mass–luminosity law: giants and supergiants are too bright for their masses, while white dwarfs are too faint.

12.4.2 Mass limits and brown dwarfs

Given the agreement between observation and theory, however, theory can in turn be used to extend the observed M–L relation to mass ranges as yet untried by binary star observations (the maximum measured at $83M_{\odot}$). The highest-luminosity main sequence dwarfs should have masses of around $100–120M_{\odot}$ (though there are arguments for even greater). The lowest luminosity is unknown. Below the hydrogen-fusing limit of $0.073M_{\odot}$, *brown dwarfs* shine by a combination of deuterium burning (the deuterium derived from the stars' birthclouds) and gravitational contraction, at least down to 13 jupiter masses (1/80 solar mass), at which point even deuterium fusion ceases. Unlike ordinary main sequence dwarfs, luminosities of brown dwarfs depend on both mass and age. Ages can be determined roughly through lithium abundance (as lithium is destroyed at low interior temperatures when it is cycled downward by convection), allowing some measure of age, hence mass [13.2.4]. Brown dwarfs can also be found from Doppler techniques [11.4.3]. Their masses may even overlap those of planets.

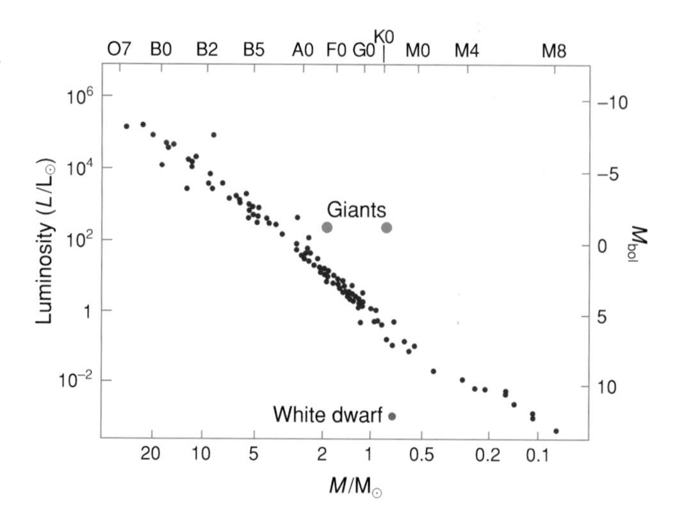

Fig. 12.13 **The mass–luminosity relation** Luminosities of main sequence stars climb steadily and quickly with increasing mass. Giants and white dwarfs, however, do not fit the curve at all. (Kaler, J.B. (1994). *Astronomy!* New York: HarperCollins, reprinted by permission of Pearson Education, Inc.)

12.4.3 Nuclear reactions and interior constructions

Other than mass itself, the most important variation along the main sequence is probably the energy-generating mechanism. While main sequence energy comes from both the p–p reaction and the carbon cycle, the H-to-He reaction rates of the two have very different dependencies on temperature. While the exponents are averages over a large range, energy generation in the p–p chain goes roughly as T^4, while that in

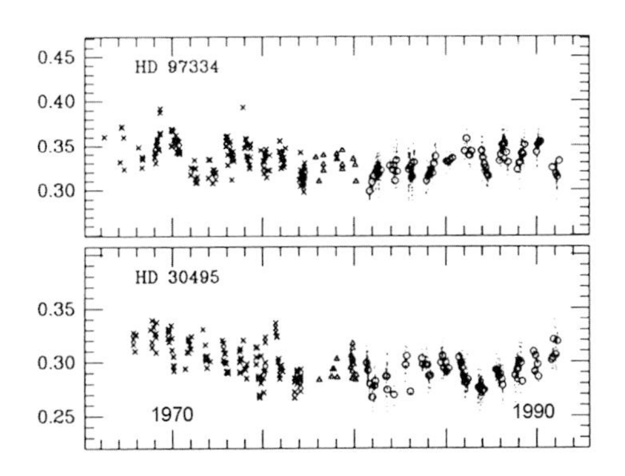

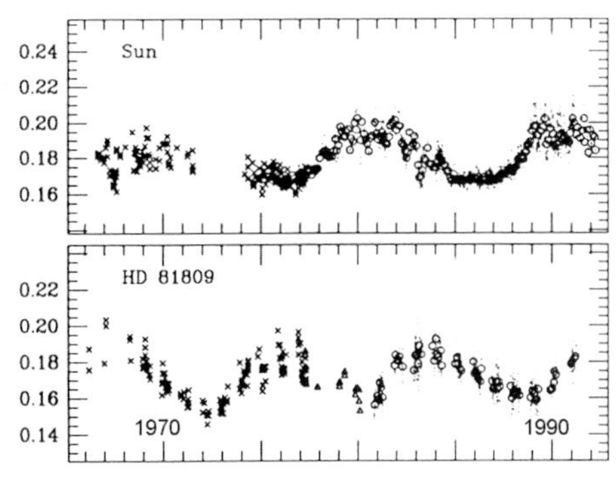

Fig. 12.14 **Stellar activity cycles** The activity of the Sun and three other stars (as expressed by a Ca II emission line index) is shown between 1965 and 1995. The stars on the left show erratic behavior, while the one at lower right has a solar-type cycle of 8 years. (HD 30495 is 58 Eridani.) (Baliunas, S.L. *et al.* (1995). *Astrophysical Journal*, **438**, 269.)

the CN cycle goes as T^{18} or so. As a result, low-mass stars operate entirely from the p–p chain, which becomes insignificant for high-mass stars that run on the CN cycle.

For solar composition, the crossover (where the two are of equal importance) takes place at around 1.25 solar masses, near class F5. From this point on up, the carbon cycle increasingly does the job, and from class A and earlier it effectively dominates; from class G0 on down, most of the work is done effectively by the p–p chain. As temperature drops, the different temperature dependencies of the individual steps of the p–p reaction become important. Below about 0.25 solar masses, later than class M4/M5, the last step ceases, and the stellar cores make ^{3}He rather than ^{4}He, as do deuterium-burning brown dwarfs.

The very high temperature dependence of the carbon cycle steepens the temperature gradient and forces the core to become convective for all early-type dwarfs. At the same time, opacity changes in the outer layers of stars hotter than the Sun cause a thinning of the convective layer. At around 1.7 solar masses (A7 and earlier), not far above where the CN cycle takes over, surface convection finally dies away, leaving the entire portion of the star surrounding the core in a state of radiative (that is, non-convective) equilibrium.

Descent of the main sequence from the Sun sees the reverse, a thickening of the convective envelope. Below around 0.3 solar masses (about class M3–M4), convection has invaded the core and the entire star becomes convective. (All these dividing lines are rather vague and subject to some controversy.)

12.4.4 Magnetism and magnetic activity

The Sun's convection combined with rotation produces the solar magnetic field, magnetic activity, and the solar cycle. We would expect no less of other dwarfs. The combination of observations of Ca II (H and K) emission line cores, ultraviolet emission lines, and X-ray fluxes show that later-type dwarfs, from class F or so on down, have chromospheres [12.1.3], transition regions, and hot million K coronae [12.1.4], from which relatively weak solar-type winds blow, though direct data are difficult to come by. Other solar phenomena are there as well. Among the M and late K dwarfs, we find the flare stars [10.7], in which solar-like flares are so powerful that they may extend over the whole star. Solar-type magnetic fields are thought to be anchored at the interface between the outer convection zone and the radiative envelope. Even though stars become fully convective below 0.3 solar masses (M3–M4), and the fields should be suppressed, some late M, even L, stars still possess magnetic fields, spots, and flares.

Cessation of convection as class F turns to class A has a dramatic effect, in that coronae and related X-ray activity cease earlier than class A5. Between A5 and B8 is a magnetic "quiet zone" that is nevertheless filled by the magnetic Ap and Bp [7.3.3] stars. The origin of these magnetically peculiar stars is still shrouded in mystery, though we suspect that the fields may be fossilized remnants from the protostellar state when the stars had primitive accretion disks [11.2.3]. Once the rotation–convection dynamo is gone, stellar winds can no longer be driven by magnetic energy, as they are in later-type stars. Winds of early dwarfs, which climb to mass loss rates of 10^{-10} $M_\odot yr^{-1}$ and then to 10^{-6} $M_\odot yr^{-1}$ among hotter O stars (eight orders of magnitude more than the flow rate of the solar wind), are driven by radiation pressure acting on absorption lines rather than by the magnetic dynamo.

Not only do solar-type stars possess magnetic activity, some two-thirds have activity cycles similar to that of the Sun, as derived from observation of the chromospherically active emission lines, the stellar-cycle periods ranging from 5 to 30 years. Some of the solar-type stars that have no obvious cycle activity have probably shut it down in their own versions of the Maunder Minimum (Fig. 12.14).

12.4.5 Rotation on the main sequence

Stellar rotation [7.2.5] correlates strongly with main sequence location (Fig. 12.15). The minimum rotation velocity of only a few kilometers per second is found among the G and K stars. Though the average rotation speed climbs some among the M dwarfs, serious rotation is found in the other direction, toward warmer stars. Upon entering class F, the mean projected rotation speeds (v_{rot} sin i) gently climb to

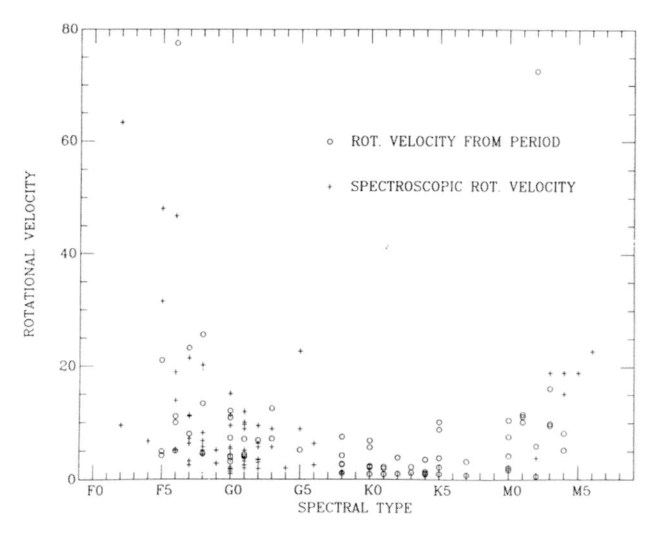

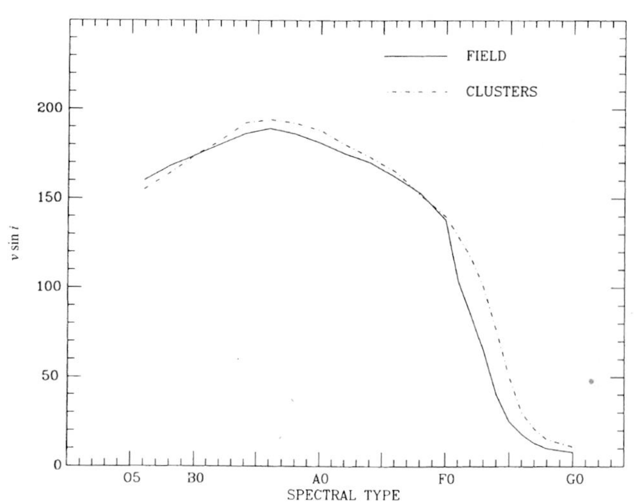

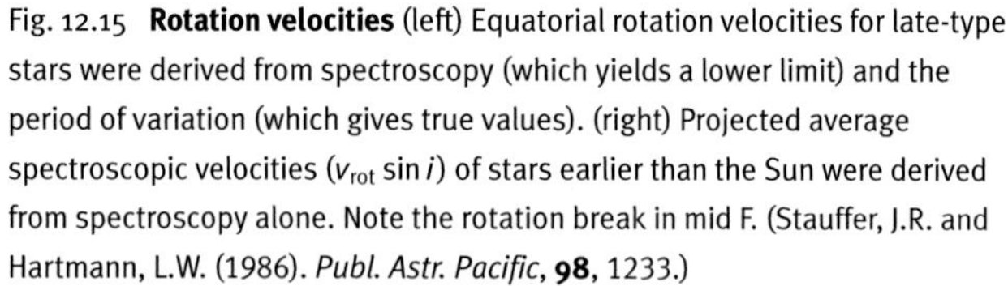

Fig. 12.15 **Rotation velocities** (left) Equatorial rotation velocities for late-type stars were derived from spectroscopy (which yields a lower limit) and the period of variation (which gives true values). (right) Projected average spectroscopic velocities ($v_{\rm rot} \sin i$) of stars earlier than the Sun were derived from spectroscopy alone. Note the rotation break in mid F. (Stauffer, J.R. and Hartmann, L.W. (1986). *Publ. Astr. Pacific*, **98**, 1233.)

30km s^{-1} at F5, which marks the *rotation break*, earlier than which the velocities launch themselves to nearly 200km s^{-1} at B5, and beyond which they decline some among the O stars. Maximum projected spin speeds, found among the Be stars [10.6.3], top out at well over 300km s^{-1}.

The dramatic difference between late and early-type stars is the result of *magnetic braking*. Among the cooler stars, stellar winds and coronal mass ejections drag the fields outward. However, they are still anchored at the stars, which as a result slow down. With less magnetic activity, hotter stars are not braked. Studies of clusters, whose ages are known, show that the G dwarfs spin down quickly within a few tens of millions of years, while the M dwarfs take some 10 times longer. Indeed, one of the ways to find the age of a cool dwarf is to measure its spin rate; a lower spin results in magnetic activity in older stars being notably suppressed. Sunspots and flares, and even terrestrial aurorae, will die away in time as the Sun prepares for its great adventure in stellar evolution and its new life as a red giant.

CHAPTER THIRTEEN

Stellar evolution

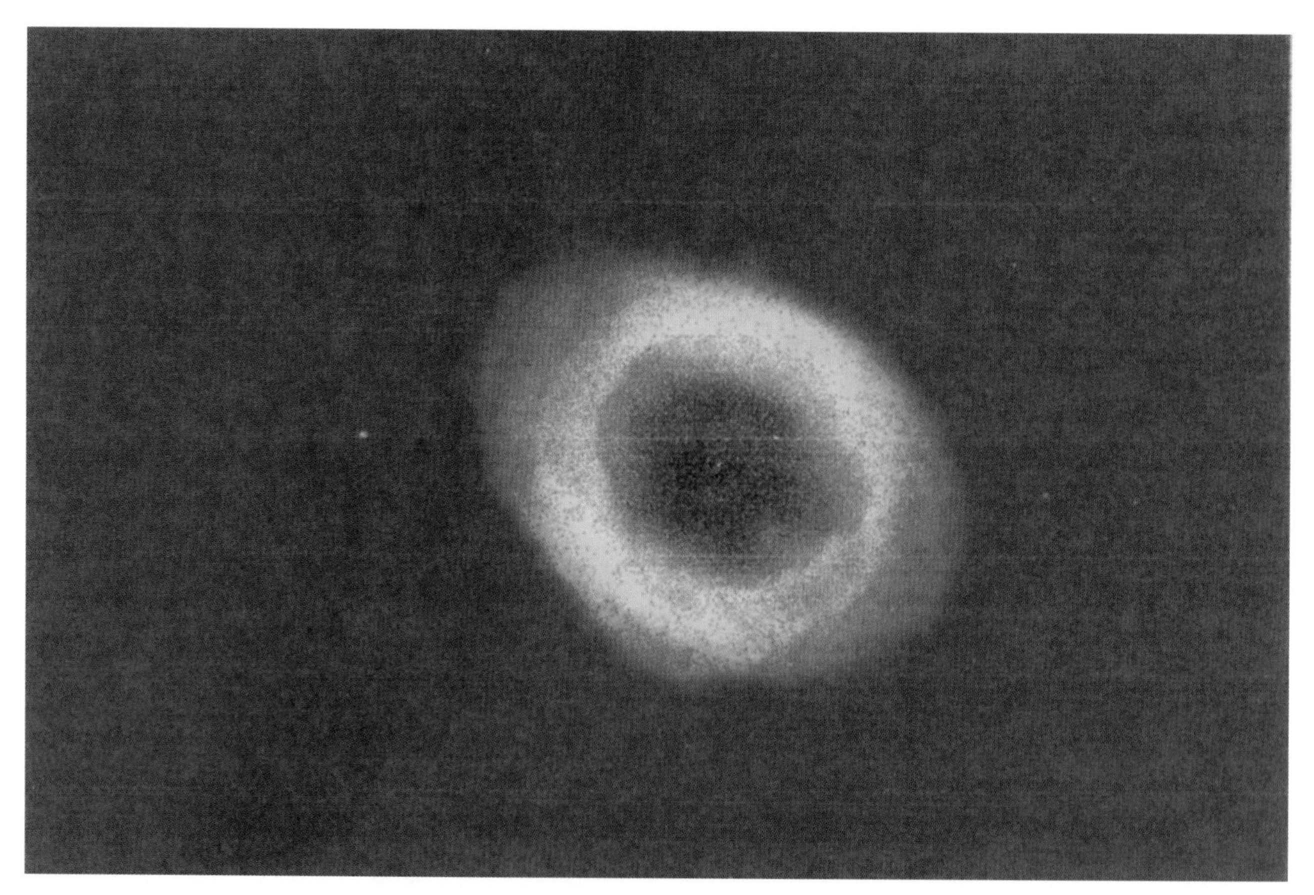

Fig. 13.1 **The Ring Nebula in Lyra, M57** A shell of gas – a "planetary nebula" – announces the transition of a giant star (the progeny of the main sequence) into its final white dwarf state. (Courtesy of Hubble Heritage Team/AURA/STScI/NASA.)

Once the kinds of stars and their collections are laid out, they can be assembled into some kind of whole within which we see how they link together, that is, how one kind of star turns into another. The driving force in *stellar evolution* – the ageing process – is gravity, which tries to make a star's nuclear-burning heart as small as possible. At different evolutionary stages there will be internal processes that slow, halt, or even reverse the contraction, giving the star some stability. At other times, it will be in rapid transition from one state to another, which can lead to spectacular, and aesthetically beautiful, effects (Fig. 13.1).

A guiding principle of stellar evolution is the *Russell–Vogt theorem* (after Henry Norris Russell and the German astronomer Heinrich Vogt, 1890–1966), which states that the characteristics of a single star (one not belonging to a binary) depend only on mass, chemical composition, and age. Extending the theorem to its conclusion, a star's final state and the time it takes to get there depend on its mass, with some input from chemical composition. The number of stars within a specific category is directly proportional to the birth rate of that particular mass, and inversely proportional to the time spent within the category. The number of stars *observed* within a category, however, may differ substantially, as we notice the bright and the odd, whereas the dim and normal may go unregarded.

At the end, all stars die:

(1) by simple contraction;
(2) by supernova explosion, as catastrophically collapsing massive singles or through action induced by close duplicity [14.2];
(3) by evaporation by a companion.

Stars thereby end their lives in one of four states that again depend on mass and duplicity:

(1) white dwarfs (simple contraction);
(2) neutron stars (catastrophic contraction [14.3.1]);
(3) black holes (catastrophic contraction [14.3.3]);
(4) nothing (the star succumbing to complete annihilation by either explosion or evaporation [14.2.5, 14.3.2]).

A triumph of twentieth-century astrophysics was that astronomers could not only recognize these processes and end-products, but could also tell through the laws of physics how they were produced.

13.1 Domains of the main sequence

The main sequence is separated into domains, either by phenomenology or by evolution. The first involves such matters as where along the sequence we find convection and radiative stability, where rapid rotation commences, where the different nuclear-burning processes become important, and the points at which magnetic fields and winds change their characteristics.

The second involves the masses at which different evolutionary scenarios become important, particularly at what point stars cease becoming white dwarfs and instead create neutron stars and black holes. Bookending the sequence are the limits to star formation, the masses of the least and greatest stars. An extensive combined list of mass divisions is presented in Table 13.1.

One of the primary (and oft under-appreciated) facets of stellar life is mass loss (Fig. 13.2). All stars lose mass through winds [7.4] from birth to death, the process crucial to stellar evolution. One must therefore differentiate between the *initial mass* (M_i) of a star, that with which it is born on the main sequence, the *current* mass (M), and the *final mass* (M_f), that with which it will die, wherein ($M_i - M_f$) goes back into the interstellar medium [11.1], is lost to a companion, or both. Along the way is a continuum with sometimes abrupt shifts in the ratio of M/M_i. The Sun will eventually lose about half of itself, while at the extreme, stars that do not annihilate completely can lose 90 percent or more of themselves back into interstellar space, leaving behind but scorched remnants of their former glory. All masses in the following are those on the zero-age main sequence [11.3.2], M_i.

Beginning at an unknown minimum up to 0.073 solar masses ($M_\odot$), or 76 jupiter-masses (M_j), are the lowly brown dwarfs [14.3.2, 12.4.2], stars that cannot reach the 3 or so million kelvin temperature required to turn hydrogen (^{1}H) into helium as at least ^{3}He [12.4.3]. This sequence is subdivided into two segments. The natural deuterium (^{2}H) given to the star by the Big Bang fuses to helium at a temperature of only 700 000 kelvin, which corresponds to a mass of $0.012M_\odot$ ($12.6M_j$), making such bodies – by the most common definition – still "stars" [11.3.2].

Below $13M_j$, however, gravitational compression is not great enough to heat the cores even to the point of deuterium fusion, the masses now falling into the realms of the planets [11.4.3]. Given that such brown dwarfs should form from the "top down" by direct contraction, they could still be called "stars," though "substellar masses" might be more appropriate. Confusing the issue, some of these bodies may be real planets that have escaped their parents; some stars may form as low-mass binary companions within the disks

Initial Mass ($M_\odot$)	Class	Kind of Star/Activity/Evolution
?	T?	↓ Lower limit to brown dwarfs unknown
1/80	T?	↓ Deuterium fusion ceases ($12.6M_j$)
0.073	*L2*	*↓ proton–proton chain ceases*
0.08	M8	↓ Core begins to become degenerate; H_2 formation
0.25	*M5*	*↓ Stars will not become red giants*
0.3	M3	↓ Fully convective
0.85	*G8*	*↓ Galaxy not old enough to allow evolution*
1.0	G2	♦ The Sun as standard
1.25	F5	↑ Carbon cycle exceeds p–p chain and core becomes convective
1.25	F5	↑ Main sequence rotation increases (rotation break)
1.7	A7	↑ Envelope becomes fully radiative
1.8	A5	↑ Coronal activity/X-ray ceases
2.0	*A2*	*↑ Helium-core flashes cease*
3	*B8*	*↑ Second dredge-up on the AGB*
3.5	B7	↑ X-ray activity from hot winds
5	*B4*	*↑ Blue loops after RGB phase*
10?	*B0.5*	*↑ Neon/oxygen white dwarfs*
10?[a]	*B0*	*↑ Core collapse supernova/neutron star creation*
40?	O4	↑ Strong winds create Wolf–Rayet stars
60?	*O3*	↑ *LBVs and creation of GRBs* and black holes
120?	O2	↑ No stars; maximum stellar mass uncertain

[a]The limit for the onset of core collapse (and for neon white dwarfs) lies in the range of 8 to $12M_\odot$.

Table 13.1 **Divisions of the main sequence**
The table lists commencement or cessation of a particular phenomenon or activity according to increasing mass (whether on the main sequence or not), where the spectral classes refer to zero age. Many critical points are only approximate, and depend on a variety of parameters, including chemical composition. Some are very uncertain and contended. The limits that directly involve stellar evolution are in italics. A down arrow refers to activity (or lack thereof) below the stated limit, the up arrow to activity (or lack thereof) above the limit.

Fig. 13.2 **OH 231.8+4.2** An aged Mira variable at the heart of the thin neck of the OH/IR star OH 231.8+4.2 is in the throes of a strong dusty "superwind" that has created a reflection nebula a third of a parsec long. The star, which seems to have a main sequence companion that helps to shape the wind, is about to spawn a planetary nebula. (Courtesy of Hubble Space Telescope, ESA and V. Bujarrabal (Obs. Astr. Nac. Spain), STScI, and NASA.)

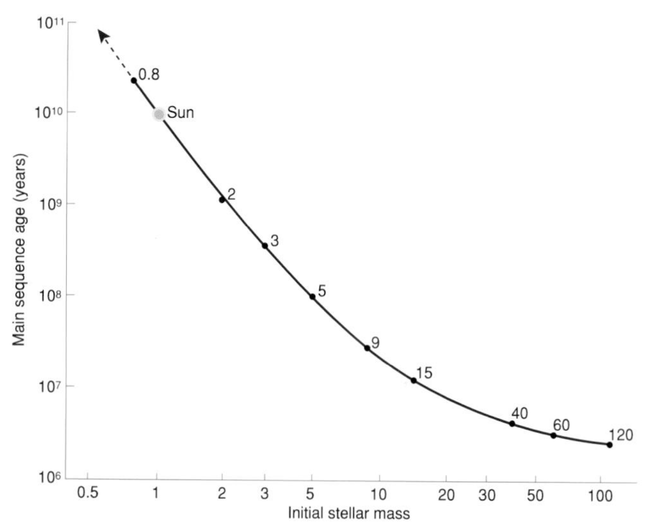

Fig. 13.3 **Main sequence ages** As stellar masses increase, main sequence ages (here for solar metal stars) decrease. For lower metal stars, the curve drops a bit.

of their bigger siblings [11.3.3], as may planets; some "planets" may form as binary companions by direct accretion from the ISM. At the end, there may be little to distinguish the two.

Next consider main sequence lifetimes versus the age of the Galaxy. The main sequence lifetime of a star (t) depends on the amount of hydrogen fuel available divided by the rate at which it is burned to helium. To a crude approximation, the fuel supply is proportional to the stellar mass (M). To a similar approximation, the burning rate relates to the luminosity (L), so $t \propto M/L$. However, from binary stars $L \propto M^{3.5}$ [12.4.1]. Therefore, $t \propto M/M^{3.5} = 1/M^{2.5}$, showing that as mass increases, t drops quickly. The exponent and its variations can be refined through detailed evolutionary models to give highly accurate values of lifetime (Fig. 13.3). Given the solar reference of 10^{10} years on the main sequence, as mass drops to about $0.85M_\odot$ (for low metals), the lifetime goes up to 12 to 13 billion years, that is, to the age of the Galaxy [5.1, 9.4.4]. Stars spend the vast majority of their lives on the main sequence, so main sequence lifetime stands in well for total lifetime. As a result, for low-metal ancient stars, none with a mass under $0.85M_\odot$ has ever had a chance to die, or even move off the main sequence. Given a 10^{10}-year lifetime for the age of the higher-metal disk [9.2.4], the limit is closer to (but still below) $1M_\odot$. Every star ever born below these limits is still there on the main sequence, including all those in the ancient globular clusters [9.4].

The Sun will fuse its internal hydrogen into helium, which through eventual core compression will be fused into a mixture of carbon and oxygen. It will ultimately die as a stable white dwarf [6.4.6] of roughly half a solar mass. From the lowest substellar limit to $10M_\odot$ (class B1 on the ZAMS), all stars proceed to the white dwarf state, but with vastly different timescales that depend critically upon mass. (The dividing mass is uncertain and contended, but seems to lie between 8 and $12M_\odot$ masses; $10M_\odot$ is adopted here.) As the main sequence mass increases, so does the white dwarf mass, M_f, and at $M_i \approx 10M_\odot$, the core mass – that which would become the white dwarf – hits a critical limit of $1.4M_\odot$, above which it cannot sustain itself, and it collapses [14.2.2]. The energy generated explodes the outer stellar layers, leaving a ball of neutrons – a neutron star – behind [14.3.1].

The main sequence therefore divides into four domains, each of which has a variety of subsets:

(1) brown dwarfs, below $0.073M_\odot$;
(2) the lower main sequence between 0.073 and $0.85M_\odot$;
(3) the intermediate main sequence between 0.85 (above which stars are observed to evolve off the main sequence in the creation of white dwarfs) and $10M_\odot$;
(4) the upper main sequence, greater than $10M_\odot$, that cannot produce white dwarfs.

(The divide between the "lower" and "intermediate" main sequences is sometimes set at 2 solar masses for reasons that involve the physics of the core.) The first three domains are the subjects of this chapter. Domain four, which is subdivided as well, will occupy the next, and last.

13.2 Early intermediate-mass evolution

As the Sun provides the fundamental reference for stellar properties and structure, so it does for evolution. The other realms of the main sequence thereby become variations on a theme, albeit some very large ones. The ageing of a star defies any analytic equations, its study more a numerical art. The theoretician constructs a stellar model [12.2.1] of a particular mass and chemical composition that has a specific content of hydrogen, helium, and heavy elements (Box 13.1). The model predicts nuclear-burning rates and thereby gives rates of change in chemical composition as a function of radius. Adoption of a time interval provides the actual change in composition, thus allowing the calculation of a new model that will have a different internal structure and different opacities, and therefore different external characteristics. This model predicts different nuclear-burning rates, and therefore allows the construction of yet another model. The series of a large number of sequential models gives the changes in stellar luminosity, effective temperature (hence spectral class or color), and radius as a function of time, which can be plotted on the theoretician's version of an HR diagram, in which the logarithm of temperature (log T) is plotted against the logarithm of luminosity (log L) to create the "log L–log T plane" [6.3.3]. Connecting the dots predicted by the succession of models allows the construction of a time-dependent *evolutionary track*.

During the chosen time interval, the nuclear-burning rates are tacitly assumed to be constant, whereas in fact they change continuously, resulting in small errors in the second model predicted from the first. Each step adds new errors and grows the previous ones, which then can combine into large errors. The trick is to make the intervals large enough so that the calculation can be run in a reasonable time (or at all), but small enough to "converge," such that successively smaller intervals make no significant difference in the outcome.

Another problem lies in the comparison of the evolutionary tracks on the log L–log T plane against the observer's HR diagram, which plots absolute magnitude in some wavelength band against spectral class or color. Comparison of theory with observation therefore requires various transformations in which temperature is related to class and color (for the chosen chemical composition) and luminosity (expressed as bolometric magnitude) is converted to visual magnitude through temperature and the bolometric correction [6.3.3]).

13.2.1 Main sequence evolution

Evolutionary tracks begin during star formation at the "birthline," where stars hover on the HR diagram (or the log L–log T plane) as they begin to burn deuterium [11.3.2]. We begin here with the stars having arrived on the zero-age main sequence, the ZAMS. The main sequence is defined most simply as a mass-dependent band on the HR diagram where stars convert the hydrogen in their cores into helium [12.4.1]. Although these cores undergo drastic changes in chemical composition (from mostly hydrogen to nearly all helium), dwarf stars do not much change their external characteristics over very long periods of time.

As a fire consumes its fuel, it dims. Stars behave quite differently. Except under extreme circumstances, the pressure exerted by a gas (P) depends on the number of particles per unit volume (N) and the temperature (which controls their speeds), or $P = NkT$, where "k" is a constant. It matters not what kinds of particles they are, whether electrons, atoms,

Box 13.1 Chemical ratios

Like all sciences, astronomy is filled with its own language and inconsistencies. The relative abundances of two elements can be expressed in terms of either the number of atoms or by mass fraction, mostly depending on whether they are given by observers (number) or theoreticians (mass). The entire periodic table is commonly also reduced to hydrogen, helium, and "metals," the last of these the collective term for everything above atomic number 2. The ratio of the number of helium atoms (N_{He}) to the number of hydrogen atoms (N_H) is expressed by N_{He}/N_H. For simplicity and to avoid subscripts, it is more commonly given as just "He/H." The mass fraction of hydrogen, helium, and metals is respectively expressed by the letters "X," "Y," and "Z." Subdwarfs and globular cluster members are sometimes referred to as "low-Z" stars.

In the Sun's atmosphere, the He/H ratio is 0.08. The alternative mass fraction of helium is N_{He} multiplied by its atomic weight of 4 divided by the numbers of all the other different kinds of atoms times their atomic weights. However, in a solar mix the metal content is very low, and to a reasonable approximation can be ignored. Therefore (given that the atomic weight of hydrogen is 1),

$$Y = 4N_{He}/(N_H + 4N_{He}) = 4He/(H + 4He).$$

Dividing the top and bottom by $N_{He} = He$,

$$Y = (4He/H)/(1 + 4He/H).$$

With $He/H = 0.08$, $Y = 0.24$. For typical solar-mix models, $Z = 0.02$ (slightly reducing Y) and rendering $X = 0.64$.

Alternatively, the conversion of Y to He/H is

$$He/H = Y/(4 - 4Y).$$

For heavily hydrogen-depleted stars, the simplified equations will obviously not work, and the actual metal composition must be taken into account.

or molecules. All things equal, four hydrogen atoms turned into one helium atom would make the pressure decline. But to hold up the star, the pressure in any given layer (as defined by mass proportion above and below the layer) must stay constant. To maintain pressure and hydrostatic equilibrium [12.2.1], gravity squeezes down on the core, increasing the pressure and density. As the core "falls" under gravity, energy is released. Some of it is radiated immediately away, while the rest goes into raising the temperature and the reaction rates [12.4.3], which also causes the core to eat slowly into the surrounding envelope, increasing its mass. As a result, the decline in the initial fuel supply is closely offset, allowing for significant stability so long as there is any hydrogen left in the center at all. The result is a densely-packed main sequence that contains some 95 percent of all stars (the remainder being those that have died or are in the process thereof).

Yet the offsetting cannot be perfect. As time proceeds, the increased reaction rates and core mass cause slow but inexorable alterations in the star's condition. The changes are complex, and not subject to simple reasoning. A "core" is not a monolithic entity. The H-to-He conversion rate is greatest at the star's center where the temperature is highest, and then drops toward the core's periphery. Moreover, different processes (the CNO cycle vs. the p–p chain and their various sub-processes) are temperature, radius, and time dependent as well [12.2.2]. As long as the core is in radiative (nonconvective) equilibrium and unmixed (below about 1¼ solar masses, all such numbers rough), the He/H ratio increases more quickly at the center than it does near the edge. For higher-mass stars, convection tends to equalize the He/H ratio throughout the core.

Models show that as the fuel is consumed, all stars slowly brighten. Those below about $1.7M_\odot$ first heat at their surfaces and then begin to cool as the stars' radii grow larger, while those of higher mass steadily cool (Fig. 13.4). For stars with masses greater than solar, as core hydrogen nears depletion (with but a few percent left), equilibrium begins to go awry. As the core of the star contracts, so does the entire star, causing it to swing back to slightly higher effective temperatures, which thus produces a prominent hook in the evolutionary track. These slow changes cause the main sequence (for a given composition) to be a band that spreads to the red (to the right) of the ZAMS (see Fig. 13.4). When the central hydrogen is finally consumed, a pure-helium core begins to advance outward, while hydrogen burns in a thick shell, dominated now by the CN cycle. The core then also begins a Kelvin–Helmholtz-like contraction [12.2.2] that (along with zero central hydrogen) makes the star significantly swell and cool, and finally signals the end of main sequence life. The process to this point takes 10 billion years at $1M_\odot$, down to only 20 million years at $10M_\odot$, the upper limit to intermediate evolution (see Fig. 13.3).

Detailed models that take all the complexities into account show that when the Sun was a ZAMS star, it was only about 70% as luminous as it is today, had 90% the current

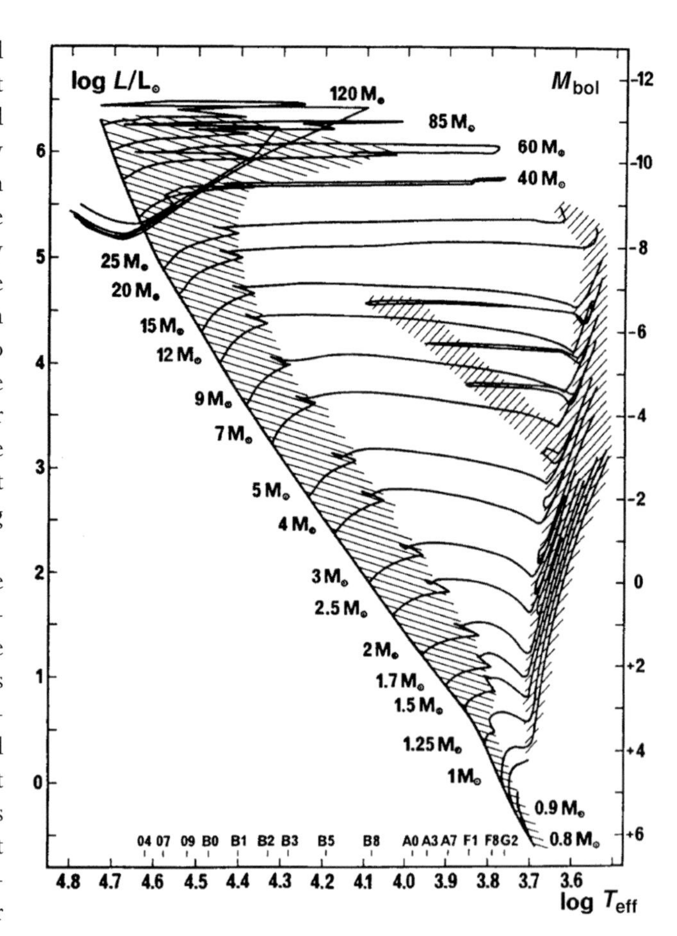

Fig. 13.4 **Evolutionary tracks** A set of evolutionary tracks (plotted by luminosity and bolometric magnitude against effective temperature) show the early ageing of solar-metal stars from under a solar mass to 120 solar masses. Up to $1.7M_\odot$, the tracks take the star to helium ignition. From $2M_\odot$ up to 5, the stars are followed into the early asymptotic giant branch before thermal pulsing begins. High-mass stars over $7M_\odot$ go to the end of carbon burning. Mass loss is included. The cross-hatched areas are regions of core nuclear fusion, hydrogen burning on the left, helium fusion on the right. (Schaller, G., Schaerer, D., Meynet, G., and Maeder, A. (1992). *Astronomy and Astrophysics*, **96**, 269.)

radius, and had an effective temperature of 140 kelvin less than now. The calculation gives rise to the "faint Sun paradox." The fossil record shows that life on Earth goes back at least 3.5 billion years. The Earth, however, should have been frozen, disallowing the formation of life. Several theories have been advanced to explain the problem, including a more rapidly rotating, hence chromospherically active [12.1.2], Sun, or a more massive greenhouse effect (in which infrared-absorbing atmospheric gases, such as carbon dioxide and methane, kept the planet's surface warm).

Over the next two billion years, the solar temperature will rise by a few tens of kelvin, while the luminosity will increase by just over 15 percent, which will probably spell the end for life on Earth (though such speculations are fraught with uncertainties). The effective temperature will maximize about two billion years from now. At the end of central hydrogen fusion, in 5.5 billion years, the Sun will be roughly 70%

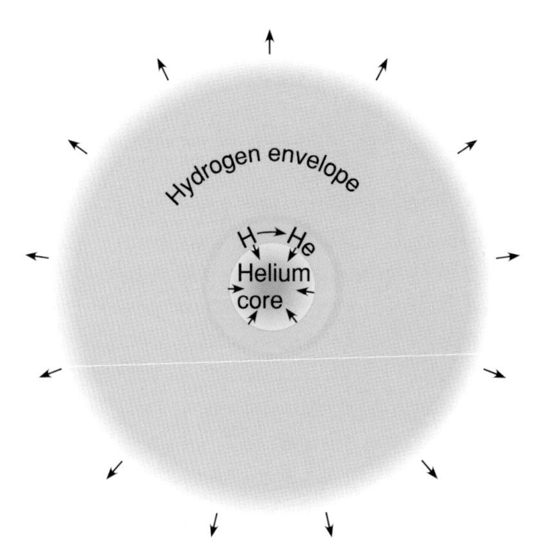

Fig. 13.5 **Subgiant** A star moves into the subgiant realm with a quiet helium core surrounded by a hydrogen-burning shell, all encased in a rapidly expanding hydrogen envelope. (Kaler, J.B. (1994). *Astronomy!* New York: HarperCollins, reprinted by permission of Pearson Education, Inc.)

brighter than now and be 30 percent larger, though about the same surface temperature. Magnetic braking [12.4.5] will slow, then probably halt, the solar-activity cycle, removing us from that energy source (whose transfer to Earth is still not understood).

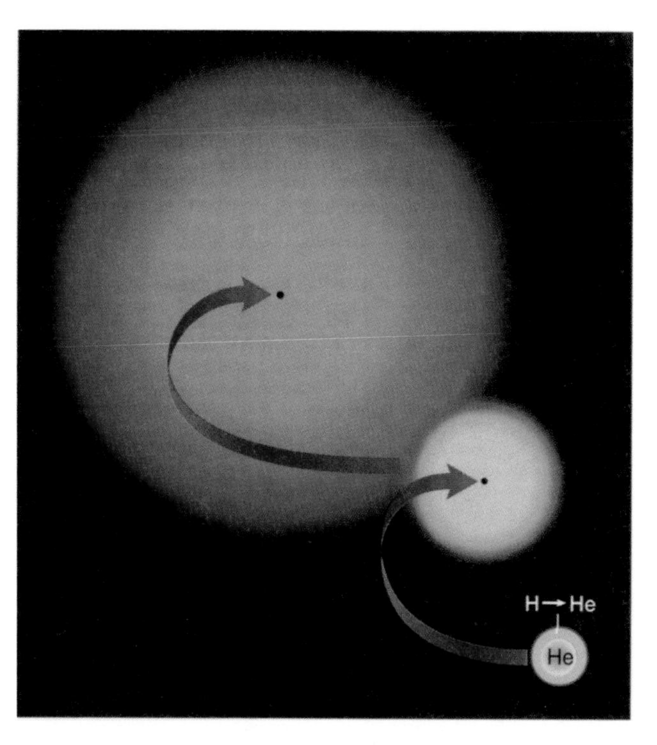

Fig. 13.6 **Red giant** A fully developed red giant has a tiny helium core (drawn approximately to scale) ready to burn helium to carbon and oxygen. (Kaler, J.B. (1994). *Astronomy!* New York: HarperCollins, reprinted by permission of Pearson Education, Inc.)

13.2.2 Red giants

When the internal density and temperature rise as a result of core contraction, much of the energy released at first causes the star as a whole to expand rather than brighten, and it swings seriously to the red on the HR diagram, to class G or K (Fig. 13.5). Though lower-mass stars (solar mass and below) do not change much in class, the changes for those of higher mass are huge: a $2M_\odot$ star passing over from class A, a $3M_\odot$ from class B. The crossing time is very short, only about one percent of the main sequence lifetime, giving rise to the observational Hertzsprung gap [Box 6.3] in which we find relatively few stars. The number of stars per unit volume of space climbs quickly downward along the main sequence. As a result there are many more crossing stars at lower masses, enough to create the distinctive subgiant branch of the observed HR diagram [6.3.2].

At a critical point, changes in the energy-generation rate of the hydrogen-burning shell (which is heating as a result of core contraction) and in the opacity of the gases of the envelope cause the star to begin to brighten at more or less constant surface temperature, and thereby to swell to much larger proportions. In theoretical jargon, the star now ascends the *red giant branch* (the RGB) of the HR diagram (see Fig. 13.4). Numerous attempts have been made to describe in analytical terms why stars behave this way. The complexity of the large variety of changing processes seems to void any simple explanation. To paraphrase a stellar modeler: "stars become red giants because that is what the calculations show they do."

At the point of entry to the RGB, if the envelope is not already convective, it will be pushed into convection by the lowered temperature and the consequent increased atmospheric opacity. Convection can then sweep gases toward the hydrogen-burning shell and into the higher regions, where carbon (^{12}C) has been partially processed into ^{13}C and nitrogen (^{14}N), and lift these creations to the top. This *first dredge-up* alters the ratio of carbon isotopes and increases the nitrogen content of giants. Even some helium made by hydrogen burning can find its way upward. At the same time, the increased luminosity and large size (and the concomitant lowered gravity) of the star allows a stronger wind to blow from its surface, which pumps some of the freshly made elements into space and begins to whittle away the stellar mass.

The show cannot go on forever, else the sky would be littered with infinitely bright red giant stars. The envelope has now expanded past the size of the orbit of Mercury, while the core has compacted to the size of Earth (Fig. 13.6). When the central temperature hits around 100 million kelvin, helium begins to form carbon. In the first step, two ^{4}He atoms collide to create ^{8}Be. This isotope of beryllium is so wildly unstable (with a half-life of 10^{-16} seconds) that it immediately falls back to two ^{4}He nuclei. Nevertheless, the process will establish an equilibrium in which at any given time there will always be a bit of ^{8}Be. When the equilibrium amount is sufficiently high, as a result of increasing temperature and density, it is inevitable that the ^{8}Be will interact with another ^{4}He nucleus to create carbon in the form of ^{12}C. In effect, three ^{4}He nuclei must collide nearly simultaneously.

In the early days of research on radioactivity, three kinds of "emissions" were seen to emerge from radioactive elements [Box 7.4], and were called "alpha," "beta," and "gamma" rays. Alpha rays turned out to be helium nuclei, which then (^{4}He only) acquired the alternative name "alpha particle." Beta rays are electrons, and gamma rays are real high-energy electromagnetic radiation [Box 7.1]. Helium nuclei therefore fuse by the three-alpha, triple alpha, or 3-α process. Another absorption of an alpha particle by carbon makes oxygen as ^{16}O.

We are now seeing deeper into the process of stellar element formation, whereby hydrogen first makes helium, then helium creates carbon and oxygen. If these elements can get out of the star – which under the right circumstances they will – they enrich the heavy element content of the interstellar medium. If all the elements are made by this and similar processes, it is clear why elements 3, 4, and 5 (lithium, beryllium, and boron) are so rare: they are skipped over in the creation of C and O. We also see why C and O are so abundant, and one reason why (given further alpha particle reactions) even-numbered elements are more common than odd-numbered elements [7.3.2, Fig. 7.17].

At this point, the Sun (and stars up to about 2.5M$_\odot$) will have reached a luminosity above 1000L$_\odot$; higher masses will hit 10 000. All have cooled to roughly 3500 kelvin, which places them in early class M. The new source of nuclear energy, a helium-burning core surrounded by a hydrogen-burning shell (which feeds more helium into the core and actually provides most of the luminosity), not only stops the contraction of the core, but causes it to expand. The star stabilizes, the envelope's radius shrinks, and the luminosity declines to a point very roughly midway (in the logarithm) down the RGB, for the Sun, to somewhat under 100L$_\odot$. Now, another divide. As stars climb the RGB, core densities become enormous, approaching a million grams per cubic centimeter. As a result, below about 2M$_\odot$ the cores become *degenerate* (Box 13.2) as the stars begin to grow white dwarfs within them [7.2.4]. The pressure of a degenerate gas has no temperature dependence. When the temperature hits the 3-α point, the increased heat from the reaction does not immediately produce expansion, consequent cooling,

Box 13.2 Degeneracy

Werner Heisenberg (1901–1976) is famed for his *uncertainty principle*. Given a subatomic particle, the uncertainty in its position (Δx) multiplied by the uncertainty in its momentum (Δp, where p = mass × velocity) $= h/2\pi$, where h is Planck's constant (the same as in the equation for the energy of a photon, $E = h\nu$ [Box 7.1]). The uncertainties are not those of measurement: they are the limits to which you can actually *know* position and momentum. If you know x exactly (such that $\Delta x = 0$), you therefore have no idea of the momentum, and vice versa. The origin of the principle is that subatomic particles behave as much like waves [Box 7.4] as they do "little balls of matter," and you cannot really pin down a wave. Planck's constant therefore relates to a profound minimum of scale.

Subatomic particles are characterized by a variety of "quantum numbers" that give energy within an atom (if bound), spin, and so on. All take on only specific values, that is, are "quantized," again the result of the particles' wave natures. A fundamental tenet of atomic physics is the *Pauli exclusion principle*. Define a three-dimensional box of size Δx, Δy, Δz, whose spatial volume is thus $\Delta x\Delta y\Delta z$. Then devise another box made of "momentum space" in which the sides are Δp_x, Δp_y, Δp_z and whose "volume" is $\Delta p_x\Delta p_y\Delta p_z$. Next, combine the two into a six-dimensional *phase space* whose sides are Δx, Δy, Δz, Δp_x, Δp_y, Δp_z. Such a six-sided "box" is impossible to visualize, but presents no more problem mathematically than a three-sided box. The "volume" of the phase space is the product of all six sides, or $\Delta x\Delta y\Delta z\Delta p_x\Delta p_y\Delta p_z$. The Pauli principle states that identical particles (those with the same quantum numbers) cannot exist simultaneously in a phase space box with a minimum volume of h^3, that is, the box can contain only one particle with a particular set of numbers.

Free electrons can spin in one direction or in the other; that is, their spins can be either "plus" or "minus" (from which originates the 21-cm radio line [11.1.4, Box 11.2]). Therefore, in a phase space box of volume h^3, there can be but two electrons, and their spins must be different. More electrons can always be packed into a specific spatial volume, but they can be added only at higher and higher momenta, that is, velocities. If the density is so high (the order of 10^6gm cm^{-3} for a star) that the little boxes become filled up to any specific momentum, the gas is said to be completely degenerate; partial filling leads to partial degeneracy. In a non-degenerate electron gas, P (pressure) $= NkT$ [13.2.1], where N is directly proportional to the density of electrons in grams per cubic centimeter (ρ), that is, $P \approx \rho T$. In a fully degenerate gas, however, $P \approx \rho^{5/3}$, with no dependence on temperature at all.

The Pauli exclusion principle is the reason why electrons are arranged around a nucleus in shells. In the hydrogen atom, only two electrons fill the first shell, 8 the second, and so on. Electron degeneracy provides outward support for white dwarfs, as there is a limit to the spatial density at any given momentum. Note though, that the heavier protons in a white dwarf are *not* degenerate. They, and neutrons, can reach degeneracy only at much higher densities, degenerate neutrons supporting neutron stars [14.3.1].

When the density increases to the point where the highest-velocity particles approach the speed of light and the theory of relativity must be used, the support disappears. Such densities are reached in white dwarfs at the *Chandrasekhar limit* (after Subrahmanyan Chandrasekhar, 1910–1995) of 1.4 solar masses. The results are catastrophic, and lead to supernovae, neutron stars, and black holes.

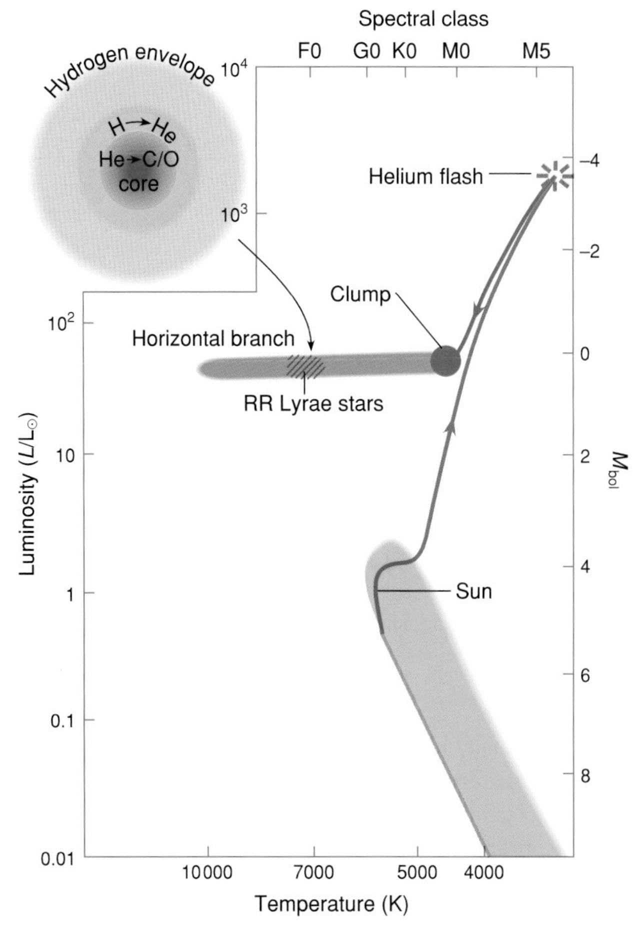

Fig. 13.7 **Red giant evolution** When core hydrogen fusion is over, the Sun will climb the red giant branch. After the ascent is terminated by the helium flash, the Sun will descend back down the giant branch with a helium-burning core and a hydrogen-burning shell (inset), and then along with similar stars will settle into the stable clump. Low-metal horizontal branch stars spread to the left. (Kaler, J.B. (1994). *Astronomy!* New York: HarperCollins, reprinted by permission of Pearson Education, Inc.)

and partial quenching of the reaction. Instead, as the heat increases further, the reaction quickly runs away with itself, and becomes violently explosive, producing a *helium flash* that must first remove the state of degeneracy before the star can stabilize (Fig. 13.7). Most of the energy is absorbed by the star and never reaches the surface. Above $2M_\odot$, helium burning begins non-explosively. (Some astronomers consider $2M_\odot$ to be the true dividing point between the "intermediate" and "lower" main sequences rather than the artificial divide set by the age of the Galaxy, making the Sun and its kind a "lower main sequence" star: watch the context.)

13.2.3 The clump and blue loops

The majority of core helium-burning stars are in the neighborhood of a solar mass with approximately solar composition. As a result, they pile up on the HR diagram in a "clump" within late G or early K near $M_V = 0$, and are in fact called *clump stars* (see Fig. 13.7). Almost every late G or early K giant (Arcturus, Aldebaran, etc.) is among their number. Dropping the metal content decreases the opacity, and makes the star bluer, as does increased mass loss, which thins the stellar envelope. As a result of differing mass loss rates among the stars in globular clusters, their helium-burning stars spread out to the left to form the horizontal branch [9.4.4] of the color–magnitude diagram to an extent that depends on metallicity, as well as on the infamous "second parameter" [9.4.9].

Beginning at around $5M_\odot$, stars do not join a clump, but begin to make *blue loops* of increasing sizes back toward the left in the HR diagram (see Fig. 13.4). At $7M_\odot$, they loop to class F, and at $9M_\odot$ back to class A, earlier for yet higher masses. The major phases of helium burning lie in and at the bottoms of the tracks that descend from the RGB peaks and around the ends of the loops for the higher-mass stars. The result is a Y-shaped helium-burning "main sequence" of sorts that runs up the right of the HR diagram and then branches at about $5M_\odot$ to run up both to the left and right.

As the more massive stars cross the HR diagram to become giants, and as they loop back and forth, they cross the Cepheid instability strip where their structures create the valving mechanism responsible for Cepheid variability [10.3, Box 10.1]. Other kinds of variables – Delta Scuti stars [10.4.1], Beta Cephei stars [10.4.2] – are produced when evolution brings them through their respective instability zones.

Nuclear fusion reactions are exothermic (energy generating) until they reach iron and nickel (which actually happens in supergiants [14.1.3]). These elements have the most tightly bound nuclei [Box 7.4], and since the heat of fusion comes mostly from the release of binding energy, further fusion must be endothermic (energy absorbing). Most of the energy from hydrogen to iron (some 80 percent) is released in hydrogen burning; most of the remainder is emitted in helium burning. As a result, a star is on the main sequence for 80 percent of its life, and in the helium-burning clump for (factoring in other stages of stellar life) somewhat under 20 percent. The result is a thickly populated giant branch that contains around 10 to 20 percent of the dwarfs old enough to have evolved, in qualitative agreement with observation.

Note that the term "red giant branch" means two very different things, depending on one's point of view. To the theoretician, it means the ascent stage with a dead helium core and a helium-burning shell, otherwise called the RGB. The observer's giant branch, however, refers to the distinctive zone of giants as seen on the HR diagram. Since most of the evolving giants are coming from roughly solar-mass dwarfs, the giants will concentrate lower on the HR diagram, collectively appearing as a band that spreads to the right of main sequence classes F and G. Such stars include both ascending and descending giants, some second-ascent AGB stars (13.3.1), and clump stars. More specifically, a "giant" is a star so classified (as II or III) by its spectrum against MKK standards, irrespective of any kind of evolutionary or physical properties. Again, context is all.

13.2.4 Isochrones and ages

Evolutionary theory leads directly to cluster ages. Begin with a newly-born cluster with a chosen chemical composition and a complete zero-age main sequence. Evolve a sample of masses over a selected time period, convert the resulting L and T to spectral class (or color) and visual magnitude, and plot on an HR (or color–magnitude) diagram. Connect the dots, and you have the locus of stars as they appear at the chosen age, an *isochrone* [9.2.4; Figs. 9.6 and 9.7]. Select additional time periods, and watch the cluster's HR diagram evolve. O stars first leave the main sequence to become supergiants, then B stars depart, and so on down through the dwarfs as they become more ordinary giants.

The theoretician thereby calculates the exact ages at which the main sequence ends, and determines the shape of the giant branch. The application of sets of isochrones [9.2.4] of increasing age applied to the observed color–magnitude diagram (corrected for interstellar reddening [3.5.2]) to find or interpolate the best fit immediately yields the age of the cluster. The technique can be applied to all systems whose stars were born at about the same time, including both open clusters [9.2.4], globular clusters [9.4.8], and OB associations [9.3]. One must only be careful to pick the right chemical composition, especially for the low-metal globulars, and to apply the proper transformations, luminosity to M_V, etc. [6.3.3]. The oldest-known open cluster determines the age of the Galaxy's disk, and the oldest globular cluster the age of the Galaxy itself [5.1, 9.4.8].

Isochrones can also be used for single stars, in which ages (along with masses) are estimated from the stars' precise positions on the main sequence relative to the ZAMS. The method works best for hot stars, and the chemical compositions again must be well known. Since stellar rotations slow as a result of magnetic braking [12.4.5], ages can also be estimated from rotation velocities [7.2.5] and chromospheric activity [12.4.4], the method working best for cooler stars. Lithium content provides another indicator for cooler stars with convection envelopes. Lithium is destroyed by nuclear reactions at a temperature of only 2.5 million kelvin, well below the temperature limit of the p–p chain. If the convection goes deeply enough, lithium will be depleted with age. Unfortunately, stellar models have difficulty replicating the observations.

13.2.5 Low mass stars

From $0.85 M_\odot$ down to the p–p limit at $0.073 M_\odot$, calculation of evolution is moot, since no one will be around to see

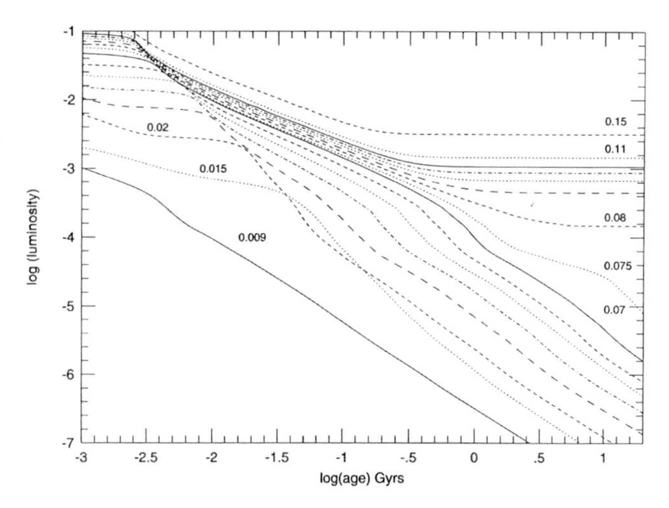

Fig. 13.8 **Brown dwarf evolution** Hydrogen-burning stars above the brown dwarf limit stabilize at nearly constant luminosity (the graph showing luminosity plotted against age expressed in billions of years, "gigayears"). Below the limit, however, they quickly dim (and cool) with time. (Courtesy of the Tucson research group, Reid, I.N. and Hawley, S. (2000). *New Light on Dark Stars*, Chichester: Praxis.)

what actually happens. Nevertheless, curiosity compels that we at least dream of what might be seen. Down to a quarter of a solar mass, stars should behave something like the Sun and produce giants. Below $\frac{1}{4} M_\odot$, however, they may not even do that. Maintaining their convection throughout, they will simply turn themselves into helium white dwarfs [6.4.6, 13.5].

Below the p–p limit, evolutionary calculations again become observationally important. As "real stars" (those above the p–p limit) are born they dim along pre-main-sequence "Hayashi tracks" [11.3.2], then stabilize as more or less constant luminosity dwarfs. Brown dwarfs, however, while at first seeming to behave as their higher-mass brethren, continue to fade with time as their surface temperatures cool (Fig. 13.8). The mass of a hydrogen-burning main sequence star is found directly from luminosity and temperature with only a small consideration for age. Unlike red dwarfs, however, brown dwarfs fade relatively quickly, so that one must have some idea of how old the stars are to be able to interpret luminosities in terms of masses. Age might be found from a higher-mass binary companion, better by the brown dwarf's membership in a cluster, or by lithium content (since lithium is destroyed as the star gets older; see Section 13.2.4). It would be much better to get masses from binary orbits, but we have yet to observe brown dwarfs long enough to do so.

13.3 Late intermediate-mass evolution

Early evolution leads to and includes stable helium burners. In the late stages of evolution, all stability is lost until death as a white dwarf. In the process, the dying stars produce magnificent red giants, carbon stars, wildly unstable variables, and stunningly beautiful planetary nebulae, while at the same time seeding the interstellar medium with newly-born chemical elements. From stellar death comes life, as without these late stages, our Sun and selves would not exist.

13.3.1 The second ascent: the AGB

A star's helium-burning core eventually converts entirely to carbon and oxygen, and the nuclear fire shuts down. The result is similar to the ascent of the star on the red giant branch, but more complex. Upon consuming its helium, the core again contracts toward degeneracy (see Box 13.2), while helium burning moves out into a shell around the core, and hydrogen burning (on the CNO cycle) moves yet farther out. With the release of gravitational energy (plus the shell burning), the star once again increases in luminosity, expands, and cools to become an even brighter and chillier red giant. The second ascent of the HR diagram is in a loose sense asymptotic to the first ascent along the RGB (an "asymptote" a mathematical curve that approaches a line or another curve but never meets it). Such *second ascent stars* are thus more commonly known as *asymptotic giant branch stars*, or simply as AGB stars, to distinguish them entirely from the RGB and clump stars (Fig. 13.9; see also Fig. 13.4). More rarely, they are called *double-shell burning stars*.

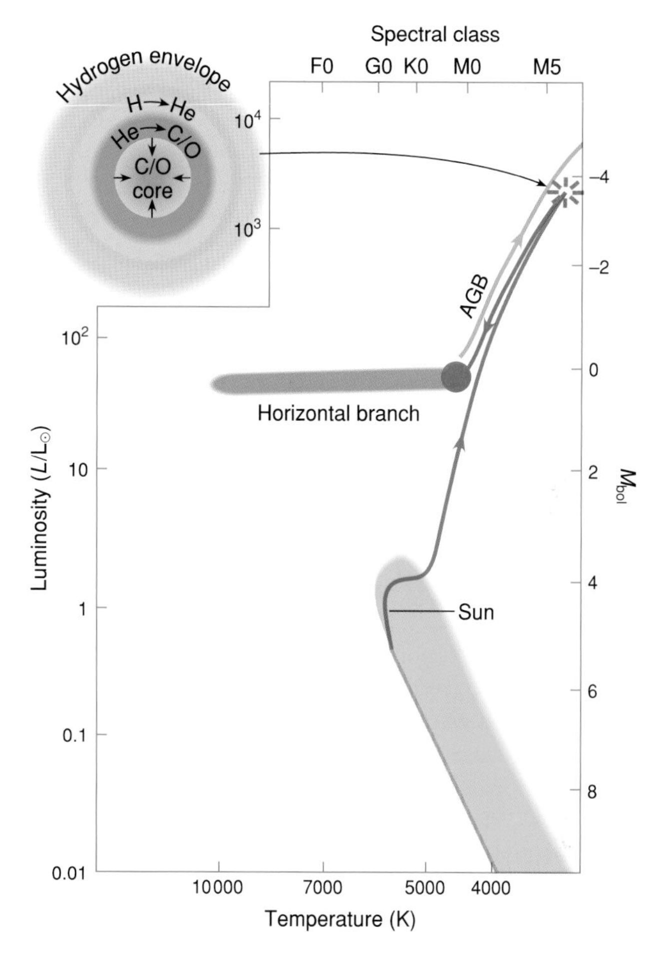

Fig. 13.9 **Asymptotic giant branch** From the clump or horizontal branch, stars brighten for the second time along the asymptotic giant branch (AGB) with carbon–oxygen cores surrounded by shells of helium and hydrogen (inset) that turn on and off in sequence. Helium shell burning begins explosively near the top of the AGB. (Kaler, J.B. (1994). *Astronomy!* New York: HarperCollins, reprinted by permission of Pearson Education, Inc.)

The hydrogen and helium shells do not burn simultaneously, but switch on and off in sequence. As the new AGB star begins to expand, the hydrogen-burning shell expands along with it, and cools, to the extent that it shuts down, leaving only the interior helium-burning shell, which continues to feed mass into the carbon–oxygen core. At this point, if the star has a mass greater than 2–3$M_\odot$, envelope convection sweeps into the region where the star had burned hydrogen to helium (the extinguished hydrogen-burning shell), and in a *second dredge-up* lofts a mixture rich in nitrogen and helium (and depleted in carbon) to the surface. As the star climbs in luminosity, the helium in the helium-fusing shell eventually runs out, which causes the old hydrogen-burning zone to contract, heat, and to fire up. This set of activities defines the lower-luminosity segment of the AGB, and is known as the *early AGB* (E-AGB), which lasts the order of a million years (depending on mass).

As the hydrogen shell burns, it adds fresh helium back into the zone between it and the carbon–oxygen core. Eventually the helium burning turns back on, but this time explosively, in a *helium shell flash* or *thermal pulse*, which pumps energy into the star, and again shuts down the hydrogen-burning shell. With the star now increasing considerably in luminosity along the *thermally pulsing AGB* (or the TP-AGB), the alternation of shell burning continues with increasing violence and luminosity, each pulse separated by the order of 100 000 years. At the same time, the C–O core is contracting, increasing in density (to the order of 10^6 gcm^{-3}), and becoming degenerate. The degeneracy stops the core contraction, which stabilizes the core, effectively turning it into a very hot (10^8K) white dwarf. Near the top of the AGB the most massive stars may undergo carbon burning to create degenerate neon–oxygen cores.

Box 13.3 End of the Earth?

Given the dramatic evolution of stars, what will happen to Earth, not to mention the rest of the Solar System? Life will likely become impossible within a couple billion years as the Sun brightens in luminosity. Our world, however, will keep turning without us. The first planetary victim will be Mercury, which will probably not survive the expansion of the Sun on the RGB. Enjoy the little planet while you can. Indeed, RGB expansion will take the Sun close to the orbit of Venus. Growing larger on the AGB, the solar gases will pass the Venusian orbit, and most likely even that of the Earth; our seeming fate is to be vaporized, our metals to be incorporated into the then-brilliant red giant, while Mars, more or less safe, looks down upon us.

But wait. Giant stars lose vast amounts of mass, which will loosen the gravitational grip that the Sun has on its planets, and they will move outward, quite possibly saving us and maybe even Venus. Since we cannot yet predict the mass loss rates of such stars, we cannot actually know whether or not we will escape in time before the future solar surface impinges upon us. But wait again. If the orbits of planets change as a result of mass loss, then their gravitational interactions change as well, which could move the planets into chaotic paths that send one or more into the Sun or may even eject them from the system altogether. Too bad we will not be around to find out.

As early evolution produces Cepheids and other variables, late evolution creates the valving conditions required to make long-period Miras along the TP-AGB [10.5.1], as well as semi-regular SRa and SRb stars [10.5.3], which are really "mini-Miras," their characteristics based mostly on their masses. When applied to the Sun, none of this activity is going to do the Earth any good (Box 13.3).

13.3.2 Mass loss

Stars begin to lose serious mass on the first ascent of the giant branch, the RGB, their winds promoted by greater luminosities and larger radii, which means lower gravities. While on the main sequence, winds are magnetically controlled as they are from the Sun [12.1.4], but as the stars climb the RGB, the winds are increasingly driven by the pressure of radiation, some "hybrid stars" doing both [7.4]. In the accepted formula, the mass loss rate is proportional to luminosity times radius divided by mass (LR/M), all in solar units. The flow rate at the tip of the RGB is thousands of times that of the Sun, reaching between 10^{-7} and 10^{-8} solar masses per year. By the time the Sun reaches the tip of the RGB, it will have lost 10 percent or more of its mass, and then even more while it descends the RGB as a helium burner. Horizontal branch stars, having begun life at around $0.8M_{\odot}$ to $0.9M_{\odot}$, weigh in at 0.5 to 0.7 solar.

Ascent of the more luminous AGB promotes even stronger winds that are driven by radiation pressure acting on dust grains that have condensed in the atmosphere [10.5.2]. Advancing TP-AGB Miras can hit 10^{-5} solar masses per year, whence those of class M turn themselves into OH/IR stars [10.5.2]; see Fig. 13.2. There is no fundamental theory for such mass loss. From speculation and observation, as a star ascends the height of the AGB, a stronger *superwind* seems to kick in that strips the star, nearly to its degenerate C–O core. The superwind effectively stops the ascent of the AGB. The luminosity of the tip depends on mass, falling at 3000 or more $L_{\odot}$ for the Sun, up to 40 000$L_{\odot}$ for a $9M_{\odot}$ star. At this point, the stars have become much redder than they were at the end of the RGB state, cooling to around 2500K (or lower) and late class M. At the extreme, the stars have radii of two or more AU, which overlap the radii of the supergiants. The TP-AGB phase also lasts about a million years (shorter for higher mass), and the white dwarf is about to be revealed.

13.3.3 The third dredge-up and chemical change

When the star reaches the pulsing AGB state, each helium flash induces convective motion in the gases that lie above it, allowing freshly made carbon to be lofted upward. If the mass is in the right range (3 to $5M_{\odot}$), envelope convection from above can tap into this carbon-rich layer and raise helium and carbon to the mass-losing stellar surface. The set of such episodes together mark the *third dredge-up*. If we could watch, the carbon-to-oxygen ratio might be reversed from the solar value ($C/O < 1$), and the star could change from class M to class S (for which C/H about equals O/H) and then to class C (wherein $C/O > 1$); that is, it could become a carbon star [6.4.3]. This process combined with high mass loss rates creates carbon-rich shrouds like that surrounding the archetypal carbon star IRC+10 216 [10.5.2; Fig. 10.8]. In the higher-mass range, carbon is converted to nitrogen, and the amount of carbon dredged to the surface decreases with mass, while nitrogen goes up. Additional CNO burning may also take place at the base of the convective envelope, this *hot bottom burning* producing more ^{13}C and nitrogen. Thus the observed chemistry of stars is at least qualitatively explained, even though the detailed calculations do not fit quite so well, as a result of uncertain treatment of convection and ignorance of the correct nuclear-burning rates.

Carbon creation is only the start. Higher-mass stars manufacture elements through nuclear burning up to and including iron and nickel [14.1.3, 14.2.3]. Chemical elements heavier than the iron peak [7.3.2] are generally built by successive capture of free neutrons, which, without an electrical barrier, can easily penetrate atomic nuclei. There are two neutron capture modes, *slow neutron capture* (the *s-process*) and *rapid neutron capture* (the *r-process*). It is not the neutrons that are fast or slow, but the rate at which they are captured. The s-process takes place in the helium-burning shells of AGB stars, while the site of the r-process is still unknown, but is somehow related to high-energy supernovae or their progeny. The s-process creates elements up to bismuth, while the r-process [14.2.3]

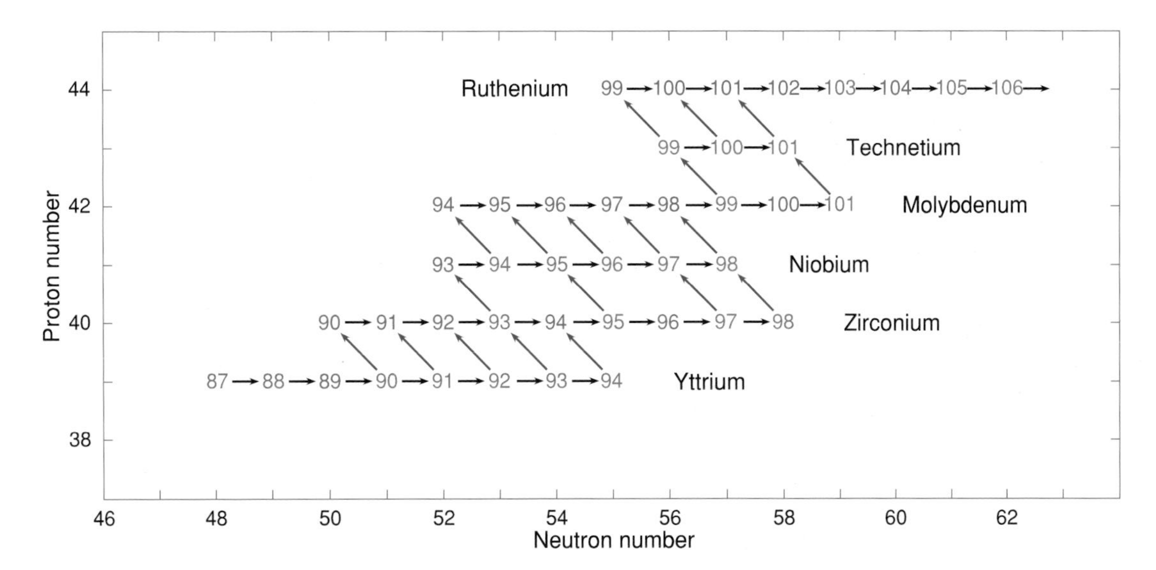

Fig. 13.10 **The s-process** Isotopes of yttrium through ruthenium are in blue (stable) and red (unstable). For any given element, the successive capture of neutrons increases the atomic weight until an unstable isotope is reached, at which point it may or may not decay with the ejection of an electron into the next element up. (Kaler, J.B. (1994). *Astronomy!* New York: HarperCollins, reprinted by permission of Pearson Education, Inc.)

goes far beyond, past uranium. (Direct proton capture, the *p-process*, helps a bit, too.)

Only the s-process is really relevant here. To make it work, a source of neutrons is required that has yet to be entirely pinned down. The most likely is a pair of reactions involving isotopes of carbon and neon, $^{13}C + \alpha \rightarrow\ ^{16}O + n$ (where α is an alpha particle and n is a neutron) and $^{22}Ne + \alpha \rightarrow\ ^{25}Mg + n$. Next, note that all elements exist in a specific range of isotopes, some stable, some not. If the instability is too great, then the isotope is said not to exist. Within a nucleus, protons and neutrons have energy levels just as do electrons in orbit about nuclei [7.1.4]. Whether an isotope is stable depends on how the protons and neutrons are arranged within the nucleus, how they are distributed within their levels, and on their between-level transition probabilities.

Start with the nucleus of a stable isotope ^{n}X (Fig. 13.10). If this nucleus absorbs a neutron, its atomic weight increases by 1 to become ^{n+1}X. If ^{n+1}X is also stable, another absorption makes it ^{n+2}X, and if that is stable, yet another gives ^{n+3}X. If ^{n+2}X is not stable, however, there are two possibilities. It may be so highly unstable that one of its neutrons immediately ejects an electron to become a proton, which converts the nucleus to $^{n+2}(X + 1)$. Since electrons were once called "beta particles," the ejection process is referred to as *beta decay*. (The ejection of a positron, as happens in the first step of the p–p reaction [12.2.2], is an example of *inverse beta decay*.) At this point, no more absorptions are possible, and the chain for that element ends. In the second possibility, ^{n+2}X is only modestly unstable, which allows it to go either to the next isotope of element X or to decay to the next element up. That is, ^{n+2}X *does* have time to create ^{n+3}X, which may or may not be stable. But there is also a chance that before ^{n+2}X grabs another neutron, it will beta decay to $^{n+2}(X + 1)$. What is statistically likely to happen is given by a *branching ratio* that depends on the half-life of the isotope.

AGB stars demonstrate the s-process taking place particularly through the observation of technetium (element 43), which has no stable isotopes; it is so radioactive that none appears on Earth. Yet there it is in the spectra of a variety of M, S, and C stars, notably Miras, vividly showing that stars *do* make chemical elements (see Fig. 13.10). Zirconium isotope ratios in S stars (derivable from minute isotope shifts in the wavelengths of ZrO lines) are also in the correct proportion to that calculated by the s-process.

In the r-process, neutron capture takes place so quickly that the chain can skip over unstable isotopes until one is reached that is so very unstable that it beta decays immediately to sufficiently high atomic number, where stability is again found. Elements are thereby characterized as "s-process" or "r-process," and most of those in the Sun (placed there during birth from the interstellar medium) have been made by a combination of the two [see Appendix 4]. Zirconium, niobium, tin, barium are largely s-process elements (from giant stars), whereas rhodium, silver, iodine, and gold are mostly r-process that are somehow related to supernovae [14.2.3]. However, compared to nuclear processes that create energy and elements into the iron peak, neutron-capture processes are minor, explaining the great drop in abundances at higher atomic number [7.3.2]. The saw-tooth pattern seen in Fig. 7.17, in which even elements are more abundant than their odd neigbors, is in part the result of alpha particle capture, but is caused mostly by the somewhat higher binding energies [Box 7.4] of the evens, the result having pairs of protons and neutrons that fulfill the Pauli exclusion principle (see Box 13.2).

13.4 Planetary nebulae

The massive change in the appearances of stars as the end approaches warrants a new name, *post AGB* (PAGB) *evolution*, which incorporates one of the most remarkable stages in evolutionary progress, the *planetary nebula* (PN, used here in the singular and plural), that phase of evolution in which giant stars are converted into white dwarfs (see Fig. 13.1). PN, discovered by William Herschel in 1785, were named from their disk-like appearances as seen through modest telescopes. Observationally, they appear as complex rings and shells of gas that surround hot blue central stars. Presented in an immense range of sizes that extend to light years across, they are the compressed, ionized ejecta of stars that were once on the AGB and that have lost nearly all their outer envelopes through winds to reveal the hot white dwarfs within. As such, these stellar leavings provide profound clues to mass loss processes in giants and, through their chemical compositions, to internal nuclear-burning processes and convection. Represented by a set in Table 13.2, they are among the most striking objects in the Universe.

Fig. 13.11 **Planetary nebula spectrum** The spectrum of the Ring Nebula (see Fig. 13.1) consists of a variety of emissions. The red image at far right is a blend of Hα and strong [N II] lines. The weaker red images are from [O I]. Just left of center are a quartet, 5007 [O III], 4959 [O III], Hβ (4861Å), and 4686 He II. Farther along to the violet, Hγ shows up. (Courtesy of Y. Norimoto, Okayama Astrophysical Observatory/NAOJ.)

13.4.1 Emission spectra

PN exhibit a rich spectrum of emission lines that overlie an emission continuum (Fig. 13.11). The spectra of diffuse nebulae [11.1.3] are similar [Fig. 11.5], so the arguments below apply to both. There are two broad kinds of emissions: *recombination lines* and *forbidden lines*. The strongest recombination lines (caused by ionization and subsequent recapture of electrons) are those of hydrogen, topped by Hα. Also observed are recombination lines of helium, oxygen, carbon, neon, and others. The strongest forbidden lines (indicated by square brackets) are usually those of [O III] (produced by O^{+2}), though [O II], [N II], [O I], and numerous others are strongly present as well. The visual color of a nebula depends on the ions, and thus the emission lines, that are present, which in

Catalogue Name	Popular Name	α (2000) δ h	m	°	′	Const.	T (star)	Remarks
NGC 2440	...	07	42	−18	13	Pup	220 000	Strongly bipolar
NGC 3242	...	10	25	−18	39	Hya	90 000	FLIERS, strong double shell
NGC 3587	Owl, M97	11	15	+55	01	UMa	110 000	Descending; bipolar
NGC 6543	Cat's Eye	17	59	+66	38	Dra	47 000	Complex FLIERS; huge outer halo
NGC 6720	Ring, M57	18	54	+33	01	Lyr	150 000	Descending; most famed; large outer halo
NGC 6853	Dumbbell, M27	20	00	+22	43	Vul	160 000	Descending, bipolar
NGC 7009	Saturn	21	04	−11	22	Aqr	90 000	Discovery object; strong FLIERS
NGC 7027	...	21	07	+42	14	Cyg	180 000	Most studied; very dusty
NGC 7293	Helix	22	30	−20	50	Aqr	110 000	Closest (300ly); descending, dusty knots
NGC 7662	Blue Snowball	23	26	+42	32	And	110 000	Classic double shell

[a]On horizontal evolutionary tracks except as indicated; FLIERS are powerful low-ionization bipolar flows.

Table 13.2 **Some famous planetary nebulae**[a]

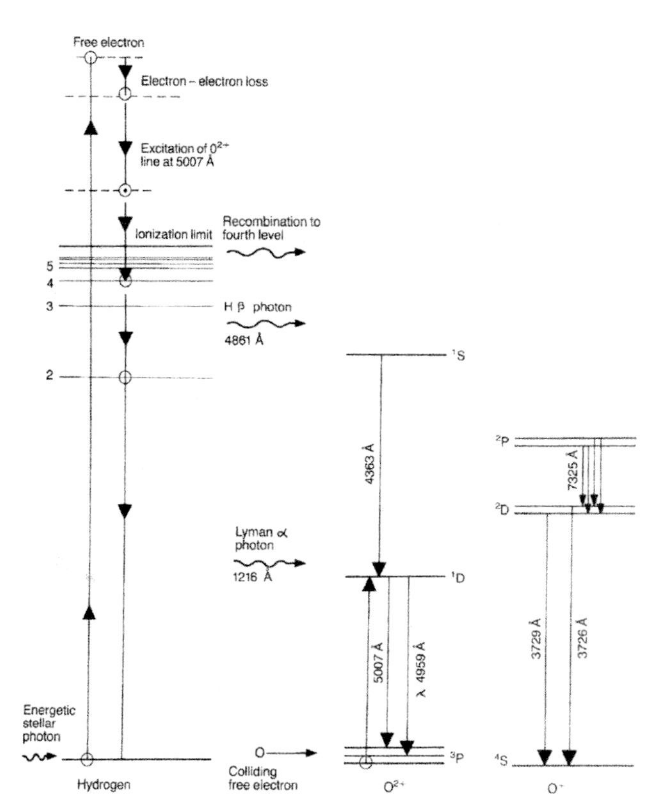

Fig. 13.12 **Nebular excitation** Left: An ultraviolet photon ionizes a hydrogen atom. After some energy losses, the electron combines with another atom, by happenstance on the fourth energy level, whereupon it jumps to the second (producing an Hβ photon) and then to the ground state (ejecting a Lyman α photon). Middle: While in the free state, the electron collides with an O^{2+} ion in the ground state (labelled ^3P), kicking it into a metastable state labelled ^1D. A drop back down to the ground state yields the forbidden lines at 4959 or 5007Å. If the electron is bumped higher, to the ^1S state, it can make another emission at 4363Å. Right: Collisional excitation of O^+ makes more forbidden lines. (Kaler, J.B. (1989). *Stars and their Spectra*, Cambridge: Cambridge University Press.)

turn depends on the temperature of the central star (25 000 K to over 200 000K) as well as on position within the object.

Recombination theory was worked out in 1927 by the Dutch astronomer Hermann Zanstra (1894–1972), and is called the *Zanstra mechanism* (Fig. 13.12). The central stars of planetary (and diffuse) nebulae are hot enough to produce ultraviolet photons with energies greater than the hydrogen ionization (Lyman) limit at 912Å [7.1.7]. A neutral hydrogen atom in the ground state absorbs a Lyman photon and is ionized to a bare proton, while the electron goes flying away with an energy E (electron) = E (photon) − E (ionization energy). After losing energy through collisions, the electron will eventually pass close enough to a proton to be captured on some level n_{cap}, with the emission of a photon.

If the recapture level $n_{cap} = 1$, it is as if the ionization and recombination never took place, and the process starts anew. If $n_{cap} > 1$, the lifetime of the electron in orbit is so short that no ionization is possible, so the electron drops down to a lower level with the emission of a photon, whose energy (hence wavelength) is given by the energy difference between the levels. If $n > 2$, the electron has a choice of levels below it. If it goes directly to $n = 1$, it emits a Lyman photon that can be absorbed by a neutral atom in the ground state, resulting in the electron being "returned" to its original state in n_{cap} (though in a different atom; the argument is statistical). If the electron jumps down to some level $n > 1$, however, it emits a photon that can escape. The process continues until the electron finally lands on $n = 2$ with the creation of a Balmer photon, from which it can go only to $n = 1$ with the emission of Lyman α. A new ionization starts the process all over again.

The Zanstra mechanism is responsible for the observed Balmer, Paschen (level 3 on the bottom), Brackett (level 4), etc. series of lines, as well as a huge array of "radio lines" (especially from diffuse nebulae) with orbits with n the order of 100 or greater. Each series consists of a *decrement* in which the longest-wave line is the strongest, followed in sequence by the others. The theory allows the calculation of the relative intensities of the Balmer (and other) lines, which among other things can be used to find the degree of interstellar reddening [3.5.2].

The forbidden lines (originally called "nebulium lines" for a presumably unknown element) were identified in 1928 by I.S. Bowen (1898–1973). Chief among them are the prominent [O III] lines at 5007 and 4959Å (confusingly still known as N1 and N2, having nothing to do with nitrogen). An electron on an upper level of hydrogen will within 10^{-7} seconds make a downward jump. Heavier atoms, however, have a number of orbital levels that do not easily interact with one another. Transitions are possible, but can take a long time, minutes, even hours, and from simplified atomic theory were called "forbidden" even though they do take place. If a level has no other choice but to produce a "forbidden photon" it is called *metastable*. The O^{+2} ion has two metastable states just above absolute ground, which itself is divided into three sub-levels.

Collisions with energetic electrons pump the bound electrons into the metastable states (see Fig. 13.12). If an excited electron can avoid a collision that takes it back to the ground state, it will eventually jump down via photon emission. Photons emitted by transitions from the lower of the two metastable states to the divided ground state make the two bright [O III] lines. An electron transition from a second, higher metastable state to the one that creates N1 and N2 makes another line at 4363Å. Forbidden transitions from O^+ produce a pair of strong [O II] lines at 3727Å, and a pair of [N II] lines flank Hα. Such lines flood the spectrum. Far too weak to be seen in terrestrial laboratories, forbidden lines are observed in nebular spectra only because of huge radiating masses. Similarly-produced forbidden lines from very highly ionized species of iron and nickel ("coronium lines") are observed in the solar corona [12.1.4].

Energies of free electrons are not quantized. As a result, the array of recaptures on different levels create emission continua to the short wave ends of the limits of each of the hydrogen series. A free electron will also radiate if it passes close enough to a proton to be forced to slow down, which adds to the emission continuum (the process called *free–free* radiation, or *brehm strahlung*, from the German for "braking radiation"). Finally, a jump from level 2 to 1 can emit *two* photons (*two-quantum emission*), whose only constraint is that their energies must add to that of Lyman α. The process that dominates the continuum depends on wavelength, two-quantum (and recapture) strongest in the optical, free–free in the radio (from which distant, dust-buried diffuse nebulae can be found).

13.4.2 Nebular conditions

With the formation mechanism known, the emission lines can easily be used to determine nebular conditions. Like the solar corona [12.1.4], planetary and diffuse nebulae are not blackbodies [7.1.2]. PN temperatures do not indicate luminosities, only the velocities of the atoms and electrons [Box 7.3]. *Electron temperatures* range from 8000K to nearly 20 000K, depending on the electron heating rate (from the central star) relative to the electron cooling rate (mostly from collisional excitation of the forbidden lines). Electron densities range from as high as $10^6 cm^{-3}$ for the most compact nebulae to as low as $10 cm^{-3}$ for the largest objects. Once temperatures and densities are known, the spectra plus theory readily yield chemical compositions. Many nebulae (those from higher mass stars) are enriched in helium, nitrogen, and carbon, as would be expected given their AGB-star parents. Though there is decent qualitative agreement between nebular chemical abundances and predictions from dredge-up theory, the discrepancies are large enough to show that there is still a great deal to learn about the workings of giant stars.

The emission lines from a planetary nebula are split in two by the Doppler effect, showing that one side is moving away from us (relative to the gross radial velocity), while the other is coming at us. PN typically expand at a rate of a few tens of kilometers per second. At that expansion velocity they last for perhaps 50 000 or so years before they blend into the interstellar medium [11.1], taking along a bounty of heavy elements derived from the dredge-up processes in the predecessor AGB stars.

13.4.3 Structure and formation

The shapes of planetary nebulae are templates against which to test theories of AGB mass loss. PN display two broad forms: *bipolar* (with two distinct lobes) and more round or elliptical (Fig. 13.13; see also Fig. 13.15). Surrounding a bright inner structure, there may be outer halos, and surrounding these are occasional huge distended envelopes, the outer structures filled with molecular hydrogen. Blasting through the nebulae on either side of the central star there may be narrow high-speed bipolar *FLIERS* that display shocks [Box 11.3] at their ends.

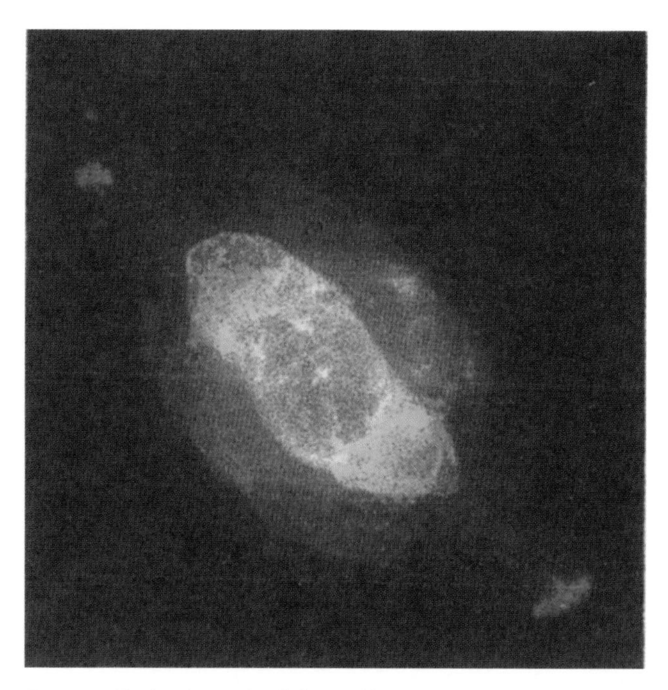

Fig. 13.13 **Bipolar planetary nebula** A Hubble image of NGC 7009, Herschel's discovery nebula, reveals a well-defined bipolar structure, as well as outlying high speed "FLIERS" that give it the name "Saturn Nebula." Compare with Figs. 13.1 and 13.15. (Courtesy of B. Balick *et al.*, STScI, and NASA.)

Unfortunately, all but a handful of PN are too far away for direct parallax measure [4.2]. Distances are estimated from expansion parallaxes [4.5], from the degree of interstellar reddening [3.5.2], and by a variety of other indirect (and mostly inaccurate) methods. The best solution is to study those in the Magellanic Clouds [Fig. 4.11, Box 5.1], whose distances are well known. Ionized masses range a few hundredths to half a solar mass, and neutral/molecular masses to a few $M_\odot$ (as might be expected for AGB ejecta).

The *interacting winds theory* explains the gross shapes. As a dying star is whittled away by the slow (15 km s^{-1}) superwind, the remaining star shrinks and heats at the surface, the escape velocity goes up, and the mass loss rate dramatically falls. The fleeing mass then detaches from the heating star, which then goes from class M (or C) back through the sequence to class B. At this point, the star is still hidden by the dusty shroud and is not directly visible. Instead, it manifests itself by heating the ejected dust, which re-radiates the energy in the infrared to produce a *protoplanetary nebula* that may also be visible through scattered starlight to make an associated reflection nebula (Fig. 13.14).

Once the heating star is detached from the expanding cloud, it begins to blow a hotter, faster, though less massive wind. The mass from the fast wind catches up with that from the earlier slow wind, shocks it, and shovels it into a dense ring that lies within an outer unshocked halo. When the interior star hits 25 000K, it begins to radiate ultraviolet photons beyond the Lyman limit at 912Å, and starts to ionize the inner edge of the compressed ring, creating the first glimmer of a real planetary nebula. As the nebula expands and the

Fig. 13.14 **Protoplanetary nebula** The protoplanetary "Egg Nebula" contains an F supergiant that is too cool to ionize the gas and that is buried inside a thick, dusty, disk filled with molecules (indicated in red in the inset). The star's light escapes through the poles to scatter from shells of dusty matter, lost to the star on the AGB. Its predecessor may have looked something like the nebula in Fig. 13.2. (Courtesy of R. Sahai and J. Trauger (JPL), the WFPC2 Science Team, and NASA.)

star continues to heat, the ionized bubble works its way into the outer neutral matter, and the nebula grows. Eventually, the expanding ionization front hits the edge of the superwind ejecta, and radiation escapes to illuminate earlier aeons of mass loss (Fig. 13.15). Dust and molecules survive in dense knots and around the nebula's periphery.

The winds of AGB stars are largely spherically symmetric. Planetary nebulae, however, usually are not. The shaping mechanism is not known. Protoplanetary nebulae exhibit the bipolarity, however, so it must somehow begin in the late stages of the superwind (as vividly displayed in Fig. 13.2). The winds may be structured by the effects of an orbiting companion (even by planets), rotation, magnetic fields, or a combination of all. Bipolar nebulae are strongly concentrated to the Milky Way, however, where we find massive AGB stars, so stellar mass is clearly related to shape.

13.4.4 Central stars

The central stars of planetary nebulae (CSPN) are so hot that the optical portions of their spectra are insensitive to

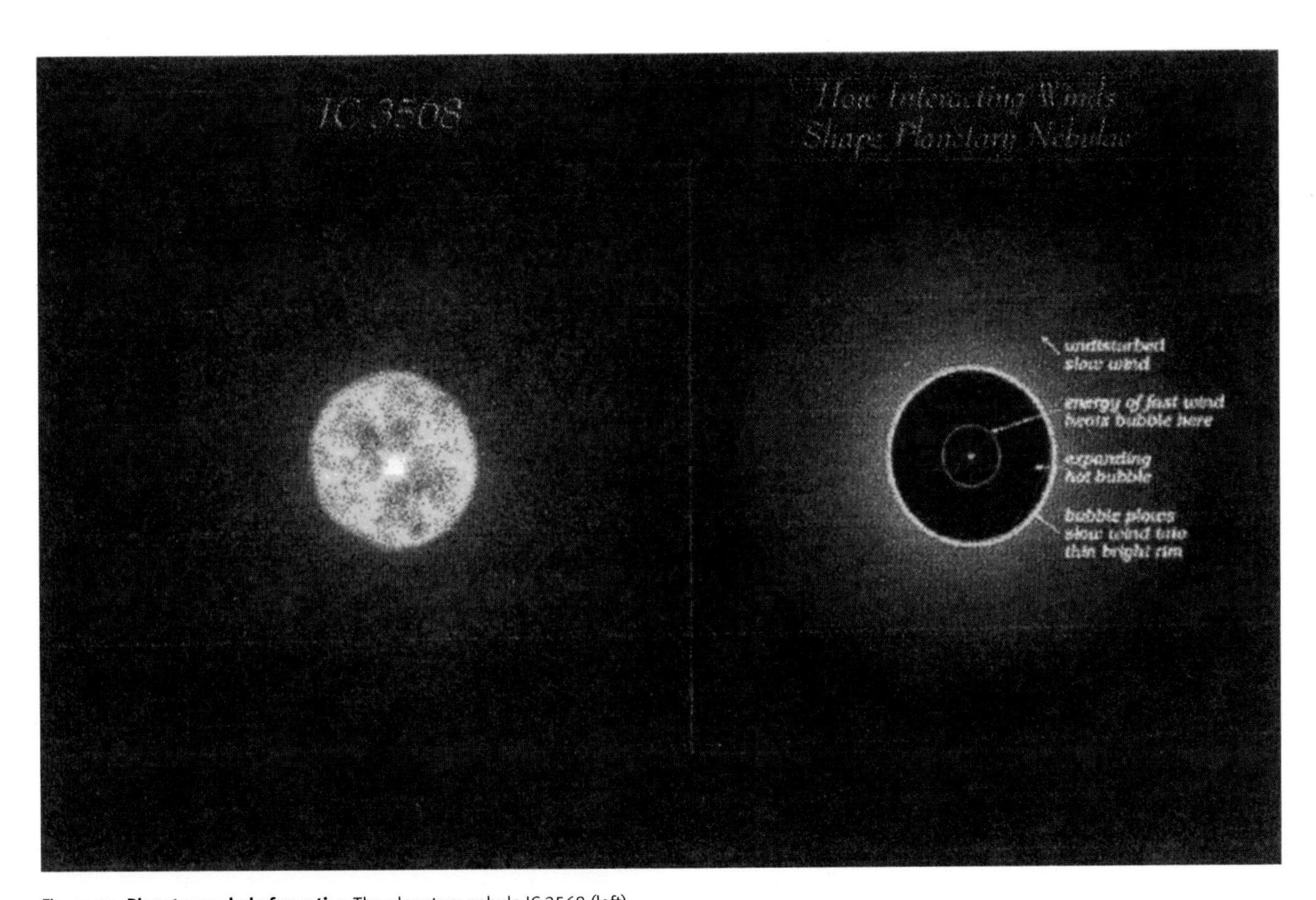

Fig. 13.15 **Planetary nebula formation** The planetary nebula IC 3568 (left) illustrates how nebulae are formed (right) when a fast wind from the heating star plows into and shocks the slow AGB wind. (Image courtesy of H. Bond (STScI) and R. Ciardullo (Penn. St. U.), STScI, and NASA; diagram by Bruce Balick from *Planetary Nebulae*, slide set, J. B. Kaler and B. Balick, courtesy of the Astronomical Society of the Pacific.)

temperature. Instead, we make use of the surrounding nebulae. Every UV photon from the star is responsible for one hydrogen ionization. For each ionization there must be one recombination. Every recombination eventually produces a transition that winds up on the second level, and therefore must produce a Balmer photon (either in the lines, or in the Balmer continuum, which is produced by direct recombination onto the second level). Given equilibrium, the number of Balmer photons radiated by the nebula per second therefore equals the number of stellar UV photons shorter than 912Å emitted per second by the star. The ratio of the total number of UV photons to the number of visual photons from a hot blackbody is very sensitive to temperature. Therefore, all we need do is measure the ratio of the sum of the nebular Balmer-line photon fluxes (number of photons per unit area per second) at the Earth to the visual stellar photon flux, which is found from the visual magnitude. Since theory allows us to relate the total Balmer-line photon flux to the Hβ energy flux, F(Hβ)/F(*V*) gives the *Zanstra temperature*. A similar measure derives from the ionized helium lines, which are created by photons beyond the helium double-ionization limit at 228Å.

Though many nebulae will not catch all the UV photons, and the stars may not exactly be blackbodies, the method shows that temperatures range from the expected lower limit near 25 000K to above 250 000K. Since Hβ fluxes reflect stellar UV fluxes, if we know distances (and temperatures), we find stellar luminosities, which range from a few times that of the Sun to over 10 000$L_\odot$. CSPN are among the most luminous stars in the Galaxy. Their seeming faintness at Earth is the result of their radiating most of their energy in the ultraviolet where we cannot see it; that is, the result of huge bolometric corrections [3.3].

With temperature and luminosity known, we can plot the CSPN on the log L–log T plane and compare with theory (Fig. 13.16), which agrees nicely with observation. A star emerges from its protoplanetary shroud puffed up with a low-mass hydrogen (or perhaps even helium) envelope, which keeps being thinned by the stellar wind. At the same time, nuclear fusion (either hydrogen or helium burning) eats away at the envelope from below. As the envelope shrinks, the star heats at constant luminosity to a maximum temperature (on a *horizontal track*, not to be confused with the "horizontal branch"). Then, as the core is nearly exposed and nuclear burning shuts down, the star dims as a hot, nascent white dwarf along a *descending track*. As masses increase, so do the luminosities on the horizontal tracks and the maximum turnaround temperatures. Along the descending tracks, higher mass stars are the *less* luminous, because of higher gravities and greater compressions, and therefore smaller radii. As the stars evolve, their nebulae inexorably expand, making those with stars on horizontal tracks small and bright, those on descending tracks increasingly larger and dimmer until there are no nebulae left to see at all, only the ancient fading central stars, the white dwarfs revealed.

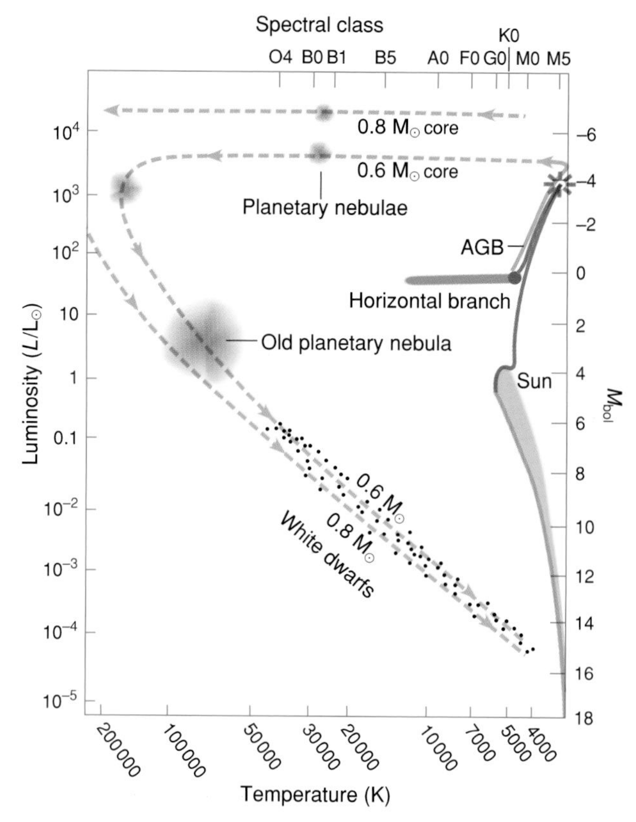

Fig. 13.16 **Evolution to the white dwarf state** Planetary nebula central stars of 0.8 and 0.5 solar masses evolve off the AGB along horizontal tracks. When the stars hit 25 000 kelvin, they illuminate their wind-compressed nebulae. Shrinking, they heat to mass-dependent maximum temperatures, and then cool as their nebulae expand to invisibility, leaving white dwarfs behind. (Kaler, J.B. (1994). *Astronomy!* New York: HarperCollins, reprinted by permission of Pearson Education, Inc., based on the work of I. Iben, Jr.)

CSPN fall into a variety of spectral categories that depend on temperature and chemical composition [6.4.5]. Along the horizontal tracks, we see "O-type" absorptions, Ofp types with helium emissions [Box 6.1], Wolf–Rayet types [6.4.4] (with one exception class WC only), the emissions of the latter two classes indicating hot winds. Like "classical" massive Wolf–Rayet stars, their hydrogen envelopes are gone, and are replaced by carbon-rich helium envelopes. At the high temperature end, near the turnaround in the evolutionary tracks, we find luminous stars with powerful O VI emissions. Some of these hot stars fall into the hydrogen-poor class of PG 1159 stars, while some are also GW Vir variables [10.4.3]. During descent, the stars mostly display absorption lines that reveal either hydrogen-rich or helium-rich atmospheres, showing us that they are the precursors of the DA and non-DA white dwarfs.

As a last gasp, a fading CSPN may undergo a final helium-shell flash that drives it back onto the AGB, from which it will repeat its evolution. Half a dozen or so candidates exist, including FG Sagittae, which over a period of a few years went from class B to G and began to exhibit huge quantities of s-process rare earth elements. R Coronae Borealis stars [10.6.2], which are very low in hydrogen but rich in helium and carbon, may derive from such an event. (Their origins are contended. They may have also lost their outer hydrogen-rich envelopes through winds, rather like Wolf–Rayet stars that reveal the by-products of helium burning, and may be on their way to *becoming* planetary nebulae. An additional scenario has them formed by the mergers of helium and carbon white dwarfs.)

The rare *subdwarf O* (sdO) *stars* have many of the same characteristics as the CSPN, except they have no surrounding nebulae. Their status is unclear. They may have lost their outer masses on the RGB before the helium began to burn. They may simply be very blue horizontal branch stars (as are their counterparts, the *subdwarf B*, sdB *stars*) that are starting to ascend to the AGB [9.4.4]. They may come from a late thermal pulse, and their nebulae have dissipated. There are almost as many theories as there are stars.

13.5 White dwarfs at last

In spite of the many uncertainties, what *is* certain is that the CSPN and their kind will become extremely dense, degenerate (Box 13.2) DA or non-DA white dwarfs [6.4.6]. Their only fate is slowly to cool and dim as they cross the HR diagram below the ordinary dwarfs, going through their stages of variability [10.4.3] and finally crystallizing near the end, their degeneracy keeping them from squeezing down any farther. Their radiative energy comes from the tail of high-speed electrons that are not degenerate. From theory, we can determine the age of a white dwarf at any point along its evolutionary track. No white dwarf has ever had time to cool sufficiently to disappear from sight in the history of the Galaxy. Every one ever made is still there to be seen. The coolest (3400 kelvin) and dimmest ($M_V = 16$) of them yield the age of the Galaxy, which is in accord with the HR diagrams of globular clusters [9.4.8]. Theory combined with observation also yields the birthrate of white dwarfs, which is of the order of $2 \times 10^{-12} pc^{-3} yr^{-1}$, in rough accord with the PN birthrate, as it ought to be.

The end-product of the studies is the relation between final mass and initial mass. There is, however, no firm theory of AGB mass loss, so we cannot easily evolve stars of known initial mass to calculate what the white dwarf masses would be. Instead, we must calibrate uncertain theory against hard-to-make observations of white dwarf masses in non-interacting binaries and estimate limits to the initial masses from the mass of the remaining dwarfs. The results, with large uncertainties, show that a $1 M_\odot$ star will develop a white dwarf of $0.55 M_\odot$, and that at $M_i = 7 M_\odot$, M_f hits a solar mass. As 10 initial solar masses is approached, rare neon–oxygen white dwarfs approach the Chandrasekhar limit of $1.4\ M_\odot$ (see Box 13.2). Very roughly, within the large errors of theory and observation, up to 7 or $8 M_\odot$,

$$M_f = 0.47 + 0.078 M_i.$$

Higher masses are subject to very uncertain, and contended, theory. Since the majority of stars are of lower mass, so are most white dwarf masses, which average about $0.6 M_\odot$, in accord with masses of CSPN derived from their evolutionary tracks. High-mass white dwarfs become increasingly rare.

The origins of the different kinds of white dwarfs [6.4.6] are yet unclear. Non-DA stars seem to have lost their outer hydrogen skins through winds. The DA white dwarfs are probably the result of diffusion, whereby the heavier helium sinks out of sight, much as particular elements do in a variety of chemically peculiar dwarfs [7.3.3]. While the precursors of the different kinds are seen among the planetary nebula nuclei [6.4.5], the variations of the ratios of DA to non-DA stars, and in particular the origin of the "non-DA desert" between 30 000 and 45 000K, are not understood. Magnetic white dwarfs may be by-products of magnetic A and B dwarfs, the Ap and Bp stars [7.3.3], whose fields are squeezed down as a result of evolutionary contraction. These in turn may be fossil fields left over from star-forming accretion disks [12.4.4], these white dwarfs still clinging to the tattered remains of their births.

13.6 Binary evolution

All ends well, except that most stars are not single, but are in binaries [Chapter 8], and the effects of the companions on each other cannot be ignored. Unfortunately, since there is almost an infinite number of combinations of binary separations and mass ratios, there is no generalized theory of binary evolution. Wide binaries, in which tidal action [8.5.1] is small, are no problem, as each star evolves on its own. The result may be a simple white dwarf orbiting an ordinary dwarf or giant, and finally a double white dwarf (providing neither star initially has so much mass that it explodes). If the two are close enough, however, mutual tides can disturb the standard ageing process. Evolution can cause the more massive star to swell and to fill and overflow its Roche lobe [8.5.1], resulting in mass transfer from one star to the other, which can produce a host of phenomena that include dwarf carbon stars [6.4.3], barium stars [8.5.10], and Algol-type binaries [8.5.3]. If the mass transfer occurs early enough in a star's history, before it starts fusing helium, it may even fizzle and turn into a low-mass helium white dwarf. The possibilities seem endless.

Expansion through evolution and mass loss can wrap the two stars in a common envelope. Friction between the stars and the envelope then dissipates orbital energy, makes them spiral together, and perhaps even merge. If the envelope dissipates before merger, then the white dwarf that develops from the evolving star can find itself stuck very close to the remaining dwarf, allowing the conditions needed for novae [8.5.4], recurrent novae [8.5.5], dwarf novae [8.5.6], and symbiotic stars [8.5.8].

Inspiralling and mergers can be produced in a variety of other ways. If the stars are close enough, mutual tides will force a state of *synchronous rotation*, that is, make the stellar rotation periods of the stars equal to the orbital period. The Moon, which keeps one face always pointed toward Earth, is in such a state. Under the right conditions, if the orbit is perturbed, the continued attempt at synchronization can dissipate orbital angular momentum [Box 8.3] and lower the orbital radius. One star may also have a strong magnetic field. Magnetic braking through winds [12.4.5] will try to slow the star's rotation, which the mutual tides are trying to keep synchronized. The result again is a lowering of orbital size. To these add gravitational disturbances from close-passing stars [9.4.10] and even radiation of energy through gravity waves [14.3.2].

No white dwarf more massive than $M_f = 1.4M_\odot$ can exist (Box 13.2). Single stars initially more massive than $10M_\odot$ grow cores that meet or exceed the limit, while binary evolution can, through mass transfer or mergers, push white dwarfs past the limit as well. What happens above the limiting mass is catastrophic death – and in one of the great dichotomies of the Universe – renewal and life.

Chapter Fourteen

High-mass evolution

High-mass stars are so rare they might seem worth but a footnote to stellar evolution. Above an adopted lower limit of 10 solar masses [13.1], we deal with the evolution of stars earlier than class B2, effectively the O dwarfs, this hottest class constituting under a millionth of the main sequence population [Box 6.3]. In spite of the paucity of its members, however, the set of high-mass stars is perversely among the most important. Unable to make stable white dwarfs, high-mass stars explode as supernovae (Fig. 14.1). Massive white dwarfs that are pushed into exceeding the Chandrasekhar limit [Box 13.2] explode as well. The two kinds of supernova together create the shock waves that compress the interstellar medium to make new stars [11.3.1], expel cosmic rays that are crucial to the chemistry of molecular clouds [11.1.1; Box 11.1] and help slow their rotations (thus allowing star-forming collapse [11.3.1]), and create a good fraction of the nuclei of the periodic table, some exclusively.

Fig. 14.1 **The Crab Nebula, M1** The exploded remnant of the supernova of 1054, the Crab epitomizes high-mass stars and their fates. Expanding into interstellar space, a spinning neutron star at its heart (the lower right of the pair at center), it carries heavy elements made in the supernova's nuclear furnace. The nebula is powered by the neutron star's wind, which produces readily visible shell-like structures. Though looking like any other star, the pulsar turns on and off 30 times per second. (Courtesy of Jay Gallagher (U. Wis.), N. A. Sharp (NOAO)/WIYN/NOAO/NSF.)

14.1 Supergiants

As the term "giant" means different things to different people, so does "supergiant." To a spectroscopist, a supergiant is a star so classified on the basis of its spectrum alone [6.3.2]. Calibration shows supergiants from class F to M to have absolute visual magnitudes brighter than around −4. These divide into lesser Ib supergiants (average $M_V \approx -5$), a brighter Ia variety ($M_V \approx -7$), and an extraordinary "hypergiant" class (called "0," "zero," $M_V \approx -7$), their detailed loci illustrated in Fig. 6.12.

To the theoretician, supergiants and giants are distinguished by their evolutionary endpoints. Giants become white dwarfs, while supergiants explode to make either neutron stars or black holes. The separation in initial mass between giants and supergiants as so defined is around $10M_\odot$ [13.1], which agrees rather well with the spectroscopic divide (though some luminous lesser-mass stars are still classified as supergiants). By further coincidence, the spectroscopic divide near class B1 between the dwarfs that become giants and those that become supergiants also roughly separates the stars that make reflection nebulae from those that make diffuse nebulae [11.1.3]. Diffuse nebulae thereby serve as flags for stars that will someday become supergiants and supernovae.

14.1.1 Main sequence evolution

Zero-age O and early B dwarfs (for simplicity, call them all "O stars") range in luminosity from about $6\,000L_\odot$ at $10M_\odot$ (class B1) to two million $L_\odot$ at $120M_\odot$ (O2–O3). While still called "dwarfs," these stars are large: the radius of a $100M_\odot$ dwarf more than 15 times solar [Fig. 7.14]. In spite of the great masses, maintenance of their radiative power requires very short main sequence lifetimes, 20 million years at $10M_\odot$ to only 2.5 million at $120M_\odot$. The more massive O stars have evolved from birth to death since humans first walked the Earth. Decreasing lifetimes, in combination with lower chance of formation, means that higher masses become increasingly rare, until there are none at all.

Standard evolutionary tracks for O stars begin with fully formed stars of specific masses that begin life on the ZAMS, as shown in Fig. 13.4. In reality, however, stars of such high mass (especially toward the upper reaches) begin rapid internal evolution before they have achieved full growth. Since the course of high-mass star formation is not well known [11.2.5], real evolutionary paths for main sequence dwarfs are impossible to calculate, as the stars move off the ZAMS even as they are moving onto it, and increasing their luminosities as a result of mass accretion. There really is no such thing as a high-mass ZAMS star.

With this caveat, the evolution of high-mass dwarfs is similar to that of their lower mass cousins [13.2.1]. Like the stars of lower masses, as core hydrogen fuses to helium (by the CNO cycle [12.2.2; Box 12.2]), O dwarfs brighten, swell, and cool at their surfaces. Above $20M_\odot$, the surface cooling during core hydrogen burning becomes progressively greater, resulting in a distinct broadening of the main sequence toward higher luminosities. Overall contraction as core hydrogen nears zero produces the same hook in the evolutionary track that we see for lower-mass stars. When core hydrogen finally runs out, the stars reverse paths once again, and main sequence life is over. The relative changes in luminosity (L) and temperature (T) are enough to make significant changes in radius (R). On the theoretical ZAMS, a 25-solar-mass star has a radius $6.5R_\odot$; by the end of hydrogen burning it has grown to $17.5R_\odot$. Between such large changes and the speed of evolution, habitable planets (if there are any planets at all in the fierce stellar environment) would be impossible.

Temperatures of newly formed O dwarfs are so high that upon formation, O stars immediately ionize their surroundings. Those deeply embedded within their birth clouds create dense, small "ultra-compact H II regions" [11.1.3] whose radio emissions immediately identify pockets of high-mass star formation.

14.1.2 Early supergiant evolution

When the central hydrogen runs out, the cores of massive stars (like those with lower masses) rapidly contract and heat, which sends the stars evolving with nearly constant luminosities and increasing sizes through the blue supergiant range to the red, where internal temperatures are high enough to sustain helium fusion by the 3-α process (see Fig. 13.4). The passage through the extended Hertzsprung gap, which widens as masses increase, is remarkably fast, a $25M_\odot$ star making the journey in a bit over 10 000 years, 0.2 percent of its main sequence lifetime. As a result, cooling mid-temperature "yellow supergiants" are rare. Up to around $25M_\odot$ or more (depending on the models), the stars stabilize as helium-burning red supergiants with immense sizes, the $25M_\odot$ star growing to a radius of some 1000AU,

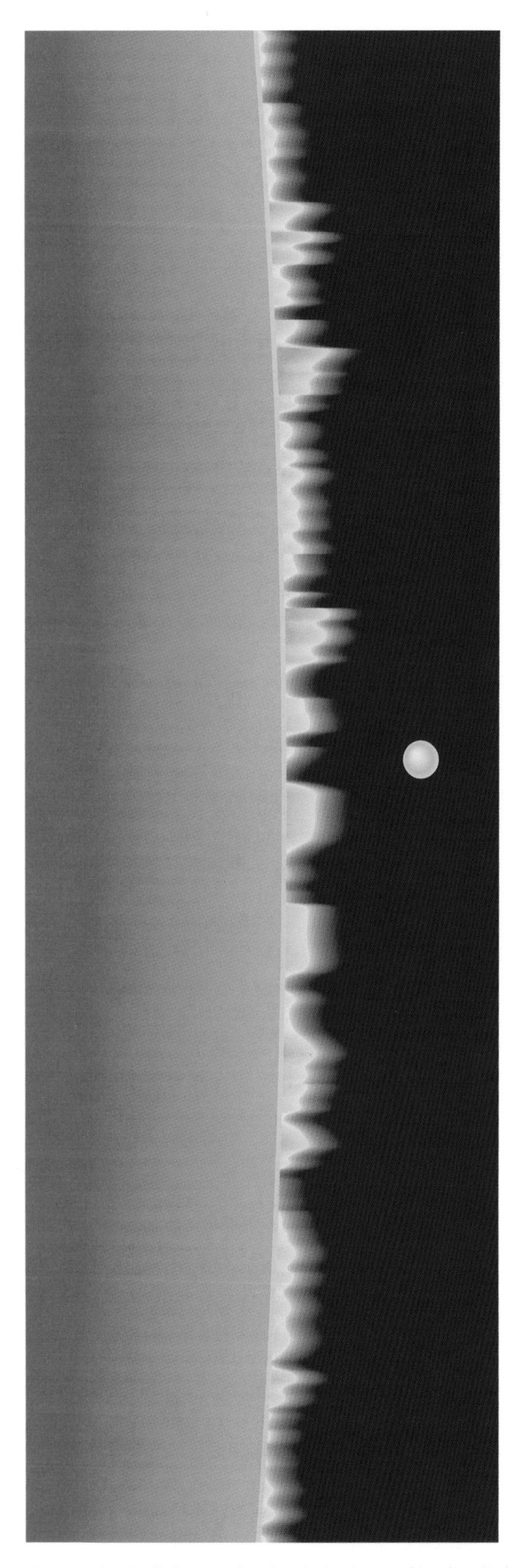

Fig. 14.2 **Mu Cephei** Mu Cephei dwarfs the Sun. A white dwarf is but a hundredth the solar size, and a neutron star is a ten thousandth the size of that. On this scale, a neutron star would be 800 angstroms across. (Kaler, J.B. (1994). *Astronomy!* New York: HarperCollins, reprinted by permission of Pearson Education, Inc.)

comparable with Jupiter's orbit. Two of the largest stars known, Mu Cephei and VV Cephei, are M2 supergiants whose radii lie in the 7 to 9 AU range, which approaches the size of Saturn's orbit (Fig. 14.2). Stabilization temperatures up to about 25 $M_\odot$ fall into the 3500–4000 kelvin range, appropriate to early class M and late K, warmer and earlier than the AGB stars [13.3.1] that can overlap supergiant luminosities, thus separating the two.

Mass loss is a profound part of high-mass life. Supergiant winds are fierce, with loss rates in the range of 10^{-5} to 10^{-4} $M_\odot$ (and on rare occasion far greater). O dwarf and blue supergiant winds are driven by the pressure of radiation on gas (via electrons and absorption lines), whereas at the cool end of the HR diagram, absorption of light by dust grains again plays a role, as it does for Mira giants [13.3.2]. The 25-solar-mass star of Fig. 13.4 is estimated to be down to 15$M_\odot$ by the end of helium burning. Higher-mass stars are even more extreme, an 80-solar-mass star whittling itself down to under 10$M_\odot$.

In part as a result of mass loss, from about 25–40 solar masses and up (the value depending on the specific model), the stars do not go all the way to the right side of the HR diagram to become red supergiants, but stop short in class G or F or even earlier, and then turn around to move back to the left, to become blue supergiants once more as they continue to fuse their helium into carbon and oxygen. These, as well as stars of higher mass, can become so hot at their surfaces (in part as a result of mass loss) that they even cross to the other side of the main sequence.

At the very highest end, mass loss is so great that stars not only do not make it to the M supergiant range, but never go past late class B, staying as blue supergiants before they make the turnaround. The effect is illustrated by the *Humphreys–Davidson* (HD) limit of the observed HR diagram. There are very few stars to the red of a sloping line in early class B that goes from high-mass limit to around 50$M_\odot$ (Fig. 14.3). Below the HD limit, stars slide most or all the way across the HR diagram as they evolve. Above it, they hit the limit and bounce back to the blue side. The effect is apparently related to the *Eddington limit*, the luminosity at which radiation pressure acting on electrons effectively balances the inward gravitational pull. Here is the domain of the "luminous blue variables," the LBVs [10.6.1], stars that are so unstable that they can erupt huge quantities of matter, as exemplified by P Cygni and Eta Carinae. As massive stars loop back across the main sequence, they have lost so much matter that the by-products of nuclear burning become visible, creating first the nitrogen-rich WN version of Wolf–Rayet stars and then the carbon WC stars (though the actual progression is argued), the lower-mass version making the B[e] stars [6.4.4]. Consistent with mass loss, Wolf–Rayet stars are sometimes surrounded by thick "Ring Nebulae" that are akin to the planetary nebulae [13.4] created by mass loss from lower-mass stars.

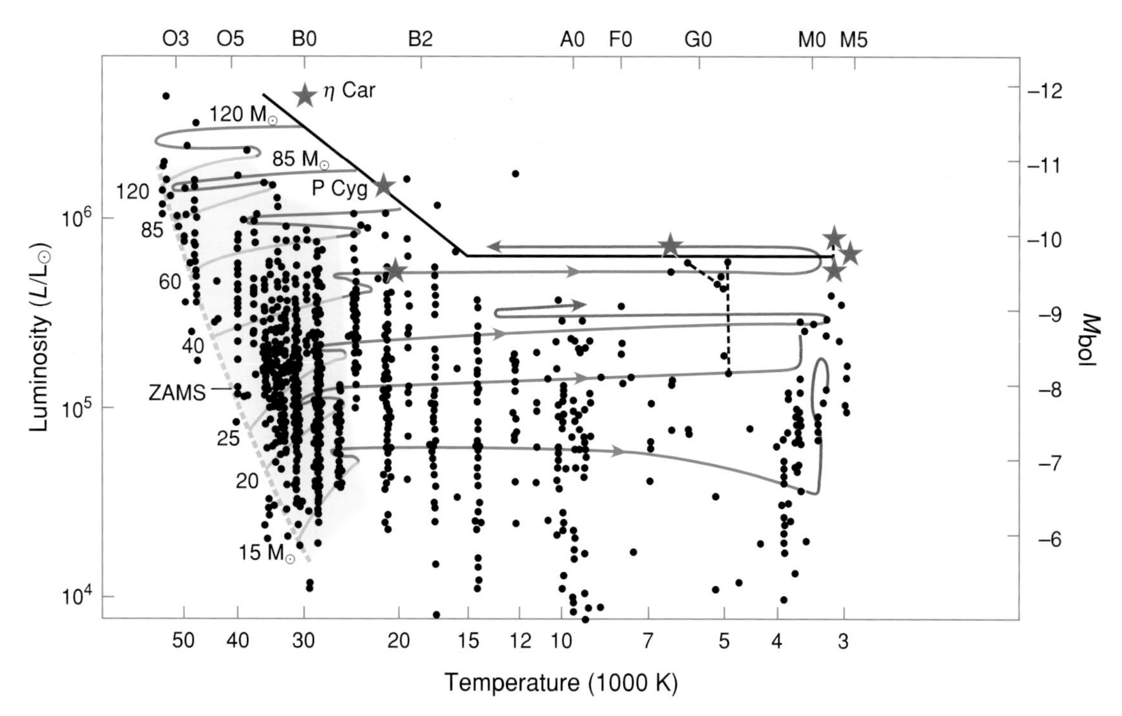

Fig. 14.3 **Supergiants** Supergiants are plotted with, and nicely agree with, theoretical evolutionary tracks that are identified by stellar mass at the left (and that differ some from those in Fig. 13.4). Mass loss prevents supergiants from crossing the HD limit, the sloping line at the top of the diagram. (Kaler, J.B. (1994). *Astronomy!* New York: HarperCollins, reprinted by permission of Pearson Education, Inc.) Observations by R. Humphreys and K. Davidson, theory by A. Maeder and G. Meynet.)

H → He
He → C → O
C/O → Ne/Mg/O + Na, Al
Ne → O/Mg
O/Mg → Si/S + Cl, K, P
Si/S → Fe

Table 14.1 **Nuclear fusion stages in a developing supergiant**

14.1.3 Carbon burning and beyond

For intermediate mass stars, the termination of helium fusion and the completion of the carbon–oxygen core generally spells the beginning of the end, as mass loss in the AGB state clears the core to expose the developing white dwarf. The exception lies at the high-mass end around $10M_\odot$, where continued alpha particle (helium nucleus) capture can create neon–oxygen white dwarfs. When their helium burning is complete, all supergiants do the same (Table 14.1). Near a temperature of 10^9 kelvin, the carbon and oxygen will fuse by alpha particle capture and by direct carbon burning to a mixture that features not just neon and oxygen, but magnesium as well, e.g.

$$^{12}\mathrm{C} + {}^{12}\mathrm{C} \rightarrow {}^{20}\mathrm{Ne} + \alpha,\ {}^{24}\mathrm{Mg},\ {}^{23}\mathrm{Mg} + \mathrm{neutron} + \cdots,$$

with sodium and aluminum on the side. The neon fuses to additional oxygen and magnesium:

$$\left({}^{20}\mathrm{Ne} + {}^{20}\mathrm{Ne} \rightarrow {}^{24}\mathrm{Mg} + {}^{16}\mathrm{O}\right),$$

with some phosphorus. Meanwhile, helium- and hydrogen-burning shells move outward to surround the core (Fig. 14.4).

When carbon burning ceases, the core again contracts and heats until the Mg-O fuses to a yet heavier mix that yields silicon and sulfur as per

$$^{16}\mathrm{O} + {}^{16}\mathrm{O} \rightarrow {}^{28}\mathrm{Si} + \alpha,\ {}^{31}\mathrm{S} + \mathrm{neutron},\ {}^{32}\mathrm{S}, \ldots,$$

along with some other nuclei, chiefly chlorine, potassium, and phosphorus. The whole thing now becomes nested in carbon, helium, and hydrogen-burning shells. When the Si–S core is complete, contraction then heats the mix to over 3×10^9 kelvin and a density in excess of 10^6 gcm^{-3}, whence the silicon, through a variety of reactions and counter-reactions, burns to create iron. At the end of the line, a Chandrasekhar-mass ($1.4M_\odot$ or greater) iron core about the size of Earth is surrounded by onion-like shells that contain all the earlier generations of nuclear fusion. As iron (and nickel) have the most tightly bound nuclei [13.2.3], no further core burning is possible.

Most of the binding energy released in the progression from hydrogen to iron is in the fusion of hydrogen to helium, which thereby takes the longest time: hence the very existence of the main sequence [13.2.3]. Most of the remainder is released in helium burning, which therefore dominates the supergiants and creates the HR diagram's helium-burning sequences. Progressively less is released in Ne–Mg fusion and in silicon burning. Yet these stages must still provide enough

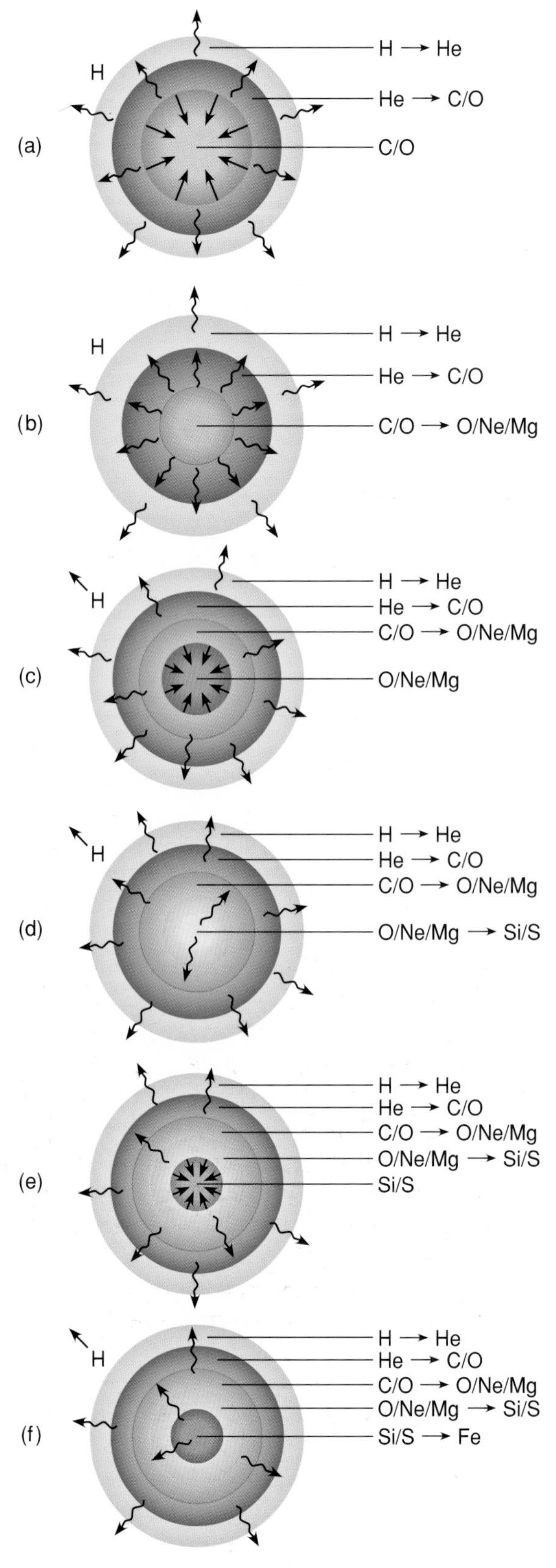

Fig. 14.4 **Nuclear burning in supergiants** The sequence shows a simplified view of the progression of fusion stages, beginning with the contracting carbon and oxygen core. At the end, the star creates a ball of iron surrounded by shells that contain earlier core-burning stages. (Kaler, J.B. (1994). *Astronomy!* New York: HarperCollins, reprinted by permission of Pearson Education, Inc.)

energy to help support the star, which speeds up later burning stages. Then factor in neutrino emission, which saps energy required for support. As a result, advanced burning stages take very little time. Silicon burning to the iron core may take just a mere week in a star that has been around for millions of years. Though the odds are that the core of any supergiant is fusing helium, it is impossible to tell exactly what state a given supergiant may be in until the iron core collapses and explodes the star in the creation of a grand supernova.

14.2 Supernovae

The modern story of supernovae (SN, singular or plural) begins with the Danish astronomer Tycho Brahe (1546–1601), the last and greatest of the naked-eye observers. While he did not discover the supernova that erupted in Cassiopeia in 1572, he followed its progress so carefully that it became known as "Tycho's Star" (Fig. 14.5). It reached apparent magnitude −4, close to the maximum brilliance of Venus, was visible in the daytime sky, and did not fade into the cosmic gloom for two years. Johannes Kepler [8.2.3], who for a time served as Tycho's assistant, repeated his mentor's feat with detailed observations of "Kepler's Star," which graced Ophiuchus in 1604. Reaching apparent magnitude −2.5, it was visible to the naked eye for over a year. Studies of ancient documents (particularly those

Fig. 14.5 **Tycho's Star** Though the supernova observed by Tycho in 1572 had long since faded away, it was nevertheless memorialized three decades later by Johannes Bayer in his *Uranometria* of 1603. (Courtesy of the Rare Book and Special Collections Library, University of Illinois.)

Year	Const.	Peak V	Type	Remnant	Dist. (kpc)	Remarks
185	Cen	−2?	. . .	RCW 86	2.5	Questionable
1006	Lup	−8?	Ia	PKS 1459–41	2.2	Brightest on record
1054	Tau	−4	II	Crab Nebula	2.0	Messier 1, Chinese "Guest Star"
1181	Cas	0	II	3C 58	5.0	. . .
1572	Cas	−4	Ia	3C 10	3.5	Tycho's Star
1604	Oph	−3	Ia[a]	3C 358	5.5	Kepler's Star
1667?	Cas	+6	II	Cas A	3.4	Brightest radio source beyond Solar System. Highly questionable sighting by Flamsteed
1987	Dor	+3	II	1987A	50	Supernova 1987A in LMC

[a]Possibly Type II.

Table 14.2 **Some famous supernovae**

from China) turned up more such events, the best known being the "Guest Star" of 1054, which rivalled Tycho's Star and created the Crab Nebula (see Fig. 14.1). The brightest ever recorded was the supernova of 1006 in Lupus, which may have reached a (vigorously argued) apparent magnitude of −8 to −10! None has been observed visually in our Galaxy since John Flamsteed may briefly (and again arguably) have spotted the supernova of the seventeenth century whose exploded remnant was much later discovered glowing in the radio spectrum as Cassiopeia A (see Fig. 14.16 below). While they are thought to occur at a rate of one or two per century, most are hidden by the dust in the Milky Way (Table 14.2).

The study of supernovae has therefore been relegated to those in other galaxies [Box 5.1]. The first of these was S Andromedae, which popped off in the Andromeda Galaxy (M31) in 1885 and reached almost to naked-eye visibility. Only when Edwin Hubble learned the distance to M31 (now known to be 780kpc) could astronomers realize the brilliance of SN and separate them from ordinary novae [8.5.4]. Even at a low explosion rate per galaxy, however, there are so many galaxies that any dedicated search easily picks them up, allowing a great number for examination. The most studied extragalactic supernova is without question the Large Magellanic Cloud's [Fig. 4.11] SN 1987A, which in that year reached third magnitude, even though 50kpc away (Fig. 14.6a; Box 14.1). With some exceptions, supernovae are named after their year of eruption, followed by one or two lower-case letters that indicate discovery order.

Box 14.1 SN 1987A

Supernova 1987A was discovered glowing in the Large Magellanic Cloud [Fig. 4.11] on February 24, 1987 by Ian Shelton, who was taking images of the LMC at Las Campanas Observatory and caught an intruder (Fig. 14.6a). It was actually seen rising in brightness the day before by the Australian amateur, Robert McNaught. SN 1987A came from the core collapse of a blue B1 Ia supergiant catalogued as Sk −69°202. Though 50 000pc away, at its peak of third magnitude it was easily visible to the naked eye. While actually an unusually faint supernova, the result of small progenitor size caused by the low metallicity of the LMC, it has been a treasury of information.

Above all else perhaps, SN 1987A provided proof of core collapse. Each neutron of the neutron star created from core collapse produces a neutrino. The number of neutrinos therefore must equal the core mass (assume $1.5M_\odot$) times the mass of the Sun (2×10^{33}g) divided by the neutron mass (1.7×10^{-24}g), resulting in an astonishing 2×10^{57} of them. All these expand into a sphere of ever-increasing size. At the Earth, a sphere centered on SN 1987A has a surface area of $4\pi R(\text{distance})^2 = 4\pi \times [5 \times 10^4$ (distance of the LMC) $\times$ 3×10^{18} (number of centimeters in a parsec)$]^2 = 3 \times 10^{47}$ square centimeters. The flux of these neutrinos at the Earth is therefore $2 \times 10^{57}/3 \times 10^{47}$ or nearly 10^{10} – ten billion! – per square centimeter.

More than that: "thermal neutrinos" are produced by the weak nuclear interactions that take place in the ultra-hot nuclear-fusing environment, notably by the annihilation of electron–positron pairs. Within a few seconds after you would have seen the internal collapse (which you cannot), the number of neutrinos that passed through you were roughly comparable to those coming from the Sun [12.2.3]. Eleven neutrinos were caught by neutrino detectors on Earth coming from the direction of the LMC, about the right number expected from the physics of the collapse. And, they were "seen" about three hours before McNaught watched the star visually brighten, the delay about right given the time needed for the shock wave to make it to the stellar surface. If you could have been at the surface of the star and seen the bath of neutrinos flooding outward, you would know what was to come.

Over the years we have watched SN 1987A interacting with its environment. Radiation from the supernova first hit the huge expanding shell of gas that the progenitor supergiant had sloughed off through its wind and that created an eerie hourglass-shaped figure. Years later, the ejecta and shock wave arrived to cause numerous spots on the ring, and we see the development of a "supernova remnant" before our very eyes (Fig. 14.6b). Oddly, however, there is no evidence for the neutron star that should have been left behind.

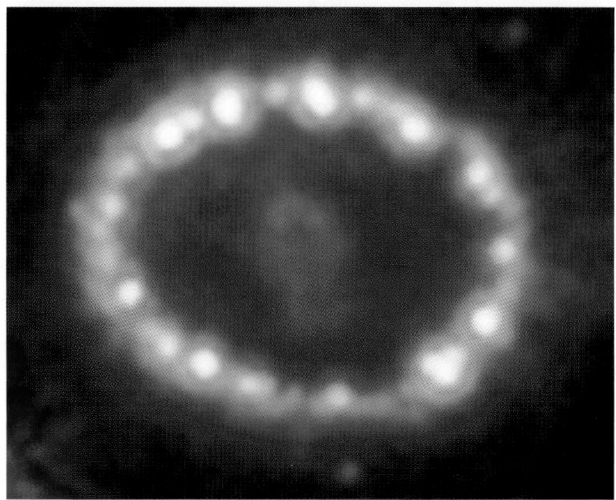

Fig. 14.6 **Supernova 1987A** (Top) A B1 supergiant (center star in the left-hand image) erupted in 1987 as the visually brightest supernova (right) since Kepler's Star of 1604. (Bottom) Some 13 years later, the shock wave hit a ring of matter ejected by the progenitor supernova. (Top) Courtesy of the Anglo-Australian Observatory, David Malin Images; (Bottom) Courtesy of NASA, P. Challis, R. Kirshner (Harvard-Smithsonian CfA), and B. Sugerman (STScI).)

Type	Spectrum	Mean M_B	Light Curve
Ia	No H; Si II	−19.5	Sharply peaked, 3 mag first 50 days; white dwarf overflow
Ib	No H; no Si II; He	−18.1[a]	Sharply peaked; WR core collapse
Ic	No H; No Si II; Weak or absent He	−18.1[a]	WR or helium star core collapse
II-L	Hydrogen	−18.0[b]	Core collapse; linear light curve
II-P	Hydrogen	−17.0	Core collapse; plateau in light curve, circumstellar shell
IIn	Hydrogen	−19.2	Core collapse, dense circumstellar shell

[a]Ib and Ic combined; normal range −17.6 to −20.1.
[b]Normal range −17.6 to −19.3.

Table 14.3 **Supernova types**

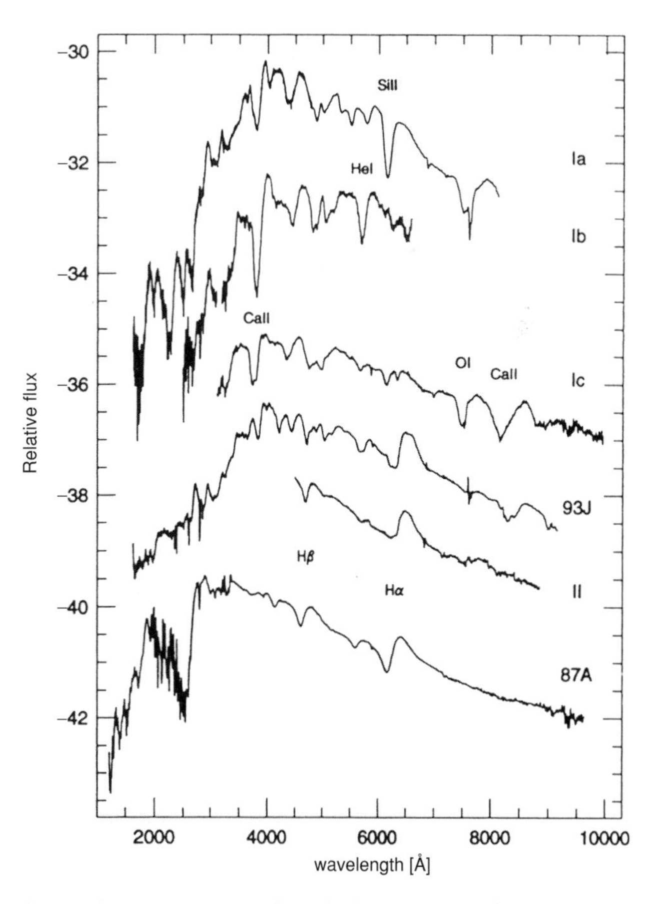

Fig. 14.7 **Supernova spectra** Different kinds of supernovae (as indicated at right) are discriminated by their spectra, seen here at maximum light. SN 1993J was a core-collapse oddball. SN 1987A was an unusual Type II. The left axis is a logarithmic flux scale. (Wheeler, J.C. and Benetti, S. (2000). *Astrophysical Quantities*, Cox, A.N., ed., American Institute of Physics Press, Springer.)

14.2.1 Kinds of supernova

From planets to galaxies, nothing in the Universe seems to come in only one form. SN are no exception. They were seen early on to divide into two principal classes based on the presence of hydrogen lines (Table 14.3). As in the spectral sequence, morphology comes first, explanation second. Hydrogen is absent from the spectra of Type I supernovae, while present in Type II spectra. Further examination allows the division of hydrogen-less Type I into three subtypes: Type Ia has a strong line of ionized silicon, Ib has no silicon but does have helium absorptions, while Ic (really a variety of Ib) has neither H nor Si, and weak or absent He (Fig. 14.7).

The types are readily correlated with galaxy class [Box 5.1], which leads to their origins. Types II, Ib, and Ic are strictly

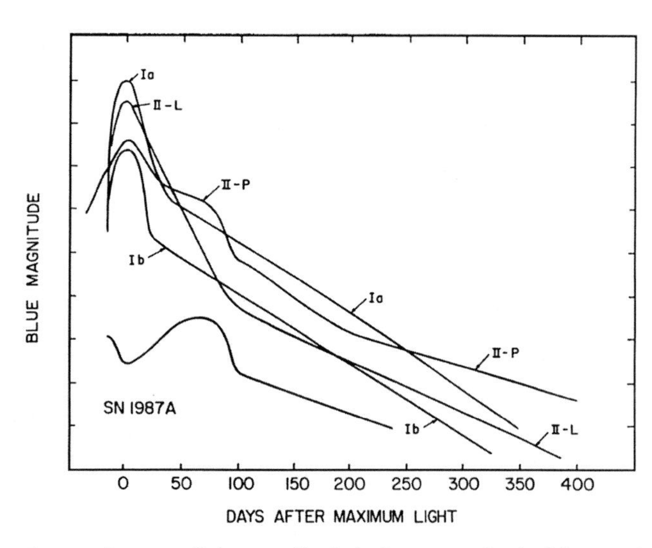

Fig. 14.8 **Supernova light curves** No single diagram can give the full range of light curves, but these are at least representative of their classes. (Wheeler, J.C. (1990). *Supernovae*, Wheeler, J.C., Piran, T., and Weinberg, S., eds., Singapore: World Scientific.)

confined to the disks of spiral galaxies and to smaller systems with active star formation. They never appear in elliptical galaxies. These three kinds must therefore be related to massive stars, since the existence of such stars is a chief distinguishing factor between the two kinds of galaxy (ellipticals having none). Type Ia SN, however, can occur anywhere, in disks of spirals, in galactic halos, or in ellipticals, showing that they are produced by much lesser stars that have masses comparable to that of the Sun. To the eternal confusion of students, Type Ia SN belong to Population II, while Type II (and Ib and Ic) belong to Population I [5.1].

The types are also discriminated by their light curves (Fig. 14.8). SN of Type Ia are on the average the brightest. They are remarkably uniform from one to the next, each of them hitting absolute blue magnitudes around −19.0, which makes them excellent "standard candles" for measuring the distances to galaxies in which they appear, and for determining the nature of the expansion of the Universe. They reach a quick peak, plunge rapidly in brightness by some 3 magnitudes after the first 50 days, and then decline at a slower pace.

More heterogeneous Type II divides into three subgroups, II-L, II-P, and IIn. Though the II-L SN can rival Type Ia, they are generally over a magnitude fainter. Plotted by magnitude vs. time, they have linear light curves that drop rapidly, by some 5 magnitudes over the first 100 days, and then change to a lower slope. Type II-P are a magnitude fainter, and display a plateau in their light curves that finally become more or less linear after the first 100 days. The two versions may simply be the tail ends of a continuous distribution.

Types Ib and Ic SN are rather like the II-P in brightness, but with more-linear light curves. The Type In can be quite luminous. SN 1987A, several magnitudes fainter than the rest, is a Type II, but rather in its own sub-category (Box 14.1).

14.2.2 Core collapse

There seems no alternative but to believe that all classes of SN, other than Type Ia, derive from massive stars. Not only do they all share similar galactic locations, but theory clearly reveals that the only fate of massive stars is indeed to blow up. As far back as the 1930s, long before we understood anything much about stellar evolution, Fritz Zwicky (1898–1974) suggested that the only viable energy source for a supernova was the gravitational collapse of a "normal" star into a tight, tiny ball of neutrons, a *neutron star*. And that is exactly what happens.

Because of their extreme binding, adding more protons or neutrons to iron or nickel can liberate no more energy [12.2.2; 13.2.3]. Instead, reactions to make heavier elements become endothermic, requiring energy instead of releasing it. Further burning can no longer provide support for the star. And all that gravitational energy is waiting to be used.

The pre-supernova core is effectively an ultra-dense iron white dwarf supported briefly by degenerate electrons [13.2], with a temperature near 10^{10} kelvin. The gamma ray photons flying around have sufficient energies to destroy the iron nuclei and shatter them back to alpha particles (helium nuclei), then to the protons and neutrons that started the process in the first place. Given the huge densities, the protons combine with free degenerate electrons that provided the original support. The rapidly increasing proton–electron reaction creates first a trickle, then a flood, of neutrons and neutrinos. Since energy was liberated in the building of the iron, its breakdown sucks energy up. Reduced to its building blocks, support removed, the core suddenly and violently collapses at a speed as high as a quarter that of light, which liberates vast amounts of stored gravitational energy.

As the collapsing core nears a radius of around 20km, the density at the center approaches that of nuclear matter (3×10^{14} gcm^{-3}), matter with no "spaces" between the particles (the word "space" is not well defined, since the "particles" also behave like waves). The violent contraction actually pushes the dense neutron core past that critical density, at which point the neutrons are totally degenerate and the strong force becomes repulsive. The growing neutron core then rebounds back, which creates a powerful shock wave that tears outward into the infalling matter. Releasing over 10^{53} ergs of energy (a rate of energy release of 10^7 ergs per second equalling one watt), the collapse and rebound prepares to blow the non-degenerate stellar envelope apart to make a Type II supernova. The Sun would have to radiate at its current level of 4×10^{33} ergs per second, $10^{53}/(4 \times 10^{33}) = 3 \times 10^{19}$seconds $= 10^{12}$ years to equal the energy release of a supernova: 100 solar lifetimes! No wonder supernovae have such extraordinary effects on their surroundings.

This awesome statistic must be moderated some, however, as 99 percent of the energy release is in the form of neutrinos, which are produced by the mergers of protons and electrons

Fig. 14.9 **Gamma ray burst afterglow** A powerful core-collapse supernova in a faint, distorted galaxy some 10 billion light years away produced a gamma ray burst whose afterglow swamps the light from the galaxy itself. (Courtesy of Andrew Fruchter (STScI) and NASA.)

into neutrons (one for each neutron produced, yielding 10^{57} of them) and by thermal neutrinos made by nuclear reactions in the heated stellar envelope (see Box 14.1). Once they escape the exploding star, they are for the most part never heard from again, since as part of the weak force they (along with the more abundant thermal neutrinos) pass through anything, with practically no effect. The result is that the mechanical energy of a supernova is 100 times less, "only" 10^{51} ergs, and the radiative energy is even less (10^{50} ergs or so), but together still the equivalent of all the radiation released by the Sun over its main sequence lifetime.

For all the apparent success of the idea, calculations show that the shock is really not strong enough to explode the stellar envelope. Mass from outside the core collapses too, and as it rains down, the shock stalls within it and cannot get out. Some other mechanism has to help it, and no one knows quite what. Various theories have been put forth. While a neutrino created by the p–p chain leaves the Sun immediately and can pass through the entire Earth as well, at nuclear densities matter becomes "optically thick" [7.2.2] to them, and they can be absorbed. Neutrino pressure may thereby act to help push the shock outward to make the rest of the star explode. Energy generated by violent convection in the forming neutron star surface might be responsible as well.

Another possibility is that the increasing spin of the collapsing star (through the conservation of angular momentum) and the increasing concentration of the stellar magnetic field can emplace the energy into violently expanding bipolar jets. Ultra-energetic supernovae – *hypernovae* – seem to relate to the jets postulated to be involved in the gamma ray bursts observed from distant galaxies (Box 14.2; see also Fig. 14.16). Such explosions may be responsible for highly magnetic neutron stars ("magnetars") [14.3.1] or even black holes [14.3.3].

Box 14.2 Gamma ray bursts

A burst of celestial gamma rays smacks the Earth about once a day. Not correlating with the distribution of stars in our Galaxy, they come at us randomly from all over the sky, showing that their origins are far away. There are two kinds. Short bursts range in duration from milliseconds to two seconds, while long bursts start at two seconds and go to many minutes. The long bursts are identified with hypernovae in galaxies a billion or more light years away. Following long bursts are "afterglows" that can last for hours, even days (Fig. 14.9). If the bursts are isotropic, the distances of the galaxies imply energies that are far too great to be accounted for by any kind of core collapse. The bursts therefore most likely emerge in narrow cones, which drops the energy requirement by a factor of 100.

These *gamma ray bursts* (GRBs) are most likely produced by the core collapse of very massive stars, of the kinds that make luminous blue variables [10.6.1] and Wolf–Rayet stars [6.4.4], in which increasing rotation acts to focus the subsequent events into bipolar jets. The gamma rays are created by the shock wave of the collapse acting on the stellar envelope, and break through the stellar surface before the shock brightens the star into the ultra-energetic hypernova. The shock acting on circumstellar matter resulting from earlier mass loss produces the afterglow.

Though the short bursts are not understood, they may come from the rare mergers of closely orbiting neutron stars. Given the rarity of such events, like the long bursts, we can see them only if we invoke their existence in an enormous number of galaxies, which again implies great distances and energetics.

14.2.3 Element building

However the outbound shock is moved along, as it roars away, it heats the interior of the star to 10^{10} kelvin, which produces explosive nuclear burning in the expanding stellar matter (particularly in the silicon- and oxygen-rich shells) that tries to run the fusion process to the most tightly bound nucleus, one that also has an equal number of protons and neutrons, ^{56}Ni (which under normal conditions is highly radioactive). However, because the dying star is expanding and cooling, *explosive nucleosynthesis* cannot maintain equilibrium burning over a sufficiently long period, so a pure nickel–iron state cannot be reached; that is, the raging fusion-storm can bring only a portion of the nuclei to ^{56}Ni. When the temperature drops below a critical value, the remaining distribution of elements (lying in something of a bell curve around the atomic numbers of iron and nickel) "freezes out," making the chemical composition depend on the rate of cooling, which in turn depends on distance from the stellar center.

Somewhere within the milieu, nucleons may undergo rapid neutron capture, the r-process [13.3.3]. (Note, however, that the site of the r-process is highly controversial. Another theory posits that the r-process takes place in neutron star winds [14.3.1].) High temperature and a great supply of neutrons allow a given chemical element to grow very quickly to heavy isotope numbers that, under milder conditions, would have time to decay into the element with the next higher proton number (i.e., the s-process). During r-process operation, only when the isotopes become violently unstable can they decay upward, and then they jump to very high values of atomic number, creating much in the way of heavy elements, including silver, gold, and uranium.

As a result of explosive nucleosynthesis and the r-process (providing it takes place in the star), the ejecta that make it out of the star contain all the elements of the periodic table, including about a tenth of a solar mass of nickel. Since the chemical composition of the expanding stellar envelope depends on temperature and density, hence on radial distance from the center, the gross composition of the ejecta will depend critically on the "mass cut," the radius at which the star separates into the mass that gets blown away and that which falls back to add to the mass of the growing neutron star (or even black hole) in the middle.

14.2.4 Core-collapse light curves

Taking about a day to reach the stellar surface of a typical red supergiant, the shock expels the chemically enriched outer envelope at speeds of tens of thousands of kilometers per second, far beyond escape velocity, rending the star apart forever. At the same time, it deposits a percentage of its energy into heat and light, and a visible supernova is born. At first, the temperature at the expanding surface – the photosphere – is so high, half a million kelvin or so, that most of the radiation is in the ultraviolet. As the expanding envelope cools and thins, and the opacity drops, more radiation is released in the optical, and the star brightens. After a time, the thinning photosphere cools to the point where the optical brightness declines as well, and the supernova starts to fade.

The most common (real) metal of the Universe is not nickel, but iron. The target of explosive nucleosynthesis, ^{56}Ni, is (outside of the hot, dense stellar interior) highly radioactive. With a half-life of only 6 days, it decays to cobalt as ^{56}Co, which, with a longer half-life of 77 days, in turn decays to a tenth of a solar mass of stable ^{56}Fe. This conversion of isotopes produces gamma rays that heat the ejecta. Some 100 days after the explosion, as the heat from the shock fades away, the dimming becomes controlled by radioactive decay. Many other radioactive isotopes are included as well. As ^{56}Co disappears to iron, for example, longer-lived radioactive isotopes such as ^{44}Ti take over the job.

The progenitors of core-collapse SN are as heterogeneous as are the supernovae themselves. While not without controversy, the matches that can be made between the kind of supergiant and the resulting kind of supernova are at least reasonable. Most highly-evolved massive stars are expected to be huge red supergiants at the cool edge of the HR diagram. But even these are not all the same. The mass of the envelope that contains the collapsing iron core (the neutron star to be), through which the shock wave must pass, depends both on initial mass and on mass lost through winds, which though not yet well understood likely depends on rotation, magnetic fields, binarity, and metallicity. II-L and II-P supernovae probably arise from red supergiants, the latter from stars with the more massive envelopes, as these store the energy that dims the peak brightness and creates the plateau.

Sufficiently low metallicity can even cause a star with a massive envelope to shrink to becoming a smaller blue supergiant, like the class B1 Ia predecessor of the LMC's SN 1987A, whose relatively small size resulted in an optically dimmer explosion and a rather different sort of light curve. Mass loss also creates surrounding shells into which the shock wave and expanding ejecta must plow, as beautifully exhibited by SN 1987A (see Fig. 14.6b).

From 25 to 40$M_{\odot}$ or so, the red supergiants evolve back across the HR diagram to become smaller blue supergiants, even Wolf–Rayet stars [6.4.4, 14.1.2], before exploding. With little or no hydrogen at the moment of internal disaster, these are thought to create hydrogen-less class Ib supernovae. Class Ic SN may come from stripped helium stars in binary systems. While still fairly massive, a WR star is only several times the size of the Sun, hardly what one would think of as "supergiant." As in SN 1987A, the shock wave cannot expand these stars to sufficiently large sizes to create the brilliance of class II-L supernovae. The shock simply dissipates too early, resulting in dimmer explosions. As a result, Ib and Ic supernovae are powered almost entirely by radioactive decay.

14.2.5 Type Ia supernovae

That leaves Type Ia, hydrogen-less SN that occur where there are no massive stars now to be found (Fig. 14.10). These bright supernovae must therefore anomalously involve lower-mass stars. Their striking uniformity, however, implies that there is but one scenario for producing them. Evolutionary theory leaves no room for single low-mass stars to erupt violently, so the events must involve doubles. The only available possibility is a white dwarf in a binary that has been forced to overflow the Chandrasekhar limit [Box 13.2]. Given that the majority of stars are low mass to start with, the most probable companions are lower main sequence stars.

Type Ia supernovae, consistent with their name, are extensions of the nova phenomenon [8.5.4]. Two stars are born sufficiently close that when the more massive of them evolves, tides [8.5.1] raised in the larger by the smaller will cause mass to flow toward the smaller [8.5.3; 13.6]. They may even become encased in a common envelope which, along with tidal or magnetic interactions, can draw the two together [13.6]. By the time the more massive star has evolved to become a carbon–oxygen white dwarf, the two may be close enough for the white dwarf to create tides in (and draw

Fig. 14.10 **Type Ia supernova** A white dwarf overflows its limit in a galaxy over 5 billion light years away to create a Type Ia supernova (the deep red star at center, the reddening compliments of the expansion of the Universe), whose light curve in turn gives the galaxy's distance. (Courtesy of NASA and J. Blakeslee (JHU).)

matter from) its companion, which is still an ordinary dwarf. The outcome is a variety of cataclysmic variables that include novae and recurrent novae [8.5.5], which are effectively the same.

If the greater of the two original stars was massive to begin with, just under the limit of the stars that create supergiants, the resulting white dwarf will be reasonably close to the $1.4M_\odot$ Chandrasekhar limit [13.5]. Such massive progenitors once inhabited galactic halos, and have now all been converted to the required massive white dwarfs. Mass transfer from a close companion will produce frequent nova eruptions. Each cycle of mass transfer and surface explosion adds a bit more mass to the white dwarf. When the white dwarf hits the limit, it erupts into a Type Ia supernova. An alternative theory posits the merger of two white dwarfs (one or both of which must still be fairly massive) into a super-Chandrasekhar mass, with a similar result.

While there is little doubt that most Ia SN are created in either or both of these two ways, the details are still not well understood. Carbon–oxygen white dwarfs seem to be the culprits (as opposed to the rare neon versions). The gravitational collapse that follows the overflow of the limit may cause unregulated carbon burning in the core that advances outward in a subsonic *deflagration*. It may also cause a supersonic runaway *detonation*, or more likely a combination of the two, the former converting to the latter. In either case, explosive nuclear fusion will advance to about a third of a solar mass of radioactive nickel (^{56}Ni), while the cooler outer parts of the star will create a mix of other elements. All of this nuclear debris quickly gets blasted out into space at speeds of 10 000 km s^{-1} as the white dwarf destroys itself, leaving no dense remnant behind, the ordinary dwarf instantly losing its life-long companion. The Ia supernova then glows entirely through the heating of the ejecta caused by radioactive decay of ^{56}Ni through ^{56}Co to ^{56}Fe, resulting in a third of a solar mass of iron ejected into the cosmos, triple the yield of core-collapse supernovae. The numbers of Type Ia (white dwarf) and non-Ia (core collapse) SN are roughly comparable, with non-Ia having a bit of an edge.

While Type Ia SN are relatively uniform, there are variations on the theme. Some are brighter than others. There is a strong relation, however, between decay time and peak visual brilliance (the brighter being the slower to fade), which allows correction of all to the same effective peak magnitude. As a result, Ia SN make amazingly good standard candles with which to determine the accurate distances of galaxies into the billions of light years [4.4.5].

14.2.6 Chemical enrichment of the Galaxy

All kinds of supernovae act to enrich the interstellar medium with freshly made chemical elements. Most of the iron comes from the Type Ia variety, with Type II SN filling in the rest. The lowest-metal stars have chemical compositions expected of the r-process, consistent with a Population III made of massive stars that were created, burned out, and exploded shortly after the Big Bang [5.4]. Since Type Ia supernovae are made from white dwarfs, they could not contribute until the Galaxy was old enough for an 8 to 10-solar-mass progenitor to have evolved to a white dwarf and then wait for the first ones to overflow the Chandrasekhar limit. Type Ia SN do not make r-process elements (though that is argued), but do make more iron than Type II (and related) supernovae. As result, when the Type Ia variety entered the scene, the abundance ratios of various elements to iron dropped [5.4]. Eventually – coupled to the fresh elements made by giant stars – we achieve the chemistry of the Sun [7.3.2], including an explanation of the detailed patterns of abundances with respect to atomic number (13.3.3).

14.3 The Remains

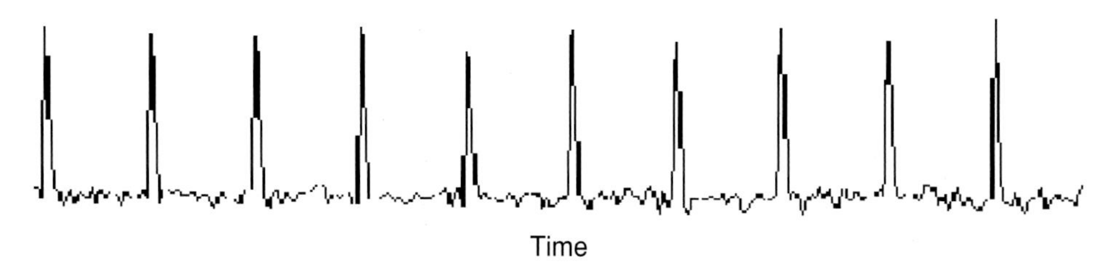

Fig. 14.11 **Pulsar pulses** Radio bursts (averaged over 500 times) from the Vela pulsar are about a tenth of a second part, with nothing at all in between. (Courtesy of Hartebeesthoek Radio Astronomy Observatory.)

All supernovae leave reminders of their fleeting existence. There are two distinctly different kinds of remains. The most obvious and first discovered is the *supernova remnant* (SNR) caused by the rapidly expanding debris of the exploded star, coupled with the illumination of the ambient surrounding medium by the supernova shock wave. All types of SN produce them. The best known is the Crab Nebula of Taurus (Fig. 14.1), which lies at the site of the Chinese Guest Star of 1054.

The other reminder is a collapsed, dense core. Since Type Ia supernovae annihilate themselves, they produce SNR but no collapsed cores, which belong exclusively to Type II, Ia, Ib, etc. Again, there are two possibilities: extremely dense neutron stars that are supported by neutron degeneracy pressure, and more massive black holes, for which support has run out, causing them to collapse to such small radii that light cannot escape. While no one yet knows how the progenitors of neutron stars differ from those of black holes, it is usually assumed that the highest-mass end of the main sequence makes the latter.

14.3.1 Single neutron stars and pulsars

Given a rate of one core-collapse supernova per century in our Galaxy, an age of the Galaxy of 10^{10} years, and discounting rare black-hole formation, the Galaxy should contain around 10^8 neutron stars. Considering the Galaxy to be a simple cylinder 20 000pc in radius and 1000pc thick gives a volume of 10^{12}pc^3, and a neutron-star density of one per 10 000pc^3, which is the volume contained in a box only about 20 pc on a side. The nearest neutron star should be fairly close. And that is about what we find, the nearest known around 100pc away. While some have popular names, most neutron stars are identified by "PSR" plus coordinates (right ascension and declination) preceded by "B" for "Besselian" (equinox 1950) or "J" (J2000), as example PSR B1957+50 [2.4].

Neutron stars divide into those that are single and those that are in binary systems. They also divide into *pulsars* and non-pulsars. Because of their radio brightness and notable behavior, pulsars were not only the first of the kind to be discovered, but also dominate in number known, even though on a global scale they are by far the rarer of the two (selection in action). The first pulsar known was discovered in 1967 by Jocelyn Bell (1943–) using a fast-response radio telescope developed by the English astronomer Anthony Hewish (1924–). In the radio spectrum, it emits brief radio bursts with a precisely known period of 1.33701 . . . seconds (Fig. 14.11). While the pulses occasionally disappear, when they return they occur on schedule according to the period. More such objects followed quickly, until well over 1000 are now known, with periods that range from 33 milliseconds to a bit over 4 seconds. The warm ionized interstellar medium [11.1.4] refracts (slows) the radio waves in accord with the frequency, higher frequencies travelling the faster. The pulses therefore arrive at frequency-dependent times. From the arrival-time difference at two frequencies and an estimate of the electron density averaged over the line of sight, a pulsar's distance can be found.

The explanation for these seemingly bizarre observations was not long in coming. The periods are too fast for pulsation (the term "pulsar" a misnomer). Only rotation could produce bursts that regular. White dwarfs are far too large to spin that fast: the emitter's radius must be only in the tens of kilometers. Any star shrunk to that size has to be a neutron star, exactly as postulated for the creation of supernovae as far back as the 1930s. Pulsars are the neutron star remains of the collapsed cores of Type II (and Ib and Ic) supernovae.

Neutron stars are effectively balls of nuclear matter. Those in binary systems reveal masses of about 1.5$M_\odot$, close to the Chandrasekhar limit (though masses may range to 2 or 3$M_\odot$). Packed into a sphere of radius 10 to 15km, the densities are of the order of 10^{14}gcm^{-3}, that of nuclear matter. All stars, even supergiants, are rotating. As the core of a supergiant collapses, conservation of angular momentum spins it up to a rotation period of under a second. At the same time, the rotating star's magnetic field (whether the fossil field or one newly

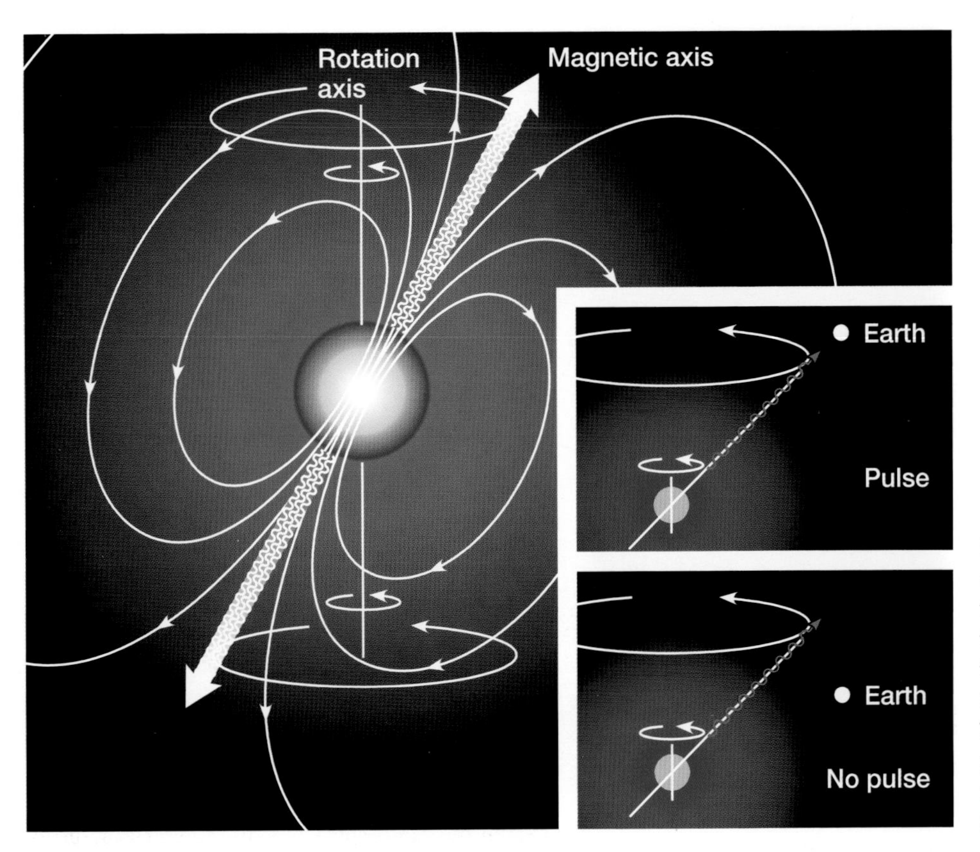

Fig. 14.12 **Rotating pulsar** A spinning neutron star beams radiation along a tilted magnetic field axis. If Earth is in the line of sight (upper inset) we see the "pulse," if not, we see nothing. (Kaler, J.B. (1994). *Astronomy!* New York: HarperCollins, reprinted by permission of Pearson Education, Inc.; adapted from Greenstein, G. *Frozen Star*, Freundlich Books.)

generated by the increasing spin) is typically concentrated to a field strength the order of 10^{12} times that of Earth. From their surfaces blow powerful winds at speeds approaching that of light. As in nearly all magnetic bodies, the magnetic axis is inclined to the spin axis. Acceleration of electrons by the magnetic field in the direction of the field axis generates opposed, cone-shaped radio beacons that wobble around the rotation axis. If the geometry of the star is such that the magnetic axis points more of less toward the Earth, we get a blast of radiation. If not, the pulsar will be invisible except for its thermal radiation, which for a star this small will be weak and undetectable unless it is very close to us. If the orientations are right, we might also see an "interpulse" as the cone from the opposite axis grazes us. A pulsar is therefore something of a rotating celestial lighthouse or airport beacon, though one so energetic that you would not want to be drawn to it (Fig. 14.12).

The pulsar's magnetic field – which is anchored to the star – and its associated wind and beamed energy are ultimately generated by the star's rotation. The wind, which is coupled to the magnetic field, and the radiation therefore act as a brake that slows the pulsar, typically at a rate of 10^{-15} seconds per second of time, small, but enough that it can be measured [12.4.5]. The spin-down rate depends on the period, the magnetic field, and time, which allows the determination of age (since the pulsar was born) and field strength (such ages sometimes in strong disagreement with the ages of the associated supernovae). A slowly spinning pulsar has the ability to radiate only in the radio spectrum, while one that spins rapidly can generate not only optical pulses, but X-ray and even gamma ray pulses. As the small star ages, the pulses disappear from high energy on down. By the time the rotation period declines to only once a second, only radio radiation appears, and after the period is down to 4-plus seconds, nothing is seen at all and the little body disappears from sight.

That pulsars are the remains of core-collapse supernovae was demonstrated by the pulsar within the confines of the Crab Nebula (see Fig. 14.1). Spinning with the short period of 0.034 . . . seconds, it is so energetic that the pulses are seen throughout the entire electromagnetic spectrum. Because the star's optical spectrum contains no absorption or emission lines, it had already been identified as a possible neutron star that resulted from the collapse that formed the Crab Nebula.

However, of all the SNR and neutron stars, only a small percentage of each is found associated with the other (witness Supernova 1987A). Several explanations are available. Some collapsing iron cores may simply never turn into pulsars, becoming black holes instead. If they do become neutron stars, their magnetic field axes may not sweep past Earth, and the beamed energy will not be visible. Some may just not produce pulses; none are observed from the visible neutron star associated with the Cas A remnant (see Fig. 14.16b). A more likely explanation involves dynamics. Many pulsars have very high proper motions, showing that they are moving at high speeds relative to the Sun, into the hundreds of kilometers per second. The core collapse is apparently off-center,

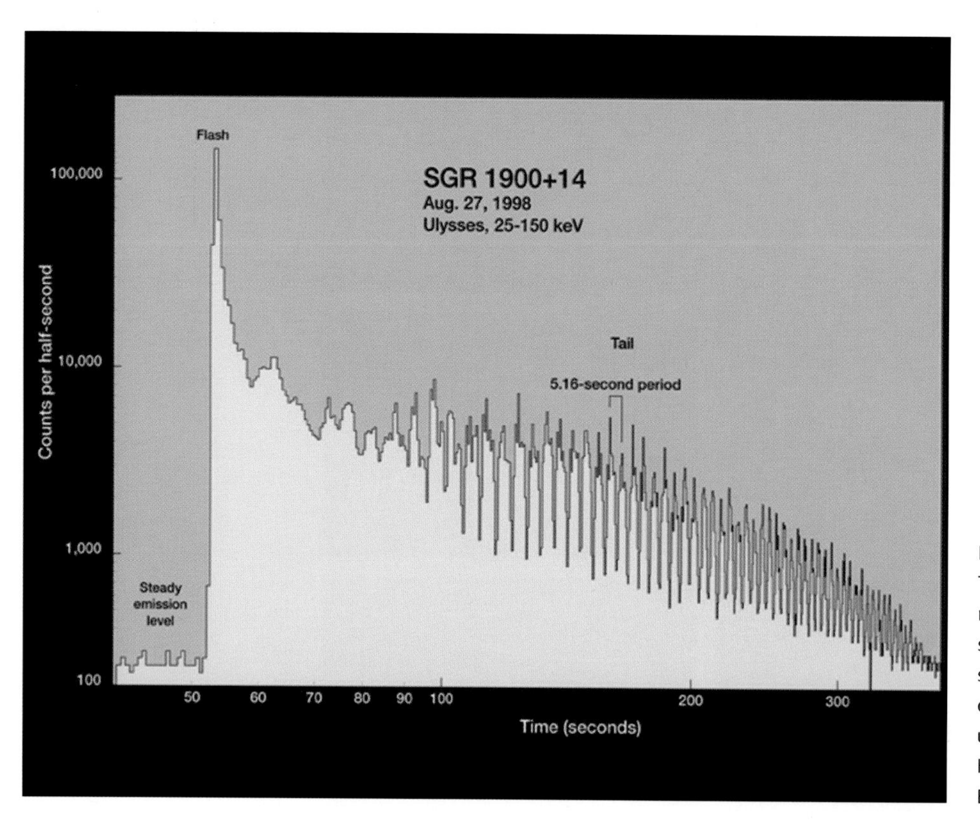

Fig. 14.13 **SGR 1900+14 superburst** The burst from this soft gamma ray repeater (recorded by the Ulysses solar satellite) overloaded orbiting satellites and affected terrestrial radio communication by ionizing the Earth's upper atmosphere. (Courtesy of Kevin Hurley, University of California at Berkeley, and NASA/Marshall.)

which gives the resulting neutron star enough energy to kick it quickly out of its own supernova remnant.

The constructions of neutron stars are quite unlike those of other stars. The surface is a crystallized solid crust about a kilometer thick, whose upper layer is made of iron atoms. Inward of the surface, the stuff turns into a neutron-rich gas. In the interior, where the density is over 10^{14}gcm^{-3}, is a liquid neutron "superfluid," one in which the viscosity has disappeared. (Or so we think.) Rotation of the neutron star produces rows of vortices that come out from the center and become caught by the more normal crust. When they are released, the neutron star undergoes a sudden "glitch" that causes a mild spin-up of the pulsar, making it rotate faster, from which it again slows.

A very small subset of pulsars visible at X-ray wavelengths, known as *anomalous X-ray pulsars* (AXPs), spin with much longer periods of 5 to 10 seconds or so, far longer than those of radio pulsars. They are apparently related to (and may be predecessors of) *soft gamma ray repeaters* (SGRs), which radiate in the X-ray but also release – like bullets – lower energy gamma rays only a hundredth of a second long. Theory and spin-down rates tell of magnetic fields 100 or more times that of more normal pulsars. Such *magnetars* have fields approaching 10^{15} times that of Earth. The powerful magnetic fields stress the crusts, which in turn causes sudden magnetic field rearrangements to produce some of the strongest bursts of energy seen anywhere. Power outputs can range to 10 or more times that of entire galaxies. The famous 1998 superburst from SGR 1900+14 partially ionized the Earth's atmosphere, even though the perpetrating body is 20 000 light years away (Fig. 14.13). Such immense fields require an active dynamo during core collapse, one driven perhaps by the same violent convection in the forming neutron star that may help drive the supernova shock wave. Extremely massive hypernovae may be involved, events that could also create black holes [14.3.3]. The winds from magnetars (and even the more ordinary pulsars) may well be the sites of rapid neutron capture (the r-process) and the creation of heavy elements beyond bismuth [13.3.3, 14.2.6].

Non-pulsing (that is, non-beaming) neutron stars are visible through their blackbody radiation. Even though intensely hot, with surface temperatures the order of 10^6K, the small radii yield low luminosities, only a few tenths that of the Sun. As a result, they are extremely difficult to find, even though they must dominate the neutron star population.

14.3.2 Neutron stars in binary systems

Since most stars, including high-mass stars that blow up, are in binary systems, we should expect a large fraction of neutron stars to have binary companions [8.5.9]. They do not. Of all the pulsars known, only a small fraction reveal evidence of having them. Neutron stars have no Doppler-shifting absorption lines. Instead, their orbits produce variations in pulse-time arrivals. When the pulsar is coming at you (relative to the center of mass), the pulses speed up, and when it is receding, they slow down (a "Doppler shift" in pulse timing, not in wavelength). The extremely regular pulse periods make even very low-mass companions detectable, so the lack of binarity is real, not a selection effect. The off-center collapse that gives a pulsar its high speed kick can also give a companion a kick in the other direction, ejecting it from the system, helping to create one of the two sets of runaway stars, the other set caused by binary interaction [8.6].

Nevertheless, a few neutron stars have retained the mates with which they started their lives, and as a result have formed some of the more bizarre characters in the celestial population. At the heart of the story are those that emit X-rays, of which there are two broad kinds: *high-mass X-ray binaries* (HMXBs) and *low-mass X-ray binaries* (LMXBs). In an HMXB, a normal stellar companion to a magnetic neutron star has a mass greater than about $5M_\odot$. The wind from the high-mass companion is accreted by the neutron star, but not uniformly. Instead, it is controlled by the wobbling magnetic field, such that the system produces X-ray pulses as viewed from Earth (yet another kind of X-ray pulsar).

An LMXB has a companion with a mass under $1M_\odot$. For unknown reasons, there are none between about 1 and $5M_\odot$. In an LMXB, mass is accreted through Roche lobe overflow [8.5.1] into an accretion disk around the neutron star surface, from which it falls onto the compact object. Heated to temperatures the order of 10^7 kelvin, the infalling gas spreads over the neutron star surface and radiates X-rays. LMXBs are neutron star analogues of white dwarf novae [8.5.4]. As the accreted hydrogen layer thickens, it heats to the point of fusion, which creates a degenerate helium layer at the bottom. When the helium becomes hot and dense enough, a runaway 3-α reaction – a helium flash – ensues, resulting in a powerful *Type I X-ray burst* with energies in excess of 100 000 times the solar luminosity. Such *X-ray bursters* pop off every few hours. The rise time is the order of a second, whereas the decline may take minutes to hours. Much rarer *Type II* X-ray bursters are caused by instabilities in the accretion disks that suddenly dump matter onto the neutron stars, which releases vast amounts of gravitational energy. Or so go the theories.

As witnessed by their considerable population in globular clusters, LMXBs are ancient systems [9.4.12]. They are the purported predecessors of the *millisecond pulsars*, which, like "normal" pulsars, are observed in the radio domain. These madly spinning systems have periods between 1 and 100 milliseconds. At the extreme end is a pulsar with a period of 0.00156 . . . seconds. Rotating 640 times a second, were there anything to see on the surface, it would be but a blur to the human eye. Given a radius of 15km, the equatorial surface velocity is 65 000 km s^{-1}, over 20 percent the speed of light. From their periods and spin-down rates, they have relatively low field strengths (10^8 to 10^9 Earth fields) and, like the LMXBs, are very old.

The short period of a millisecond pulsar can be explained by mass transfer during an earlier LMXB state, wherein the matter falling from the accretion disk hits the neutron star to the side, which spins the aging neutron star back up again, way past the period with which it was born as a normal pulsar. Most millisecond pulsars are indeed in binary systems. A few, however, are not, which seems to violate the theory. Support comes to us from the "Black Widow" (PSR B1957+20), in which an ultra-short-period (0.0016 second) millisecond pulsar is orbited (and eclipsed) by a star the size of the Sun, but with a mass only 0.02 solar. The pulsar is caught in the act of evaporating its companion, which is still donating itself to speeding up the pulsar (Fig. 14.14).

The debris of an evaporated star seems to have accumulated into a pair of planets orbiting PSR B1257+12, which whirls around with a period of only 0.006 seconds. From the pulse variations, the inner has a lower-mass limit of 3.4 Earth masses and orbits at 0.36AU from the neutron star with a period of 67 days, while the outer contains at least 2.8 Earth masses with a period of 98 days. Two more planets might also be in the system. A tiny set of similar systems has been found. Given the chance, apparently, planets will form nearly anywhere.

The ultimate pulsar binary has *two* neutron stars in orbit about each other, the result of the successive explosions of the components of a massive predecessor double, perhaps one similar to Eta Carinae. The pulse variations in PSR B1913+16 (the *Hulse-Taylor pulsar*) demonstrate a mutual orbit with a neutron star companion. The two are a mere 2 million kilometers apart, apparently drawn together by the emission of energy by gravity waves, as predicted by the general theory of relativity (Box 14.3). The rate at which the orbit decays provides a stunning confirmation of the existence of gravity waves, even if they have yet to be directly detected. Mergers of binary neutron stars are a candidate mechanism for ultra-distant short gamma ray bursts (see Box 14.2).

14.3.3 Black holes

As there is a limiting mass of $1.4M_\odot$ at which electron degeneracy can support a white dwarf [Box 13.2], so is there a mass limit to support by degenerate neutrons. The equation of state – which describes the relation between density, pressure, and temperature – is completely understood for both normal gases and for white dwarfs. Because it remains somewhat mysterious for neutron stars, the upper limit to their existence is not well known, though it can be constrained to between 2 and $3M_\odot$. Beyond the limit there is no longer any mechanism that can support the star, and its only fate is to collapse forever to become a *black hole*.

Black holes are remarkably simple objects. Every body has an escape velocity (v_{esc}), the speed needed such that a launched projectile will not come back, which depends on the strength of the gravitational field, and therefore on the body's mass (M) and radius (R), to the tune of

$$v_{esc}^2 = 2GM/R$$

(where G is the gravitational constant). Given the Earth's parameters, $v_{esc} = 11.1$ km s^{-1}. Shrink a body of a given mass, and v_{esc} climbs according to $1/\sqrt{R}$. For any mass, there will be a critical radius (R_{bh}) at which v_{esc} hits the speed of light, $c = 2.9979\ldots \times 10^5$ km s^{-1}. Light can no

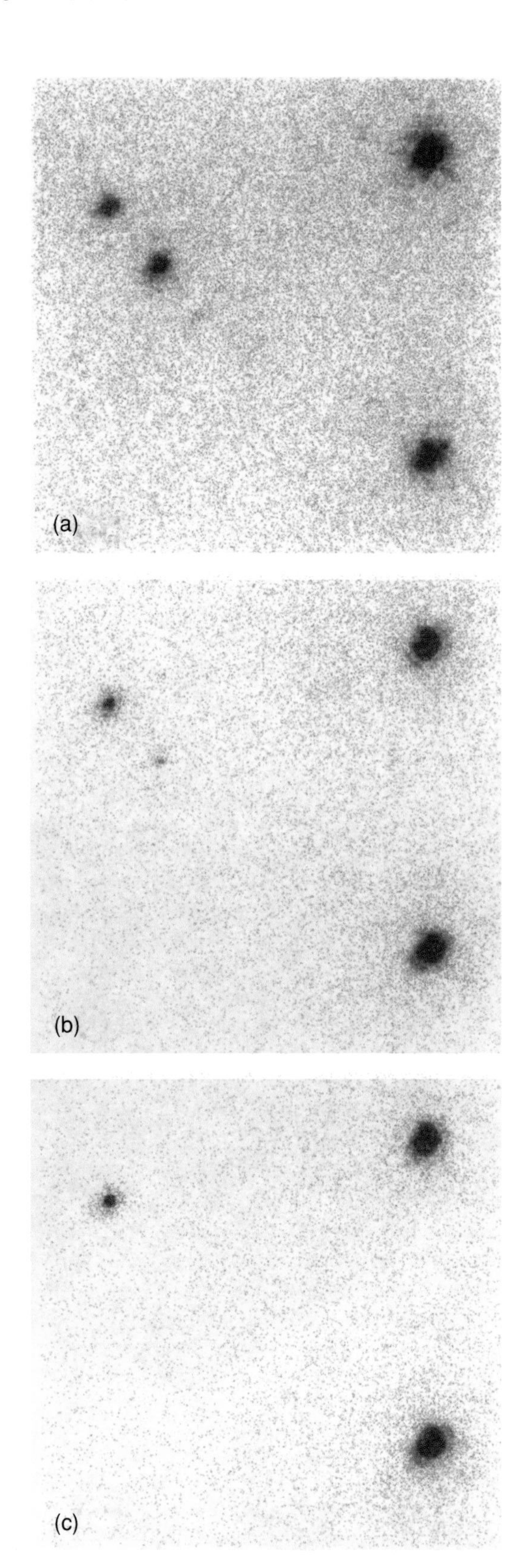

Fig. 14.14 **PSR B1957+20** The "Black Widow" pulsar is destroying its stellar companion. The pair orbit every 9.2 hours. At the top, the invisible millisecond pulsar is between us and the companion and we see the star's heated side. In the middle, the heated side is turning away, while at bottom, it faces in the other direction and the star disappears from view. (Courtesy of the Hubble Space Telescope, A.S. Fruchter, J. Bookbinder, and C.D. Baiylyn, STScI, and NASA.)

Box 14.3 Black holes and relativity

Light does not behave as does an ordinary particle. Throw a ball at 30km hr^{-1}, and the catcher receives it at 30km hr^{-1}. Run at the catcher at 10km hr^{-1}, and it is caught at 40km hr^{-1}. The speed of light, however, is independent of the speeds of either the pitcher or the catcher. You cannot catch up with a speeding photon. No matter how fast you might chase it, it always passes you at c, which is not theory but experimental fact. The only known explanation is that high relative speed distorts time, making it slow down, and thereby also distorts Einstein's amalgam of space and time, a four-dimensional *spacetime*.

Relativity is a theory of mechanics and gravity. Instead of Newton's mysterious force that draws two bodies together, relativity posits that mass bends spacetime to it such that in a sense the bodies "roll down the slopes" toward each other. For weak gravity fields and low speeds, Newton's laws of motion work fine. But as the field strength increases, and speeds become a significant part of that of light (c), the two diverge, and Einstein must be invoked.

A ball thrown upward loses speed, hence energy, as it climbs. If it loses all its energy in working against the Earth's gravity, it comes back down; if it has enough energy, however, it will not. A photon sent flying from the Earth's surface must also lose energy, but it cannot slow down. A photon's energy is inversely proportional to wavelength [Box 7.1]. As a result, it loses energy by stretching, that is, by reddening. This usually subtle *gravitational redshift* has been observed both in the laboratory and in the spectrum lines of stars. In your mind, shrink the Sun. As its surface gravity increases, its light becomes redder. When its hits a critical value that is close to that at which the Newtonian escape velocity equals c (the two are not exactly the same), the Sun's radiation simply redshifts to infinite wavelength, and the Sun disappears into its black hole.

longer escape the body, and the body disappears from view. Though black holes really involve Einstein's general theory of relativity, the classical model serves simple discussion (see Box 14.3).

Setting $v_{esc} = c$ into the above equation yields

$$R_{bh} = 2GM/c^2 = 1.485 \times 10^{-33}\,M = 2.95\,M/M_\odot,$$

where R_{bh} is in km, M in grams, and $M/M_\odot$ is the stellar mass in solar masses. That is, the Sun would turn into a black hole if it could be squeezed into a radius of only 2.95km, 4.2 millionths of its actual radius. This "radius" is not actually that of the body itself (whatever that means). It is instead that of the *event horizon*, within which nothing can get out.

The masses of white dwarfs climb in synchrony with progenitor masses. Perhaps the same is true for neutron stars. If so (and there is no certainty), at some increasing progenitor mass, the remains of a core-collapse supernova hits the neutron star limit of two to three solar masses, and instead of a neutron star, nature might make a black hole. At $2M_\odot$, R_{bh} is 5.9km, while at $3M_\odot$, R_{bh} is 8.9km. No one knows the lower limit to a black hole progenitor, but it is commonly presumed

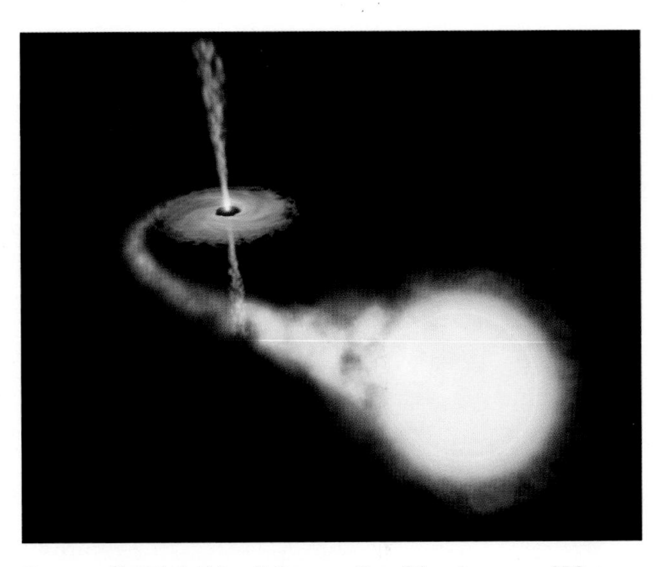

Fig. 14.15 **Black hole** This artist's conception of the microquasar GRO J1655–40 could stand in for any black hole mass-transfer binary. Here the accreting mass not only radiates X-rays, but sends out a high-speed bipolar jet. (Courtesy of ESA, NASA, and Felix Mirabel (CEA).)

to be around $40 - 60M_\odot$. The collapsing cores of such massive stars may produce the extremely powerful hypernovae related to distant gamma ray bursts (see Box 14.2). Such events are also linked to magnetars [14.3.1]. Rare stars, rare events, much speculation, little certainty.

Do black holes exist? From a theoretical point of view, they almost have to. Plus, the radius of the event horizon is not that much smaller than the radii of neutron stars, which demonstrably exist. Observation yields several candidates in the form of a handful of X-ray binaries with high masses (that are not HMXBs). Among the best is Cygnus X-1, an O9 supergiant of 33 solar masses that is a single line spectroscopic binary [8.3.2] that anomalously emits X-rays. Its invisible companion orbits in 5.6 days at a separation of about 0.2AU. The X-rays can (so far as known) come only from a heated accretion disk that surrounds, and feeds, some kind of compact object. Weighing in at around 15 solar masses, the companion almost has to be a black hole.

Other candidates – "microquasars" – eject mass in high-speed bipolar flows (Fig. 14.15). The first known was SS 433. Centered in a huge supernova remnant called W 50, the binary ejects optical jets at 26 percent the speed of light, which wildly wobble over a 164-day period, thanks to a precessing accretion disk surrounding a probable black hole that every 13 days orbits an O or B supergiant. X-ray jets over 30pc long extend into W 50 itself. While a neutron star is not ruled out for SS 433, other microquasar systems that spew bipolar flows at nearly the speed of light are more certain black hole candidates.

Curiously, the evidence for supermassive black holes of millions of solar masses that lie at the cores of galaxies is even better (as deduced from both Kepler's laws and bipolar flows). If huge ones can exist, so can small ones. Black holes are apparently as real as the normal stars you see twinkling on a clear cool night.

14.3.4 Supernova remnants

Supernova explosions not only violently eject mass at thousands of kilometers per second, but blast out shock waves with energies of the order of 10^{51}ergs [14.2.2]. The result is an expanding bubble of superheated gas, the supernova remnant, whose effects can stretch for dozens, even hundreds, of parsecs and last for tens of thousands of years (see Table 14.2). The nature of an SNR depends on the kind of supernova, on the immediate surroundings (which can include a dense interstellar medium as well as many solar masses of circumstellar matter lost through winds), and on time. At first the SNR is dominated by the stellar ejecta. But the shock wave quickly takes over. Acting like a snowplow, it sweeps up the matter of local interstellar space, which eventually dominates. The process also sweeps up the local Galactic magnetic field. Electrons accelerated to near the speed of light spiral around field lines emit *synchrotron radiation*, visible primarily in the radio spectrum (but at higher energies too) with a very characteristic spectrum.

Type Ia SN derive from old systems that uniformly eject only the mass of the white dwarf, $1.4M_\odot$, and ordinarily occur in relatively empty environments. Not only does the surrounding interstellar medium tend to be sparse, but mass lost through giant star winds has long since dissipated. As the blast wave moves outward ahead of the stellar ejecta, it sweeps the thin local ISM into a shell and heats it to X-ray producing temperatures (Fig. 14.16a). Shock waves bounce off barriers much as water waves reverse upon hitting a wall. A "reverse shock" therefore propagates in the opposite direction and slams into, and heats, the slower moving stellar ejecta. The line of sight from Earth passes through more matter at the edge than at the center, and as a result the bubble appears as a "limb-brightened" ring. The passing shock also excites neutral hydrogen atoms into creating a wispy "Balmer line SNR."

For a trio of reasons that involve the great masses of their progenitors, core-collapse (Type II, Ib, and Ic) SN create quite different SNR structures (Fig. 14.16b). First, the amount of stellar ejecta is much greater, in the tens of solar masses. Second, supergiants are surrounded by great circumstellar shells created by their powerful winds, which is especially true of Types Ib and Ic, which are related to stars (specifically Wolf–Rayet stars) that have been stripped of their hydrogen envelopes. Third, the short lifetimes of massive stars ensure that they are generally in proximity to dense molecular clouds.

The resulting SNR pass quickly from a free expansion phase, in which the supernova debris goes flying outward, to a snowplow state in which outbound shock sweeps up first the circumstellar shell and then the mass of the local, chaotic ISM. Reverse shocks heat the interior matter, the

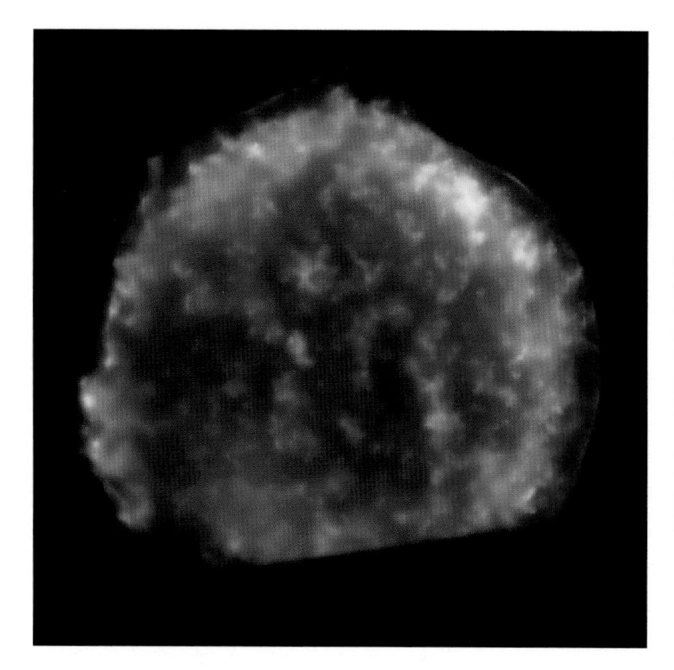

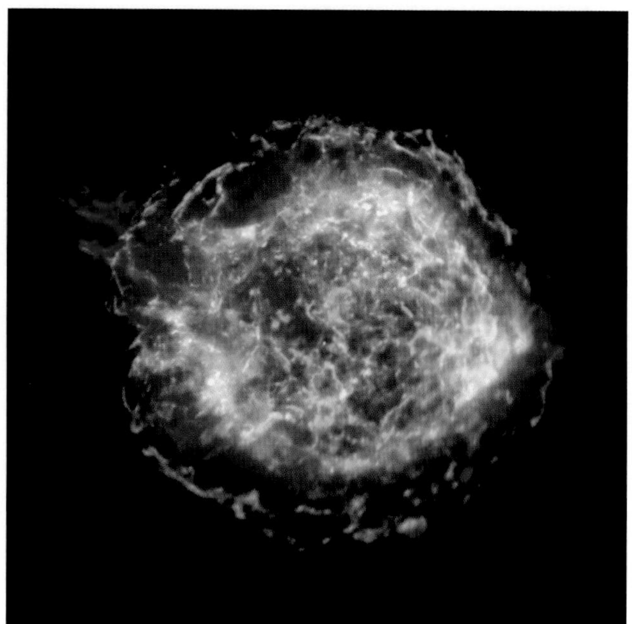

Fig. 14.16 **Tycho and Cas A remnants in X-rays** Left: Blue represents the 20 million kelvin 8-pc-wide shock wave from Type Ia Tycho's Star expanding into the interstellar medium; red and green, the stellar debris at 10 million kelvin. Right: Green now shows the shock from Type II Cas A, which is half the size of Tycho's SNR. Opposing jets rich in silicon suggest a mild hypernova of the type that makes gamma ray bursts. Blue reveals newly created iron atoms. At the center is a quiet pulsar, possibly a magnetar. (Left: Courtesy of NASA/CXC/SAO; Right: Courtesy of NASA/CXC/GSFC/U. Hwang *et al.*)

whole structure again appearing as a limb-brightened shell. Within a thousand or so years, the original ejecta – in some cases appearing as oxygen-rich knots – has become overwhelmed. As more and more mass is picked up, the shock eventually slows to the hundreds of kilometers per second. All the while, ambient ISM is being mixed with stellar ejecta that are rich in elements created by the original supernova (Fig. 14.17).

There are great variants on any simple theme. The Crab Nebula (see Fig. 14.1), which resulted from the supernova of 1054, is a filled "plerion" or *pulsar wind nebula* (PWN) energized by the high-speed relativistic wind from the magnetic pulsar within (Fig. 14.18). Consistently, it is elongated along the star's apparent axis of rotation. Other SNR exhibit a composite nature with an interior plerion and an outer shell. Since massive O stars gang into associations, so do SN and SNR. Multiple SN popping off within a small volume combine to create *superbubbles*, which are among the largest single structures in the Galaxy (Fig. 14.19).

14.3.5 Cosmic rays

The ultimate ejecta of supernovae are probably *cosmic rays* (CR). Cosmic rays are *not* electromagnetic radiation, but atomic nuclei (and electrons) that fly through space at speeds close to that of light and are accelerated into curved paths by the Galaxy's magnetic field. Upon striking the Earth's atmosphere, they smash atoms and send cascades of subatomic particles to the ground. The primary particles are observed from balloon flights and satellites. CR include the nuclei of the periodic table through iron and beyond, and are the most energetic particles known. The average energy of the electrons in a 10 000 kelvin gas is about 1 electron volt (eV), while the ionization energy of hydrogen is 13.6eV (where the eV is 1.6×10^{-12}erg [7.1.4, 14.2.2]). With strongly decreasing numbers of particles, CR energies range to 10^{18}eV, the number then flattening out slightly with increasing energy to 10^{21}eV, a billion ergs – the energy radiated in one second by a 100-watt light bulb – all in one particle.

The only phenomenon that seems to have the power to produce the immense number of CR circulating in the Galaxy, at least up to the break at 10^{18}eV, is the supernova. The chemical composition of CR is not that of the exploded debris (which is very rich in iron), however, but more or less that of the Sun or ISM, leading to the conclusion that cosmic rays are created by supernova shock waves acting on the ambient interstellar medium. Particles are sped up by the shock, scattered forward, hit again by the shock, and so on until they are flying practically at light-speed. Synchrotron X-rays coming from SNR show that the particles (including electrons) are there, being launched outward. No one knows what causes the CR at the highest energies. They come to us from magnetars or even from other galaxies. Another suggestion involves the merger of neutron stars that may make short gamma ray bursts. Cosmic rays smashing into interstellar carbon and oxygen atoms are responsible for the creation of the three "skipped" elements between helium and carbon: lithium (some of which was made in the Big Bang), beryllium, and boron.

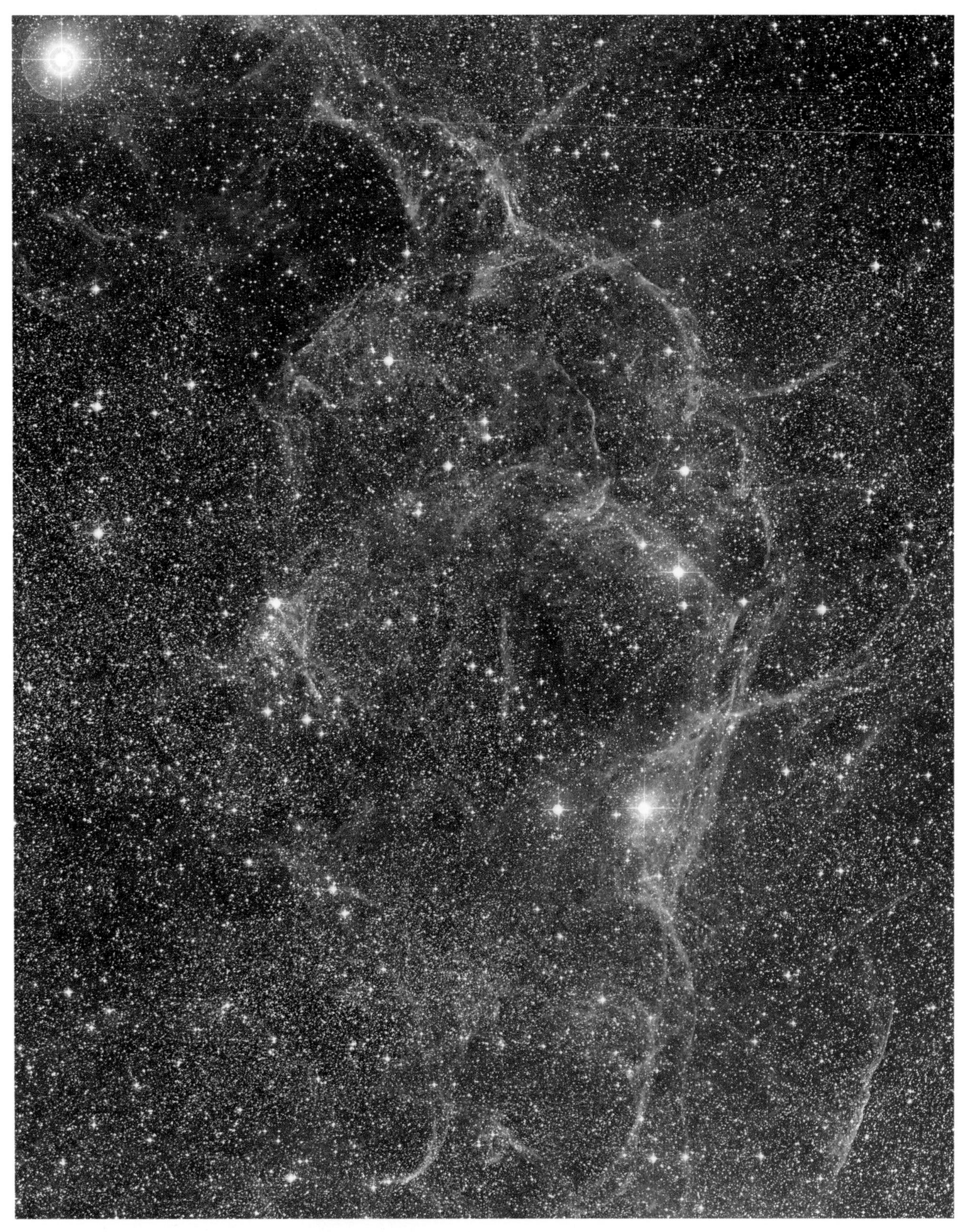

Fig. 14.17 **The Vela SNR** The huge Vela supernova remnant, some 30pc across (the northern portion of which is shown here), is the expanding shock from a supernova that blew up over 10 000 years ago and is associated with a radio pulsar (Fig. 14.11). (Courtesy of the Anglo-Australian Observatory, David Malin Images.)

Fig. 14.18 **The Crab pulsar** Overlapping X-ray (blue) and optical (red) images of the interior of the Crab Nebula reveal the pulsar's mighty wind, which creates a disk perpendicular to the star's rotation axis. (X-ray: NASA/CXC/ASU/; optical: NASA/HST/ASU/; both courtesy of J. Hester *et al.*)

Fig. 14.19 **The N44 superbubble** Multiple supernovae within the Large Magellanic Cloud have created an immense superbubble 65pc across. (Courtesy of Chris Smith and the MCELS Team, M. Hanna; NOAO/ AURA/NSF.)

14.4 Synergy

Perhaps the most remarkable property of the stars is their relation to one another, which is expressed no better than through the story of massive stars and their supernovae. Their explosions create the bulk of heavy elements, which blend with and enrich the ISM. Along with supergiant winds, their shock waves pound molecular clouds to help create the dense cores that grow into the next generation of stars of all kinds, including more of high mass that will one day do the same thing. Cosmic rays launched by shock waves partially ionize molecular clouds to help generate the cold chemistry that makes an abundance of molecules, and that allows the Galaxy's magnetic field to grab onto the cores to slow them, in turn allowing new suns to drip from the clouds' confines.

At the edges of supernova bubbles and superbubbles, new generations of massive stars and their supernovae generate yet more bubbles, making our star-forming Galaxy into a froth. The bubbles break through the Galactic disk, creating immense chimneys that send hot gas far into the Galactic halo, from which it cools and falls back, causing one part of the Galaxy to mix with another.

As you read this, you sit within a rain of particles falling on you that were ultimately generated by distant supernovae. Indeed, cosmic ray collisions with Earthly atoms make the radioactive carbon-14 in the air that is incorporated into living things, the decay of which allows us to go back in time to date ancient civilizations. At the end, we do not just admire the stars to learn the friendly constellations, we do not just study them to see how they work. They are part of our daily environment, of our very lives, their light shining upon us to illuminate the day and night, their fingers reaching out to touch us to show that all the Galaxy, indeed the assembly of galaxies, is a unit of which we are a part.

Appendix 1 The Messier Catalogue

M	NGC	Constell.	α (2000) δ				Size	Description
			h	m	°	′		
1	1952	Taurus	05	35	+22	01	5′	*Crab Nebula*, remnant of supernova of 1054.
2	7089	Aquarius	21	33	−00	50	12′	Globular cluster.
3	5272	Canes Venatici	13	42	+28	25	19′	Bright globular cluster; binocular object.
4	6121	Scorpius	16	24	−26	31	23′	Globular cluster; binocular object.
5	5904	Serpens	15	19	+02	05	20′	Globular cluster.
6	6405	Scorpius	17	40	−32	12	26′	Open cluster; easy binocular objct.
7	6475	Scorpius	17	54	−34	49	50′	Magnificent open cluster; naked-eye object.
8	6523	Sagittarius	18	04	−23	23	1°	*Lagoon Nebula*; bright diffuse nebula; naked-eye object.
9	6333	Ophiuchus	17	19	−18	31	6′	Globular cluster.
10	6254	Ophiuchus	16	57	−04	06	12′	Globular cluster; binocular object.
11	6705	Scutum	18	51	−06	18	12′	Open cluster; striking in telescope.
12	6218	Ophiuchus	16	47	−01	57	12′	Globular cluster; binocular object.
13	6205	Hercules	16	42	+36	30	23′	*Great cluster in Hercules*; magnificent globular cluster; barely naked eye; easy in binoculars.
14	6402	Ophiuchus	17	38	−03	15	7′	Globular cluster.
15	7078	Pegasus	21	30	+12	09	12′	Globular cluster.
16	6611	Serpens	18	19	−13	46	8′	Open cluster.
17	6618	Sagittarius	18	21	−16	12	40′	*Omega Nebula*; *Horseshoe Nebula*; diffuse nebula.
18	6613	Sagittarius	18	21	−17	08	7′	Open cluster.
19	6273	Ophiuchus	17	03	−26	15	5′	Globular cluster.
20	6514	Sagittarius	18	03	−23	02	30′	*Trifid Nebula*; diffuse nebula.
21	6531	Sagittarius	18	05	−22	30	12′	Open cluster.
22	6656	Sagittarius	18	36	−23	55	17′	Bright globular cluster; binocular object.
23	6494	Sagittarius	17	57	−19	01	27′	Open cluster.
24	...	Sagittarius	18	18	−18	29	1.5°	Star cloud in Milky Way; naked eye.
25	IC4725	Sagittarius	18	32	−19	15	35′	Open cluster; binocular object.
26	6694	Scutum	18	45	−09	25	9′	Open cluster.
27	6853	Vulpecula	20	00	+22	43	6′	*Dumbbell Nebula*; planetary nebula.

M	NGC	Constell.	α (2000) δ				Size	Description
			h	m	°	′		
28	6626	Sagittarius	18	25	−24	52	15′	Globular cluster.
29	6913	Cygnus	20	24	+38	32	7′	Open cluster.
30	7099	Capricornus	21	40	−23	11	9′	Globular cluster.
31	224	Andromeda	00	43	+41	15	1×2°	*Great Nebula in Andromeda*; *Andromeda Galaxy*; spiral galaxy; naked-eye object.
32	221	Andromeda	00	43	+40	51	9′	Elliptical galaxy; companion to M31.
33	598	Triangulum	01	34	+30	38	1°	*Triangulum Spiral*; spiral galaxy; binocular object and just visible to naked eye.
34	1039	Perseus	02	42	+42	47	30′	Open cluster.
35	2168	Gemini	06	09	+24	21	30′	Open cluster; easy binocular object.
36	1960	Auriga	05	36	+34	08	16′	Open cluster.
37	2099	Auriga	05	52	+32	33	24′	Open cluster; binocular object.
38	1912	Auriga	05	29	+35	51	18′	Open cluster.
39	7092	Cygnus	21	32	+48	26	32′	Open cluster.
40	...							Does not exist; star.
41	2287	Canis Major	06	47	−20	44	32′	Open cluster; binocular object.
42	1976	Orion	05	35	−05	25	1°	*Orion Nebula*; bright diffuse nebula; easy binocular object.
43	1982	Orion	05	35	−05	16	10′	Diffuse nebula at northern edge of Orion Nebula.
44	2632	Cancer	08	40	+20	00	1.5°	*Beehive* or *Praesepe Cluster*; open cluster; naked-eye object.
45	...	Taurus	03	48	+24	06	2°	*Pleiades*; *Seven Sisters*; open cluster; obvious naked-eye object.
46	2437	Puppis	07	42	−14	49	27′	Open cluster.
47	2422	Puppis	07	37	−14	29	25′	Open cluster; naked-eye object.
48	2548	Hydra	08	14	−05	47	30′	Open cluster; binocular object.
49	4472	Virgo	12	30	+07	59	4′	Elliptical galaxy.
50	2323	Monoceros	07	03	−08	20	16′	Open cluster.
51	5194	Canes Venatici	13	30	+47	11	10′	*Whirlpool Nebula*; spiral galaxy.
52	7654	Cassiopeia	23	26	+61	35	13′	Open cluster.
53	5024	Coma Berenices	13	13	+18	10	14′	Globular cluster.
54	6715	Sagittarius	18	55	−30	28	6′	Globular cluster.
55	6809	Sagittarius	19	40	−30	56	15′	Globular cluster.
56	6779	Lyra	19	17	+30	02	5′	Globular cluster.
57	6720	Lyra	18	54	+33	01	1.2′	*Ring Nebula in Lyra*; planetary nebula.
58	4579	Virgo	12	38	+11	48	4′	Spiral galaxy.

M	NGC	Constell.	α (2000) δ				Size	Description
			h	m	°	′		
59	4621	Virgo	12	42	+11	39	3′	Elliptical galaxy.
60	4649	Virgo	12	44	+11	33	4′	Elliptical galaxy.
61	4303	Virgo	12	22	+04	28	6′	Spiral galaxy.
62	6266	Ophiuchus	17	01	−30	07	6′	Globular cluster.
63	5055	Canes Venatici	13	16	+42	01	6′	Spiral galaxy.
64	4826	Coma Berenices	12	58	+21	41	6′	Spiral galaxy
65	3623	Leo	11	19	+13	07	6′	Spiral galaxy.
66	3527	Leo	11	20	+13	01	6′	Spiral galaxy.
67	2682	Cancer	08	51	+11	51	18′	Open cluster; one of oldest known.
68	4590	Hydra	12	39	−26	45	9′	Globular cluster.
69	6637	Sagittarius	18	31	−32	21	4′	Globular cluster.
70	6681	Sagittarius	18	43	−32	18	4′	Globular cluster.
71	6838	Sagitta	19	54	+18	47	6′	Globular cluster.
72	6981	Aquarius	20	53	−12	33	5′	Globular cluster.
73	6994	Aquarius	20	59	−12	38	...	Four stars.
74	628	Pisces	01	37	+15	47	8′	Spiral galaxy.
75	6864	Sagittarius	20	06	−21	56	5′	Globular cluster.
76	650–1	Perseus	01	42	+51	34	1′	Planetary nebula.
77	1068	Cetus	02	43	−00	01	2′	Spiral galaxy.
78	2068	Orion	05	47	+00	03	7′	Diffuse nebula.
79	1904	Lepus	05	24	−24	31	8′	Globular cluster.
80	6093	Scorpius	16	17	−22	59	5′	Globular cluster.
81	3031	Ursa Major	09	56	+69	04	13′	*Great Spiral in Ursa Major*; spiral galaxy; binocular object.
82	3034	Ursa Major	09	56	+69	42	7×2′	Starburst galaxy.
83	5236	Hydra	13	37	−29	52	9′	Spiral galaxy.
84	4374	Virgo	12	25	+12	53	3′	Elliptical galaxy.
85	4382	Coma Berenices	12	25	+18	11	3′	Elliptical galaxy.
86	4406	Virgo	12	26	+12	56	4′	Elliptical galaxy.
87	4486	Virgo	12	31	+12	23	3′	*Virgo A*; active elliptical galaxy.
88	4501	Coma Berenices	12	32	+14	25	6×3′	Spiral galaxy.
89	4552	Virgo	12	36	+12	33	2′	Elliptical galaxy.
90	4569	Virgo	12	37	+13	09	6×3′	Spiral galaxy.
91	4548	Coma Berenices	12	35	+14	30	5′	Barred spiral galaxy.
92	6341	Hercules	17	17	+43	08	12′	Globular cluster.
93	2447	Puppis	07	45	−23	52	18′	Open cluster.
94	4736	Canes Venatici	12	51	+41	07	5′	Spiral galaxy.
95	3351	Leo	10	44	+11	42	3′	Spiral galaxy.

M	NGC	Constell.	α (2000) δ				Size	Description
			h	m	°	′		
96	3368	Leo	10	47	+11	49	7×4′	Spiral galaxy.
97	3587	Ursa Major	11	15	+55	02	3′	*Owl Nebula*; planetary nebula.
98	4192	Coma Berenices	12	14	+14	54	8×2′	Spiral galaxy.
99	4254	Coma Berenices	12	19	+14	25	4′	Spiral galaxy.
100	4321	Coma Berenices	12	23	+15	49	5′	Spiral galaxy.
101	5457	Ursa Major	14	03	+54	21	22′	*Pinwheel Galaxy*; spiral galaxy.
102	...	...	...			...	...	Same as M101.
103	581	Cassiopeia	01	33	+60	41	6′	Open cluster.
104	4594	Virgo	12	40	−11	37	7×2′	*Sombrero Galaxy*; spiral galaxy.
105	3379	Leo	10	48	+12	35	2′	Elliptical galaxy.
106	4258	Canes Venatici	12	19	+47	18	20×6′	Spiral galaxy.
107	6171	Ophiuchus	16	32	−13	03	8′	Globular cluster.
108	3556	Ursa Major	11	12	+55	41	8×2′	Spiral galaxy.
109	3992	Ursa Major	11	58	+53	22	7′	Spiral galaxy.
110	205	Andromeda	00	40	+41	41	16′	Elliptical galaxy; companion to M31.

Appendix 2 The 51 brightest stars (to Polaris, magnitude 2.02)

The table lists the brightest stars down to Polaris with visual magnitude (V) of 2.02 with data taken chiefly from the *Bright Star Catalogue* (Fourth Revised Edition, D. Hoffleit and C. Jaschek, Yale University Observatory, New Haven, 1982) and the *Hipparcos Catalogue* [4.2.3]. If the components of double stars are brighter than $V = 2.0$ and have individual HR numbers, they are listed separately, but in order of their combined brightness. Though Castor B by itself would not make the list, it is included with Castor A. Prominent close binaries are noted, while wider companions are ignored. Though Delta Scorpii could make the list, it has been in outburst, and is not included. Proper motion, transverse velocity, radial velocity, and space velocity are represented by μ, v_t, v_r, and v_s.

Proper Name	Greek let.	α (2000) δ				V	Dist	M_V	μ	v_t	v_r	v_s	Spectrum and Remarks
		(h)	(m)	(o)	(′)		(pc)		($''yr^{-1}$)	($km\,s^{-1}$)	($km\,s^{-1}$)	($km\,s^{-1}$)	
Sirius	α CMa	06	45	−16	43	−1.46	2.64	1.43	1.339	17	−8	19	A1 V; wht. dwf. companion
Canopus	α Car	06	24	−52	41	−0.72	96	−5.63	0.031	14	21	25	F0 II
Rigil Kent	α Cen A	14	40	−60	50	−0.01	1.35	4.34	3.711	24	−25	35	G2 V; companion Proxima Cen.
	α Cen B	14	40	−60	50	1.33	1.35	5.68	3.726	24	−21	32	K1 V; combined $V = -0.29$
Arcturus	α Boo	14	16	+19	11	−0.04	11.3	−0.30	2.279	122	−5	122	K1.5 III; high velocity
Vega	α Lyr	18	37	+38	47	0.03	7.8	0.58	0.351	13	−14	19	A0 V; circumst. disk
Capella	α Aur	05	17	+46	00	0.08	12.9	−0.48	0.434	27	30	40	G8 III + G0 III close binary
Rigel	β Ori	05	15	−08	12	0.12	240	−6.8	0.002	2	21	21	B8 Iab
Procyon	β CMi	07	39	+05	13	0.36	3.50	2.64	1.258	21	−3	21	F IV–V; wht. dwf. compan.
Achernar	α Eri	01	38	−57	14	0.46	44	−2.76	0.097	20	16	26	B3 Vpe; rapid rotator
Hadar	β Cen	14	04	−60	22	0.61	102	−4.4	0.042	20	6	21	B1 III; Beta Cep var; binary.
Betelgeuse	α Ori	05	55	+07	24	0.70	130	−4.9	0.029	18	21	28	M1 Ia; semi-reg. variable
Acrux	α Cru A	12	27	−63	06	1.33	98	−3.6	0.030	17	−11	20	B0.5 IV; combined 0.76
	α Cru B					1.73	98	−3.2	0.037	17	−6	18	B1 V
Altair	α Aql	19	51	+ 08	52	0.77	5.1	2.21	0.661	16	−26	30	A7 V; rapid rotator
Aldebaran	α Tau	04	36	+ 16	31	0.85	20	−0.65	0.199	19	54	57	K5 III
Antares	α Sco	16	29	−26	26	0.96	185	−5.9	0.025	35	−3	35	M1.5 Ib; semi-reg. variable
Spica	α Vir	13	25	−11	10	1.04	80	−3.5	0.053	20	1	20	B1 III–IV + B2 V
Pollux	β Gem	07	45	+ 28	02	1.14	10.3	1.07	0.627	31	3	31	K0 III
Fomalhaut	α PsA	22	58	−29	37	1.16	7.7	1.73	0.368	13	7	15	A3 V; circumstellar disk
Deneb	α Cyg	20	41	+ 45	17	1.25	800	−8	0.002	18	−5	9	A2 Ia; no parallax
Mimosa	β Cru	12	48	−59	41	1.25	110	−3.9	0.050	26	16	30	B0.5 IV; Beta Cep var.
Regulus	α Leo	10	08	+ 11	58	1.35	24	−0.53	0.249	28	6	29	B7 V
Adhara	ϵ CMa	06	59	−28	58	1.50	132	−4.2	0.003	2	27	27	B2 II; $A_V = 0.06$
Castor	α Gem A	07	35	+ 31	53	1.94	15.8	0.95	0.254	19	6	20	A1 V; double, combined 1.57
	α Gem B					2.92	15.8	1.93	0.254	19	−1	19	A5 Vm; double

Proper Name	Greek let.	α (2000) δ (h)	(m)	(o)	(′)	V	Dist (pc)	M_V	μ ($''yr^{-1}$)	v_t ($km\,s^{-1}$)	v_r ($km\,s^{-1}$)	v_s ($km\,s^{-1}$)	Spectrum and Remarks
Gacrux	γ Cru	12	31	−57	07	1.63	27	−0.52	0.265	34	21	40	M3.5 III
Shaula	λ Sco	17	34	−37	06	1.63	215	−5.1	0.031	32	−3	32	B2 IV; Beta Cep var; $A_V = 0.1$
Bellatrix	γ Ori	05	25	+06	21	1.64	75	−2.72	0.016	6	18	19	B2 III
Elnath	β Tau	05	26	+28	36	1.65	40	−1.37	0.176	33	9	35	B7 III
Miaplacidus	β Car	09	13	−69	43	1.68	34	−1.98	0.192	31	−5	31	A2 IV
Alnilam	ϵ Ori	05	36	−01	12	1.70	410	−6.6	0.002	4	26	26	B0 Ia; $A_V = 0.26$
Alnitak	ζ Ori	05	40	−01	57	1.74	250	−5.5	0.005	6	16	17	O9.5 Ibe + B0 III; $A_V = 0.29$
Al Nair	α Gru	22	08	−46	58	1.74	31	−0.72	0.195	29	12	31	B7 IV
Alioth	ϵ UMa	12	54	55	58	1.77	25	−0.20	0.112	13	−9	16	A0p; magnetic star
Regor	γ-2 Vel	08	10	−47	20	1.78	260	−5.3	0.012	14	35	38	WC8 + O7.5e close binary
Dubhe	α UMa	11	04	+61	45	1.79	38	−1.10	0.141	25	−9	27	K0 III, F0 V companion
Mirfak	α Per	03	24	+49	52	1.79	180	−4.5	0.035	30	−2	31	F5 Iab
Wezen	δ Cma	07	08	−26	24	1.84	550	−6.9	0.004	11	34	36	F8 Iab
Kaus Aus.	ϵ Sgr	18	24	−34	23	1.85	44	−1.38	0.130	27	−15	31	B9.5 III
Alkaid	η UMa	13	48	+49	19	1.86	31	−0.59	0.122	18	−11	21	B3 V
Avior	ϵ Car	08	23	−59	31	1.86	194	−4.6	0.034	31	12	34	K3 III + B2 V close binary
Girtab	θ Sco	17	37	−43	00	1.87	83	−2.73	0.006	2	1	3	F1 II
Menkalinan	β Aur	06	00	+44	57	1.90	25	−0.10	0.056	7	−18	19	A2 IV eclipsing binary
Atria	α TrA	16	49	−69	01	1.92	127	−3.61	0.037	23	−3	23	K3 II–III; hybrid star
Alhena	γ Gem	06	38	+16	24	1.93	32	−0.60	0.067	10	−12	16	A0 IV
Peacock	α Pav	20	26	−56	44	1.94	56	−1.81	0.087	23	2	23	B2 IV close binary
. . .	δ Vel	08	45	−54	43	1.96	24	0.02	0.108	13	2	13	A1 V; multiple star
Mirzam	β CMa	06	23	−17	57	1.98	153	−3.94	0.004	3	34	34	B1 II–III; Beta Cep var.
Alphard	α Hya	09	28	−08	40	1.98	54	−1.70	0.036	9	−4	10	K3 II–III; barium star
Hamal	α Ari	02	07	23	28	2.00	20	0.47	0.240	23	−14	27	K2 III
Polaris	α UMi	02	32	89	16	2.02	132	−3.58	0.046	29	−17	33	F7 Ib–II; Cepheid var.; + F3 V

Appendix 3 The stars within four parsecs (plus 2)

The table includes all known stars confirmed as of 2004 to lie within four parsecs (13 light years) of the Earth, with data taken from the compilations of the Astronomical Data Center (CDS), University of Strasbourg and checked with the RECONS website. Distances are from the *Hipparcos Catalogue*. Krüger 60 is included as the first star just barely over the line. Flare stars' names are given in the last column. "BD" means "brown dwarf."

Name	α (2000) δ (h)	(m)	(°)	(′)	D (pc)	μ (″yr^{-1})	v_t (km s^{-1})	v_r (km s^{-1})	v_s (km s^{-1})	V	M_V	Spec.	Flare
Proxima Cen	14	30	−62	41	1.29	3.85	24	−17	−29	11.05	15.49	M5.5 Ve	V645 Cen
α Cen A	14	40	−60	50	1.35	3.71	24	−25	35	−0.01	4.34	G2 V	
α Cen B						3.73	24	−21	32	1.33	5.68	K1 V	
Barnard's Star	17	58	+04	42	1.82	10.36	89	−108	140	9.54	13.24	M4 Ve	
Wolf 359	10	56	+07	01	2.35	4.70	52	13	54	13.54	16.68	M6 V	CN Leo
BD+36 2147	11	03	+35	58	2.55	4.81	58	−85	103	7.49	10.47	M2 V	
Sirius A	06	45	−16	43	2.64	1.34	17	−8	19	−1.47	1.43	A1 V	
Sirius B					2.64	1.33	17	−8	19	8.44	11.33	DA	
Luyten 726-8 A	01	39	−17	57	2.65	3.36	42	29	51	12.57	15.45	M5.5 Ve	BL Cet
Luyten 726-8 B							42	32	53	12.52	15.40	M5.5e V	UV Cet
Ross 154	18	50	−23	50	2.97	0.67	9	−4	10	10.95	13.58	M3.5 V	V1216 Sgr
Ross 248	23	42	+44	10	3.17	1.60	24	−81	84	12.28	14.77	M5 V	HH And
ϵ Eri	03	33	−09	28	3.22	0.98	15	16	22	3.73	6.19	K2 V	
CD-36 15693	23	06	−35	51	3.29	6.90	108	10	108	7.34	9.75	M0.5 V	
Ross 128	11	48	+00	48	3.34	1.36	22	−13	26	11.08	13.46	M4 V	FI Vir
Luyten 789-6ABC	22	39	−15	18	3.33	3.26	51	−60	79	12.18	14.57	M5.5 V	EZ Aqr
61 Cyg A	21	07	+38	45	3.48	5.28	87	−64	108	5.21	7.50	K5 V	
61 Cyg B					3.50	5.17	86	−64	107	6.03	8.31	K7 V	

Name	α (2000) δ				D	μ	v_t	v_r	v_s	V	M_V	Spec.	Flare
	(h)	(m)	(°)	(′)	(pc)	(″yr^{-1})	(km s^{-1})	(km s^{-1})	(km s^{-1})				
Procyon A	07	39	+05	13	3.50	1.26	21	−3	21	0.36	2.64	F5 IV–V	
Procyon B										10.7	13.0	DA	
BD+59 1915 A	18	43	+59	38	3.57	2.24	38	−1	38	8.91	11.15	M3 V	
BD+59 1915 B	18	43	+59	37	3.52	2.31	39	6	39	9.69	11.96	M3.5 V	
BD+43 44 A	00	18	+44	01	3.57	2.92	49	14	51	8.07	10.31	M1.5 V	GX And
BD+43 44 B					3.57	2.90	49	20	53	11.04	13.27	M3.5 V	GQ And
G 051-015	08	30	+26	47	3.63	1.29	22	9	24	14.81	17.01	M6	DX Cnc
ϵ Ind A	22	03	−56	47	3.63	4.70	81	−40	90	4.89	7.00	K4.5 V	
ϵ Ind B										...	...	T1.0 BD	
ϵ Ind C										...	...	T6.0 BD	
τ Cet	01	44	−15	56	3.65	1.92	33	−16	37	3.50	5.69	G8 V	
L372–58	03	36	−44	31	3.66	0.83	14	...	...	13.03	15.21	M5.5 V	
L725–32	01	13	−17	00	3.72	1.37	24	28	37	11.6	13.7	M4.5e	YZ Cet
Luyten's Star	07	27	+05	14	3.78	3.74	66	26	71	9.82	11.98	M5 V	
Kapteyn's Star	05	11	−45	01	3.92	8.66	161	246	294	8.89	10.92	M0 V	VZ Pic
CD−39 14192	21	17	−38	52	3.95	3.45	65	23	69	6.68	8.70	K7 V	AX Mic
Krüger 60 A	22	28	+57	42	4.01	0.99	19	−24	31	9.59	11.57	M3 V	
Krüger 60 B					4.01	0.85	16	−28	32	10.30	12.28	M4 V	DO Cep

Appendix 4 The chemical composition of the Sun

The first part (A) ranks the elements according to atomic number, the second (B) by abundance. Both are scaled to 1 000 000 hydrogen atoms. To find the "standard" abundance number scaled to hydrogen at 10^{12}, multiply by one million. The abundances of elements not observed in the Sun are those found in meteorites, and are enclosed in parentheses. The references in the "Group" column are obvious except for "RE," which stands for "rare earth." Radioactive elements (noted by a * in the "Group" column) with no solar measurements of abundance are placed at the end and are arranged according to their half-lives.

The "origin" column tells whether the element was created in the Big Bang (BB), by cosmic rays in interstellar space (CR), or by slow (s) or rapid (r) neutron capture. If the s- and r-processes contribute more or less equally, within a factor of two of each other, the column gives "rs" or "sr" depending on which of the two dominates. The dollar sign "$" indicates that at least 90% of the element is made by the s- or r-processes [13.3.3], and r$$ indicates that the element is exclusively an r-process element. Other elements are created by energy-generating fusion or by explosive burning in supernovae and by some additional minor processes.

Unstable elements not observed are listed with their longest half-lives. (The abundance data are from N. Grevesse, A. Noels, and A.J. Sauval, in *Cosmic Abundances*, ASP Conference series, eds. S.S. Holt and G. Sonneborn, 1996.)

A. Elements ranked by atomic number

Rank	Element	Abundance or half-life	Atomic Number	Group	Origin
1	Hydrogen	1 000 000	1		BB
2	Helium	98 000	2		BB
58	Lithium	0.000 014	3	Li	BB, CR
59	Beryllium	0.000 014	4	Li	CR
34	Boron	0.000 40	5	Li	CR
4	Carbon	355	6	CNO	3-α
6	Nitrogen	93	7	CNO	
3	Oxygen	741	8	CNO	
24	Fluorine	0.036	9		
5	Neon	120	10		
14	Sodium	2.14	11		
7	Magnesium	38.0	12		
12	Aluminum	2.95	13		
8	Silicon	35.5	14		
18	Phosphorus	0.282	15		
10	Sulphur	21.3	16		
17	Chlorine	0.32	17		
11	Argon	3.31	18		
20	Potassium	0.132	19		
13	Calcium	2.29	20		
30	Scandium	0.001 5	21		
21	Titanium	0.105	22	Fe	
26	Vanadium	0.010 0	23	Fe	
16	Chromium	0.47	24	Fe	
19	Manganese	0.245	25	Fe	
9	Iron	31.6	26	Fe	
22	Cobalt	0.083	27	Fe	
15	Nickel	1.78	28	Fe	
25	Copper	0.016 2	29	Fe	
23	Zinc	0.040	30	Fe	
32	Gallium	0.000 76	31		sr
27	Germanium	0.002 6	32		sr
37	(Arsenic)	(0.000 23)	33		r
28	(Selenium)	(0.002 4)	34		r
33	(Bromine)	(0.000 43)	35		r
29	(Krypton)	(0.001 7)	36		sr

Rank	Element	Abundance or half-life	Atomic Number	Group	Origin
35	Rubidium	0.000 40	37		sr
31	Strontium	0.000 93	38		s
38	Yttrium	0.000 17	39		s
36	Zirconium	0.000 40	40		s
55	Niobium	0.000 026	41		s
45	Molybdenum	0.000 083	42		s
***	Technetium	42 000 000 years	43	*	s
46	Ruthenium	0.000 069	44		rs
61	Rhodium	0.000 013	45		r
49	Palladium	0.000 049	46		rs
69	Silver	0.000 008 7	47		r$
48	Cadmium	0.000 059	48		sr
50	Indium	0.000 046	49		r
43	Tin	0.000 10	50		sr
66	Antimony	0.000 010	51		r$
39	(Tellurium)	(0.000 17)	52		r
53	(Iodine)	(0.000 032)	53		r$
40	(Xenon)	(0.000 17)	54		r$
62	(Cesium)	0.000 013	55		r$
41	Barium	0.000 17	56		s$
42	Lanthanum	0.000 15	57	RE	s
51	Cerium	0.000 038	58	RE	s
74	Praseodymium	0.000 005 1	59	RE	rs
52	Neodymium	0.000 032	60	RE	rs
***	Promethium	18 years	61	*	
67	Samarium	0.000 010	62	RE	r
76	Europium	0.000 003 2	63	RE	r$
63	Gadolinium	0.000 013	64	RE	r
81	Terbium	0.000 000 8	65	RE	r$
60	Dysprosium	0.000 014	66	RE	r$
78	Holmium	0.000 001 8	67	RE	r$
70	Erbium	0.000 008 5	68	RE	r$
80	Thulium	0.000 001 0	69	RE	r$
65	Ytterbium	0.000 012	70	RE	r
73	Lutetium	0.000 005 8	71	RE	r$
72	Hafnium	0.000 007 6	72		rs
82	(Tantalum)	(0.000 000 7)	73		rs
64	Tungsten	0.000 013	74		sr
77	(Rhenium)	(0.000 001 9)	75		r$
54	Osmium	0.000 028	76		r$
56	Iridium	0.000 022	77		r$

Rank	Element	Abundance or half-life	Atomic Number	Group	Origin
47	Platinum	0.000 063	78		r$
68	Gold	0.000 010	79		r$
57	(Mercury)	(0.000 015)	80		sr
71	Thallium	0.000 007 9	81		s
44	Lead	0.000 089	82		s
75	(Bismuth)	(0.000 005 1)	83		rs
***	Polonium	102 years	84	*	
***	Astatine	8 hours	85	*	
***	Radon	4 days	86	*	
***	Francium	22 minutes	87	*	
***	Radium	1600 years	88	*	
***	Actinium	22 years	89	*	
79	(Thorium)	(0.000 001 2)	90	*	r$$
***	Protactinium	3000 years	91	*	
83	(Uranium)	(0.000 000 3)	92	*	r$$
***	Neptunium	2 100 000 years	93	*	
***	Plutonium	80 000 000 years	94	*	r$$

B. Elements ranked by abundance

Rank	Element	Abundance or half-life	Atomic Number	Group	Origin
1	Hydrogen	1 000 000	1		BB
2	Helium	98 000	2		BB
3	Oxygen	741	8	CNO	
4	Carbon	355	6	CNO	3-α
5	Neon	120	10		
6	Nitrogen	93	7	CNO	
7	Magnesium	38.0	12		
8	Silicon	35.5	14		
9	Iron	31.6	26	Fe	
10	Sulphur	21.3	16		
11	Argon	3.31	18		
12	Aluminum	2.95	13		
13	Calcium	2.29	20		
14	Sodium	2.14	11		
15	Nickel	1.78	28	Fe	
16	Chromium	0.47	24	Fe	
17	Chlorine	0.32	17		
18	Phosphorus	0.282	15		
19	Manganese	0.245	25	Fe	
20	Potassium	0.132	19		
21	Titanium	0.105	22	Fe	
22	Cobalt	0.083	27	Fe	
23	Zinc	0.040	30	Fe	
24	Fluorine	0.036	9		
25	Copper	0.016 2	29	Fe	
26	Vanadium	0.010 0	23	Fe	
27	Germanium	0.002 6	32		sr
28	(Selenium)	(0.002 4)	34		r
29	(Krypton)	(0.001 7)	36		sr
30	Scandium	0.001 5	21		
31	Strontium	0.000 93	38		s
32	Gallium	0.000 76	31		sr
33	(Bromine)	(0.000 43)	35		r
34	Boron	0.000 40	5	Li	CR
35	Rubidium	0.000 40	37		sr
36	Zirconium	0.000 40	40		s
37	(Arsenic)	(0.000 23)	33		r

Rank	Element	Abundance or half-life	Atomic Number	Group	Origin
38	Yttrium	0.000 17	39		s
39	(Tellurium)	(0.000 17)	52		r
40	(Xenon)	(0.000 17)	54		r$
41	Barium	0.000 17	56		s$
42	Lanthanum	0.000 15	57	RE	s
43	Tin	0.000 10	50		sr
44	Lead	0.000 089	82		s
45	Molybdenum	0.000 083	42		s
46	Ruthenium	0.000 069	44		rs
47	Platinum	0.000 063	78		r$
48	Cadmium	0.000 059	48		sr
49	Palladium	0.000 049	46		rs
50	Indium	0.000 046	49		r
51	Cerium	0.000 038	58	RE	s
52	Neodymium	0.000 032	60	RE	rs
53	(Iodine)	(0.000 032)	53		r$
54	Osmium	0.000 028	76		r$
55	Niobium	0.000 026	41		s
56	Iridium	0.000 022	77		r$
57	(Mercury)	(0.000 015)	80		sr
58	Lithium	0.000 014	3	Li	BB, CR
59	Beryllium	0.000 014	4	Li	CR
60	Dysprosium	0.000 014	66	RE	r$
61	Rhodium	0.000 013	45		r
62	(Cesium)	0.000 013	55		r$
63	Gadolinium	0.000 013	64	RE	r
64	Tungsten	0.000 013	74		sr
65	Ytterbium	0.000 012	70	RE	r
66	Antimony	0.000 010	51		r$
67	Samarium	0.000 010	62	RE	r
68	Gold	0.000 010	79		r$
69	Silver	0.000 008 7	47		r$
70	Erbium	0.000 008 5	68	RE	r$
71	Thallium	0.000 007 9	81		s
72	Hafnium	0.000 007 6	72		rs
73	Lutetium	0.000 005 8	71	RE	r$
74	Praseodymium	0.000 005 1	59	RE	rs
75	(Bismuth)	(0.000 005 1)	83		rs
76	Europium	0.000 003 2	63	RE	r$
77	(Rhenium)	(0.000 001 9)	75		r$
78	Holmium	0.000 001 8	67	RE	r$

Rank	Element	Abundance or half-life	Atomic Number	Group	Origin
79	(Thorium)	(0.000 001 2)	90	*	r$$
80	Thulium	0.000 001 0	69	RE	r$
81	Terbium	0.000 000 8	65	RE	r$
82	(Tantalum)	(0.000 000 7)	73	*	rs
83	(Uranium)	(0.000 000 3)	92	*	r$$
	Plutonium	80 000 000 years	94	*	r$$
	Technetium	42 000 000 years	43	*	s
	Neptunium	2 100 000 years	93	*	
	Protactinium	33 000 years	91	*	
	Radium	1600 years	88	*	
	Polonium	102 years	84	*	
	Actinium	22 years	89	*	
	Promethium	18 years	61	*	
	Radon	4 days	86	*	
	Astatine	8 hours	85	*	
	Francium	22 minutes	87	*	

Appendix 5 Endpaper: riches at the anticenter

This story begins in Chapter 1 with a spectacular view of the Milky Way centered on the core of the Galaxy; it ends with a detailed look at the Galactic anticenter [5.1] that summarizes much of stellar astronomy. Figure 4.7 shows a broad view of the Milky Way in Taurus and Auriga around Galactic longitude 180° [2.3.5]. Appendix Fig. A5.1 shows the same image, but labelled with 18 open clusters (one of which probably does not exist) and 18 variable stars that run the gamut from protostars to highly evolved long-period Miras. The characteristics of each are presented in Appendix Tables A5.1 and A5.2. The clusters and variables appear in a giant ragged ring with none in the center, and thereby outline the nearby Taurus–Auriga dark cloud [11.1.1], a region of vigorous star formation whose dust hides the background and the distant stars. In front of the cloud is our own planet Saturn, which was the product of a similar, but long-dispersed, dark cloud that birthed our Sun, Earth, and our very selves.

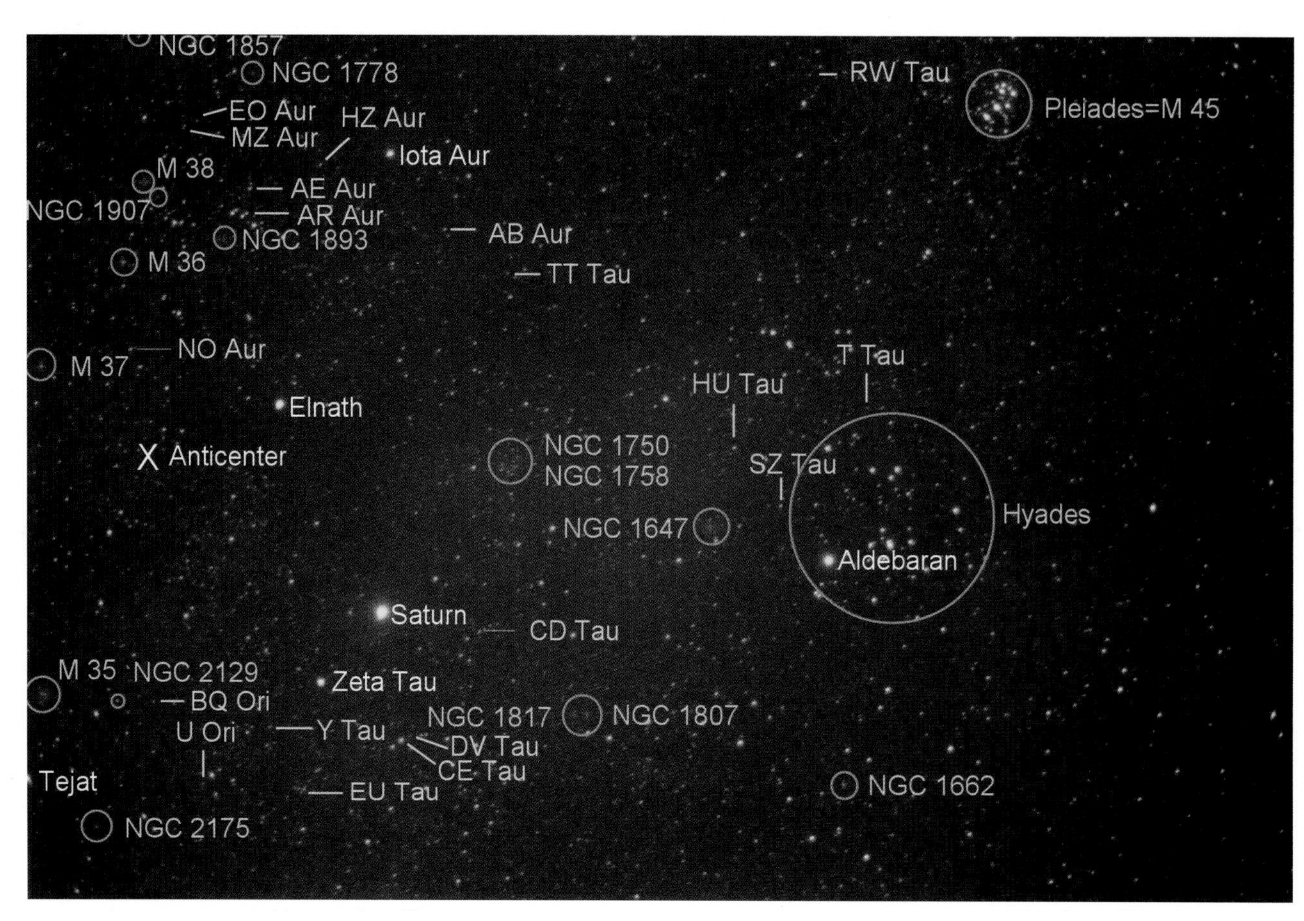

Appendix Fig. A5.1 (J.B. Kaler)

Name and Remarks	Mess.[a]	Const.	α (2000) δ				Dist.	A_V	Diameter		Age
			h	m	°	′	(pc)		(min)	(pc)	(Myr)
Pleiades	M45	Tau	03	47	24	06	132	0.13	110	4.2	130
Seven Sisters											
Hyades	...	Tau	04	26	15	51	46	0.03	330	4.4	650
Nearest large cluster											
NGC 1647	...	Tau	04	46	19	05	540	0.93	40	6	140
NGC 1662	...	Ori	04	49	10	56	410	1.09	11	1.3	300
NGC 1750	...	Tau	05	04	23	44	630	1.08	10	1.8	200
With NGC 1758, "NGC 1746"											
NGC 1758	...	Tau	05	04	23	47	2500	1.08	7	1.6	400
With NGC 1750, "NGC 1746"											
NGC 1778	...	Aur	05	05	36	59	1500	1.09	7	3	110
NGC 1807	...	Tau	05	10	16	32	...	...	12	...	...
NGC 1817	...	Tau	05	12	16	42	1700	1.06	20	10	630
NGC 1857	...	Aur	05	20	39	21	1900	2.2	10	5	180
NGC 1893	...	Aur	05	23	33	24	4000	1.57	25	28	10
With nebula IC 410											
NGC 1907	...	Aur	05	28	35	20	1500	1.38	5	2	400
NGC 1912	M38	Aur	05	29	35	51	1120	0.83	18	6	300
NGC 1960	M36	Aur	05	36	34	08	1300	0.70	16	15	32
NGC 2099	M37	Aur	05	52	32	33	1500	0.90	24	10	400
NGC 2129	...	Gem	06	01	23	18	1500	2.3	5	2.2	10
May not exist											
NGC 2168	M35	Gem	06	09	24	21	1070	0.80	30	3	150
NGC 2175	...	Ori	06	10	20	30	2200	3	5	3	1
With nebula S 252											

[a]Messier number [Appendix 1].
[b]Interstellar absorption at V [3.5.1].

Appendix Table A5.1 **Open clusters near the anticenter**

Name	α (2000) h	m	δ °	′	Class	V	Distance[a] (pc)
AB Aur	04	56	30	33	A0pe	7–7.1	140
Herbig Ae/Be protostar in the last stages of forming							
AE Aur	05	16	34	19	O9.5 Ve	6–6.1	450
AE and Mu Col are runaway stars from Iota Orionis							
AR Aur	05	18	33	46	B9.5 V	6.1–6.7	400
17 Aur, eclipsing binary, 4.13 day, primary eclipse 0.75 mag							
EO Aur	05	18	36	38	B5	7.7–8.2	370
Eclipsing binary, 4.07 day, primary eclipse 0.5 mag							
HZ Aur	05	06	33	55	B9p	7–7.1	155
Magnetic chemically peculiar star, 6.43-day period							
MZ Aur	05	20	36	38	B1.5 Ivp	8.1–8.3	1400
Beta Cephei star, small variation with multiple periods							
NO Aur	05	40	31	55	M2 IIIS	6.1–6.3	420
Zirconium-rich incipient class S irregular							
U Ori	05	56	20	15	M8 III	6–12	400
Mira variable, 373-day period							
BQ Ori	05	57	22	50	M5 III	7.1–8.0	240
Semi-regular variable							
T Tau	04	22	19	32	G5 Ve	9–11	180
Multiple low-mass protostar							
Y Tau	05	36	20	42	C5 II	6–8	300
Semi-regular carbon star, 241-day period							
SZ Tau	04	37	18	33	F7 Ib	6.5–6.8	450
Cepheid, 3.15-day period							
TT Tau	04	52	28	32	C5 II	8–9	1000
Semi-regular carbon star, 166-day period							
CD Tau	05	18	20	08	F7 V	6.8–7.3	73
Eclipsing binary, 3.44-day period							
CE Tau	05	32	18	36	M2 Iab	4.2–4.5	600
119 Tauri, semi-regular supergiant, 160-day period, similar to Betelgeuse							
DV Tau	05	31	18	34	M6?	8–9	1000
Irregular, 1 magnitude							
EU Tau	05	45	18	39	G5	8.1–8.4	1200
Cepheid, 2.10-day period							
HU Tau	04	38	20	41	B8 V	5.9–6.7	110
Eclipsing binary, 2.06-day period, 0.75 mag, total ecl.							

[a]Distances from Hipparcos parallaxes except for MZ Aur, U Aur, TT Tau, DV Tau for which absolute magnitudes were assumed spectroscopically, and EU Tau, where distance is from period–luminosity relation.

Appendix Table A5.2 **Variable stars near the anticenter**

Further reading

Aller, Lawrence H. (1971). *Atoms, Stars, and Nebulae*. Cambridge, Mass: Harvard University Press.

(1984). *The Physics of Thermal Gaseous Nebulae*. Dordrecht: Reidel.

Batten, Alan H. (1973). *Binary and Multiple Systems of Stars*. Oxford: Pergamon Press.

Binney, James, and Merrifield, Michael (1998). *Galactic Astronomy*. Princeton: Princeton University Press.

Binney, James, and Tremaine, Scott (1988). *Galactic Dynamics*. Princeton: Princeton University Press.

Böhm–Vitense, Erika (1989). *Introduction to Stellar Astrophysics: Volume 1, Basic Stellar Observations and Data*. Cambridge: Cambridge University Press.

(1989). *Introduction to Stellar Astrophysics: Volume 2, Stellar Atmospheres*. Cambridge: Cambridge University Press.

(1992). *Introduction to Stellar Astrophysics: Volume 3, Stellar Structure and Evolution*. Cambridge: Cambridge University Press.

Burnham, Robert R. Jr. (1978). *Burnham's Celestial Handbook: An Observer's Guide to the Universe Beyond the Solar System*. Three volumes. New York: Dover Publications.

Carroll, Bradley W., and Ostlie, Dale A. (1996). *Modern Astrophysics*. New York: Addison-Wesley.

Cowley, C. R. (1995). *An Introduction to Cosmochemistry*. Cambridge: Cambridge University Press.

Cox, Arthur N., ed. (2000). *Allen's Astrophysical Quantities*. Fourth edition. New York: Springer-Verlag.

Croswell, Ken (1995). *The Alchemy of the Heavens*. New York: Anchor Books.

Garrison, R. F., ed. (1984). *The MK Process and Stellar Classification*. Toronto: David Dunlap Observatory.

Golub, Leon, and Pasachoff, Jay M. (2001). *Nearest Star: The Surprising Science of our Sun*. Cambridge, Mass: Harvard University Press.

Harwit, Martin (1988). *Astrophysical Concepts*. Second edition. New York: Springer-Verlag.

Hearnshaw, J. B. (1986). *The Analysis of Starlight: One Hundred and Fifty Years of Astronomical Spectroscopy*. Cambridge: Cambridge University Press.

(1996). *The Measurement of Starlight: Two Hundred Years of Astronomical Photometry*. Cambridge: Cambridge University Press.

Hirshfeld, Alan W. (2001). *Parallax: The Race to Measure the Cosmos*. New York: Henry Holt and Company/Freeman.

Hoffmeister, C., Richter, G., and Wenzel, W. (1985). *Variable Stars*. Berlin: Springer-Verlag.

Jaschek, Carlos, and Jaschek, Mercedes (1987). *The Classification of Stars*. Cambridge: Cambridge University Press.

Kaler, James B. (1989, 1997). *Stars and their Spectra: An Introduction to the Spectral Sequence*. Cambridge: Cambridge University Press.

(1992, 1998). *Stars*. New York: Scientific American Library, Freeman.

(1997). *Cosmic Clouds: Birth, Death, and Recycling in the Galaxy*. New York: Scientific American Library, Freeman.

(2001). *Extreme Stars: At the Edge of Creation*. Cambridge: Cambridge University Press.

(2002). *The 100 Greatest Stars*. New York: Copernicus Books.

Lang, Kenneth R. (1991). *Astrophysical Data: Planets and Stars*. New York: Springer-Verlag.

(2001). *The Cambridge Encyclopedia of the Sun*. Cambridge: Cambridge University Press.

Mann, Alfred K. (1997). *Shadow of a Star: The Neutrino Story of Supernova 1987a*. New York: Freeman.

Maran, Stephen P., ed. (1992). *The Astronomy and Astrophysics Encyclopedia*. New York: Van Nostrand Reinhold.

Mihalas, D. (1978). *Stellar Atmospheres*. San Francisco: Freeman.

Mihalas, D., and Binney, J. (1981). *Galactic Astronomy, Structure, and Kinematics*. Second edition. San Francisco: Freeman.

Murdin, P., ed. (2001). *Encyclopedia of Astronomy and Astrophysics*. Four volumes. Bristol: Institute of Physics Publishing.

Ostlie, Dale A., and Carroll, Bradley W. (1996). *An Introduction to Modern Stellar Astrophysics*. New York: Addison-Wesley.

Phillips, Kenneth J. H. (1992). *Guide to the Sun*. Cambridge: Cambridge University Press.

Prialnik, Dina (2000). *Stellar Structure and Evolution*. Cambridge: Cambridge University Press.

Reid, I. Neill, and Hawley, Suzanne L. (2000). *New Light on Dark Stars: Red Dwarfs, Low-Mass Stars, Brown Dwarfs*. Chichester: Springer-Verlag/Praxis.

Verschuur, G. L. (1989). *Interstellar Matters*. New York: Springer-Verlag.

Wenzel, Donat G. (1989). *The Restless Sun*. Washington: Smithsonian Institution Press.

Index

Page numbers in *italic* refer to figures

C

N

O

P

Q

R

S

X

Y

Z